D0891904

POLYMER SYNTHESES

Volume I

This is Volume 29 of
ORGANIC CHEMISTRY
A series of monographs
Editors: ALFRED T. BLOMQUIST and HARRY WASSERMAN

A complete list of the books in this series appears at the end of the volume.

POLYMER SYNTHESES

Volume I

Stanley R. Sandler

CENTRAL RESEARCH LABORATORIES
BORDEN CHEMICAL COMPANY
DIVISION OF BORDEN, INC.
PHILADELPHIA, PENNSYLVANIA

Wolf Karo

LACTONA CORPORATION
SUBSIDIARY OF WARNER-LAMBERT COMPANY
PHILADELPHIA, PENNSYLVANIA

ACADEMIC PRESS New York and London 1974
A Subsidiary of Harcourt Brace Jovanovich, Publishers

ACADEMIC PRESS, INC.
111 Fifth Avenue, New York, New York 10003

United Kingdom Edition published by
ACADEMIC PRESS, INC. (LONDON) LTD.
24/28 Oval Road, London NW1

Library of Congress Cataloging in Publication Data

Sandler, Stanley R Date
Polymer syntheses.

(Organic chemistry; a series of monographs, v.29)
Includes bibliographic references.
1. Polymers and polymerization. I. Karo, Wolf, Date joint author. II. Title. III. Series.
QD281.P6S27 547'.84 73-2073
ISBN 0-12-618560-3 (v. 1)

PRINTED IN THE UNITED STATES OF AMERICA

CONTENTS

Chapter 5. **Polymerization of Aldehydes**

Chapter 6. **Polymerization of Epoxides and Cyclic Ethers**

Chapter 7. **Polyureas**

Chapter 8. **Polyurethanes**

Chapter 9. **Thermally Stable Polymers**

Chapter 15. **Free-Radical Initiators: Hydroperoxides**

Appendix

PREFACE

The aspects of organic polymer theory and mechanisms, polymer processes, and practical chemistry have already appeared in other books. However, the synthesis of the various classes of polymers by functional group types remained unavailable. This book aims to fill this gap and to present detailed laboratory instructions for the preparation of polymers by various functional group classes. Each chapter contains a critical review of the best available synthetic methods. The classes of polymers covered include olefin and diolefin, hydrocarbon polymers, polyesters, polycarbonates, polymerization products of epoxides, cyclic ethers, aldehydes, polyureas, polyurethanes, thermally stable polymers, acrylic–methacrylic esters, polyacrylonitriles, polyacrylamides, and organophosphorus polymers. Some of the heterocyclic polymers included in the chapter on thermally stable polymers are polyimides, polybenzimidazoles, polyquinoxalines, poly-1,3,4-oxidazoles, poly-1,2,4-triazoles, polybenzothiazoles, and polybenzoxazoles.

This book is enhanced by the chapter on phosphorus-containing polymers. It appears to be the first review that presents detailed laboratory procedures for the preparation of polymers containing phosphorus in various oxidation states. The synthesis of phosphonitrilic polymers is also included in this chapter. This chapter should be especially valuable to those chemists involved in the designing of phosphorus polymers to meet present and future fire retardant requirements of their various products.

Added details on the synthesis of peroxide and hydroperoxide free radical initiators are of special interest. This information is included to aid those polymer chemists who may have to design special initiators for unusual applications.

In all chapters the latest journal articles and patents have been reviewed. Each chapter contains tables of data to show the scope of the various methods of synthesis with references given for each entry. Some preparations are taken from the older literature because they are of a classic nature and suitably describe the polymer preparation. Most are taken from present-day literature and are included only if they appear to be the best available.

In presenting preparative details of the various techniques of polymerization, an effort was made to select, if possible, methods which would have wide application not only for the formation of polymers of the specific system cited but also for many related situations. Thus, for example, the principles involved in the preparation of emulsion polymers are applicable to the preparation of polymers of a wide variety of vinyl monomers as well as to many copolymerizations.

This book is designed only to give helpful polymer synthesis information and not to override the question of legal patentability or to suggest allowable industrial use.

We would like to take this opportunity to thank Dr. Jack Dickstein, Research Manager of the Central Research Laboratories of Borden Chemical, Division of Borden, Inc., for encouragement and support in the preparation of this manuscript. Special thanks are due to Miss Emma Moesta for her untiring devotion in the preparation of the typed manuscript. Finally, we thank our wives and children for their patience, understanding, and encouragement during all stages involved in the preparation of the manuscript.

*Stanley R. Sandler**
Wolf Karo†

* **Present address:** Pennwalt Corporation, King of Prussia, Pennsylvania.
† Currently a research consultant in Huntingdon Valley, Pennsylvania.

Chapter 1

POLYMERIZATION OF OLEFINIC AND DIOLEFINIC HYDROCARBONS

I. INTRODUCTION

Olefinic and diolefinic monomers can be polymerized using either free-radical, anionic, cationic, or coordination type initiators. These will be discussed individually in each of the four sections of this chapter. It is interesting to note that not all monomers respond equally well to each of the types of initiators.

The substituents placed on ethylene greatly affect stereochemistry, resonance, and polarity of the monomer, and have a decided effect on which initiator system works best with it. For example, propene and 1-butene can only be homopolymerized well with heterogeneous initiators, whereas isobutene responds mainly to cationic initiators. Styrene can be polymerized using any of the four types of initiators. Isoprene and 1,3-butadiene can be homopolymerized with all the initiators except the cationic type, whereas ethylene polymerization can be initiated by all except the anionic type.

The free-radical initiating system has the practical advantage that the polymerizations can be carried out in the gas, solid, and liquid phases (bulk, solution, emulsion, suspension, and precipitation techniques). Free-radical reactions can be carried out in water, whereas the other initiators usually require anhydrous conditions.

Many of the anionic and cationic polymerizations can be considered to have no inherent termination step and may be called "living polymers."

The stereochemistry of the repeating units depends on the structure of the starting monomer, the initiating system, and the conditions of the polymerization reaction. Optical isomerism, geometric isomerism, repeat unit configuration (isotactic, syndiotactic, atactic) and repeat unit orientation (head-to-tail or head-to-head) are some important aspects of the stereochemistry problem.

The mechanisms of polymerization will not be discussed here but several worthwhile references should be consulted [1]. This chapter will give mainly examples of some selected preparative methods for carrying out the major methods of polymerization as encountered in the laboratory. All intrinsic viscosities listed in this chapter have units of dl/gm.

2. FREE-RADICAL POLYMERIZATIONS

In 1838 Regnault [2] reported that vinylidene chloride could be polymerized. In 1839 Simon [3] and then Blyth and Hofmann (1845) [4] reported the preparation of polystyrene. These were followed by the polymerization of vinyl chloride (1872) [5], isoprene (1879) [6], methacrylic acid (1880) [7], methyl acrylate (1880) [8], butadiene (1911) [9], vinyl acetate (1917) [10], vinyl chloroacetate [10], and ethylene (1933) [11]. Klatte and Rollett reported that benzoyl peroxide is a catalyst for the polymerization of vinyl acetate and vinyl chloroacetate [10].

In 1920 Staudinger [12] was the first to report on the nature of olefin polymerizations leading to high polymers. A great many of his studies were carried out on the polymerization of styrene. These studies led to recognition of the relationship between relative viscosity and molecular weight [13]. The radical nature of these reactions was later elucidated by Taylor [14], Paneth and Hofeditz [15], and Haber and Willstätter [16]. The understanding of the mechanism of polymerization was greatly aided by Kharasch [17], Hey and Waters [18], and Flory [19].

No effort will be made to discuss the mechanism of polymerization in this chapter, but let it suffice to say that the polymerization is governed by the three main steps shown in Eqs. (2), (3), and (4), (5), (6), and (8).

The most common initiators are either acyl peroxides, hydroperoxides, or azo compounds. Hydrogen peroxide, potassium persulfate, and sodium perborate are popular in aqueous systems. Ferrous ion in some cases enhances the catalytic effectiveness. The use and preparation of acyl peroxides and hydroperoxides is described in Chapters 14 and 15.

Ethylene is conveniently polymerized in the laboratory at atmospheric pressure using a titanium-based coordination catalyst (see Section 5) [20]. It may also be polymerized less conveniently in the laboratory under high pressures using free-radical catalysts at high and low temperatures [21]. Other olefins such as propylene, 1-butene, or 1-pentene homopolymerize free radically only to low molecular weight polymers and require ionic or coordination catalysts to afford high molecular weight polymers [22,22a]. These olefins can effectively be copolymerized free radically.

The free-radical polymerization process which can be carried out in the laboratory is best illustrated by polymerization of styrene.

Free-radical polymerization processes [22a] are carried out either in bulk, solution, suspension, emulsion, or by precipitation techniques, as described in Sections A–E. In all cases the monomer used should be free of solvent and inhibitor or else a long induction period will result. In some cases this may be overcome by adding an excess of initiator. Table I describes some typical examples of the polymerization of various olefinic and diolefinic monomers.

TABLE I

FREE-RADICAL POLYMERIZATION OF VARIOUS OLEFINIC AND DIOLEFINIC MONOMERS

Monomer (gm or ml)	Catalyst (gm)	Solvent (ml)	Additive (gm)	Reaction conditions: Temp. (°C)	Reaction conditions: Time (hr)	% Yield polymer	Polymer mol. wt.	Ref.
Styrene (50 gm)	—	—	125	24	90	—	150,000	*a*
			125	168				
			150	48	100	—	—	*a*
(50 ml)	Benzoyl peroxide (1.0)	—	50	72	—	—	700,000	*a*
(50 gm)	α,α-Azobis(isobutyronitrile) (0.5)	Toluene (500)	100	24	—	—	—	*b*
(50 gm)	$K_2S_2O_8$ (0.05)	H_2O (100)	Na_2HPO_4 (0.05) + sodium lauryl sulfate (1.0)	79–95	2	—	—	*c*
Butadiene (77.2)	$K_2S_2O_8$ (0.4)	H_2O (150)	CCl_4 (7.5) + sodium oleate (2.3)	50	65	69	—	*d*
$CH_2{=}CH_2$ + $R{-}CH{=}CH_2$	Di-*t*-butyl peroxide + acetone oxime	—	—	130–220 at 15,000–25,000 lbs/in^2	—	—	—	*e*

Butadiene (75 gm) + styrene (25 gm)	$K_2S_2O_8$ (0.3)	H_2O (180) + soap flakes (56)	—	150	12	79	—	f
Isoprene (1000 ml)	Benzoyl peroxide (0.0488 mole)	—	—	60	336	50	—	g
$CH_2{=}C(CH_3){-}C(CH_3){=}CH_2$ (20 gm)	Diisopropylbenzene monohydroperoxide (0.08 gm)	H_2O (36)	Soap (1.0) + KCl (0.2) + K_3PO_4 (0.08) + tetraethylene-pentamine (0.08)	15	4	60	10,000[i]	h

[a] Dow Chemical Company, "The Polymerization of Styrene," Product Bulletin. 1961.
[b] M. Hunt, U.S. Patent 2,471,959 (1949).
[c] C. E. Schildknecht, "Vinyl and Related Polymers." Wiley, New York, 1952; I. M. Kolthoff and W. J. Dale, *J. Amer. Chem. Soc.* **69**, 442 (1947).
[d] G. H. Fremon and W. N. Stoop, U.S. Patent 3,168,593 (1965).
[e] L. Boghetich, G. A. Mortimer, and G. W. Dawes, *J. Polym. Sci.* **61**, 3 (1962).
[f] I. M. Kolthoff and W. E. Harris, *J. Polym. Sci.* **2**, 41 (1967).
[g] R. H. Gobran, M. B. Berenbaum, and A. V. Tobalsky, *J. Polym. Sci.* **46**, 431 (1960).
[h] M. Morton and W. E. Gibbs, *J. Polym. Sci., Part A* **1**, 2679 (1963).
[i] $\eta_i = 0.7$ (C_6H_6).

Initiator:

$$I_2 \longrightarrow 2I\cdot \quad (1)$$

Initiation:

$$CH_2{=}CHR + I\cdot \longrightarrow I{-}CH_2{-}\underset{\displaystyle R}{\underset{|}{CH}}\cdot \quad (2)$$

Propagation:

$$I{-}CH_2\underset{\displaystyle R}{\underset{|}{CH}}\cdot + CH_2{=}CHR \longrightarrow I{-}CH_2{-}\underset{\displaystyle R}{\underset{|}{CH}}{-}CH_2{-}\underset{\displaystyle R}{\underset{|}{CH}}\cdot \xrightarrow{nCH_2=CHR}$$

$$I(CH_2\underset{\displaystyle R}{\underset{|}{CH}})_{n+1}CH_2\underset{\displaystyle R}{\underset{|}{CH}}\cdot \quad (3)$$

Termination (by radical coupling, disproportionation, or chain transfer):

Radical coupling:

$$\sim CH_2\underset{\displaystyle R}{\underset{|}{CH}}\cdot + \sim CH_2{-}\underset{\displaystyle R}{\underset{|}{CH}}\cdot \longrightarrow \sim CH_2\underset{\displaystyle R}{\underset{|}{CH}}\underset{\displaystyle R}{\underset{|}{CH}}CH_2\sim \quad (4)$$

Disproportionation of two radicals:

$$\sim CH_2{-}\underset{\displaystyle R}{\underset{|}{CH}}\cdot + \sim CH_2{-}\underset{\displaystyle R}{\underset{|}{CH}}\cdot \longrightarrow \sim CH_2\underset{\displaystyle R}{\underset{|}{CH_2}} + \sim CH{=}\underset{\displaystyle R}{\underset{|}{CH}} \quad (5)$$

Chain transfer:

$$\sim CH_2\underset{\displaystyle R}{\underset{|}{CH}}\cdot + R'SH \longrightarrow \sim CH_2\underset{\displaystyle R}{\underset{|}{CH_2}} + R'S\cdot \quad (6)$$

$$R'S\cdot + CH_2{=}\underset{\displaystyle R}{\underset{|}{CH}} \longrightarrow RS'CH_2\underset{\displaystyle R}{\underset{|}{CH}}\cdot \quad \text{(start of new monomer chain)} \quad (7)$$

A. Bulk Polymerization

Bulk polymerization consists of heating the monomer without solvent with initiator in a vessel. The monomer–initiator mixture polymerizes to a solid shape fixed by the shape of the polymerization vessel. The main practical disadvantages of this method are the difficulty in removal of polymer from a reactor or flask and the dissipation of the exothermicity of the polymerization. Some typical examples are shown in Preparations 2-1 and 2-2. This method finds importance in producing cast or molded products, such as plastic scintillators, in small or very large shapes, but is difficult since the formation of local hot spots must be avoided.

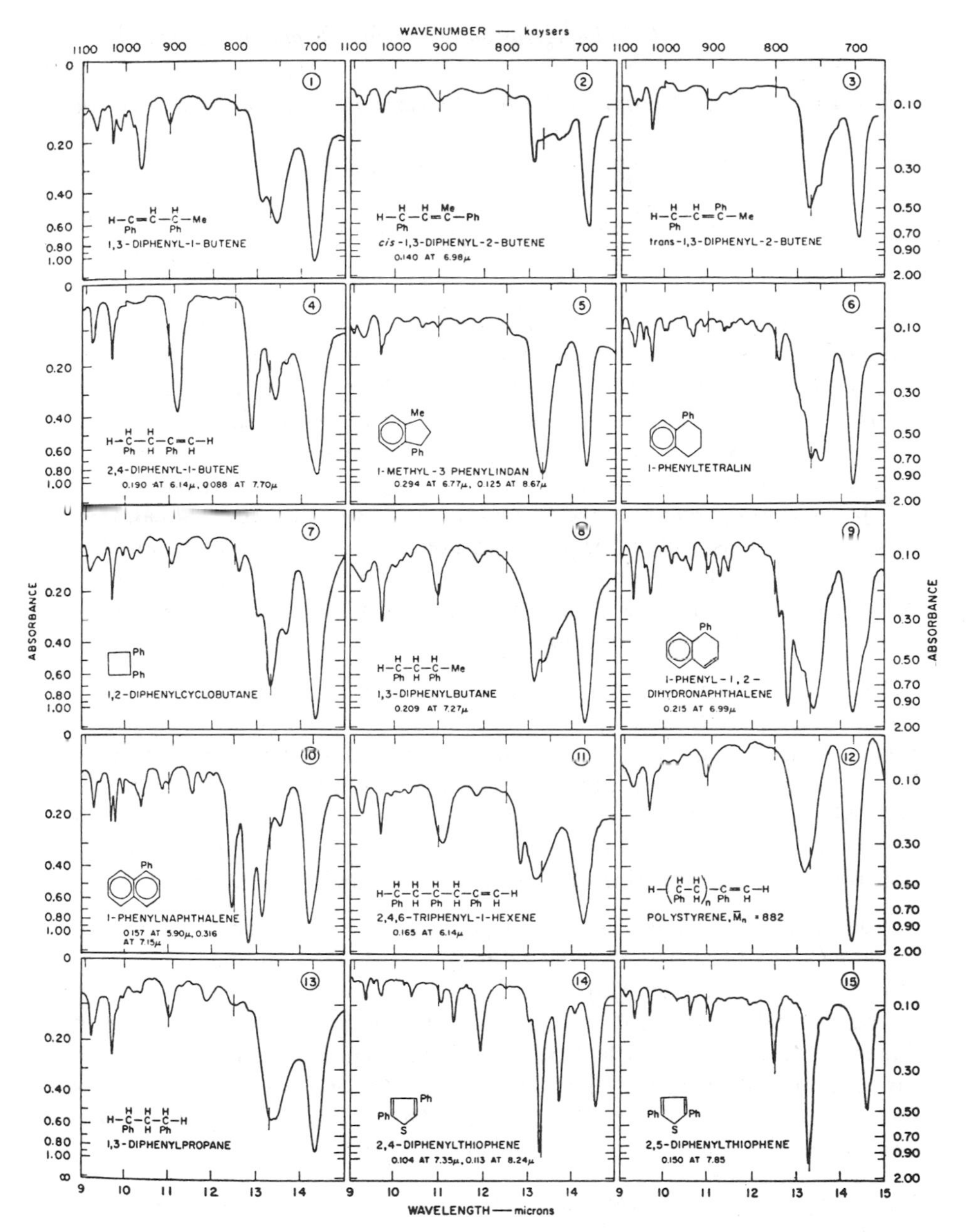

FIG. 1. Infrared absorption spectra between salt plates and identifying numbers of styrene dimers and related compounds. Other distinctive bands and their maximum absorbances on the same sample are indicated. The curves are marked at 11.00, 12.50, and 13.30 μ to bring out differences at critical points. Glpc retention times for some of the compounds are: **2** and **13**, 10.5 min; **8**, 11.3 min; **4** and **5**, 13.3 min; **1**, 15.6 min; **7**, 16.1 min; **6**, 17.9 min; **3**, 18.6 min; **9**, 20.4 min; **10**, 26.0 min. Reprinted from F. R. Mayo, *J. Amer. Chem. Soc.* **90**, 1289 (1968). Copyright 1968 by the American Chemical Society. Reprinted by permission of the copyright owner.

2-1. *Preparation of Polystyrene by the Thermal Polymerization of Styrene* [23,23a]

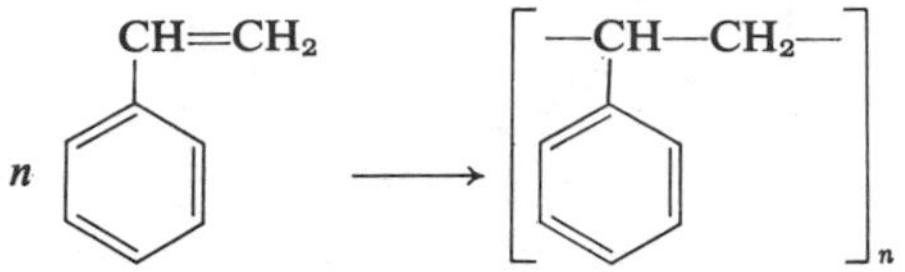

$$n\ C_6H_5CH{=}CH_2 \longrightarrow [-CH(C_6H_5)-CH_2-]_n \tag{8}$$

To a polymer tube is added 25 gm of distilled styrene monomer (preferably inhibitor-free) (see Note a), then the tube is flushed with nitrogen and placed in an oil bath at 125°C for 24 hr. At the end of this time the monomer is converted in 90% yield to the polymer. Heating for 7 days at 125°C and then 2 days at 150°C gives a 99% or better conversion to polymers. If all the air has been excluded the polystyrene will be free of yellow stains on the surface [23a]. The polystyrene is recovered by breaking open the tube and can then be purified by dissolving in benzene and precipitating it with methanol. The solid polymer is filtered, and dried at 50°–60°C in a vacuum oven to give a 90% yield (22.5 gm). The molecular weight is about 150,000–300,000 as determined by viscometry in benzene at 25°C.

NOTES: (a) The monomer should be distilled prior to use to give a faster rate of polymerization. If this is not practical then the sample should be washed with 10% aq. NaOH and dried. (b) Styrene monomer also polymerizes at room temperature over a period of weeks or months. (c) Use of 0.5% of benzoyl peroxide allows one to prepare the polymer at 50°C over a 72 hr period to give $\eta_i = 0.4$ (0.5 gm/100 ml C_6H_6 at 25°C). This type of combination is suitable for preparing castings or molds of polystyrene objects.

Additional information on the polymerization of styrene and related monomers has been described by Boundy and Boyer [24] and also by the Dow Chemical Company [25].

The thermal polymerization of styrene at 150°C also gives a mixture of dimers of which eleven C_{16} hydrocarbons have been identified by Mayo [26]. For example, after 0.18 hr at 155°C (no catalyst or solvent used) there is formed 0.2% dimer and 13.7% polymer. The structure of some of these dimers and their infrared spectra are shown in Fig. 1.

2-2. *Preparation of Polybicyclo[2.2.1]hepta-2,5-diene (Polynorbornadiene)* [27a]

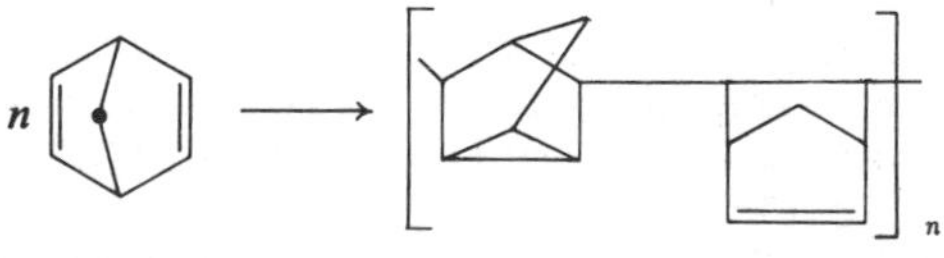

$$n\ \text{(norbornadiene)} \longrightarrow [\ \ ]_n \tag{9}$$

To a 250 ml three-necked round-bottomed flask equipped with a reflux condenser, Teflon magnetic bar, and gas inlet and outlet tube is added nitrogen over a 1 hr period to purge the system. Then a slurry of 0.1804 gm (0.1 mole %) of azobis(isobutyronitrile) in 100 gm of norbornadiene is added under nitrogen. The reaction mixture is purged for several minutes using nitrogen and then the mixture is heated for 6 hr at 70°C. The polymer is isolated by precipitating it in hexane, dissolving it in benzene, and then reprecipitating it in hexane. The polymer (mol. wt. 17,000 by light scattering in benzene) is freeze-dried to afford 5.7 gm (5.7%). The polymer darkens at 225°C and softens at 238°C, at which point it thermosets. The heat distortion temperature is 186°C. Using 1 mole % of initiator and heating for 24 hr at 70°C gives 56% of benzene-insoluble polymer (colorless gel). Norbornadiene (M_2) copolymerizes with vinyl acetate (M_1) to give the following reactivity ratios as calculated by the Fineman-Ross equation: $r_1 = 1.28$; $r_2 = 0.82$ [27].

B. Solution Polymerization

Solution polymerization has the advantage of allowing heat dissipation and ease of removal of polymer from the vessel [27a]. The main disadvantages are that the solvent acts as a chain transfer agent to lower the molecular weight and the solvent is difficult to remove from the free polymer. Properly chosen solvent systems allow one to prepare polymer solutions which can be used to cast films or be spun into fibers. In most cases dilute solutions have to be prepared in order to avoid high viscosity buildup and stirring problems.

Aqueous solution polymerization (emulsifier-free latex system) is of technological importance and has recently been reviewed [27b].

C. Precipitation Polymerization

If the polymer formed is insoluble in its own monomer of a monomer–solvent combination and precipitates out as it is formed, the process is termed "precipitation polymerization" [27a]. Some examples are bulk polymerization of vinyl chloride [28] and vinylidene chloride [29], styrene in alcoholic solutions [30], and methyl methacrylate in water [31].

D. Suspension Polymerization

In suspension polymerization a catalyst is dissolved in the monomer, which is then dispersed in water. A dispersing agent is added separately to stabilize the resulting suspension. The particles of the polymer are 0.1 to 1 mm in size. The rate of polymerization and other characteristics are similar to those found in bulk polymerization. Some common dispersing agents are polyvinyl alcohol, polyacrylic acid, gelatin, cellulose, and pectins. Inorganic dispersing agents are phosphates, aluminum hydroxide, zinc oxide, magnesium silicates, and kaolin. Several worthwhile references should be consulted for more details [27a,32,32a].

2-3. Preparation of Polystyrene by Suspension Polymerization [32a]

To a 350 ml glass bottle is added 100 gm of styrene (see Note), 0.3 gm of benzoyl peroxide, 150 ml of water, and 3.0 gm of zinc oxide. Then conc. aqueous ammonium hydroxide is added to adjust the pH to 10. The bottle is then sealed and rotated (30 rpm) in an oil bath for 7 hr at 90°C and 5 hr at 115°C. At the end of this time the pH is 7.25. The polymer granules are filtered, suspended in water, and acidified to pH approx. 2 with 10% aqueous hydrochloric acid in order to remove the zinc oxide. The polymer granules are then washed with water and dried.

NOTE: This process works well for other olefins such as vinyltoluene, vinylxylene, and *tert*-butylstyrene.

E. Emulsion Polymerization

Emulsion polymerization is discussed more fully in Chapter 10 and only a brief review will be given here.

The system basically consists of water and 1–3% of a surfactant (sodium lauryl sulfate, sodium dodecyl benzenesulfonate, or dodecylamine hydrochloride) and a water-soluble free-radical generator (alkali persulfate, hydroperoxides, or hydrogen peroxide–ferrous ion). The monomer is added gradually or is all present from the start. The emulsion polymerization is usually more rapid than bulk or solution polymerization for a given monomer at the same temperature. In addition the average molecular weight may also be greater than that obtained in the bulk polymerization process. The particles in the emulsion polymerization are of the order of 10^{-5} to 10^{-6} mm in size. It is interesting to note that the locus of polymerization is the micelle and only one free radical can be present at a given time. The monomer is fed into the locus of reaction by diffusion through the water where the reservoir of the monomer is found. If another radical enters the micelle then termination results because of the small volume of the reaction site. In other words, in emulsion polymerization the polymer particles are not formed by polymerization of the original monomer droplets but are formed in the micelles to give polymer latex particles of very small size. For a review of the roles of the emulsifier in emulsion polymerization see a recent review by Dunn [33]. Other references with a more detailed account of the field should be consulted [34,34a]. These references are only a sampling of the many that can be found in *Chemical Abstracts*.

The polymer in emulsion polymerization is isolated by either coagulating or spray-drying.

2-4. Emulsion Polymerization of Styrene [35]

To a resin kettle equipped with a mechanical stirrer, condenser, and nitrogen inlet tube, is added 128.2 gm of distilled water, 100 gm of styrene, 0.070 gm of potassium persulfate, and 100 ml of 3.56% soap solution (see Note). The system is purged with nitrogen to remove dissolved air. Then the temperature is raised to 50°C and kept there for 2 hr to afford a 90% conversion of polymer. The polymer is isolated by freezing–thawing or by adding alum solution and boiling the mixture. The polystyrene is filtered, washed with water, and dried.

NOTE: In place of soaps such as sodium stearate one can use 1.0 gm of either sodium dodecyl benzenesulfonate or sodium lauryl sulfate.

Price and Adams [36] earlier reported on the effect of changes in catalyst concentration, type of soap, and monomer ratio on the emulsion polymerization of styrene at 50°C. Figures 2, 3, and 4 and Tables II and III summarize their data.

Styrene emulsion polymerizations have been studied in greater detail by several workers. Recently Williams and Bobalek [37] studied the application of molecular weight and particle growth measurements in continuously uniform lattices to kinetic studies of styrene emulsion polymerization.

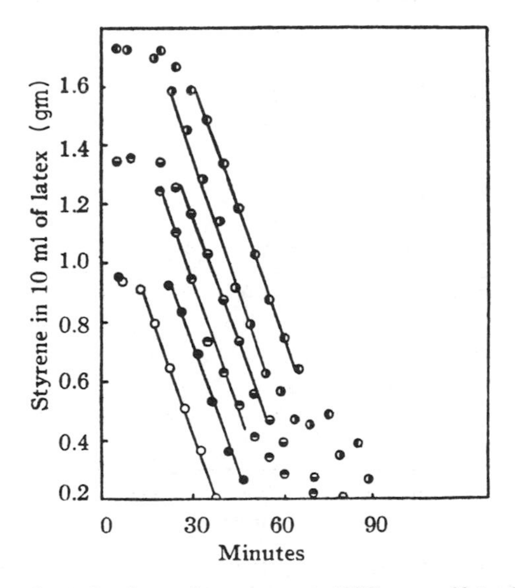

FIG. 2. Rate of polymerization of styrene at 50°C, persulfate 0.00203 mole/liter: expt. 5, ○; expt. 10, ●; expt. 28, ◒; expt. 29, ◓; expt. 30, ◐; expt. 31, ◑. [Reprinted from C. C. Price and C. E. Adams, *J. Amer. Chem. Soc.* **67**, 1674 (1945). Copyright 1945 by the American Chemical Society. Reprinted by permission of the copyright owners.]

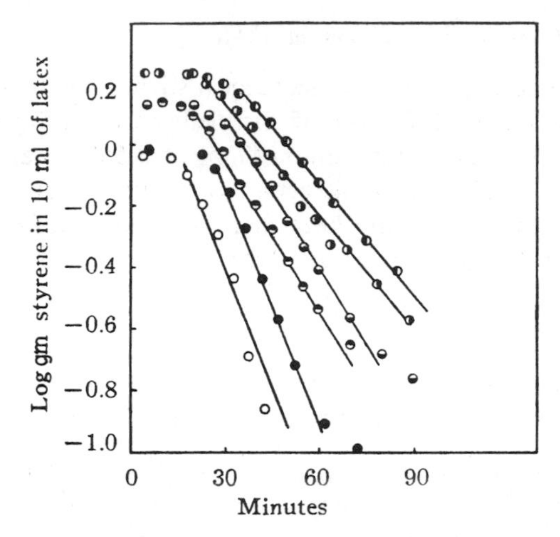

FIG. 3. Rate of polymerization of styrene at 50°C (plotted as a first-order reaction), persulfate 0.00203 mole/liter: expt. 5, ○; expt. 10, ●; expt. 28, ◓; expt. 29, ◒; expt. 30, ◐; expt. 31, ◑. [Reprinted from C. C. Price and C. E. Adams, *J. Amer. Chem. Soc.* **67**, 1674 (1945). Copyright 1945 by the American Chemical Society. Reprinted by permission of the copyright owners.]

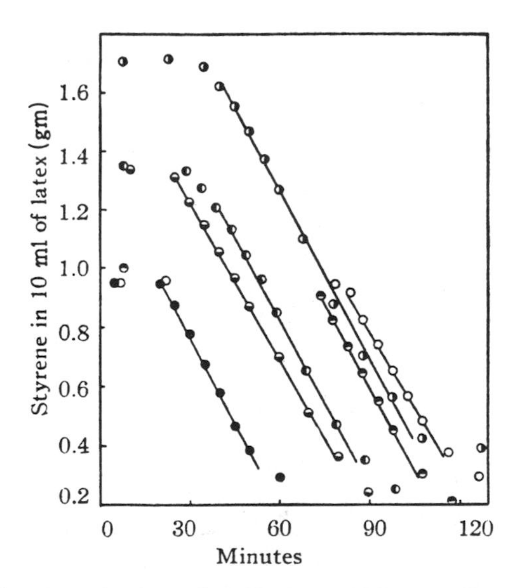

FIG. 4. Rate of polymerization of styrene at 50°C, persulfate 0.00101 mole/liter: expt. 22, ●; expt. 23, ○; expt. 24, ◓; expt. 25, ◒; expt. 26, ◐; expt. 27, ◑. [Reprinted from C. C. Price and C. E. Adams, *J. Amer. Chem. Soc.* **67**, 1674 (1945). Copyright 1945 by the American Chemical Society. Reprinted by permission of the copyright owners.]

TABLE II

EFFECT OF VARIABLES ON THE EMULSION POLYMERIZATION OF STYRENE AT 50°C BATH TEMPERATURE[a]

Expt.	Soap	Catalyst[c] (mole/liter)	Styrene		Induction period (min)	k_0 (mole liter^{-1} min^{-1})	Mol. wt.
			cm^3	Purity			
1	Na[b]	0.00406	50	Comm.[d]	105	0.0259	185,200
2	Na[b]	0.00448	100	Comm.	89	0.0436	222,300
3	Na[b]	0.00211	100	Comm.	292	0.0219	226,000
4	K	0.00203	50	Res.[e]	35	0.0265	257,300
5	Na[b]	0.00203	50	Chem.	13	0.0281	290,700
6	Na[b]	0.00203	50	Comm.	230	0.0192	248,200
7	Na[b]	0.00203	50	Comm.	159	0.0194	231,100
8	Na	0.00203	50	Comm.	316	0.0251	237,400
9	Na	0.00203	50	Comm.	46	0.0240	256,700
10	K	0.00203	50	Res.	23	0.0282	269,500
11	K	0.00203	50	Res.	1461 (air)	—	—
12	K	0.00203	50	Res.	1675 (air)	—	—
					40 (N_2)	0.0306	218,400
13	Na, neut.	0.00203	50	Comm.	129	0.0200	205,300
14	Na, 10% excess alk.	0.00203	50	Comm.	33	0.0206	228,700
15	Na, 10% excess acid	0.00203	50	Comm.	72	0.0238	218,400
16	Na	0.00203	50	Res.	22	0.0280	158,500
17	K	0.00203	50	Res.	20	0.0282	219,700
18	Na, 848 ppm act. oxygen	0.00203	50	Comm.	37	0.0226	223,700
19	Na, 3.3 ppm act. oxygen	0.00203	50	Comm.	36	0.0246	246,700
20	K	0.00305	50	Res. + 0.5% dodecyl mercaptan	3	0.0298	15,400
21	K dehydroabietate	0.00305	50	Res.	13	0.0115	168,600
47	K dehydroabietate	0.00305	50	Res.	14	0.0125	171,000
48	K hydro- and dehydroabietate	0.00305	50	Res.	17	0.0100	145,300

[a] Reprinted from C. C. Price and C. E. Adams, *J. Amer. Chem. Soc.* **67**, 1674 (1945). Copyright 1945 by the American Chemical Society. Reprinted by permission of the copyright owners.

[b] This soap was prepared in aqueous solution previous to experiments, the other samples were prepared *in situ*. The soap is prepared from stearic acid and either NaOH or KOH.

[c] $K_2S_2O_8$ (100% pure). [d] Commercial. [e] Research.

TABLE III

RATES OF EMULSION POLYMERIZATION OF STYRENE AT VARYING MONOMER AND CATALYST CONCENTRATIONS[a]

Expt.	Catalyst[b] (mole/liter)	Ratio styrene: water	Max. temp. (°C)	k_0	k^c (cor.)		Mol. wt.
42	0.00203	1:8	40.8	0.0101	0.207	40°C	235,300
43	0.00406	1:8	41.0	0.0130	0.185		292,700
44	0.00406	1:8	41.0	0.0122	0.174		379,400
45	0.00609	1:8	41.5	0.0199	0.221		306,600
46	0.00812	1:8	42.0	0.0218	0.199		—
22	0.00101	1:8	51.3	0.0186	0.520	50°C	314,000
23	—	1:8	51.0	0.0164	0.471		352,000
24	—	1:8	51.4	0.0183	0.507		290,500
25	—	1:6	51.0	0.0168	0.482		301,000
26	—	1:6	51.0	0.0177	0.509		332,000
27	—	1:4	51.2	0.0180	0.508		375,000
5	0.00203	1:8	52.0	0.0282	0.523		290,700
10	—	1:8	52.5	0.0283	0.502		269,500
28	—	1:6	52.5	0.0290	0.514		321,000
29	—	1:6	52.2	0.0287	0.523		299,300
30	—	1:4	52.3	0.0283	0.509		299,300
31	—	1:4	52.7	0.0294	0.512		305,200
32	0.00305	1:8	52.9	0.0352	0.512		279,500
33	—	1:8	52.7	0.0344	0.488		250,500
34	0.00406	1:8	53.0	0.0394	0.471		269,900
35	—	1:8	53.2	0.0394	0.463		238,400
36	—	1:6	53.1	0.0396	0.464		226,800
37	—	1:4	53.4	0.0404	0.466		258,100
38	0.0406	1:8	55.8	0.216 (?)	0.638		196,200
39	0.00203	1:8	64.3	0.0642	0.987	60°C	219,300
40	0.00101	1:8	63.7	0.0526	1.205		256,700
41	0.00051	1:8	62.6	0.0344	1.218		271,300

[a] Reprinted from C. C. Price and C. E. Adams, *J. Amer. Chem. Soc.* **67**, 1674 (1945). Copyright 1945 by the American Chemical Society. Reprinted by permission of the copyright owner.

[b] $K_2S_2O_8$.

[c] $k = k_0/[\text{cat.}]^{1/2}$. The values reported have been corrected for the observed temperature rise.

Robb [38] has studied the emulsion polymerization of styrene at 40°C [styrene (30 gm), water (300 gm), sodium dodecyl sulfate (2.1 gm), and potassium persulfate (0.60 gm)]. After 2 hr a 92% conversion was obtained. Robb determined the number of particles per unit volume of latex during the emulsion polymerization of styrene and described their properties.

2-5. *Emulsion Polymerization of Styrene to Give Uniform Monodisperse Polymer Particles* [37]

Polymerization experiments are carried out in a bottle polymerizer maintained at 60° ±1°C. To the bottle is added 155 gm of water, 100 gm (0.95 mole) of styrene monomer, 0.5 gm of potassium persulfate (0.014 mole) dissolved in 10 gm of water, and 0.78 gm (0.0035 mole) of potassium laurate soap. The bottle is sealed and through a hyperdermic cap opening is added 2.35 gm (0.0105 mole) of soap dissolved in 15 gm of water in 1 ml increments every 12 min after the reaction starts. The kinetic data, molecular weight, particle size, and other data are shown in Table IV.

In contrast, α-methylstyrene has been polymerized in only low yields (24.1%) by emulsion polymerization techniques and even lower yields by other radical processes [39].

2-6. *Preparation of Butadiene–Styrene Copolymers by the Emulsion Polymerization Technique* [40]

$$n\mathrm{CH_2{=}CH{-}CH{=}CH_2} + n\mathrm{C_6H_5{-}CH{=}CH_2} \longrightarrow \left[\mathrm{{-}CH_2{-}CH{=}CH{-}CH_2{-}CH_2{-}\underset{\displaystyle C_6H_5}{\underset{|}{CH}}{-}}\right]_n \quad (10)$$

To an ice–water–salt-cooled 275 ml stainless steel pressure bomb (or 4 oz bottles with metal caps containing a rubber gasket for a total charge of only 100 gm) is added 180 gm of distilled water, 5 gm of Proctor & Gamble SF Flakes (see Note a), 0.3 gm of potassium persulfate (reagent grade) (Note b), and 0.5 gm of dodecyl mercaptan dissolved in 25 gm of styrene. Then the butadiene, 75.0 gm, is weighed into an ice-chilled 4 oz bottle and one additional 0.5 gm is added to allow for transfer loss in pouring into the stainless steel bomb. The bomb is sealed and placed into a shaker–oil bath at 110°C–125°C for 1 hr. The bomb is cooled in a stream of air and then finally cooled in an ice bath. The latex is poured out into a beaker containing phenyl-β-naphthylamine (see Note c) and coagulated with a saturated salt–dilute sulfuric acid solution to give 75 gm (75%) of 96% solubility in benzene.

NOTES: (a) Sodium soap of hydrogenated tallow fatty acid (anhydrous) may also be used. (b) A mixture of 0.08 gm of 100% active *p*-menthane hydroperoxide (Hercules Powder C.), 0.08 gm of sodium formaldehyde sulfoxylate ($FeSO_4 \cdot 7H_2O$), and 0.02 gm of Versene Fe-3 (100%) may be used if one wants the polymerization to proceed at 5°C [41]. In addition, amines and peroxides have been reported to give electron-transfer complexes which decompose at room temperature or below [42]. (c) CAUTION: carcinogenic.

TABLE IV

KINETIC DATA AND PARTICLE SIZE UNIFORMITY FOR PREPARATION 2-5[a]

Soap conc. per charge (mole)	Experiment No.	Time (min)	% Conversion	Particle size diameter (Å)	Particles converted	Uniformity ratio	Intrinsic viscosity	Mol. wt. $\times 10^{-6}$
0.014	1	15	2.36	370	7	1.031	0.903	0.291
0.014	2	45	19	835	15	1.024	2.405	1.412
0.014	3	75	35.8	925	19	1.009	2.631	1.632
0.014	4	105	60.5	1090	14	1.011	2.560	1.556
0.014	5	135	80.5	1250	19	1.010	2.544	1.546
0.014	6	180	90.2	1245	17	1.010	2.306	1.319
0.014	Final sample	—	100	1290	21	1.008	2.23	1.257

[a] Data taken from Ref. [37].

2-7. Emulsion Polymerization of 1,2-Dimethylenecyclohexane to Give an All-cis-Diene Polymer [43]

$$\begin{array}{c} H_2C{=}C{-}C{=}CH_2 \\ H_2C\langle \quad \rangle CH_2 \\ H_2C{-}CH_2 \end{array} \longrightarrow$$

$$\begin{array}{c} {-}CH_2 \quad\quad CH_2{-} \\ C{=}C \\ H_2C\langle \quad \rangle CH_2 \\ H_2C{-}CH_2 \end{array} \left[\begin{array}{c} CH_2 \quad\quad CH_2 \\ C{=}C \\ H_2C\langle \quad \rangle CH_2 \\ H_2C{-}CH_2 \end{array} \right] \begin{array}{c} CH_2 \quad\quad CH_2{-} \\ C{=}C \\ H_2C\langle \quad \rangle CH_2 \\ H_2C{-}CH_2 \end{array} \quad (11)$$

In a 2 oz screw-cap bottle are placed 10.0 gm of 1,2-dimethylenecyclohexane, 0.50 gm of sodium stearate, 0.03 gm of potassium persulfate, 0.05 gm of lauryl mercaptan, and 18.0 gm of water. The bottle is rotated end over end in a 55°C water-bath for 24 hr, at which time the emulsion has broken. The mixture is acidified and the solid filtered off, washed thoroughly, and dried under vacuum to produce 8.5 gm (85% conversion) of the polymer. The 8.5 gm of polymer is dissolved in 1 liter of boiling toluene containing a trace of *N*-phenyl-β-naphthylamine and the solution poured with rapid stirring into 3 liters of cold methanol to reprecipitate the polymer. The precipitate is filtered, washed with a solution of *N*-phenyl-β-naphthylamine, and dried under vacuum to yield 7.5 gm of purified poly-1,2-dimethylenecyclohexane, m.p. 164.5°–165.0°C. Additional purification can be effected by dissolving the polymer in boiling toluene, filtering to remove any extraneous material, cooling to allow the polymer to crystallize out, filtering the precipitated polymer, and treating as before. The melting point is determined by placing the polymer in a thin-walled capillary and heating the capillary in a melting point bath, in the same way that the melting point of any crystalline organic compound is determined. No visible change is noted below 160°C, but at approximately this temperature shrinking occurs. Shrinking continues without any visible appearance of melting until the temperature reaches 164.5°C. At this temperature the polymer consists of a solid white column with about half the original diameter but otherwise with very nearly the original appearance. Within a half of one degree the polymer passes from this white solid to a thin, water-white liquid which runs down the side of the capillary. [From W. J. Bailey and H. R. Golden, *J. Amer. Chemical Soc.* **76**, 5418 (1954).

3. CATIONIC POLYMERIZATIONS

Cationic polymerization has a history dating back about 183 years and has been extensively investigated by Plesch [44], Dainton [45], Polanyi [46], Pepper [47,47a], Evans [48], Heiligmann [49], and others [50]. Whitmore [51] is credited with first recognizing that carbonium ions are intermediates in the acid-catalyzed polymerizations of olefins. The recognition of the importance of proton-donor cocatalysts for Friedel-Crafts catalysts was first reported by Polanyi and co-workers [46]

$$MX_n + SH \rightleftharpoons [MX_nS]^-H^+ \tag{12}$$

(SH and RX = Lewis Base)

$$MX_n + RX' \rightleftharpoons [MX_nR]^-E^+ \tag{13}$$

Some common initiators for cationic polymerization reactions are either protonic acids, Friedel-Crafts catalysts (Lewis acids), compounds capable of generating cations, or ionizing radiation.

Of all the acid catalysts used [52a–c] sulfuric acid is the most common. Furthermore, sulfuric acid appears to be a stronger acid than hydrochloric acid in nonaqueous solvents. Some other commonly used catalysts are HCl, H_2SO_4, BF_3, $AlCl_3$, $SnCl_4$, $SnBr_4$, $SbCl_3$, $BeCl_3$, $TiCl_4$, $FeCl_3$, $ZnCl_2$, $ZnCl_4$, and I_2.

For alkenes the reactivity is based on the stability of the carbonium ion formed and they follow the order: tertiary > secondary > primary. Thus olefins react as follows: $(CH_3)_2C{=}CH_2 \simeq (CH_3)_2C{=}CHCH_3 > CH_3CH{=}CH_2 > CH_2{=}CH_2$. Allylic and benzylic carbonium [53a,b] ions are also favored where appropriate.

The cationic polymerization process has been reviewed and several references are worthwhile consulting [47,54,55].

Certain aluminum alkyls and aluminum dialkyl halides in the presence of proton- or carbonium-donating cocatalysts act as effective polymerization catalysts. Table V gives several examples used to polymerize isobutene–isoprene (3 wt %) mixtures with $AlEt_2Cl$ catalyst in several solvents.

In the absence of monomers, trimethylaluminum (0.5 mole) reacts with *t*-butyl chloride (1.0 mole) at −78°C to give a quantitative yield of neopentane [55]. Kennedy found that aluminum trialkyls ($AlMe_3$, $AlEt_3$, $AlBu_3$) in the presence of certain alkyl halides are efficient initiators for the cationic polymerization of isobutylene, styrene, etc. [56].

$$AlR_3 + R'X \longrightarrow [AlR_3X]^- + [R]^+$$

All the experiments described in Table V were carried out under a nitrogen atmosphere in stainless steel equipment [57]. The purification of the reagents has been described previously.

Some typical examples of the experimental conditions for the cationic polymerization of various olefinic and diolefinic monomers are shown in Table VI.

It should be recognized from the results in Table VI that cationic polymerizations are usually initiated at low temperatures in order to suppress chain-terminating reactions and also to keep the reaction from becoming explosive in nature. These low temperatures thus favor high molecular weight polymer formation.

Substituted olefins which are capable of forming secondary or tertiary carbonium ion intermediates polymerize well by cationic initiation but are polymerized with difficulty or not at all free radically.

3-1. Preparation of Polyisobutylene [58]

$$n\mathrm{CH_3{-}\overset{\overset{\displaystyle CH_3}{|}}{C}{=}CH_2} \longrightarrow \left[\mathrm{{-}CH_2{-}\overset{\overset{\displaystyle CH_3}{|}}{\underset{\underset{\displaystyle CH_3}{|}}{C}}{-}} \right]_n \qquad (14)$$

A flask containing 10 gm of isobutylene and 20 gm of carbon tetrachloride is cooled to −78°C and boron trifluoride gas is bubbled into the mixture. The isobutylene polymerizes and the whole mass apparently solidifies. On warming to room temperature the carbon tetrachloride melts away, leaving 5.0 gm (50%) of polymer, mol. wt. 20,000.

NOTE: Ethylene dichloride can also be used in place of carbon tetrachloride, and other suitable solvents for the polymerization of isobutylene are either ethylene dichloride or a mixture of 2 volumes of liquid ethylene and 1 volume of ethyl chloride. Propane is not as good as the latter solvents but gives a higher molecular weight product than for the case of carbon tetrachloride above.

A good account of the early development of high polymers and copolymers of isobutylene is given by Schildknecht [59].

Thomas, Sparks, and Frolich have also described the early work and patents on the polymerization of isobutylene with acidic catalysts ($TiCl_4$, $AlCl_3$, BF_3) [60]. In particular, Thomas and Sparks described the early work on the $AlCl_3$–CH_3Cl-catalyzed copolymerization of isobutylene with isoprene at low temperatures and gave several worthwhile examples in their patent [60a]. Additional examples are later described by Calfee and Thomas [60b].

TABLE V

POLYMERIZATION OF ISOBUTENE–ISOPRENE (3 MOLE %) MIXTURE WITH $AlEt_2Cl$ CATALYST[a,b]

Solvent	ml	Temp. (°C)[c]	Reaction conditions	Catalyst solution	Observation, remarks	Yield	
						gm	%
CH_3Cl	100	−50	Monomer added dropwise to catalyst and solvent charge	None	No polymerization (90 min)	—	—
CH_3Cl	75	−75	Monomer and catalyst stirred for 10 min	1 ml of CH_3Cl containing 4×10^{-3} mole HCl	Explosive polymerization on HCl addition	—	100
C_2H_5Cl	100	−50	Monomer feed added dropwise to catalyst and solvent charge, stirred for 30 min	Undetermined amount of HCl in C_2H_5Cl	Slow polymerization	54.5	72.0
C_2H_5Cl	100	−50	As in 3rd entry except C_2H_5Cl treated with KOH to remove HCl	C_2H_5Cl treated with KOH	Very slow polym[d] for 100 min	11.0	16
C_2H_5Cl	250	−45	As in 3rd entry	As in 3rd entry	Slow polym for 6 min	34.5	48.5
n-Pentane	100	−50	As in 3rd entry	0.07 ml *n*-pentane containing 1.5×10^{-4} mole HCl	Explosive polym on HCl addition	—	100
CS_2	100	−50 ↑	As in 3rd entry	Small amount of HCl in 0.07 ml CH_3Cl	Immediate reaction on HCl addition	—	High

CH_3Cl	100	$-75\uparrow$	Monomer feed + catalyst stirred for 60 min	0.025 mole HF in 20 ml CH_3Cl	Immediate polym on HF addition run for 2 min	56	80
CH_3Cl	100	$-50\uparrow$	As in 3rd entry	About 1 ml CH_3Cl containing 1.3×10^{-4} mole (approx.) HBr	Immediate polym on HBr addition. Run for 2 min	7.0	9.9
CH_3Cl	100	-50	As in 3rd entry. No polym for 300 min	0.4 ml CH_3Cl containing 2.2×10^{-4} mole CH_3COOH added	Immediate slow polym started. Run for 90 min	13.0	18.3
CH_3Cl	100	-50	As in 3rd entry	25 ml CH_3Cl containing 1.87×10^{-3} mole CH_3OH	Immediate interaction (haziness) on CH_3OH addition by slow polym	0.81	1.1
CH_3Cl	87.6	-100	As in 3rd entry	H_2O saturated stream of N_2	Slow polym on H_2O–N_2 introduction	15.0	21.0
CH_3Cl	100	-50	As in 3rd entry	25 ml CH_3Cl containing 1.36×10^{-2} mole CH_3COCH_3	Slow initiation on CH_3COCH_3 addition (polym for 180 min)	0.06	0.1

[a] Conditions: Monomer mixture (97 ml isobutene + 3 ml isoprene) stirred in the presence of 0.01 mole $AlEt_2Cl$ for about 30 min. Cocatalyst solution added to the quiescent mixture. Polymerization terminated by introducing precooled methanol.

[b] Reprinted from J. P. Kennedy, *Polym. Prepr., Amer. Chem. Soc., Div. Polym. Chem.* 7, 485 (1966). Copyright 1966 by the American Chemical Society. Reprinted by permission of the copyright owner.

[c] ↑ Indicates temperature run-away, consequently no molecular weights are given.

[d] Polymerization.

TABLE VI

CATIONIC POLYMERIZATION OF VARIOUS OLEFINIC–DIOLEFINIC MONOMERS

Monomer (gm or ml)	Catalyst (gm or mole)	Solvent (ml)	Additives (gm)	Reaction conditions		% Yield of polymer	Polymer properties		Ref.
				Temp. (°C)	Time (hr)		Mol. wt.	PMT (°C)	
$C_6H_5C(CH_3){=}CH_2$ (200 gm)	$AlCl_3$ (1.0 gm or mole in 100 gm C_2H_5Cl)	C_2H_5Cl (2800)	—	−130	—	80–90	—	>200	*a*
$(CH_3)_2C{=}CH_2$ 100 (gm)	BF_3 (0.5)	—	Dry Ice (50)	−80	—	—	—	—	*b*
(20 gm)	BF_3 (0.2)	CH_3CH_3 (80)	—	−88	—	—	—	—	*c*
(31.9 gm)	$AlCl_3$ in CH_3Cl (32 ml of 4.13×10^{-2} mole/liter)	—	—	−78	2	100	—	—	*d*
(8.9 gm)	BF_3 (slow stream)	Ether	—	−50 to 25	4	—	183,000 to 341,000	—	*e*
(14.41 gm)	BF_3 (slow stream)	ClC_6H_5 (100)	—	−20 to 0	$\frac{1}{4}$–1	—	—	—	*e*

(50 gm)	BF_3 (0.2 gm)	ClC_6H_5 (100)	—	−50 to −20	½	74	—	—	*e*
$CH_2{=}CH{-}CH{=}CH_2$ (5 ml)	$AlCl_3$ (0.12 gm) + $FeCl_3$ (0.06 gm)	C_2H_5Cl (5 ml)	—	−78	2	80–90	—	—	*f*
(100 gm)	$RhCl_3$ (0.5)	H_2O (200)	Sodium dodecylbenzenesulfonate (5 gm)	50	20	25–30[g] (98% *trans*)	—	—	*h*
$CH_2{=}CH{-}CH{=}CH_2$[i] (200 ml) + $(CH_3)_2C{=}CH_2$ (800 ml)	$AlCl_3$ (0.5 gm in 100 ml CH_3Cl)	CH_3Cl (2000 ml)	—	−78	1/12	27 (reaction prematurely stopped)	42,000	—	*j*
$CH_2{=}C(CH_3){-}CH{=}CH_2$ (1.45, 93%) + $(CH_3)_2C{=}CH_2$ (98.55, 98%)	$AlCl_3$ (0.2% in CH_3Cl, added at the rate of 1 cc/sec)	CH_3Cl (300–400 ml)	—	−40 to −103	1/30	60–70	20,000–300,000	—	*k*
$CH_2{=}CH{-}C(CH_3){=}CH_2$ (3 ml) + $(CH_3)_2C{=}CH_2$ (97 ml)	$(C_2H_5)_2AlCl$ (0.01 mole in 75 ml CH_3Cl)	CH_3Cl (100)	0.025 mole HF in 20 ml CH_3Cl added dropwise	−103 to −60	1	80	—	—	*l*

(*cont.*)

TABLE VI (*cont.*)

Monomer (gm or ml)	Catalyst (gm or mole)	Solvent (ml)	Additives (gm)	Reaction conditions: Temp. (°C)	Reaction conditions: Time (hr)	% Yield of polymer	Polymer properties: Mol. wt.	Polymer properties: PMT (°C)	Ref.
$CH_2{=}CH{-}C(CH_3){-}CH_2$ (0.0775 mole/liter) + $CH_2{=}C(CH_3){-}CH_3$ (3.03 moles/liter)	BF_3 (1.8×10^{-5} mole)	CH_3Cl (30)	—	−78	1–2	68.6	140.2×10^3	—	*m*
$CH_2{=}CHCH{=}CH_2$ (50 gm) + $C_6H_5{-}CH{=}CH_2$ (50 gm)	$AlCl_3$ (1.0)	*n*-Hexane (100)	—	−75	1	77	—	—	*n*
(100 gm)	BF_3-ether (0.7 ml)	$CHCl_3$ (300)	—	20–30	$1\frac{1}{4}$	60–80	—	—	*o*
(0.53 mole)	$RnCl_3 \cdot 3H_2O$ (1.0 gm)	Ethanol (95%) (200)	—	78–80	6	60	—	72–90	*p*
$(CH_3)_2C{=}CH{-}CH{=}C(CH_3)_2$ (75 gm)	BF_3 (slow stream of gas for a few min)	Pet·ether (750)	—	−70 to −50	$\frac{1}{4}$	72	—	—	*q*

[a] A. B. Hersberger, J. C. Reid, and R. G. Heiligmann, *Ind. Eng. Chem.* **37**, 1075 (1945).

[b] A. J. Morway and F. L. Miller, U.S. Patent 2,243,470 (1941).

[c] M. Müller-Cumradi and M. Otto, U.S. Patent 2,203,873 (1940).

[d] J. P. Kennedy and R. M. Thomas, *J. Polym. Sci.* **46**, 233–481 (1960).

[e] H. F. Miller and R. G. Flowers, U.S. Patent 2,445,181 (1948).

[f] T. E. Ferington and A. V. Tobolsky, *J. Polym. Sci.* **31**, 25 (1958).

[g] η_i = 0.4 dl/gm at 30°C in $CHCl_3$.

[h] R. E. Rinehart and H. P. Smith, *Macromol. Syn.* **2**, 39 (1966); R. E. Rinehart, H. P. Smith, H. S. Witt, and H. Romeyn, Jr., *J. Amer. Chem. Soc.* **83**, 4864 (1961); R. E. Rinehart, H. P. Smith, H. S. Witt, and H. Romeyn, Jr., *J. Amer. Chem. Soc.* **48**, 4145 (1962).

[i] The following other dienes may be used in place of butadiene: isoprene, cyclopentadiene and dimethylbutadiene.

[j] R. M. Thomas and W. J. Sparks, U.S. Patent 2,356,128 (1944).

[k] J. D. Clalfee and R. M. Thomas, U.S. Patent 2,431,461 (1947).

[l] J. P. Kennedy, U.S. Patent 3,349,065 (1967).

[m] J. P. Kennedy and R. G. Squires, *Polymer* **6**, 579 (1965).

[n] C. S. Marvel, R. Gilkey, C. R. Morgan, J. F. Noth, R. D. Rands, Jr., and C. H. Young, *J. Polym. Sci.* **6**, 483 (1951).

[o] M. Wismer and H. P. Doerge, *Macromol. Syn.* **2**, 74 (1966).

[p] F. W. Michelotti and W. P. Keaveney, *Macromol. Syn.* **2**, 100 (1966). The structure of the polymer is as follows:

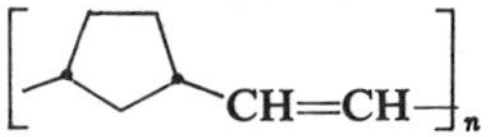

$$\text{CH=CH}\,]_n$$

[q] F. B. Moody, *Amer. Polym. Prepr., Chem. Soc., Div. Polym. Chem.* **2**, 285 (1961).

3-2. *Preparation of Polyisobutylene by the Cationic Polymerization of Isobutene with $AlCl_3$–CH_3Cl Catalyst* [61,62]

$$nCH_3-\underset{\displaystyle CH_3}{\underset{|}{C}}=CH_2 \longrightarrow \left[-\overset{\displaystyle CH_3}{\overset{|}{\underset{\displaystyle CH_3}{\underset{|}{C}}}}-CH_2-\right]_n \qquad (15)$$

To a cooled (Dry Ice–hexane) three-necked 1 liter round-bottomed flask equipped with a stirrer, nitrogen inlet–outlet tube, and low-temperature addition funnel is added a solution of 31.9 gm of isobutylene in 250 gm of *n*-pentane (see Note). Then 32 ml of 4.13×10^{-2} mole/liter aluminum chloride in methyl chloride is slowly added while keeping the temperature at −78°C. After 1 hr the polymer is isolated by evaporating the solvent to afford a quantitative yield of polyisobutylene, DP = 10,000.

NOTE: All glassware is dried at 130°C and all the reagents are anhydrous and purified as described by Kennedy [62]. The reaction can best be carried out in a stainless steel dry box under a nitrogen atmosphere.

The effectiveness of several Friedel-Crafts catalysts for the polymerization of isobutylene at −78°C is given in Table VII [63]. BF_3 is the most effective catalyst. In most cases the molecular weight of the polymer increased as the starting temperature of polymerization was reduced. Moisture had the effect of acting as a promoter [48].

TABLE VII

POLYMERIZATION OF ISOBUTYLENE BY FRIEDEL-CRAFTS CATALYSTS

Catalysts	Minimum effective cat. conc. %/% isobutene	% Conc. used	% Starting isobutylene	Reaction time (sec)	% Yield polymer	Mol. wt. $\times 10^{-2}$
BF_3	—	0.05	10	—	100	120–150
$AlBr_3$	0.01/1	0.05	20	0–5 min	70–90	120–150
$TiCl_4$	0.12/7.0	0.12	30	20–70 min	35–50	100–130
$TiBr_4$	0.50/20	1.0	30–40	12–18	30–50	70–90
BCl_3	0.8/20.0	0.9	40–50	12–18	0.5–1.5	30–50
BBr_3	—	0.6	50	12–18	0.5–1.5	20–30
$SnCl_4$	10/1.0	1.5	50	17–50	10–18	12–25

3-3. Emulsion Polymerization of Butadiene Using a Rhodium Chloride Cationic Catalyst to Give 99% trans-1,4-Polybutadiene [64]

$$nCH_2{=}CH{-}CH{=}CH_2 \xrightarrow[\substack{H_2O \\ 80^\circ C}]{RhCl_3 \cdot 3H_2O} \left[{-}CH_2{-}CH{=}CH{-}CH_2{-} \right]_n \tag{16}$$

To a stainless steel Hoke cylinder containing a pressure relief valve and a valve for venting is added 200 ml of distilled water, 1.0 gm of rhodium chloride trihydrate (Engelhard Industries, Newark, N.J.), and 5.0 gm of sodium dodecylbenzenesulfonate. The cylinder is cooled in an ice bath, weighed, and more than 100 gm of butadiene is added. The excess is allowed to evaporate off in a hood until only 100 gm remains. The cylinder is sealed closed and then placed in a shaking thermostatted water bath at 80°C for 5 hr. The cylinder is then cooled, vented, and the polymer precipitated by the addition of methanol containing 0.5 gm of *N*-phenyl-β-naphthylamine (CAUTION: carcinogenic). The polymer is washed with fresh methanol and dried to afford 72 gm (72%) of polybutadiene of greater than 98% *trans*-1,4-configuration. The crystallinity as estimated by X-ray diffraction is 37%. Further details are given in Table VIII and Figs. 5 and 6. The data in Table VIII and Fig. 6 show that the rate of polymerization is faster in the presence of emulsifier.

3-4. Preparation of Butyl Rubber (Copolymerization of Isobutylene with Isoprene Using BF_3 Catalyst) [65]

$$CH_2{=}C(CH_3)_2 + CH_2{=}CH{-}\underset{\displaystyle CH_3}{\underset{|}{C}}{=}CH_2 \xrightarrow[\substack{CH_3Cl \\ -78^\circ C}]{BF_3} \left[{-}CH_2{-}C(CH_3)_2{-}CH_2{-}CH{=}\underset{\displaystyle CH_3}{\underset{|}{C}}{-}CH_2{-} \right]_n \tag{17}$$

To an oven-dried micro resin flask equipped with a thermocouple, stirrer, and cooled jacketed dropping funnel and placed in a Dry Ice–acetone bath (−78°C) is added 9.7 ml (6.8 gm, 0.121 mole) of isobutylene and 0.3 ml of isoprene dissolved in 30 ml of methyl chloride (all volumes measured at −78°C). Then BF_3 (1.8×10^{-5} mole) dissolved in methyl chloride is added dropwise so that a temperature rise no higher than 1° is experienced. After 1–2 hr reaction at −78°C the polymerization is terminated by adding 10 ml of precooled methanol and stirring for 5 min. The unreacted monomers are evaporated and the polymer washed with methanol and dried for 48 hr at 45–50°C to afford 4.8 gm (68.6%), mol. wt. 140,200; DP = 2510. The mole percent isoprene content is 1.84 and is determined as described in Ref. [66].

TABLE VIII

POLYMERIZATION OF BUTADIENE BY RHODIUM SALTS[a,b]

Rhodium salt	Solvent	Emuls.[c] (gm)	Reaction temp. (°C)	Yield (gm/hr)	[η] Tetralin 135°C (dl/gm)	% Cryst.[d]	Ratio[e] of *trans*	vinyl	*cis*
Rhodium chloride·$3H_2O$	Water	5	5	0.02	0.5	43	99	0.2	<1
	Water	5	50	2.4	0.4	37	99	0.3	<1
	Water	5	80	21	0.1	37	>98	0.2	1.2
Rhodium nitrate·$2H_2O$	Water	None	50	0.1	0.1	37	>96	<2	<2
	Water	5	50	0.4	0.1	60	98	1	<1
	Ethanol	None	Room	0.1	0.1	41	98	0.5	<2
	Ethanol	None	80	7.0	0.1	21	>90	<8	2.7
	Dimethylformamide	None	Room	0.1	0.1	32	98	0.7	<2
Ammonium chlororhodate·$1.5H_2O$	Water	5	50	1.3	0.3	36	99	0.2	<1
Sodium chlororhodate·$18H_2O$	Water	5	50	1.5	0.3	49	>98	<1	<1

[a] Conditions: 1 gm of rhodium salt, 100 gm of butadiene, 200 ml of solvent in capped bottles.

[b] Reprinted from R. E. Rinehart, H. P. Smith, H. S. Witt, and H. Romeyn, Jr., *J. Amer. Chem. Soc.* **83**, 4864 (1961). Copyright 1961 by the American Chemical Society. Reprinted by permission of the copyright owner.

[c] Sodium dodecyl benzenesulfonate.

[d] Estimated by X-ray diffraction.

[e] Determined by the infrared absorption in KBr discs by a method similar to that described by Hampton. The ratios are normalized to 100% total unsaturation.

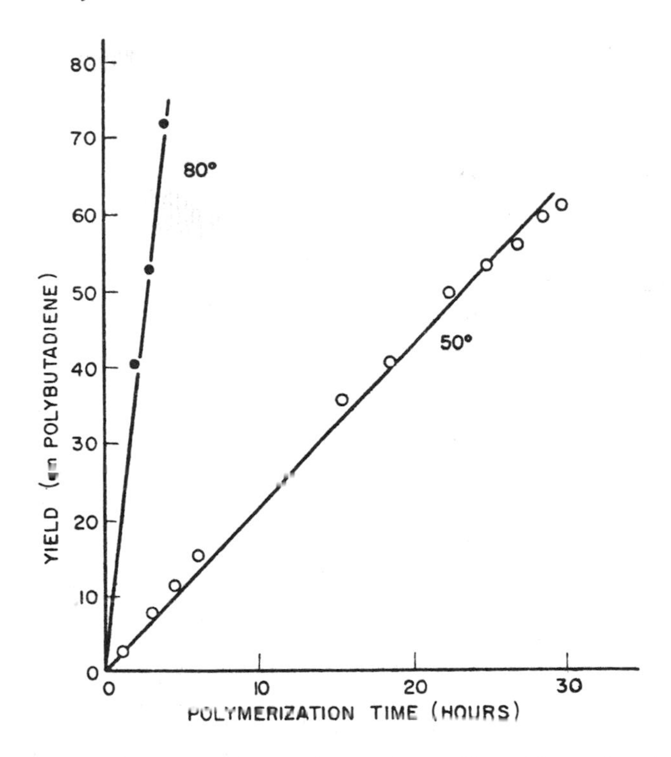

FIG. 5. Polymerization in emulsion: (1 gm $RhCl_3 \cdot 3H_2O$, 100 gm butadiene, 5 gm sodium dodecyl benzenesulfonate, 200 ml H_2O). [Reprinted from R. E. Rinehart, H. P. Smith, H. S. Witt, and H. Romeyn, Jr., *J. Amer. Chem. Soc.* **84**, 4145 (1962). Copyright 1962 by the American Chemical Society. Reprinted by permission of the copyright owner.]

Kennedy has also described the preparation of butyl rubber utilizing an AlR_2X catalyst with an HX promoter at 0° to −100°C [66a]. In this case R = C_1 to C_{12} aliphatic hydrocarbon radical and X = F, Cl, or Br. (See Table VI for an example.)

Marvel and co-workers [67] have also described the cationic polymerization of butadiene using one of several Friedel-Crafts catalysts at various temperatures and concentrations. Aluminum chloride and chlorosulfonic acid were effective at −75°C and stannic chloride, boron trifluoride etherate, boron trifluoride hydrate, sulfuric acid, and fuming sulfuric acid required higher temperatures to bring about substantial polymerization. The catalysts usually were added in chloroform or ethyl bromide as solvent. Butadiene also was found to copolymerize well with styrene at −75°C using aluminum chloride in ethyl bromide as catalyst.

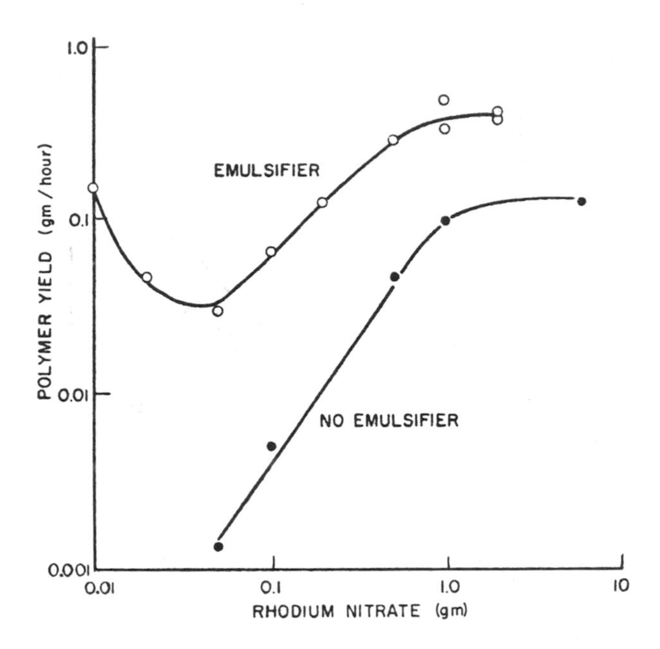

FIG. 6. Catalyst concentration effects in the polymerization of butadiene by $Rh(NO_3)_3 \cdot 2H_2O$ (100 gm butadiene, 200 ml H_2O: emulsion curve; plus 5 gm sodium dodecyl benzenesulfonate). [Reprinted from R. E. Rinehart, H. P. Smith, H. S. Witt, and H. Romeyn, Jr., *J. Amer. Chem. Soc.* **84**, 4145 (1962). Copyright 1962 by the American Chemical Society. Reprinted by permission of the copyright owner.]

4. ANIONIC POLYMERIZATIONS

The anionic polymerization of styrene was first reported in 1914 by Schlenk and co-workers [68] and more recently reinvestigated by Szwarc [69] and others [70]. Almost 60 years ago sodium and lithium metal were used to polymerize conjugated dienes such as butadiene [71a–c], isoprene [71a], 1-phenylbutadiene [71b], and 2,3-dimethylbutadiene [72]. Ziegler [73] in 1929 described the addition of organoalkali compounds to a double bond. In 1940 the use of butyllithium for the low-pressure polymerization of ethylene was described [74]. In 1952 the kinetics of the anionic polymerization of styrene using KNH_2 was reported [75]. Some anionic polymerizations have been described as living polymers in the absence of impurities) [76].

Electron-withdrawing substituents adjacent to an olefinic bond tend to stabilize carbanion formation and thus activate the compound toward anionic polymerization [77].

The relative initiator activities are not always simple functions of the reactivity of the free anion but probably involve contributions by complexing ability, ionization, or dissociation reactions [78].

Recently Waack and Doran [79] reported on the relative reactivities of 13 structurally different organolithium compounds in polymerization with styrene in tetrahydrofuran at 20°C. The reactivities were determined by the molecular weights of the formed polystyrene. The molecular weights are inversely related to the activity of the respective organolithium polymerization initiators. Reactivities decreased in the order alkyl > benzyl > allyl > phenyl > vinyl > triphenylmethyl as shown in Table IX.

The structure–reactivity behavior found for similar organosodium polymerization initiators of styrene [80] or that for addition reactions with 1,1-diphenylethylene [80a] is almost identical with that found for the lithium initiators of Table IX. It is interesting to note from Table IX that the reactivity of lithium naphthalene, a radical anion type initiator, is between that of the alkyl lithiums and the aromatic lithium initiators.

The anionic polymerization of styrene using the organolithium initiators can be described as a termination-free polymerization, as shown in Eqs. (18) and (19).

$$\text{RLi} + \text{St} \longrightarrow \text{RStLi} \quad (18)$$

$$\text{RStLi} + n\text{St} \longrightarrow \text{RSt}_n\text{StLi} \quad (19)$$

The degree of polymerization (DP) is determined by the ratio of the overall rate of propagation to that of initiation.

"Living" polymers have been extensively investigated by Szwarc and coworkers [81], who have shown that the formed polymer can spontaneously resume its growth on addition of the same or different fresh monomer. Block copolymers are easily synthesized by this technique. The lack of self-termination is overcome by the addition of proton donators, carbon dioxide, or ethylene oxide.

Earlier examples of "living" polymers were reported by Ziegler for the sodium-initiated polymerization of butadiene [82]. The anionic polymerization of ethylene oxide also gives a "living" type polymer [83]. Cationic polymerizations such as the $BF_3 \cdot H_2O$ complex-catalyzed polymerization of isobutylene is considered a "living" type polymer [84].

Alfin (alkoxide–olefin) catalysts have been developed by Morton and coworkers [85,85a] and they are special combinations of sodium salts (typically 1.5 mole sodium chloride, 1.0 mole sodium alkoxide, and 0.39 mole allylsodium) which cause the rapid polymerization of butadiene to high molecular weight polymer. The reagents are also insoluble aggregates of ions whose behavior is affected by the ions in the aggregate [85]. Table X shows some

TABLE IX
"STANDARD POLYMERIZATIONS" OF STYRENE IN TETRAHYDROFURAN SOLUTION AT 20°C

Organolithium catalyst	Mol. wt. of polymer[a] (temp, °C)
t-Butyl	3,200[b] (−66)
	3,200 (−40)
sec-Butyl	3,500[b] (−69)
Ethyl	3,500
n-Butyl	3,600
α-Methylbenzyl	3,700
Crotyl	6,500
Benzyl	6,700
Allyl	9,600
p-Tolyl	9,900
Phenyl[c]	12,000
Phenyl[d] (LiX)	24,000 (LiCl)
	22,000 (LiBr)
Methyl	19,000
Vinyl	23,000
Triphenylmethyl[e]	66,000
Triphenylmethylsodium	53,000
Lithium naphthalene	6,000

[a] Average values.
[b] At higher temperatures there is rapid reaction with THF.
[c] Salt-free.
[d] Contains equimolar lithium halide (see Ref. 12).
[e] Contains equimolar LiCl. Lithium halides are indicated to have little effect on the reactivity of such resonance stabilized species. Reprinted from R. Waack and M. A. Doran, *J. Org. Chem.* **32**, 3395 (1967).

examples of alfin catalysts and their effectiveness in the polymerization of polybutadiene. Other examples are described in Ref. [86,86a].

Replacement of 85% of the sodium ions by potassium ions in the alkoxide portion and then replacement of 2/3 of NaCl by NaF causes butadiene to give more 1,2- than 1,4-polybutadiene (ratio 1.4 to 1.2 = 0.86). If only Na ion was present then the ratio of *trans*-1,4 to *trans*-1,2 structures is approx. 3.0 [85]. This difference indicates that the nature of the catalyst surface is important for the alfin catalyst system.

TABLE X

GENERAL AND SPECIFIC FORMULAS FOR SOME ALFIN CATALYSTS AND VISCOSITIES OF CORRESPONDING POLYBUTADIENES[a,b]

Alcohol	Hydrocarbon	[η]
$RCH(CH_3)ONa$	$CH_2{=}CHCH(R)Na$	—
	or $HC{=}C{-}CH_2Na$ (H, H / H H)	—
$(CH_3)_2CHONa$	$CH_2{=}CHCH_2Na$	11–13
n-$C_3H_7(CH_3)CHONa$	$CH_2{=}CHCH_2Na$	12–13
$C_6H_6(CH_3)CHONa$	$CH_2{=}CHCH_2Na$	6–7
$(CH_3)_2CHONa$	$[CH_2{\cdots}CH{\cdots}CHCH_3]Na$	6–7
$(CH_3)_2CHONa$	$C_6H_5CH_2Na$	6–7
$(CH_3)_2CHONa$	$[CH_2{=}CHCHCH_2CH{=}CH_2]Na$	11–12
n-$C_3H_7(CH_3)CHONa$	$[C_2H_5CH{\cdots}CH{\cdots}CH_2]Na$	<2

[a] All catalysts contain NaCl.

[b] Reprinted from A. A. Morton, *Ind. Eng. Chem.* **42**, 1488 (1950). Copyright 1950 by the American Chemical Society. Reprinted by permission of the copyright owner.

The use of xylylsodium greatly affects the stereochemistry of the polymer [86b].

The alfin catalyst can also polymerize styrene and it can be used in styrene–butadiene copolymerizations.

It is interesting to note that with ordinary organosodium reagents such as amylsodium, styrene polymerizes much faster than butadiene. However, with the alfin catalyst butadiene polymerizes nearly five times faster. In addition, the intrinsic viscosity of the alfin polybutadiene (11–13) is much higher than that of the alfin polystyrene (1.1–3.6) depending on the amount of isopropoxide present.

Reference [87] is a general review of the anionic polymerization process and is worthwhile consulting.

Some typical examples of the anionic polymerization of various olefinic–diolefinic monomers are given in Table XI and Preparations 4-1, 4-2, 4-3, and 4-4.

4-1. Preparation of Polystyrene by the Polymerization of Styrene Using Sodium Naphthalene Catalyst [88]

$$\text{Styrene} + \text{naphthalene}^- \longrightarrow \text{styrene}^- \xrightarrow{\text{styrene}} \text{polystyrene}^- \quad (20)$$

$$\text{Polystyrene}^- \begin{cases} \xrightarrow{CH_3OH} \text{polystyrene} \\ \xrightarrow{\text{styrene}} \text{further polymerization} \end{cases} \quad (21)$$

TABLE XI

ANIONIC POLYMERIZATION OF VARIOUS OLEFINIC AND DIOLEFINIC MONOMERS

Monomer (gm, ml, or moles)	Catalyst (gm)	Solvent (ml)	Reaction conditions		% Yield of polymer	Polymer properties	
			Temp. (°C)	Time (hr)		Mol. wt.	Ref.
$C_6H_5C(CH_3)=CH_2$ (72.5 gm)	K metal (2.0)	—	15	12	35–40 gm	[a]	*b*
$C_6H_5CH=CH_2$ (10 ml)	Sodium naphthalene (50 gm)[c]	Ethylene glycol dimethyl ether (50)	−70	1–2	100	[d]	*e*
(16.9 gm)	Sodium naphthalene (0.00034 mole)	THF (60)	−80	1	100	—	*f*
(0.190 mole)	K in liq. NH_3 (0.0340 mole/liter)	NH_3 (900)	−33	1–3	32+	4450	*g*
(11.0 moles)	*n*-BuLi (0.25 mole/liter, 1.0 mole)	THF	−66	½-1	100	2500	*h*
$CH_2=CH-CH=CH_2$ (30 gm)	Alfin catalyst (30 ml)	Hexane or *n*-pentane (150)	−20 60	3	70–90	229,000 to 5×10^6	*i*

(0.72 gm)	Amylsodium (0.66 gm, 0.007 mole)	Pentane (60)	−10 to 30	1½	89	—	*j*
$CH_2{=}C(CH_3){-}CH{=}CH_2$ (84 ml)	*n*-BuLi (1.0 molar, 3.0 ml in *n*-pentane)	*n*-Pentane (180)	30	13	100	—	*k*
(90 ml)	Li metal (0.04–0.1 gm, 35% dispersion in petroleum oil)	—	30–40	½ or ½–72	100	—	*l*
$CH_2{=}CH{-}CH{=}CH_2$ (60–70 gm)	Na metal (0.15)	Toluene (10)	50	24	100	—	*m*

[a] $\eta_i = 0.7$–0.8 (0.5% in toluene at 25°C).

[b] T. E. Werkema, U.S. Patent 2,658,058 (1953).

[c] Before adding the required amount of Na naphthalene catalyst a few drops of this reagent is added until the greenish black of the reagent persists.

[d] $\eta_i = 1.0$ to 1.5 at 0.5% conc. in toluene at 25°C.

[e] R. Waack, A. Rembaum, J. D. Coombes, and M. Szwarc, *J. Amer. Chem. Soc.* **79**, 2026 (1957).

[f] M. Szwarc, M. Levy, and R. Milkovich, *J. Amer. Chem. Soc.* **78**, 2656 (1956).

[g] N. S. Wooding and W. C. E. Higginson, *J. Chem. Soc., London* p. 1178 (1952).

[h] R. Waack and M. A. Doran, *J. Org. Chem.* **32**, 3395 (1967).

[i] H. Greenberg and L. Grinninger, *Macromol. Syn.* **3**, 117 (1969). See also A. A. Morton R. P. Welcher, F. Collins, S. E. Penner, and R. D. Coombs, *J. Amer. Chem. Soc.* **71**, 481 (1949); A. Morton, *Ind. Eng. Chem.* **42**, 1488 (1950).

[j] A. A. Morton and R. L. Letsinger, *J. Amer. Chem. Soc.* **69**, 172 (1947).

[k] German Patents 1,040,795 and 1,040,796 (1958); M. Morton, R. D. Sanderson, and R Sakata, *J. Polym. Sci., B* **9**, 61 (1970).

[l] F. W. Stavely *et al.*, *Ind. Eng. Chem.* **48**, 778 (1956); R. S. Stearns and L. E. Forman, *J. Polym. Sci.* **41**, 381 (1959).

[m] C. S. Marvel, W. J. Bailey, and G. E. Inskeep, *J. Polym. Sci.* **4**, 275 (1946).

(*a*) *Preparation of Sodium Naphthalene* [88a]. A 2-liter three-necked flask

$$\text{(naphthalene)} + \text{Na} \xrightarrow{\text{THF}} \text{Na}^+ \text{(naphthalene}^-\text{)} \tag{22}$$

which has been dried at 150°C and then assembled warm with a nitrogen gas stream going through it is equipped with a mercury-sealed stirrer (or ground glass type). Then 1 liter of a molal solution of naphthalene in pure dry tetrahydrofuran is added followed by 25 gm of sodium sticks (2–3 cm long × 3–5 cm square on the end). After adding the sodium the mixture is stirred rapidly at the start. The reaction starts in 1–3 min and some cooling may be necessary to keep the temperature from exceeding 30°C. At 20°–25°C the reaction takes about 2 hr, and then an aliquot of the mixture is added to methanol and titrated with dilute standard HCl using methyl red.

(*b*) *Polymerization of styrene* [88]. To a flask as described previously but with a side arm with a serum cap is added 60 ml of pure dry THF containing 3.3×10^{-4}mole of sodium naphthalene (positive nitrogen pressure maintained throughout reaction). Then the contents are cooled to −80°C and 9.2 gm of dry pure distilled styrene is injected. Polymerization proceeds rapidly and the reaction mixture turns a bright red color due to the formation of the styrene anion. After completion of the reaction the solution is warmed to room temperature and the viscosity (see Note) is 1.2–1.5 sec. The solution is recooled to −80°C, and an additional 7.7 gm of styrene in 50 ml of THF is injected. After 1–2 hr the color of the reaction is still bright red and the viscosity (see Note) of the reaction mixture is 18–20 sec at room temperature. The polymerization is quenched (red color disappears) by the addition of cold methanol, filtered, washed with methanol, and dried to afford 16.6 gm (98%) of polystyrene, inherent viscosity in toluene at 0.5% concentration is 1 to 1.5. This preparation proves that the living ends of the polystyrene were able to initiate further polymerization when the second addition of styrene was made.

NOTE: Viscosity taken without removing sample from reaction vessel.

Szwarc [88] has shown that styrene initiated as described above can also react with isoprene to give block copolymers.

4-2. *Preparation of Polybutadiene by the Amylsodium-Catalyzed Polymerization of Butadiene* [89]

$$n\text{CH}_2\text{=CH—CH=CH}_2 \xrightarrow[\text{pentane}]{C_5H_{11}Na} [\text{—CH}_2\text{—CH=CH}_2\text{—CH}_2\text{—}]_n \tag{23}$$

A 500 ml creased Morton flask [89a] that has been dried at 150°C is assembled warm with a mechanical stirrer, dropping funnel, and nitrogen inlet–outlet. While a positive nitrogen atmosphere is maintained a 300 ml pentane suspension of amylsodium (see Note) (0.083 mole) is added. The flask is cooled to 0°–10°C and 1.00 mole of butadiene is added over a 1 hr period. After stirring for an additional hour the reaction mixture is carbonated by pouring the contents of the flask on solid carbon dioxide. When the solid carbon dioxide evaporates 200 ml of water is dropwise added. Then 0.4 gm of hydroquinone is added and the pentane is removed by warming on a steam bath. The polymer is removed from the aqueous layer, acidified with hydrochloric acid, and steam-distilled in order to remove any decane. The rubbery mass is dried by distilling chloroform from it and then dried under reduced pressure to afford 45.4 gm (74%).

NOTE: Amylsodium is prepared in 75–80% yield by the dropwise addition (1 hr) of the halide (0.25 mole) to two equivalents of finely divided sodium in pentane at −10°C.

4-3. Preparation of Polybutadiene Using the Alfin Catalyst [90]

$$n\mathrm{CH_2{=}CH{-}CH{=}CH_2} \xrightarrow{\text{alfin catalyst}} [\mathrm{-CH_2{-}CH{=}CH{-}CH_2-}]_n \quad (24)$$

Mixture of *cis*- and *trans*-1,4 and vinyl

(*a*) *Catalyst preparation* [86,90]

CAUTION: Alfin catalysts are pyrophoric and can ignite the pentane if spilled in the air.

$$\mathrm{C_5H_{11}Cl + Na \xrightarrow{\text{pentane}} C_5H_{11}Na + NaCl \xrightarrow{\textit{i}\text{-}C_3H_7OH}}$$

$$\left\{\begin{array}{c}\mathrm{C_5H_{11}Na + NaCl}\\ +\\ \textit{i}\text{-}\mathrm{C_3H_7ONa}\end{array}\right. \xrightarrow{\mathrm{CH_2{=}CH{-}CH_3}} \left\{\begin{array}{c}\mathrm{CH_2{=}CH{-}CH_2Na}\\ +\\ \mathrm{NaCl}\\ +\\ \textit{i}\text{-}\mathrm{C_3H_7ONa}\end{array}\right. \quad (25)$$

To a 1 liter four-necked flask attached to a high-speed stirrer (see Note b) is added 500 ml of dry *n*-pentane and 23 gm (1.0 atom) of sodium sand under a nitrogen atmosphere. The flask is cooled to −10°C and 63 ml (0.522 mole) of amyl chloride is added dropwise. The addition is stopped after 10–15 ml of the amyl chloride is added and the mixture stirred vigorously until the reaction is initiated as evidenced by a dark purple coloration. The addition is then continued over a 45–60 min period while keeping the temperature at −20° to −10°C. The mixture is warmed to room temperature and then stirred for an additional ½ hr.

Isopropanol (0.20 mole) is added at 0°C just prior to the addition of propylene. The propylene is bubbled into the reaction mixture over a 0.5–2½ hr period while the temperature is kept at −10°C using high-speed stirring. The propylene is bubbled in until there is a steady reflux of it from the Dry Ice–methanol condenser. The propylene addition is stopped, and the catalyst stirred for an additional 1 hr before use or may be kept overnight without any deleterious effect.

(*b*) *Polymerization.* To a 16 oz oven-dried soft drink bottle containing a serum pressure cap (see Note a) is added 150 ml of dry *n*-pentane using a nitrogen atmosphere. Then 30 gm of polymerization grade (Matheson) butadiene is condensed into the calibrated bottle at −20°C using a two-hole rubber stopper and weighed. At −20°C 30 ml of the Alfin catalyst is added, the bottle securely capped (Note b), and then shaken well. Polymerization starts at −20°C as evidenced by a temperature rise to 40°C in about 10–15 min. The bottle is agitated at 40°–60°C until the contents completely solidify (less than 1 hr). The bottle is allowed to stand for 1–2 hr and then the seal slowly opened to allow pressure equalization. The solid polymer is removed and added to 500 ml of ethanol containing 1.0 gm of *N*-phenyl-2-naphthylamine antioxidant. The mixture is stirred in a blender and then filtered. The solid crumbly polymer is washed twice with 500 ml of ethanol in the blender and then in the same way twice with acetone (1.0% *N*-phenyl β-naphthylamine per 500 ml of acetone). The polymer is filtered and dried at 40°C under reduced pressure to afford 21–27 gm (70–90%) or more of polybutadiene (mixture of *cis*, *trans*, and vinyl). The intrinsic viscosity in toluene at 25°C is approx. 11–15 dl/gm (mol. wt. about 2–5 $\times$ 10^6 based on $KM^a = [\eta]$ where $K = 1.1 \times 10^{-3}$ and $a = 0.62$) [86].

NOTES: (a) Serum cap catalog #5300 Harshaw Scientific Co. or other suitable type. (b) "Slip-Seal" bottle closure device may be used as available in local hardware stores for resealing beverage bottles.

The use of 40 ml of a catalyst (prepared by passage of propylene into 1.0 mole of amylsodium which had been first been reacted with 0.2 mole of methylphenylcarbinol caused 75% conversion of 30 ml of butadiene over a 1 hr period to give a polymer with an intrinsic viscosity of 5.2 [90].

The catalyst prepared without use of an alcohol gives a slow rate of polymerization of butadiene.

The molecular weights of the alfin polybutadienes are very high and are difficult to process. Greenberg and Hansley [86,86a] found that the molecular weight can be controlled to give products of mol. wt. 50,000 to 1,250,000 by adding dihydro aromatic compounds such as 1,4-dihydrobenzene to the polymerization mixture, as shown for example [86], in the following tabulation.

Butadiene (gm)	*n*-Hexane (gm)	Alfin cat. (0.065 gm/ml)	1,4-Dihydrobenzene (gm)	Temp. (°C)	Time (hr)	Yield (%)	Mol. wt.	η_i (toluene)
30	100	17.5	2.0	25	2	96	537,000	3.9
360	300	210	10	25	6	95	229,000	2.3

In place of a soft drink bottle the polymerization can be carried out in a three-necked flask equipped with a Dry Ice–acetone-cooled condenser. The dry butadiene can be metered into the stirred mixture of alfin catalyst and solvent, etc. [86].

4-4. *Preparation of Poly(cis-1,4-isoprene)* [91]

$$n\mathrm{CH_2{=}C(CH_3){-}CH{=}CH_2} \longrightarrow \left[\begin{array}{ccc} \mathrm{CH_3} & & \mathrm{H} \\ & \mathrm{C{=}C} & \\ \mathrm{{-}CH_2} & & \mathrm{CH_2{-}} \end{array} \right]_n \tag{26}$$

To a polymerization bottle cooled to 0°C is added 100 gm of isoprene, 300 gm of *n*-pentane, and 0.04 gm of a fine dispersed lithium metal. The bottle is sealed with a crown cap and slowly warmed from 0° to 50°C over a 1–2 hr period to give 100% conversion to polymer. The polymer is purified by dissolution in toluene followed by precipitation in acetone. Such a polymer was analyzed by infrared spectroscopy and shown to contain 94.2% as 1,4 groups and 5.8% as 3,4 groups. The inherent viscosity is 6.85.

The use of ethyllithium or butyllithium in *n*-pentane in place of lithium metal in the above preparation gave similar results.

Surprisingly, ethyl ether in place of *n*-pentane gave little or no *cis*-1,4-polyisoprene but a mixture of *trans*-1,4-, *trans*-1,2-, and 3,4-polyisoprene.

In addition sodium and alkyl sodium compounds, potassium and alkyl potassium compounds, and rubidium, cesium, or alfin (sodium) catalysts in *n*-pentane gave mixtures of *cis*-1,4-, *trans*-1,4-, *trans*-1,2-, and 3,4-polyisoprenes.

5. COORDINATION CATALYST POLYMERIZATIONS

The earliest reported coordination polymerization was the cationic polymerization of vinyl isobutyl ether [92]. Other early references reported the polymerization of ethylene using titanium tetrachloride–aluminum–aluminum chloride [93] and the polymerization of styrene using phenylmagnesium bromide–$Ti(OBu)_4$ [94].

Ziegler [95] first recognized the importance of the catalyst and the novel type of polymerization it induced, especially for ethylene monomer. The polyethylene produced was very linear (m.p. 135°C), whereas the high-pressure radical process gives a short-chain branched product (m.p. 120°C). Natta [96] almost at the same time reported on a similar catalyst useful for a wide range of olefin, diolefin, and acetylenic monomers. These new catalysts have now become known as the Ziegler-Natta catalysts. These catalysts are complexes formed by the reaction of halides or other derivatives of transition metals of groups IV–VIII and alkyls of metals of groups I–III, for example, as shown in Eq. (27).

$$AlR_3 + MCl_n \longrightarrow R_2AlCl + RMCl_{n-1} \tag{27}$$

The catalysts which are based on insoluble transition metal compounds or complexes afford little information as to the active species present. Many of the catalysts have metal salts in varying oxidation states. These catalysts therefore have a structure which is not known with certainty and leads to difficulties in trying to reproduce results.

Recently various well-defined π-allylnickel halides [97,98] have been reported to initiate diene polymerization. However, the polymerization is slow and also affords low molecular weight products. The addition of Lewis acids ($AlCl_3$, VCl_4, or $TiCl_4$) to these π-allylnickel complexes speeds up the rates and gives products with high molecular weights. Again, the structure of these new catalysts is unknown.

The factor that influences the control of the stereoregularity of the polymers obtained by polymerizing monomers with coordination catalysts is not fully understood. However, it is probable that the monomer is precomplexed with the catalyst center in the configuration which is ultimately retained in the polymer chain. For example, *cis* attachment of the monomer double bonds in 1,3-pentadiene to the catalyst site ($AlEt_2Cl/CoCl_2$) leads to syndiotactic *cis*-1,4-polymer from the *trans* isomer of the monomer and no polymer for the *cis* isomer (Eq. 28) [99].

R
Cat
CH_3 ⟶ R Cat CH_3 (28)

Some data on using soluble catalysts for the polymerization of butadiene are shown in Tables XII and XIII.

TABLE XII

POLYMERIZATION OF BUTADIENE BY SOLUBLE CATALYSTS[a,b]

Catalyst[c]	[C][d] $(ml^{-1} \times 10^4)$	[M] $(ml)^{-1}$	Temp. (°C)	R_p, $10^4\ ml^{-1}\ sec^1$
$Co(Np)_2/AlEt_2Cl/H_2O$	0.28	2.26	25	1.75
$Co/Al/H_2O = 1/1000/100$				
$Co(acac)_3/AlEt_2Cl/H_2O$	2.5	2.0	2	8.5
$Co/Al/H_2O = 1/400/8$				
$Co(acac)_3/Al_2Cl_3Et_3$	0.32	0.9	41	32.4
$Al/Co = 1700$				
$Cr(acac)_3/AlEt_3$	43	1.0	20	0.5
$Al/Cr = 11.6$				
π-C_3H_5NiCl	100	2.0[b]	50	0.3

[a] Monomer concentration not given but calculated assuming polymerization temperature, also not stated, is 50°C.

[b] Reprinted from W. Cooper, *Ind. Eng. Chem., Prod. Res. Develop.* **9**, 457 (1970). Copyright 1970 by the American Chemical Society. Reprinted by permission of the copyright owner.

[c] Np = naphthenate.

[d] Transition metal concentration.

Cooper has reviewed the coordination catalyst mechanism of polymerization of conjugated dienes and is worthwhile consulting [100]. Other earlier reviews, papers, and texts should also be consulted [101].

Some typical examples of the monomers polymerized using these catalysts are shown in Table XIV and Preparations 5-1, 5-2, 5-3, and 5-4.

TABLE XIII

COMPARISON OF SOLUBLE Cr AND SOLUBLE Co CATALYSTS FOR BUTADIENE POLYMERIZATION[a]

Catalyst	Polymerization rate	Mol. wt.	Increase in Al/[M] ratio on mol. wt.	Polymer structure	ROH termination
$Cr(acac)_3/AlEt_3$	Low	Low	Falls	1,2	+
$Co(acac)_3/AlEt_2Cl$	High	High	Independent	1,4	−

[a] Reprinted from W. Cooper, *Ind. Eng. Chem., Prod. Res. Develop.* **9**, 457 (1970). Copyright 1970 by the American Chemical Society. Reprinted by permission of the copyright owner.

TABLE XIV

POLYMERIZATION OF A REPRESENTATIVE GROUP OF OLEFINIC AND DIOLEFINIC MONOMERS WITH COORDINATION TYPE CATALYSTS

Monomer	Catalyst	Product	Ref.
Ethylene	$Al(C_2H_5)_3$–$Ti(i\text{-}C_4H_9)_4$	Polyethylene	*a*
	$Al(C_2H_5)_3$–$TiCl_4$ or $AlBr_3$–VX_n–$Sn(C_6H_5)_4$	High-density polyethylene	*b*
Allene	$Al(i\text{-}C_4H_9)_3$–$VOCl_3$ or other catalysts[d]	Polyallene	*c*
Propylene	$i\text{-}Bu_2AlCl$–VCl_4	Syndiotactic polypropylene	*e*
Isoprene	$Al(i\text{-}C_4H_9)_3$–$TiCl_4$	Polyisoprene	*f*
	$Al(i\text{-}C_4H_9)_3$–$Ti(OC_6H_{13})_4$–VCl_3	*trans*-1,4-Polymer	*g*
	$Al(C_2H_5)_3$–$ViCl_3$	*trans*-1,4-Polymer	*h*
	$Al(C_2H_5)_3$–$Ti(OC_3H_7)_4$	3,4-Polyisoprene	*i*
2-Butene	$Al(C_2H_5)_3$–$TiCl_3$	Polymer with repeating units similar to that produced from the polymerization of 1-butene	*j*
Isobutylene	$Al(CH_3)_3$–$t\text{-}C_4H_9Cl$	Polyisobutylene	*k*
Butadiene	$(C_4H_7NiCl)_2$	90% *cis*-1,4-Polybutadiene	*l*
	$Al(C_2H_5)_3$–$TiCl_4$	90% *cis*-1,4-Polybutadiene	*m*
	$Al(C_2H_5)_2Cl + H_2O$ + cobalt octoate	90% *cis*-1,4-Polybutadiene	*n*
	R_2Hg–$TiCl_4$	90% *cis*-1,4-Polybutadiene	*o*
	$Al(C_2H_5)_3$–$CoCl_2$	Syndiotactic 1,2-polybutadiene	*p*
	$Al(C_2H_5)_3$–$TiCl_4$–VCl_4	*trans*-1,4-Polybutadiene	*q*
3-Methyl-1-pentene	$n\text{-}BuLi$–$TiCl_4$	Isotactic polystyrene	*r*
4-Methyl-1-pentene	$LiAl(C_{10}H_{21})_4$–$TiCl_4$	Polymers	*s*
3-Methyl-1-butene			
4-Methyl-1-hexene			
4-Phenyl-1-butene			
1,3-Pentadiene	$C_2H_5AlCl_2$–cobalt diacetylacetonate	*cis*-1,4-Syndiotactic poly-1,3-pentadiene	*t*
	$C_2H_5AlCl_2$–$Ti(OC_4H_9)_4$	*cis*-1,4-Isotactic poly-1,3-pentadiene	*t*
Hexatriene	$(C_2H_5)_2AlCl$–$Ti(OC_6H_5)_4$	Amorphous 1,6-polyhexatriene	*u*
	$Al(i\text{-}Bu)_3$–$ViCl_3$	Crystalline 1,6-polyhexatriene	*u*
Styrene	$Al(i\text{-}C_4H_9)_3$–$TiCl_4$	Isotactic polystyrene	*v*
	$Al(C_2H_5)_3$–$TiCl_3$ or $Al(C_2H_5)_3$–$ViCl_3$	Polystyrene	*w*
	$n\text{-}BuLi$–$TiCl_4$	Isotactic polystyrene	*x*
Norbornene	$LiAl(alkyl)_4$–$TiCl_4$	Polynorbornene	*y*

[a] C. E. H. Baron and R. Symcox, *J. Polym. Sci.* **34**, 139 (1959); A. Orzechowski, *ibid.*, p. 65.

[b] W. L. Carrick, *Macromol. Syn.* **2**, 33 (1966); G. W. Phillips and W. L. Carrick, *J. Polym. Sci.* **59**, 401 (1962).

[c] W. P. Baker, Jr., *J. Polym. Sci., Part A* **1**, 655 (1963).

[d] S. Otsuka, *J. Amer. Chem. Soc.* **87**, 3017 (1965).

[e] A. Zambelli, G. Natta, and I. Pasquon, *J. Polym. Sci., Part C* **4**, 411 (1963).

[f] H. E. Adams, R. S. Stearns, W. A. Smith, and J. L. Binder, *Ind. Eng. Chem.* **50**, 1507 (1958); British Patent 776,326 (1957); S. E. Horne, J. P. Kiehl, J. J. Shipman, V. L. Fult, C. F. Gibbs, E. A. Wilson, E. B. Newton, and M. A. Reinhart, *Ind. Eng. Chem.* **48**, 784 (1956); W. M. Saltman, W. E. Gibbs, and J. Lal, *J. Amer. Chem. Soc.* **80**, 5616 (1958); W. M. Saltman and E. Schoenberg, *Macromol. Syn.* **2**, 50 (1966).

[g] J. S. Lasky, U.S. Patents 3,254,754 (1962); 3,114,744 (1963); J. S. Lasky, H. K. Garner, and R. H. Ewart, *Ind. Eng. Chem., Prod. Res. Develop.* **1**, 82 (1962).

[h] G. Natta, L. Porri, P. Corradini, and D. Morero, *Chim. Ind. (Milan)* **40**, 362 (1958).

[i] G. Natta, L. Porri, and A. Carbonaro, *Makromol. Chem.* **77**, 126 (1964).

[j] T. Otsu, A. Shimizu, and M. Imoto, *J. Polym. Sci.* **4**, 1579 (1966).

[k] J. P. Kennedy and G. E. Milliman, *Advan. Chem. Ser.* **91**, 287 (1969).

[l] V. A. Kormer, B. D. Babitskiy, and M. I. Lobach, *Advan. Chem. Ser.* **91**, 308 (1969).

[m] G. Natta, L. Porri, A. Mazzei, and D. Morero, *Chim. Ind. (Milan)* **41**, 398 (1959).

[n] M. Grippin, *Macromol. Syn.* **2**, 42 (1966); *Ind. Eng. Chem., Prod. Res. Develop.* **1**, 31 (1962); **4**, 160 (1965).

[o] W. S. Anderson, U.S. Patent 2,943,987 (1960).

[p] E. Susa, *J. Polym. Sci., Part C* **4**, 399 (1964).

[q] G. J. Van Amerongen, *Advan. Chem. Ser.* **52**, 11 (1966).

[r] W. J. Bailey and E. T. Yates, *J. Org. Chem.* **25**, 1800 (1960).

[s] T. W. Campbell and A. C. Haven, Jr., *J. Appl. Polym. Sci.* **1**, 79 (1959).

[t] G. Natta and L. Porri, *Advan. Chem. Ser.* **52**, 24 (1966).

[u] V. L. Bell, *J. Polym. Sci., Part A* **2**, 5291 (1964).

[v] C. G. Overberger, F. Ang, and H. Mark, *J. Polym. Sci.* **35**, 381 (1959).

[w] F. D. Otto and G. Parravano, *J. Polym. Sci., Part A* **2**, 5131 (1964).

[x] K. C. Tsou, J. F. Megee, and A. Malatesta, *J. Polym. Sci.* **58**, 299 (1962).

[y] W. L. Truett, D. R. Johnson, I. M. Robinson, and B. A. Montague, *J. Amer. Chem. Soc.* **82**, 2337 (1960).

5-1. *Preparation of High-Density Polyethylene* [102]

$$nCH_2{=}CH_2 \xrightarrow[\text{cyclohexane}]{AlBr_3,\ VCl_4,\ Sn(C_6H_5)_4} [-CH_2-CH_2-]_n \qquad (29)$$

CAUTION: The reaction should be carried out in a well-ventilated hood behind a safety shield.

The polymerization is carried out with stirring or near atmospheric pressure in a 3 liter flask (see Note a) attached to a manometer and sealed from the atmosphere. The reactor is thermostatted at 65°C, purged with nitrogen and then ethylene, and charged with 1 liter of purified dry cyclohexane, 5 mmoles of $Sn(C_6H_5)_4$ (Note c), 0.026 mmole of VCl_4 (Note c), and 15 mmoles of $AlBr_3$ (Note c) in this order. Upon addition of the $AlBr_3$, polymerization starts and the pressure drops to 200 mm Hg (Note d). Ethylene is fed in at the rate of 1 liter/min and after 1 hr the reaction is quenched with methanol, the polymer washed with methanol, filtered, and dried to afford 40.8 gm of polyethylene, m.p. 136°C, intrinsic viscosity dl/gm is 2.6 in tetralin at 130°C (mol. wt. 135,000).

NOTES: (a) The three-necked flask is oven-dried at 150°C, assembled while warm, and immediately purged with dry nitrogen. The joints should be lightly greased and the stirrer must be sealed to prevent air from entering the reactor flask. A bubbler should be attached to the exit from the condenser to show that an excess of ethylene is passing through the system. A serum cap is attached to one neck of the flask.

(b) The ethylene is obtained from Union Carbide and contains 96% ethylene, 4% inert hydrocarbons, and 0.003% oxygen. The ethylene is passed through Linde molecular sieves and Ascarite before use. Oxygen concentrations in the 0.003 to 0.05% range were obtained either by injecting oxygen into the flow metering system or by blending with Matheson C.P. commercial ethylene containing 0.05% oxygen. In some cases only Matheson C.P. commercial ethylene is used. Note that 400–600 ppm of oxygen is essential for this polymerization catalyst system.

(c) The catalyst components are added as cyclohexane solutions through a serum cap via a hypodermic syringe. Tetraphenyltin may be added as a solid since it is stable in air.

(d) CAUTION: Do not allow air to be sucked into the flask. Keep the ethylene flow going steadily into the flask as evidenced by the gas bubbler attached to the exit of the condenser.

The use of $TiCl_4$ (4.6 mmoles)–$Al(C_2H_5)_3$ (2.0 mmoles) catalyst does not require oxygen in the system. A more branched polyethylene is obtained and the reaction is carried out in a similar manner [103].

5-2. Preparation of Polyethylene Using a Ni–BuLi Catalyst [104]

$$n\mathrm{CH_2{=}CH_2} \xrightarrow{\text{Ni–BuLi}} [\mathrm{—CH_2—CH_2—}]_n \tag{30}$$

To a dry steel pressure bomb is added under a nitrogen atmosphere 0.5 gm of an active hydrogenation catalyst (Note a) and 189 gm of benzene solution containing 3.72 gm of *n*-BuLi. The bomb is sealed and then charged with ethylene to a pressure of 990 lbs/in^2. The temperature rises to 48°C upon introduction of the ethylene. The mixture is agitated for 6 hr under a pressure of 800 to 1010 lbs/in^2 which is maintained by occasional introduction of ethylene to a total of 126 gm added. During this reaction time the temperature drops to 35°C and the bomb upon standing for an additional 16 hr gives a reaction temperature of 22°C. The pressure of the reaction mixture drops from 995 to 700 lbs/in^2. The excess ethylene is vented and the semisolid product transferred to a flask containing 500 gm of aq. 5% HCl. The volatile material is removed by steam distillation and the residue dried to afford 36 gm (28.8%) of a light gray waxy solid, m.p. 106–110°C. The polymer is insoluble in cold solvents but dissolves in hydrocarbon solvents boiling approximately 106°–110°C or higher. The polymer shows a crystalline pattern by X-ray diffraction analysis.

Omitting the nickel catalyst gave only 0.6 gm of crude polymer.

NOTE: The nickel catalyst is an active hydrogenation catalyst on kieselguhr. It is prepared by reaction of 276 gm of 25% $Ni(NO_3)_4$ solution containing 34 gm of kieselguhr in suspension with 1700 gm of 6% $NaHCO_3$. The precipitate of basic nickel carbonate on kieselguhr is reduced by hydrogen at 450°–475°C. The reduced catalyst is stabilized by exposure at room temperature to air. The catalyst contains 30–35% elementary nickel and the balance mainly consists of nickel oxide.

5-3. Preparation of Polypropylene [105]

$$n\mathrm{CH_3—CH{=}CH_2} \xrightarrow[\mathrm{TiCl_4}]{\mathrm{Al(isobutyl)_3}} \left[\begin{array}{c}\mathrm{—CH_2—CH—} \\ \quad\quad | \\ \quad\quad \mathrm{CH_3}\end{array}\right]_n \tag{31}$$

To a 200-ml three-necked oven-dried flask equipped with a condenser, nitrogen inlet–outlet tube, mechanical stirrer, and syringe cap on one neck is added 100 ml of dry xylene and 2.0 ml of triisobutylaluminum via a hypodermic syringe and 10 drops of titanium tetrachloride. The flask is swept with nitrogen and then heated in an oil bath to 59°–62°C for $\frac{1}{2}$ hr. Propylene gas, dried by passage through an activated alumina column, is bubbled through the mixture for $1\frac{1}{2}$ hr while the temperature is kept at 60°–65°C. The reaction mixture is then cooled, poured into 50:50 conc. HCl–acetone solution, filtered, and extracted with boiling methanol to afford 2.0 gm of isotactic polypropylene, m.p. 156°–160°C.

Other olefins are also polymerized in an analogous manner using this same catalyst system, as shown in Table XV.

TABLE XV
COORDINATION CATALYST-INITIATED POLYMERIZATION OF OLEFINS [105]

Olefin (gm)	Reaction conditions		% Polymer	M.p. (°C)	Mol. wt.
	Temp. (°C)	Time (hr)			
1-Octene (10.6 gm)	93–96	27	13	—	1400
d,l-3-Methyl-1-pentene (4.85 gm)	120–132	168	12	229–237	—
d-3-Methyl-1-pentene (6.28 gm)	125–130	168	11	271–278	—

5-4. Preparation of trans-1,4-Polyisoprene (Synthetic Balata) [106]

$$nH_2C{=}CH{=}\underset{\displaystyle CH_3}{\underset{|}{C}}{=}CH_2 \xrightarrow[\text{Al(isobutyl)}_3]{VCl_3} \left[-CH_2-CH{=}\underset{\displaystyle CH_3}{\underset{|}{C}}-CH_2- \right]_n \qquad (32)$$

To a stainless steel bomb containing 350 gm of dry benzene is added under an argon atmosphere, 0.5 gm of a supported VCl_3 catalyst (see Note) (contains 55 mg VCl_3) and 0.495 ml (0.382 gm) of triisobutylaluminum as a 10% solution in benzene. The bomb is sealed and then 60 gm of isoprene is injected into the bomb. The bomb is shaken for 72 hr at 50°C using a shaking machine. Then the bomb is cooled, opened, and the polymer precipitated with about 800 ml of methanol. The polymer is washed with methanol and dried to afford 26 gm (43.3%), intrinsic viscosity in benzene 3.47 dl/gm. The X-ray diffraction pattern is substantially identical with that of natural balata.

NOTE: The catalyst is prepared as follows: 7.0 gm of TiO_2 (specified surface area, 5.5 sq. meters/gm) and 1.53 gm VCl_4 in 50 ml of dry benzene are stirred and irradiated for 3 hr with a UV light source (quartz mercury arc) under a nitrogen atmosphere. The reaction mixture is filtered, washed with benzene, and vacuum-dried to give a catalyst containing 7.48% of chlorine (Volhard method) corresponding to 11% VCl_3 [106].

$$TiO_2 + VCl_4 \xrightarrow{UV} \underbrace{TiO_2 + VCl_3}_{\text{catalyst}} + Cl_2 \uparrow \qquad (33)$$

The use of the above catalyst in the presence of tetra(2-ethylbutyl)titanate gave increased yields at lower reaction times [106a].

cis-1,4-Polyisoprene has been prepared by Saltman using a $TiCl_4$-$Al(i\text{-}C_4H_9)_3$ catalyst system [107]. Some typical reaction rate data at 10°C are shown in Fig. 7.

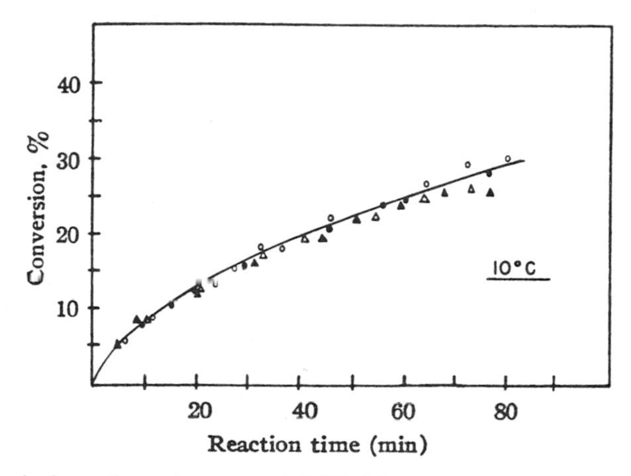

FIG. 7. Typical reaction rate curve at 10°C. Monomer concentration 1.00 mole/liter, TIBA concentration 7.96 × 10^{-3} mole/liter, $TiCl_4$ concentration 6.64 × 10^{-3} mole/liter. Different symbols represent replicate runs. Reprinted from W. M. Saltman, W. E. Gibbs, and J. Lal, *J. Amer. Chem. Soc.* **80**, 5615 (1958). Copyright 1958 by the American Chemical Society. Reprinted by permission of the copyright owner.

6. MISCELLANEOUS METHODS

1. *cis–trans* Isomerization of polybutadienes and polyisoprenes [108].
2. Heterogeneous radical polymerization of ethylene [109].
3. Preparation of polyethylene in a continuous process using free-radical-forming catalysts [110].
4. Preparation of divided expandable polystyrenes [111].
5. Preparation of foamed polystyrene [112].
6. Effect of dimethylaniline on polystyrene preparation [113].
7. Preparation of polystyrene in liquid sulfur dioxide [114].
8. Polymerization of cycloolefin derivatives [115].
9. Trialkylboron catalysts for the polymerization of olefins [116].
10. Polymerization of substituted 1,6-heptadiene [117].
11. Preparation of norbornene–ethylene copolymers [118].
12. Preparation of crystalline polyacetylene [119].
13. Polymerization of acetylenes to polyenes [120].

14. Polymerization of bicyclo[2.2.2] heptadiene to give a polymer with nortricyclene units [121].

15. Thermal polymerization of 2,6-diphenyl-1,6-heptadiene [122].

16. Polymerization of 1,5-hexadiene to afford a polymer containing cyclopentane rings [123].

17. Polymerization of cyclooctadiene using Al(iso-C_4H_9)–$TiCl_4$ at $-20°$ to 70°C to give five-membered ring-fused polymer systems [124].

18. Polymerization of diisopropenyl monomers [125].

19. Ring-forming polymerizations [126].

20. Preparation of a butadiene oligomer: *trans*-3-methylhepta-1,4,6-triene [127].

21. Polymerization of isobutylene via photoinitiation with vacuum ultraviolet radiation [128].

22. Cyclooligomerization of butadiene on polymerization catalysts such as diethylbis(dipyridyl)iron and diethyldipyridylnickel [129].

23. Polymerization of cyclopentadiene by free ions [130].

24. Polymerization of *trans*-1,3-pentadiene by γ radiation [131].

25. Radiation-induced polymerization of isobutylene [132].

26. Polymerization of optically active 2- and 1,2-substituted butadienes [133].

27. Polymerization of styrene catalyzed by clay-cracking catalysts [134].

28. Radiation-induced cationic polymerization of ethylene [135].

29. Radical-initiated homopolymerization of α-methylstyrene [136].

30. Free-radical copolymerization of α-olefins with ethylene [137].

31. Photopolymerization of ethylene [138].

32. Free-radical polymerization of ethylene at low temperature [139].

33. Polymerization of methylstyrenes and the effect of dimethylaniline on their polymerization [140].

34. Copolymerization and terpolymerization in continuous nonideal reactors [141].

35. Ring-opening polymerization of cycloolefins [142].

36. Polymerization of ethylene or propylene using cation-exchanged zeolite [143].

37. Preparation of amorphous polybutadiene (96% 1,2-units) using $MoCl_5$ and $Al(C_2H_5)_2OC_2H_5$–ethyl acetate catalyst [144].

38. Catalytic polymerization of ethylene using a 1:1 complex of bis(1,5-cyclooctadiene)nickel(0) and *o*-(diphenylphosphine)benzoic acid [145].

39. Chromium oxide–TiO_2 catalysts for the polymerization of olefins[146].

40. Crystalline polypropylene using $Al(C_2H_5)_2Cl$–$TiCl_3$–cinnamates or cinnamic amides [147].

41. Organometallic halide initiators for the polymerization of butadiene [148].

42. Ionic-coordinated catalysts for the initiation of butadiene polymerization [149].

43. Perfluorocarboxylato nickel salts as initiators for the stereospecific polymerization of 1,3-butadiene [150].

44. Kinetics of Ziegler propylene polymerization [151].

REFERENCES

1. C. Walling, "Free Radicals in Solution." Wiley, New York, 1957; J. C. Bevington, "Radical Polymerization." Academic Press, New York, 1961; P. E. M. Allen and P. H. Plesch, *in* "The Chemistry of Cationic Polymerization" (P. H. Plesch, ed.). Macmillan, New York, 1963; P. J. Flory, "Principles of Polymer Chemistry." Cornell Univ. Press, Ithaca, New York, 1953; R. W. Lenz, "Organic Chemistry of Synthetic High Polymers." Wiley (Interscience), New York, 1967; C. L. Arcus, *Progr. Stereochem.* **3**, 264 (1962); M. L. Huggins, G. Natta, V. Derreux, and H. Mark, *J. Polym. Sci.* **56**, 153 (1962); G. Natta, L. Porri, P. Corradini, G. Zanini, and F. Ciampelli, *ibid.* **51**, 463 (1961); N. Beredjick and C. Schuerch, *J. Amer. Chem. Soc.* **78**, 2646 (1956); C. S. Marvel and R. G. Wollford, *J. Org. Chem.* **25**, 1641 (1960); G. B. Butler and R. W. Stackman, *ibid.*, p. 1643 (1961); M. Farina, M. Peraldo, and G. Natta, *Angew. Chem., Int. Ed. Engl.* **4**, 107 (1965).
2. V. Regnault, *Ann. Chim. Phys.*, [2] **69**, 151 (1838).
3. E. Simon, *Ann. Pharm.* **31**, 265 (1839).
4. J. Blyth and A. W. Hofmann, *Ann. Chem. Pharm.* **53**, 289 and 311 (1845).
5. E. Baumann, *Ann. Chem. Pharm.* **163**, 312 (1872).
6. G. Bouchardat, *C. R. Acad. Sci.* **89**, 1117 (1879).
7. R. Fittig and F. Engelhorn, *Justus Liebigs Ann. Chem.* **200**, 65 (1880).
8. G. W. A. Kahlbaum, *Ber. Deut. Chem. Ges.* **13**, 2348 (1880).
9. S. W. Lebedev and N. A. Skavronskaya, *J. Russ. Phys. Chem. Soc.* **43**, 1124 (1911).
10. F. Klatte and A. Rollett, U.S. Patent 1,241,738 (1917).
11. E. W. Fawcett, British Patent 471,590 (1937).
12. H. Staudinger, *Ber. Deut. Chem. Ges.* **53**, 1073 (1920).
13. H. Staudinger, *Ber. Deut. Chem. Ges.* **62**, 241 and 292 (1929); "Die Hochmolekularen Organischen Verbindungen." Springer-Verlag, Berlin and New York, 1932.
14. H. S. Taylor, *Trans. Faraday Soc.* **21**, 560 (1925).
15. F. Paneth and W. Hofeditz, *Ber. Deut. Chem. Ges.* **62**, 1335 (1929).
16. F. Haber and R. Willstätter, *Ber. Deut. Chem. Ges.* **64**, 2844 (1931).
17. M. S. Kharasch, H. Engelmann, and F. R. Mayo, *J. Org. Chem.* **2**, 288 (1937).
18. D. H. Hey and W. A. Waters, *Chem. Rev.* **21**, 169 (1937).
19. P. J. Flory, *J. Amer. Chem. Soc.* **59**, 241 (1937); "Principles of Polymer Chemistry, Cornell Univ. Press, Ithaca, New York, 1953.
20. K. Ziegler, Belgian Patents 540,459 and 543.837 (1956).
21. R. A. V. Raff and K. W. Doaks, eds., "Crystalline Olefin Polymers," Parts 1 and 2. Wiley (Interscience), New York, 1964; H. V. Boening, "Polyolefins—Structure and Properties." Amer. Elsevier, New York, 1966; A. Renfrew and P. Morgan, eds., "Polythene: The Technology and Uses of Ethylene Polymers." Wiley (Interscience), New York, 1960.

22. E. F. Lutz and G. M. Bailey, *J. Polym. Sci., Part A-1* **4**, 1885 (1966); G. Natta, *Makromol. Chem.* **35**, 93 (1960); A. Turner-Jones, *ibid.* **71**, 1 (1964).
22a. J. A. Faucher and E. P. Reding, *in* "Crystalline Olefin Polymers" (A. V. Raff and K. W. Doak, eds.), Chapter 13. Wiley (Interscience), New York, 1965.
23. Author's Laboratory (S.R.S.); S. R. Sandler, P. J. McGonigal, and K. C. Tsou, *J. Phys. Chem.* **66**, 166 (1962).
23a. S. R. Sandler, *Int. J. Appl. Radiat. Isotop.* **16**, 473 (1965).
24. R. H. Boundy and R. F. Boyer, eds., "Styrene—Its Polymers Copolymers and Derivatives," Amer. Chem. Soc., Monogr. No. 115. Van Nostrand-Reinhold, Princeton, New Jersey, 1952.
25. Dow Chemical Company, "Styrene-Type Monomer," Technical Brochure No. 114-151-69 (1969); Dow Chemical Co., "Storage and Handling of Styrene-Type Monomers," Technical Brochure No. 170-280-6M-1067 (1967).
26. F. R. Mayo, *J. Amer. Chem. Soc.* **90**, 1289 (1968).
27. J. Pellon, R. L. Kugel, R. Marcus, and R. Rabinowitz, *J. Polym. Sci., Part A* **2**, 4105 (1964).
27a. G. C. Eastmond, *Encycl. Polym. Sci. Technol.* **7**, 361 (1967).
27b. S. R. Palit, T. Guha, R. Das, and R. J. Konar, *Encycl. Polym. Sci. Technol.* **2**, 229 (1965).
28. H. S. Mickley, A. S. Michaels, and A. L. Moore, *J. Polym. Sci.* **60**, 121 (1962).
29. W. I. Bengough and R. G. W. Norrish, *Proc. Roy. Soc., Ser. A* **218**, 149 (1953).
30. A. Chapiro, *J. Chem. Phys.* **47**, 747 (1950).
31. B. Atkinson and G. R. Cotten, *Trans. Faraday Soc.* **54**, 877 (1958).
32. E. Farber, *Encycl. Polym. Sci. Technol.* **13**, 522 (1970); G. J. Gammon, C. T. Richards, and M. J. Symes, British Patent 1,243,057 (1971); H. Nishikawa *et al.*, Japanese Patent 71/06,423 (1971); K. Wilkinson, British Patent 1,226,959 (1971); W. N. Maclay, *J. Appl. Polym. Sci.* **15**, 867 (1971); J. R. Hiltner and W. F. Bartoe, U.S. Patent 2,264,376 (1941); A. H. Turner and W. D. Bannister, British Patent 873,948 (1960); Kanegafuchi Chemical Industry Co., Ltd., French Patent 1,373,240 (1964); H. K. Chi, U.S. Patent 3,258,453 (1966); Shell Internationale Research Maatschappij N.V., French Patent Appl. 2,020,845 (1970).
32a. Dow Chemical Company, Netherlands Patent Appl. 6,514,851 (1966).
33. A. S. Dunn, *Chem. Ind.* (*London*) p. 1406 (1971).
34. F. A. Bovey, I. M. Kolthoff, A. F. Medalia, and E. J. Meehan, "Emulsion Polymerization." Wiley (Interscience), New York, 1955; S. N. Sautin, P. A. Kulle, and N. I. Smirnov, *Zh. Prikl Khim* (*Leningrad*) **44**, 1569 (1971).
34a. A. W. De Graff, "Continuous Emulsion Polymerization of Styrene in a One Stirred Tank Reactor." Lehigh Univ. Press, Bethlehem, Pennsylvania, 1970; O. Gellner, *Chem. Eng.* (*New York*) **73**, 74 (1966); A. G. Parts, D. E. Moore, and J. G. Waterson, *Makromol. Chem.* **89**, 156 (1965); E. W. Duck, *Encycl. Polym. Sci. Technol.* **5**, 801 (1966).
35. I. M. Kolthoff and W. J. Dale, *J. Amer. Chem. Soc.* **69**, 441 (1947).
36. C. C. Price and C. E. Adams, *J. Amer. Chem. Soc.* **67**, 1674 (1945).
37. D. J. Williams and E. G. Bobalek, *J. Polym. Sci., Part A-1* **4**, 3065 (1966).
38. I. D. Robb, *J. Polym. Sci., Part A-1* **7**, 417 (1969).
39. G. G. Lowry, *J. Polym. Sci.* **31**, 187 (1958).
40. I. M. Kolthoff and W. E. Harris, *J. Polym. Sci.* **2**, 41 (1947); N. Rabjohn, R. J. Dearborn, W. E. Blackburn, G. E. Inskeep, and H. R. Snyder, *J. Polym. Sci.* **2**, 488 (1947).
41. J. A. Rozmajzl, *Macromol. Syn.* **2**, 57 (1966).

42. L. Horner and J. Junkerman, *Justus Liebigs Ann. Chem.* **591**, 53 (1955).
43. W. J. Bailey and H. R. Golden, *J. Amer. Chem. Soc.* **76**, 5418 (1954).
44. P. H. Plesch, ed., "Cationic Polymerization and Related Complexes." Academic Press, New York, 1954; "The Chemistry of Cationic Polymerization." Pergamon Press, New York, 1964.
45. F. S. Dainton and G. B. B. M. Sutherland, *J. Polym. Sci.* **4**, 37 (1949).
46. A. G. Evans, B. Holden, P. H. Plesch, M. Polanyi, H. A. Skinner, and M. A. Weinberger, *Nature (London)* **157**, 102 (1946); A. G. Evans, G. W. Meadows, and M. Polanyi, *ibid.* **158**, 94 (1946).
47. D. C. Pepper, *Quart. Rev., Chem. Soc.* **8**, 88 (1954).
47a. D. C. Pepper, *in* "Friedel-Crafts and Related Reactions" (G. A. Olak, ed.), Vol. II, p. 123. Wiley (Interscience), New York, 1964.
48. A. G. Evans and G. W. Meadows, *J. Polym. Sci.* **4**, 359 (1949).
49. R. G. Heiligmann, *J. Polym. Sci.* **6**, 155 (1950).
50. J. A. Bittles, A. K. Chandhuri, and S. W. Benson, *J. Polym. Sci., Part A* **2**, 1221 (1964); G. F. Endres and C. G. Overberger, *J. Amer. Chem. Soc.* **77**, 2201 (1955); D. O. Jordan and A. R. Mathieson, *J. Chem. Soc., London* p. 611 (1952); J. A. Bittles, A. K. Chandhuri, and S. W. Benson, *J. Polym. Sci., Part A* **2**, 1221 (1964)
51. F. C. Whitmore, *Ind. Eng. Chem.* **26**, 94 (1964).
52a. R. Simha and L. A. Wall, *in* "Catalysis" (P. H. Emmett, ed.), Vol. VI, p. 266. Van Nostrand-Reinhold, Princeton, New Jersey, 1958.
52b. J. Hine, "Physical Organic Chemistry," p. 219. McGraw-Hill, New York, 1956.
52c. Y. Tsuda, *Macromol. Chem.* **36**, 102 (1960).
53a. G. A. Olah, ed., "Friedel-Crafts and Related Reactions," Vol. I. Wiley (Interscience), New York, 1963.
53b. D. N. P. Satchell, *J. Chem. Soc., London* pp. 1453 and 3822 (1961).
54. A. M. Eastham, *Encycl. Polym. Sci. Technol.* **3**, 35 (1965).
55. J. P. Kennedy, *J. Org. Chem.* **35**, 532 (1970).
56. J. P. Kennedy, *in* "Polymer Chemistry of Synthetic Elastomers" (J. P. Kennedy and E. Tarnquist, eds.), Part 1, Chapter 5A, p. 291. Wiley (Interscience), New York, 1968; Belgian Patent 663,319 (1965).
57. J. P. Kennedy and R. M. Thomas, *Advan. Chem. Ser.* **34**, Chapt. 7 (1962).
58. A. J. Morway and F. L. Miller, U.S. Patent 2,243,470 (1941).
59. C. E. Schildknecht, "Vinyl and Related Polymers," Chapter 10. Wiley, New York, 1952.
60. R. M. Thomas, W. J. Sparks, and P. K. Frolich, *J. Amer. Chem. Soc.* **62**, 276 (1940).
60a. R. M. Thomas and W. J. Sparks, U.S. Patent 2,356,128 (1944).
60b. J. D. Calfee and R. M. Thomas, U.S. Patent 2,431,461 (1947).
61. J. P. Kennedy and R. M. Thomas, *J. Polym. Sci.* **46**, 481 (1960).
62. J. P. Kennedy and R. M. Thomas, *J. Polym. Sci.* **46**, 233 (1960).
63. P. H. Plesch, M. Polanyi, and H. A. Skinner, *J. Chem. Soc., London* p. 257 (1947).
64. R. E. Rinehart, H. P. Smith, H. S. Witt, and H. Romeyn, Jr., *J. Amer. Chem. Soc.* **83**, 4864 (1961); **84**, 4145 (1962).
65. J. P. Kennedy and R. G. Squires, *Polymer* **6**, 579 (1965).
66. S. G. Gallo, H. K. Wiese, and J. F. Nelson, *Ind. Eng. Chem.* **40**, 1277 (1948).
66a. J. P. Kennedy, U.S. Patent 3,349,065 (1967).
67. C. S. Marvel, R. Gilkey, C. R. Morgan, J. F. Noth, R. D. Rands, Jr., and H. C. Young, *J. Polym. Sci.* **6**, 483 (1951).
68. W. Schlenk, J. Appenrodt, A. Michael, and A. Thal, *Ber. Deut. Chem. Ges.* **47**, 473 (1914).

69. M. Szwarc, M. Levy, and R. Milkovich, *J. Amer. Chem. Soc.* **78**, 2656 (1956); M. Szwarc, "Carbanions Living Polymers and Electron Transfer Processes." Wiley (Interscience), New York, 1968.
70. D. J. Worsfold and S. Bywater, *J. Phys. Chem.* **70**, 162 (1966); F. S. Dainton *et al.*, *Makromol. Chem.* **89**, 257 (1965).
71a. C. Harries, *Justus Liebigs Ann. Chem.* **383**, 213 (1911).
71b. W. Schlenk, J. Appenrodt, A. Michael, and A. Thal, *Ber. Deut. Chem. Ges.* **47**, 473 (1914).
71c. K. Ziegler and K. Bähr, *Ber. Deut. Chem. Ges.* **61**, 253 (1928).
72. W. Schlenk and E. Bergmann, *Justus Liebigs Ann. Chem.* **479**, 42 (1930).
73. K. Ziegler, F. Crossman, H. Kliener, and O. Schafter, *Justus Liebigs Ann. Chem.* **473**, 1 (1929).
74. L. M. Ellis, U.S. Patent 2,212,155 (1940).
75. W. C. E. Higginson and N. S. Wooding, *J. Chem. Soc., London*, p. 760 (1952).
76. M. Szwarc, *Nature* **178**, 1168 (1956); J. Smid and M. Szwarc, *J. Polym. Sci.* **61**, 31 (1962); M. Szwarc, "Carbanions, Living Polymers and Electron Transfer Processes." Wiley (Interscience), New York, 1968; M. Szwarc, *Encycl. Polym. Sci. Technol.* **8**, 303 (1968); T. Shimomura, J. Smid, and M. Szwarc, *J. Amer. Chem. Soc.* **89**, 5743 (1967).
77. D. J. Cram, *Chem. Eng. News* **41**, 92 (1963).
78. R. Woack and M. Doran, *Polymer* **2**, 365 (1961); D. J. Cram, *Chem. Eng. News* **41**, 92 (1963); A. A. Morton and E. Grovenstein, Jr., *J. Amer. Chem. Soc.* **74**, 5434 (1952); K. Yoshida and T. Morikawa, *Sci. Ind. Osaka* **27**, 80 (1953); W. E. Goode, W. H. Snyder, and R. C. Fettes, *J. Polym. Sci.* **42**, 367 (1960); W. H. Puterbaugh and C. R. Hauser, *J. Org. Chem.* **24**, 416 (1969).
79. R. Waack and M. A. Doran, *J. Org. Chem.* **32**, 3395 (1967).
80. A. A. Morton and E. Grovenstein, Jr., *J. Amer. Chem. Soc.* **74**, 5434 (1952).
80a. A. A. Morton and E. J. Lanpher, *J. Polym. Sci.* **44**, 239 (1960).
81. M. Szwarc, *Encycl. Polym. Sci. Technol.* **8**, 303 (1968); "Carbanions, Living Polymers and Electron Transfer Processes." Wiley (Interscience), New York, 1968.
82. K. Ziegler and K. Bähr, *Ber. Deut. Chem. Ges.* **61**, 253 (1928).
83. G. Gee, W. C. E. Higginson, and G. T. Merrell, *J. Chem. Soc., London* p. 1345 (1959).
84. A. G. Evans and G. W. Meadows, *Trans. Faraday Soc.* **46**, 327 (1950).
85. A. A. Morton, *Encycl. Polym. Sci. Technol.* **1**, 629 (1964).
85a. A. A. Morton, G. H. Patterson, J. J. Donovan, and E. L. Little, *J. Amer. Chem. Soc.* **68**, 93 (1946); A. A. Morton, *Ind. Eng. Chem.* **42**, 1488 (1950); A. A. Morton, R. P. Welcher, F. Collins, S. E. Penner, and R. D. Coombs, *J. Amer. Chem. Soc.* **71**, 481 (1949).
86. H. Greenberg and V. L. Hansley, U.S. Patent 3,067,187 (1962).
86a. V. L. Hansley and H. Greenberg, *Rubber Age* **94**, 87 (1963).
86b. J. L. R. Williams, J. Van Den Berghe, W. L. Dulmage, and K. R. Dunham, *J. Amer. Chem. Soc.* **78**, 1260 (1956); J. L. R. Williams, J. Van Den Berghe, K. R. Dunham, and W. L. Dulmage, *ibid.* **79**, 1716 (1957); A. A. Morton and L. Taylor, *J. Polym. Sci.* **38**, 7 (1959).
87. C. G. Overberger, J. E. Mulvaney, and A. M. Schiller, *Encycl. Polym. Sci. Technol.* **2**, 95 (1965); M. Morton and L. J. Fetters, *Macromol. Rev.* **3**, 71 (1967).
88. M. Szwarc, M. Levy, and R. Milkovich, *J. Amer. Chem. Soc.* **78**, 2656 (1956).
88a. N. D. Scott, J. F. Wilken, and V. L. Hansley, *J. Amer. Chem. Soc.* **58**, 2442 (1936).
89. A. A. Morton, M. L. Brown, and E. Magat, *J. Amer. Chem. Soc.* **68**, 161 (1946).
89a. A. A. Morton and R. L. Letsinger, *J. Amer. Chem. Soc.* **69**, 172 (1947).

90. A. A. Morton, R. P. Welcher, F. Collins, S. E. Renner, and R. D. Coombs, *J. Amer. Chem. Soc.* **71**, 481 (1949).
91. R. S. Stearns and L. E. Forman, *J. Polym. Sci.* **41**, 381 (1959).
92. C. E. Schildknecht, A. O. Zoss, and C. McKinley, *Ind. Eng. Chem.* **39**, 180 (1947).
93. M. Fischer, German Patent 874,215 (1953).
94. D. F. Herman and W. K. Nelson, *J. Amer. Chem. Soc.* **75**, 3877 (1953).
95. K. Ziegler, Belgian Patent 533,362 (1954); K. Ziegler, E. Holzkamp, H. Breil, and H. Martin, *Angew. Chem.* **67**, 541 (1955).
96. G. Natta, *Atti Accad. Naz. Lincei, Cl. Sci. Fis., Mat. Natur., Rend.* [8] **4**, 61 (1955); G. Natta, P. Pino, P. Corradini, F. Danusso, E. Mantica, G. Mazzanti, and G. Moraglio, *J. Amer. Chem. Soc.* **77**, 1708 (1955); G. Natta, L. Porri, and S. Valenti, *Makromol. Chem.* **67**, 225 (1963).
97. L. Porri, G. Natta, and M. C. Gallazzi, *J. Polym. Sci., Part C* **16**, 2525 (1967); V. A. Kormer, B. D. Babitskii, M. I. Lobach, and N. N. Chesnokova, *ibid.*, p. 4351 (1969).
98. G. Wilke, *Angew Chem.* **75**, 10 (1963); Y. Tajima and E. Kunioka, *J. Polym. Sci., Part D* **5**, 221 (1967); G. Wilke, M. Kroener, and B. Bogdanovic, *Angew. Chem.* **73**, 755 (1961); B. D. Babitski, V. A. Kormer, I. M. Lapak, and V. I. Soblikova, *J. Polym. Sci., Part C* **16**, 3219 (1968).
99. G. Natta, L. Porri, A. Carbonaro, F. Ciampelli, and G. Allegra, *Makromol. Chem.* **51**, 229 (1962); G. Natta, L. Porri, A. Carbonaro, and G. Stoppa, *ibid.* **77**, 114 (1964).
100. W. Cooper, *Ind. Eng. Chem., Prod. Res. Develop.* **9**, 457 (1970).
101. W. Cooper and G. Vaughn, *Progr. Polym. Sci.* **7** (1967); J. Boor, *Macromol Rev.* **2**, 115 (1967); J. P. Kennedy and A. W. Langer, Jr., *Fortsch. Hochpolym.-Forsch.* **3**, 539 (1964); N. G. Gaylord and H. Mark, "Linear and Stereoregular Addition Polymers." Wiley (Interscience), New York, 1959; K. Ziegler, *Angew. Chem.* **76**, 545 (1964); J. P. Kennedy and G. E. Milliman, Jr., *Advan. Chem. Ser.* **91**, 287–305 (1969); V. A. Kormer, B. D. Babitskiy, and M. I. Lobach, *ibid.*, pp. 306–316; G. Natta and U. Giannini, *Encycl. Polym. Sci. Technol.* **4**, 137 (1966).
102. G. W. Phillips and W. L. Carrick, *J. Polym. Sci.* **59**, 401 (1962).
103. W. L. Carrick, *Macromol. Syn.* **2**, 33 (1966).
104. L. M. Ellis, U.S. Patent 2,212,155 (1940).
105. W. J. Bailey and E. T. Yates, *J. Org. Chem.* **25**, 1800 (1960).
106. J. S. Lasky, U.S. Patent 3,054,754 (1962).
106a. J. S. Lasky, U.S. Patent 3,111,744 (1963).
107. W. M. Saltman, W. E. Gibbs, and J. Lal, *J. Amer. Chem. Soc.* **80**, 15615 (1958); Goodrich-Gulf Chemical Company, Belgian Patent 543,292 (1955); W. M. Saltman and E. Schoenberg, *Macromol. Syn.* **2**, 50 (1966).
108. M. A. Golub and D. B. Parkinson, *Macromol. Syn.* **3**, 32 (1969).
109. B. Erusalimskii, *Polymer* **3**, 639 (1962).
110. Badische Anilin & Soda-Fabrik, German Patent 1,105,168 (1961).
111. Badische Anilin & Soda-Fabrik, German Patent 1,163,025 (1964).
112. V. Vorob'ev, *Plast. Massy* No. 9, p. 6 (1962); R. Andrianov, U.S.S.R. Patent 138,049 (1961).
113. M. Imoto, *Bull. Chem. Soc. Jap.* **34**, 186 (1961).
114. N. Tokura, *Sci. Rep. Res. Inst., Tohoku Univ., Ser.* **12**, 380 (1960).
115. T. Yamaguchi, *Kobunshi Kagaku* **22**, 835 (1965).
116. R. De Coene, U.S. Patent 3,041,324 (1962); Solvic, Belgian Patent 570,028 (1959); F. Arimota, *J. Polym. Sci., Part A-1*, **4**, 275 (1966).
117. S. Matsoyan, *Vysokomol. Soedin* **5**, 183 (1963).

118. Shawinigan Resins Corp., British Patent 1,019,069 (1966).
119. Kureha Chemical Industry Co., Japanese Patent 10,590 (1965).
120. W. E. Daniels, *J. Org. Chem.* **29**, 2936 (1964).
121. P. J. Graham, E. L. Buhle, and N. Pappas, *J. Org. Chem.* **26**, 4658 (1961); R. H. Wiley, W. H. Rivera, T. H. Crawford, and N. F. Bray, *J. Polym. Sci.* **61**, S38 (1962); N. L. Zutty, *J. Polym. Sci., Part A* **1**, 2231 (1963).
122. W. D. Field, *J. Org. Chem.* **25**, 1006 (1960).
123. H. S. Makowski, B. K. C. Shim, and Z. E. Wilchensky, *J. Polym. Sci., Part A* **2**, 1549 (1964).
124. B. Reichel, C. S. Marvel, and R. Z. Greenley, *J. Polym. Sci., Part A* **1**, 2935 (1963).
125. H. Brunner and D. J. Walbridge, British Patent 864,275 (1961).
126. R. J. Cotter and M. Matzner, "Ring-Forming Polymerizations," Part A. Academic Press, New York, 1969.
127. S. Otsuka, T. Kikuchi, and T. Takeforni, *J. Amer. Chem. Soc.* **85**, 3709 (1963).
128. E. W. Schlag and J. J. Sparapany, *J. Amer. Chem. Soc.* **86**, 1875 (1964).
129. A. Yamamoto, K. Morifuji, S. Ikeda, T. Sarto, Y. Uchida, and A. Misono, *J. Amer. Chem. Soc.* **87**, 465a (1965).
130. M. A. Bonin, W. R. Busler, and F. Williams, *J. Amer. Chem. Soc.* **87**, 199 (1965).
131. M. Farina, G. Audisio, and G. Natta, *J. Amer. Chem. Soc.* **89**, 5071 (1967).
132. R. B. Taylor and F. Williams, *J. Amer. Chem. Soc.* **89**, 6359 (1967); **91**, 3728 (1969).
133. Z. Janovic and D. Fles, *J. Polym. Sci., Part A* **19**, 1103 (1971).
134. J. A. Bittles, A. K. Chandhuri, and S. W. Benson, *J. Polym. Sci., Part A* **2**, 1221 (1964).
135. Y. Tabata, H. Shibano, H. Sobue, and K. Hara, *J. Polym. Sci., Part A* **1**, 1049 (1963).
136. G. G. Lowry, *J. Polym. Sci.* **31**, 187 (1958).
137. L. Boghetich, G. A. Mortimer, and D. W. Danes, *J. Polym. Sci.* **61**, 3 (1962).
138. M. Hagiwara, H. Okamoto, T. Kaiya, and T. Kagiya, *J. Polym. Sci., Part A-1* **8**, 3295 (1970).
139. S. Machi, T. Sakai, T. Tamura, M. Gotoda, and T. Kagiya, *Polym. Lett.* **3**, 709 (1965).
140. M. Imoto, K. Takemoto, and Y. Iikubo, *Bull. Chem. Soc. Jap.* **34**, 186 (1961).
141. T. S. Szabo and E. B. Nauman, *AIChE J.* **15**, 575 (1969).
142. G. Dall'Asta, G. Motroni, and L. Milan, *J. Polym. Sci., Part A-1* **10**, 1601 (1972).
143. T. Nishizawa, H. Hattori, T. Uematsu, and T. Shiba, *Proc. Int. Congr. Catal., 4th, 1968* Vol. 2, p. 114 (1971).
144. F. Dawans and P. Amignes, German Patent 2,157,004 (1972).
145. M. D. Singleton, W. P. Glockner, and W. Keim, German Patent 2,159,370 (1972).
146. E. R. Dietz, German Patent 2,155,884 (1972).
147. S. Tanaka, A. Kanamura, S. Yoshida, T. Maeno, T. Hori, K. Tashiro, and R. Koga, Japanese Patent 11,808 (1972).
148. T. Otsu and M. Yamaguchi, *J. Polym. Sci., Part A-1* **7**, 387 (1969).
149. L. S. Bresler, I. Y. Poddubnyi, and V. N. Sokolov, *J. Polym. Sci., Part C* **16**, 4337 (1969).
150. F. Dawans, J. P. Durand, and P. Teyssie, *Polym. Lett.* **10**, 493 (1972).
151. V. W. Buls and T. L. Higgins, *J. Polym. Sci.* **11**, 925 (1973).

Chapter 2

POLYESTERS

1. INTRODUCTION

Polyesters are polymers with repeating carboxylate groups

$$-\overset{\overset{\displaystyle O}{\|}}{C}O-$$

in their backbone chain. Polycarbonates are esters of carbonic acid and will be discussed in Chapter 3.

Polyesters are synthesized [1–11] by the typical esterification reactions, which can be generalized by the reaction shown in Eq. (1)

$$R\overset{\overset{\displaystyle O}{\|}}{C}-X + N: \rightleftharpoons \left[R-\underset{\underset{\displaystyle X}{|}}{\overset{\overset{\displaystyle \ddot{O}}{|}}{C}}-N \right] \longrightarrow R\overset{\overset{\displaystyle O}{\|}}{C}-N + X: \qquad (1)$$

where N: is a nucleophilic reagent such as $\bar{O}R'$. The rate of reaction will be dependent on the structure of R, R′, X, N and on whether a catalyst is used.

Tartaric acid–glycerol polyesters were reported in 1847 by Berzelius [12] and those of ethylene glycol and succinic acid were reported by Lorenzo in 1863 [13]. Carothers extended much of the earlier work and helped to clarify the understanding of the polyesterification reaction in light of the knowledge of polymer chemistry at that time. Polyethylene terephthalate [14] and the polyadipates [15] (for polyurethane resins) were the first major commercial fruition of polyesters.

The major synthetic methods used to prepare polyesters all involve condensation reactions as shown in Eqs. (2)–(12).

$$\text{HOR—COOH} \xrightarrow{[16]} \text{H}\left[\text{OR—}\overset{\text{O}}{\overset{\|}{\text{C}}}\right]_n\text{OH} \tag{2}$$

$$\text{R'COOR—COOH} \xrightarrow{[17]} \text{R'CO}\left[\text{OR}\overset{\text{O}}{\overset{\|}{\text{C}}}\right]_n\text{OH} \tag{3}$$

$$\text{RCOOROCOR} + \text{R(COOH)}_2 \xrightarrow{[18]} \text{RCO}\left[\text{OR}\overset{\text{O}}{\overset{\|}{\text{C}}}\right]_n\text{OH} \tag{4}$$

$$\text{HO—R—OH} + \text{R'(COOH)}_2 \xrightarrow{[18,19]} \text{H}\left[\text{ORO}\overset{\text{O}}{\overset{\|}{\text{C}}}\text{R'}\overset{\text{O}}{\overset{\|}{\text{C}}}\right]_n\text{OH} \tag{5}$$

$$\text{HO—R—OH} + \text{R'(COOR'')}_2 \xrightarrow{[20,20a]} \text{H}\left[\text{OR—O}\overset{\text{O}}{\overset{\|}{\text{C}}}\text{—R'}\overset{\text{O}}{\overset{\|}{\text{C}}}\right]_n\text{OH} \tag{6}$$

$$\text{HO—R—OH} + \text{R'(COX)}_2 \xrightarrow{[21,21a]} \text{H}\left[\text{ORO}\overset{\text{O}}{\overset{\|}{\text{C}}}\text{R'}\overset{\text{O}}{\overset{\|}{\text{C}}}\right]_n\text{X} \tag{7}$$

$$\text{NaOR—ONa} + \text{R'(COX)}_2 \xrightarrow{[22]} \text{Na}\left[\text{ORO}\overset{\text{O}}{\overset{\|}{\text{C}}}\text{R'}\overset{\text{O}}{\overset{\|}{\text{C}}}\right]_n\text{X} \quad (\text{X} = \text{Cl, OCH}_3) \tag{8}$$

$$\text{HO—R—OH} + \text{R'(CO)}_2\text{O} \xrightarrow{[23]} \text{H}\left[\text{ORO}\overset{\text{O}}{\overset{\|}{\text{C}}}\text{R'}\overset{\text{O}}{\overset{\|}{\text{C}}}\right]_n\text{OH} \tag{9}$$

$$\text{HO—R—OH} + \text{R'O}\overset{\text{O}}{\overset{\|}{\text{C}}}\text{OR'} \xrightarrow{[24]} \text{H}\left[\text{ORO}\overset{\text{O}}{\overset{\|}{\text{C}}}\right]\text{OR'} \tag{10}$$

$$\text{Br—R—Br} + \text{R'(COOAg)}_2 \xrightarrow{[25]} \text{Br}\left[\text{ORO}\overset{\text{O}}{\overset{\|}{\text{C}}}\text{R'}\overset{\text{O}}{\overset{\|}{\text{C}}}\right]_n\text{Ag} \tag{11}$$

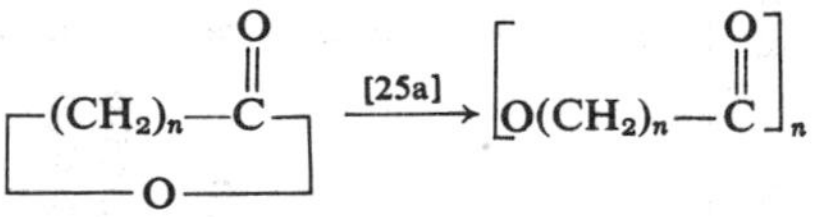

$$\underbrace{\overset{}{\text{—(CH}_2)_n\text{—}\overset{\text{O}}{\overset{\|}{\text{C}}}\text{—}}}_{\text{O}} \xrightarrow{[25a]} \left[\text{O(CH}_2)_n\text{—}\overset{\text{O}}{\overset{\|}{\text{C}}}\right]_n \tag{12}$$

The use of triols or tricarboxylic acids leads to cross-linked or network polyesters. For example, an alkyd resin is formed by the reaction of glycerol with phthalic anhydride [26].

Some typical examples of the preparation and properties of some representative polyesters are shown in Table I.

Polyesters [27] find use in fibers [poly(ethylene terephthalate), poly(ethylene oxybenzoate), poly(ester-ethers), poly(ester-amides), etc.] [27a], coatings (especially unsaturated polyesters) [28], plasticizers, adhesives, polyurethane base resins, films, etc. Cross-linked polyesters prepared from glycerol and phthalic anhydride (alkyd resins) have been reviewed [29]. High-melting polyaryl esters have been investigated for high-temperature applications.

Polyesters usually have good thermal and oxidative stability (up to 200°C) but have poor hydrolytic stability at elevated temperatures.

2. CONDENSATION REACTIONS

Polyesters are essentially prepared by the typical procedures used for the preparation of the monoesters described in an earlier volume by the present authors [30]. Catalysts are used to increase the speed of esterification. Aromatic acids and aromatic alcohols afford high-melting polyesters as compared to the viscous liquids prepared from aliphatic starting materials. The use of stoichiometric amounts affords high molecular weight products whereas an excess of one reactant lowers the molecular weight. In many cases hydroxy acids have a great tendency toward forming cyclic dimers, especially if five- or six-membered rings can be formed as in the case of hydroxyacetic acid [31].

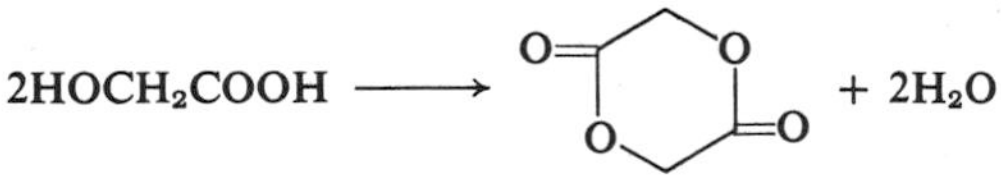

$$2HOCH_2COOH \longrightarrow \text{(glycolide)} + 2H_2O \tag{13}$$

(glycolide)

Heating glycolide with zinc chloride affords a linear polyester, $HO(CH_2COO)_nH$. Heating lactide at 250°–275°C affords a linear polyester of a molecular weight of about 3000. Use of potassium carbonate allows the same reaction to occur at 140°–150°C [32]. β-Hydroxy acids often undergo dehydration on heating [33]. γ-Hydroxy acids afford γ-lactones [34] and

TABLE I

PREPARATION OF POLYESTERS BY CONDENSATION REACTIONS

Reactants		Reaction conditions					
Alcohol	Diacid or derivative	Catalyst (gm)	Temp. (°C)	Time (hr)	M.p. (°C)	Mol. wt.	Ref.
$HO(CH_2)_nOH$ $n = 2,3,6,10$	Aliphatic type: carbonic, oxalic, succinic, glutaric, adipic, pimelic, and sebacic acids	—	—	—	—	—	*a*
$HO(CH_2)_nOH$ $n = 2,3,6,10$	Phthalic acid	—	—	—	—	—	*a*
$HO(CH_2)_4OH$	Isophthaloyl chloride	—	40–218	$\frac{1}{2}$–$\frac{3}{4}$	140–142	—	*b*
$HO(CH_2)_nOH$ $n = 2,3,6,8,10,18$	Terephthalic acid	—	—	—	—	—	*c*
$HOCH_2(CF_2)_3CH_2OH$	$(CH_2)_4(COOH)_2$	$ZnCl_2$ (0.01)	150–215	240	Visc. liq.	4000	*d*
$HOCH_2(CF_2)_3CH_2OH$	$(CF_2)_4(COCl)_2$	—	—	21.5	Approx. 35	M_n, 6570	*d*
$HO(CH_2)_{10}OH$	1,1-Dichlorocarbonylferrocene	—	—	—	25–65	500–4000	*e*
$HO(CH_2)_4OH$	Dimethyl sebacate	PbO (0.0625)	183–259	3	64–65	—	*f*
$HOCH_2HC{=}CH{-}CH_2OH$	Dimethyl sebacate	PbO (0.0625)	100–142 (2 mm Hg)	18–24	49	—	*f*
$HOCH_2(CF_2)_3CH_2OH$	$C_2H_5OC(=O)-C_6H_3(OC_5H_{11})-COOC_2H_5$	—	150–300	29	310	—	*g*

	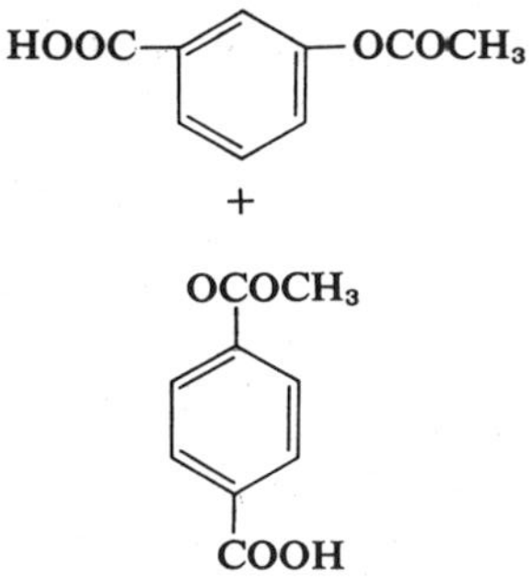 $HOOC-C_6H_4-OCOCH_3$ (meta) + $CH_3OCO-C_6H_4-COOH$ (para)	—	220–300	—	—	—	*h*
$HOCH_2CH_2OH$	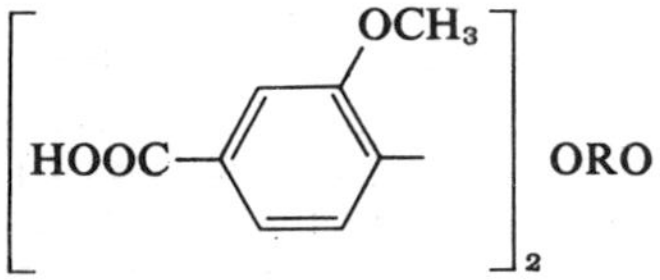 $[HOOC-C_6H_3(OCH_3)-]_2ORO$	—	—	—	200–210	—	*i*
Phenolphthalein	Isophthaloyl chloride	—	10	1–2	355	—	*j*
Hydroquinone	Terephthaloyl chloride	—	25–30	1–2	Infusible	—	*k*
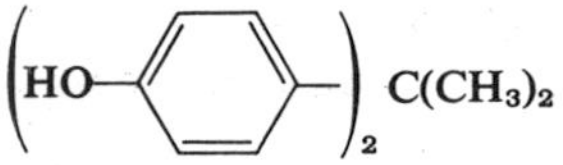 $(HO-C_6H_4-)_2C(CH_3)_2$	$(CH_2)_4(COOH)_2$	—	—	—	Yield point 45	—	*l*
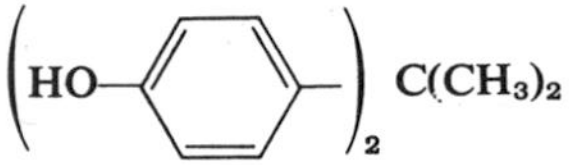 $(HO-C_6H_4-)_2C(CH_3)_2$	Dimethyl terephthalate	—	—	—	350	—	*l*
	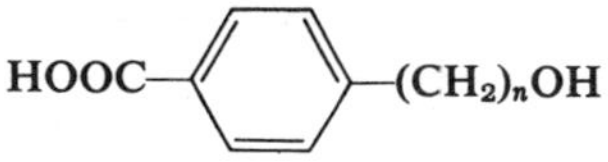 $HOOC-C_6H_4-(CH_2)_nOH$, $n = 0, 1, 2$	—	—	—	—	—	*m*
2,2′-Dihydroxydinaphthyl-1	Terephthaloyl chloride	—	—	—	420 (dec.)	—	*n*

(*cont.*)

TABLE I (*cont.*)

Reactants		Catalyst (gm)	Reaction conditions		M.p. (°C)	Mol. wt.	Ref.
Alcohol	Diacid or derivative		Temp. (°C)	Time (hr)			
CH_2—C=O, CH_2—O (four-membered ring)	—	$FeCl_3$	150	—	—	—	*o*
CH_2—CH_2—CH_2—C(=O)—O—CH_2—CH (ring)	—	Diol	—	—	—	—	*p*

[a] W. H. Carothers, U.S. Patent 2,071,053 (1937).
[b] P. J. Flory and F. S. Leutner, U.S. Patent 2,623,034 (1952).
[c] J. R. Whinfield and J. T. Dickson, British Patent 578,079 (1946); E. F. Izard, *J. Polym. Sci.* **8**, 503 (1952); J. R. Caldwell and R. Gilkey, U.S. Patent 2,891,930 (1959).
[d] G. C. Schweiker and P. Robitschek, *J. Polym. Sci.* **24**, 33 (1957).
[e] C. U. Pittman, Jr., *J. Polym. Sci., Part A-1* **6**, 1687 (1968).
[f] C. S. Marvel and J. H. Johnson, *J. Amer. Chem. Soc.* **72**, 1674 (1950).
[g] R. C. Evers and G. F. L. Ehlers, *J. Polym. Sci., Part A-1* **7**, 3020 (1969).
[h] R. Gilkey and J. R. Caldwell, *J. Polym. Sci.* **2**, 198 (1959).
[i] L. H. Bock and J. K. Anderson, *J. Polym. Sci.* **28**, 121 (1958).
[j] P. W. Morgan, *J. Polym. Sci., Part A* **2**, 437 (1964).
[k] W. M. Eareckson, III, *J. Polym. Sci.* **50**, 399 (1959).
[l] M. Levine and S. C. Temin, *J. Polym. Sci.* **28**, 179 (1958).
[m] J. R. Whinfield, *Nature (London)* **158**, 930 (1946); J. G. Cook, J. T. Dickson, A. R. Lowe, and J. R. Whinfield, U.S. Patent 2,471,023 (1949).
[n] Z. Jedlinski and D. Sek, *J. Polym. Sci., Part A-1* **7**, 2587 (1969).
[o] T. L. Gresham, J. E. Jansen, and F. W. Shaver, *J. Amer. Chem. Soc.* **70**, 998 (1948).
[p] F. Hostettler and D. M. Young, U.S. Patent 3,169,945 (1965); H. Cherdron, H. Ohse, and F. Korte, *Makromol. Chem.* **56**, 187 (1962).

δ-hydroxy acids afford unstable lactones which are easily converted to linear polyesters [32,34]. When the hydroxyl is more than six positions away from the carboxyl group only linear polyesters are obtained [35]. ω-Hydroxydecanoic acid polymerizes to afford molecular weights of 780–25,000 depending on the polymerization conditions [16].

Functionality greater than two in the acid or alcohol affords three-dimensional or cross-linked polymers often referred to as alkyd resins [36].

Preparation of polyesters using a diacid and an excess of diol leads first to low molecular weight hydroxy-terminated polyesters (see Eq. 14). These polymers are useful for the preparation of polyurethane resins or they may be converted to high molecular weight polyesters by further heating under reduced pressure to eliminate the excess diol by ester exchange reactions.

$$n\mathrm{HOOC(CH_2)_4COOH} + (n+1)\underset{\mathrm{OH}}{\underset{|}{\mathrm{CH_2}}}\!-\!\underset{\mathrm{OH}}{\underset{|}{\mathrm{CH_2}}} \longrightarrow \mathrm{HO}\left[\mathrm{CH_2CH_2}\!-\!\mathrm{O\overset{O}{\overset{\|}{C}}(CH_2)_4\overset{O}{\overset{\|}{C}}O}\right]_n\mathrm{CH_2CH_2OH} + 2n\mathrm{H_2O} \tag{14}$$

Mol. wt. 2000–4000

Polymerization to high molecular weight products requires that an equal number of reacting groups be present at all times. Since some glycol or diol may be lost during the liberation of the water as steam, a 10–20 mole % excess of diol should be used. This should also lead to a hydroxy-terminated polyester capable of reacting with other functional groups such as diisocyanates.

The direct esterification usually is self-catalyzed but helpful catalysts include (1–2.5% by weight of the reactants) *p*-toluenesulfonic acid or camphorsulfonic acid. Some typical examples are described in Ref. [37]. The starting materials are either heated at 170°–230°C under reduced pressure or the water is removed with high-boiling solvents by azeotropic distillation.

The direct esterification procedure works well for diols containing primary or secondary hydroxyl groups but is not effective for tertiary hydroxy groups or most phenols except *p*-hydroxybenzoic acid. Oxalic and malonic acids cannot be heated too long otherwise decarboxylations occur. Kinetic studies indicate that the functional groups have reactivities independent of the molecular chain size; the reaction is third order for $p = 0.8$ to $p = 0.98$ $[k(\mathrm{COOH})^2(\mathrm{OH})]$ and the reaction is catalyzed by strong acids [1,38]. The rate of polyesterification of substituted phenyl itaconic acids with ethylene glycol was found to follow a second-order rate equation [39].

The ester-interchange reaction (Eq. 15) is used commercially to prepare poly(ethylene terphthalate) [40] and related polyesters [41]. Griehl and Schnock [42] found that the rate of the ester-interchange reaction varied considerably with the type of diacid but hardly at all with the type of glycol. Terephthalic acid ester reacted more rapidly than phthalic or isophthalic acid esters. Benzoic acid esters were the least reactive, whereas sebacic acid esters

$$\left[HOCH_2CH_2O\overset{O}{\overset{\|}{C}}-\right]_2R \xrightarrow{\text{heat}} \left[OCH_2CH_2O\overset{O}{\overset{\|}{C}}R-\overset{O}{\overset{\|}{C}}-\right]_n + HOCH_2CH_2OH \quad (15)$$

were as reactive as terephthalic acid esters. Exchange catalysts such as sodium alkoxides, and cadmium, zinc, and lead acetates are more effective ester-interchange catalysts than sulfuric acid, *p*-toluenesulfonic acid, and sodium, cobalt, barium, and magnesium acetates [42].

Polyesterification using acidic catalysts is useful when preparing polyarylates of aliphatic [43] and ring-containing [44] dicarboxylic acids. For example, the diacetate of the dihydroxybenzene compound is heated with the dicarboxylic acid in the presence of *p*-toluenesulfonic acid (or magnesium metal, antimony oxide, zinc or sodium acetate). At 208°C acetic acid is eliminated and then the reaction is heated further under reduced pressure at 280°C.

A complicating factor in some polyester preparations is isomerization of the unsaturated alcohols or acids. For example, maleic acid esters isomerize to fumarates at elevated temperatures [45]. Ethylene glycol reacts with maleic anhydride at 160°–175°C to give a polyester with 55.5% fumarate and 44.5% maleate structures [45].

The polycondensation of diols with diacid chlorides is effected either by directly heating the components in a nitrogen atmosphere, or by a variation of the Schotten-Baumann reaction using trialkylamine dissolved in an inert solvent (acetone) [46]. The latter reaction can also be carried out interfacially by using a bisphenol dissolved in aqueous base (NaOH) and then adding the diacid dichloride in a water-immiscible solvent. Using the latter method the yields are approximately 80–100% and the molecular weights are as high as 80,000–90,000. The interfacial polycondensation procedure gives higher molecular weight polyesters than the melt polymerization procedure. Monofunctional phenols or acid chlorides act as molecular weight regulators.

a. Melting Point–Chemical Structure Relationship

Wilfong [19] has summarized many studies describing the relationship between chemical structure and physical properties. The following factors afford high-melting point polymers: (a) hydrogen bonding, dipole interaction, and polarization [resonance forces lead to high melting points, as is the case in poly(ethylene terephthalate), m.p. 265°C, and poly(ethylene *p,p'*-diphenylate, m.p. 355°C], (b) stiff interchain bonds, (c) molecular symmetry or regularity, (d) linear polymer chains capable of close packing. The following factors afford low-melting amorphous polymers: (a) bulky side chains and irregularity and (b) flexible interchain bonds.

Several other references describing studies of the effect of melting point on the structure of polyesters are worthwhile consulting [47].

b. Thermal Stability of Polyesters

Polyesters usually degrade via a cyclic transition state to afford the free acid and an olefin [48]. The presence of a β-hydrogen aids this degradation,

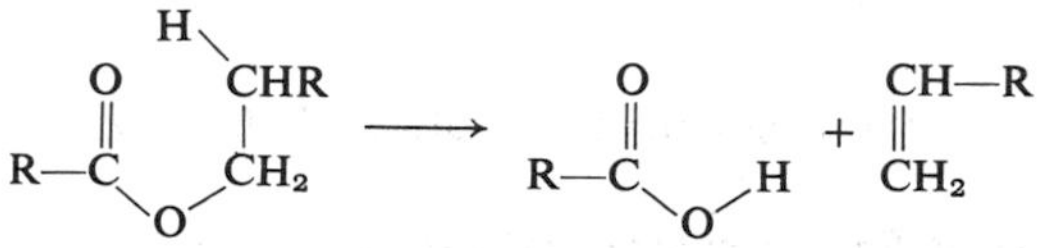

$$R{-}C({=}O){-}O{-}CH_2{-}CHR{-}H \longrightarrow R{-}C({=}O){-}O{-}H + CH_2{=}CH{-}R \quad (16)$$

as is seen in Eq. (16). Substitution of the H by F raises the decomposition temperature and the degradation probably occurs by a free-radical mechanism. Removal of the α-hydrogen of the carboxylic acid makes the free-radical decomposition difficult, as would be true for poly(hexafluoropentamethylene isophthalate), which degrades at 480°C (in helium, based on TGA data) [49].

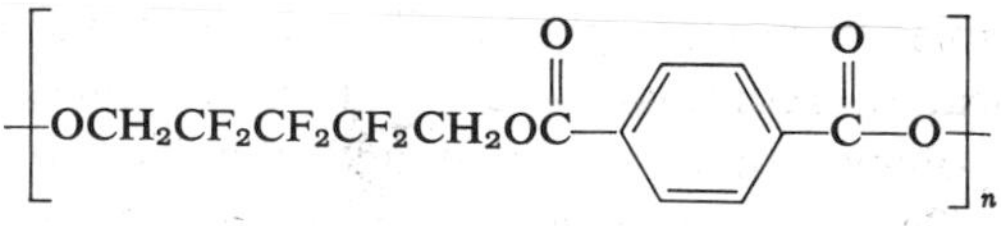

$$\left[OCH_2CF_2CF_2CF_2CH_2OC({=}O){-}C_6H_4{-}C({=}O){-}O \right]_n$$

The reaction of terephthaloyl chloride with dihydroxyanthraquinone affords polyesters with melting temperatures of about 500°C [50].

Some polyesters which are thermally stable are shown below [51].

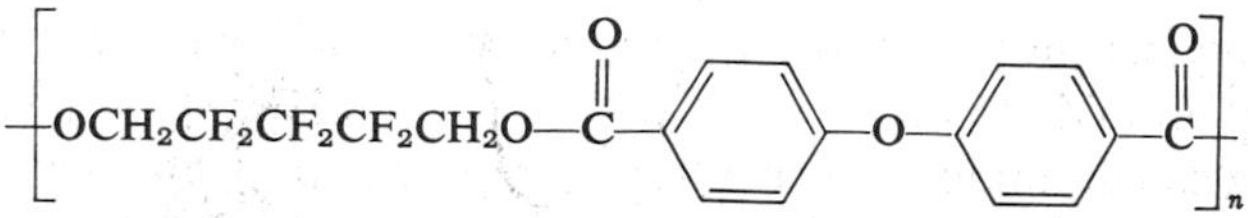

$$\left[OCH_2CF_2CF_2CF_2CH_2O{-}C({=}O){-}C_6H_4{-}O{-}C_6H_4{-}C({=}O) \right]_n$$

Decomposition temperature, 500°–525°C (TGA)

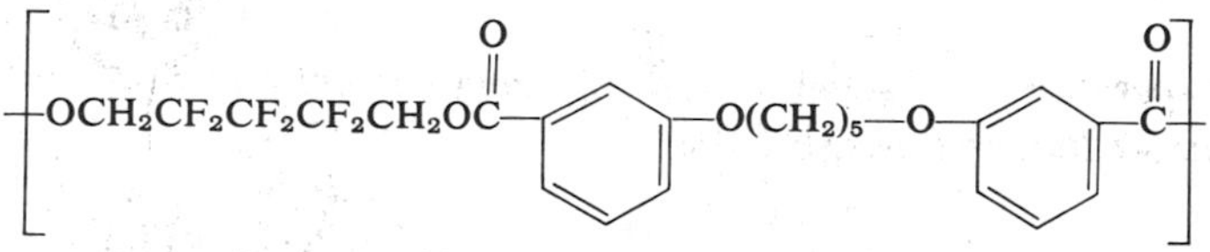

$$\left[OCH_2CF_2CF_2CF_2CH_2OC({=}O){-}C_6H_4{-}O(CH_2)_5{-}O{-}C_6H_4{-}C({=}O) \right]$$

Decomposition temperature, 480°C (TGA)

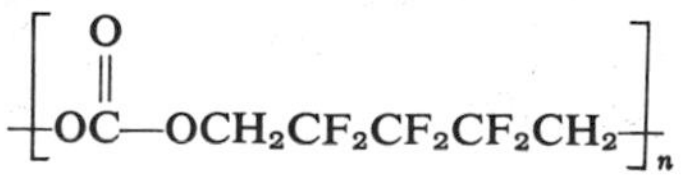

$$\left[OC({=}O){-}OCH_2CF_2CF_2CF_2CH_2 \right]_n$$

Decomposition temperature, 450°C (TGA)
(oxidatively stable polymer)

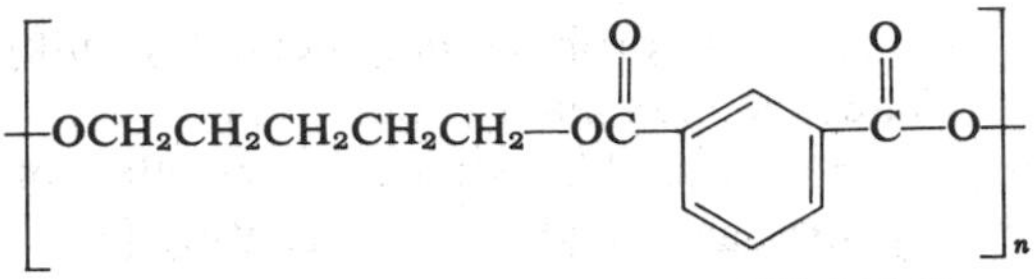

$$\left[\!-OCH_2CH_2CH_2CH_2CH_2-O\overset{O}{\overset{\|}{C}}-C_6H_4-\overset{O}{\overset{\|}{C}}-O-\right]_n$$

Decomposition temperature, 150°C (TGA)

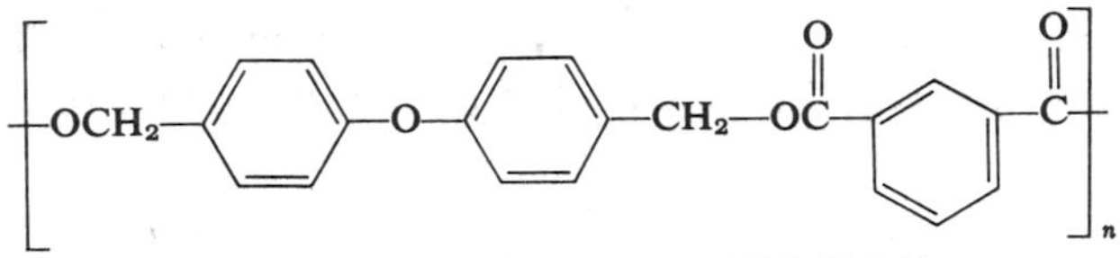

$$\left[\!-OCH_2-C_6H_4-O-C_6H_4-CH_2-O\overset{O}{\overset{\|}{C}}-C_6H_4-\overset{O}{\overset{\|}{C}}-\right]_n$$

Decomposition temperature, 150°C (TGA)

2-1. Preparation of Poly(ethylene maleate) [52]

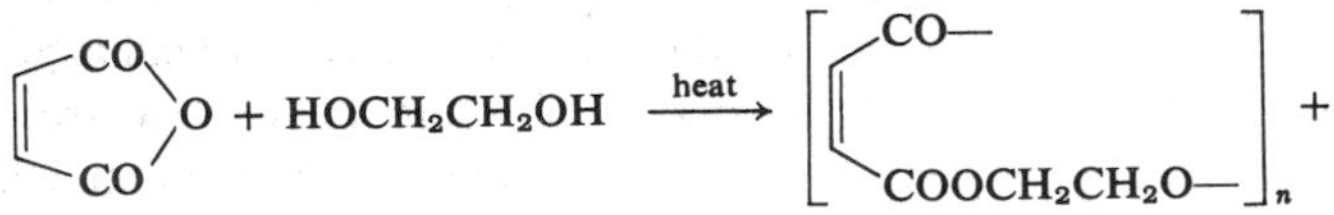

$$\text{(maleic anhydride)} + HOCH_2CH_2OH \xrightarrow{\text{heat}} \left[-CO-CH=CH-COOCH_2CH_2O-\right]_n + 2H_2O \quad (17)$$

To a polymer tube is added 32.5 gm (0.33 mole) of maleic anhydride and 18.6 gm (0.30 mole) of ethylene glycol. The tube is heated to 195°–200°C for 4 hr. Then the heating is continued at 210°–215°C at reduced pressure (0.2 mm) for 3 hr. The polymer tube is cooled and ethylene chloride added. The polymer 40 gm (94%) separates as an oil which later changes to a white powder when cooled for 2 hr at 5°–10°C, m.p. 88°–95°C. After drying the polymer has m.p. >250°C and is insoluble in most common solvents including ethylene dichloride.

This polymer has also been obtained by reacting silver maleate with ethylene dibromide [53].

In a related manner Carothers prepared poly(ethylene phthalate), poly(trimethylene phthalate), poly(hexamethylene phthalate), poly(decamethylene phthalate), and poly(ethylene fumarate) [52].

2-2. Preparation of All-trans Polyesters from Diethyl Fumarate and trans-2-Butene-1,4-diol [54]

$$\underset{trans}{nHOCH_2CH{=}CHCH_2OH} + \underset{trans}{nC_2H_5O\overset{O}{\overset{\|}{C}}CH{=}CH-\overset{O}{\overset{\|}{C}}-OC_2H_5} \xrightarrow{-2C_2H_5OH}$$

$$\underset{\text{all } trans}{\left[-O-CH_2CH{=}CHCH_2O\overset{O}{\overset{\|}{C}}-CH{=}CH-\overset{O}{\overset{\|}{C}}-\right]_n} \quad (18)$$

To a resin kettle with a nitrogen atmosphere and containing 137.6 gm (0.80 mole) of diethyl fumarate and 70.4 gm (0.80 mole) of *trans*-2-butene-1,4-diol is added 0.4 gm of freshly cut sodium metal. The mixture is warmed at 45°–58°C and becomes a clear orange color after 1 hr. The temperature is gradually raised and the pressure lowered in order to remove the ethanol by distillation. After 13 hr the temperature is finally raised to 116°C and the pressure reduced to 1 mm Hg. The contents are heated for 60 hr at the latter conditions to give an 80% yield of ethanol and an opaque yellow polymer at room temperature. A polyester prepared in this way had an intrinsic viscosity of 0.094 at 25°C in cyclohexanone and a Rast mol. wt. of 837.

2-3. *Preparation of All-cis Polyesters from cis-2-Butene-1,4-diol and Maleic Acid* [54]

$$n\,HOCH_2\underset{cis}{CH{=}CH}{-}CH_2OH + n\,HO\overset{O}{\overset{\|}{C}}{-}\underset{cis}{CH{=}CH}{-}\overset{O}{\overset{\|}{C}}OH \longrightarrow$$

$$\Big[{-}OCH_2CH{=}CHCH_2{-}O\overset{O}{\overset{\|}{C}}{-}CH{=}CH{-}\overset{O}{\overset{\|}{C}}{-}\Big]_n \quad \text{(all } cis\text{)} \qquad (19)$$

To a resin flask containing 176.0 gm (2.0 moles) of *cis*-2-butene-1,4-diol is added 232 gm (2.0 mole) of maleic acid. The mixture is heated for 5 hr at 55°–95°C and then the pressure reduced to 100 mm Hg. After 2 hr the pressure is reduced to 1 mm Hg and the contents heated for 77 hr at 90°–105°C. The polyester is a clear amber color and has an intrinsic viscosity of 0.055 at 25°C in cyclohexanone. The polymer is also soluble in acetone.

NOTE: Whether or not any of the maleic acid is converted to fumaric acid units under these conditions is questionable. The authors at that time used only infrared spectra to establish their conclusions. Reinvestigation of this problem using NMR should give more conclusive data.

Similar polyesters were prepared from *cis*- or *trans*-2-butene-1,4-diol and sebacic acid [55]. The *cis* polyester was obtained in 90% yield, m.p. 58°–59°C, inherent viscosity 0.56; and the *trans* polyester was obtained in 86% yield, m.p. 68°–69.5°C, inherent viscosity 0.30.

2-4. *Preparation of Poly(1,4-butylene sebacate)* [20]

$$HOCH_2CH_2CH_2CH_2OH + (CH_2)_8(COOCH_3)_2 \xrightarrow[-2CH_3OH]{PbO} H[OCH_2CH_2CH_2CH_2OCO(CH_2)_8CO]_nOH \qquad (20)$$

To a polymer tube containing 4.95 gm (0.055 mole) of butane-1,4-diol and 11.5 gm (0.05 mole) of dimethyl sebacate is added 0.1 gm of litharge. The mixture is heated at atmospheric pressure to 183°C for 2 hr, then raised to 259°C, quickly lowering the pressure to 0.01 mm Hg. The melt becomes very viscous and after $\frac{1}{2}$ hr heating the polymerization is discontinued. The polymer is dissolved in 100 ml of hot chloroform, filtered, precipitated with 100 ml of cold acetone to afford the solid polyester, m.p. 64°–64.5°C; intrinsic viscosity, $[\eta] = 0.61$ (measured as a 0.4% solution in chloroform at 25.5°C). The crude and purified polyesters can be drawn into fibers.

The same experiment was carried out under modified conditions to give the same polyester of slightly different properties, as shown in Table II.

TABLE II
EFFECT OF CATALYST AND REACTION CONDITIONS ON POLYMER PROPERTIES[a,b]

Inhibitor added	Catalyst used (gm)	Temp. (°C)	Time (hr)	Pressure (mm Hg)	M.p. (°C)	$[\eta]$
—	PbO (0.1)	155	3	760		
		155	1	0.03	64–64.5	0.33
Di-*t*-butyl hydroquinone (0.1 gm)	PbO (0.1)	172	2	760		
		172	4	0.05		
		215	18	0.05	—	1.16
—	Pyrophoric lead (0.05)	172	2	760		
		172	6	1.0		
		215	18	1.0	—	0.98
—	PbO (0.1)	172	6–7	760	—	0.26

[a] From Marvel and Johnson [20].
[b] Reactants the same as in Preparation 2-4.

2-5. *Preparation of Poly(ethylene terephthalate)* [40]

$$CH_3OC(=O)-C_6H_4-C(=O)-OCH_3 + 2HOCH_2CH_2OH \xrightarrow{-2CH_3OH} HOCH_2CH_2OC(=O)-C_6H_4-C(=O)-OCH_2CH_2OH$$

$$\xrightarrow[\text{heat}]{-HOCH_2CH_2OH} \left[-OCH_2CH_2OC(=O)-C_6H_4-C(=O)-\right]_n \qquad (21)$$

CAUTION: This reaction should be carried out in a hood behind a protective shield.

To a weighed thick-walled glass tube with a constricted upper portion for vacuum tube connection and equipped with a metal protective sleeve is added 15.5 gm (0.08 mole) of dimethyl terephthalate (see Note a), 11.8 gm (0.19 mole) of ethylene glycol (Note b), 0.025 gm of calcium acetate dihydrate, and 0.006 gm of antimony trioxide. The tube is warmed gently in an oil bath to melt the mixture and then a capillary tube connected to a nitrogen source is placed in the melt. While heating to 197°C a slow stream of nitrogen is passed through the melt to help eliminate the methanol. The tube is heated for 2–3 hr at 197°C, or until all the methanol has been removed (Note c). The side arm is also heated to prevent clogging by the condensation of some dimethyl terephthalate. The polymer tube is next heated to 222°C for 20 min and then at 283°C for 10 min. The side arm is connected to a vacuum pump and the pressure is reduced to 0.3 mm Hg or less while heating at 283°C for 3½ hr. The tube is removed from the oil bath, cooled (Note d) behind a safety shield, and then weighed. The yield is quantitative if no loss of dimethyl terephthalate has occurred. The polymer melts at about 270°C. (The crystalline melting point is 260°C.) The inherent viscosity of an 0.5% solution in *sym*-tetrachloroethanol/phenol (40/60) is approx. 0.6–0.7 at 30°C.

NOTES: (a) Dimethyl terephthalate is the best grade or is recrystallized from ethanol, m.p. 141°–142°C. (b) The ethylene glycol is anhydrous reagent grade or prepared by adding 1 gm of sodium/100 ml, refluxing for 1 hr in a nitrogen atmosphere, and distilling, b.p. 196°–197°C. (c) Failure to remove all the methanol leads to low molecular weight polymers. (d) On cooling the polymer contracts from the walls and this may cause the tube to shatter.

2-6. Preparation of Poly(1,4-butylene isophthalate) [21]

$$HOCH_2CH_2CH_2CH_2OH + ClC(=O){-}C_6H_4{-}C(=O)Cl \xrightarrow{-2HCl} \left[-O(CH_2)_4OC(=O){-}C_6H_4{-}C(=O){-} \right]_n \qquad (22)$$

To a polymer tube equipped with a capillary tube for a nitrogen inlet extending below the surface of the reaction mixture is added 6.3 gm (0.0310 mole) of isophthaloyl chloride and 2.8 gm (0.0311 mole) of 1,4-butanediol

(1% excess). The reactants are added in a nitrogen atmosphere and the nitrogen flow is slowly continued during the reaction. The reaction is exothermic and warms up to 40°C. After 10 min the reaction mixture is heated to 218°C and kept there for 35 min to afford a clear viscous polymer with a melt viscosity of 2500 poises. A clear film of the amorphous material crystallizes on standing to form a white opaque polymer, m.p. 140°–142°C.

2-7. *General Procedure for the Preparation of Polyphenyl Esters by the Interfacial Polycondensation Process* [50a]

$$\text{HOR—OH} + \text{R}'(\text{COCl})_2 \longrightarrow \text{H}\left[\text{OROC}(=\text{O})\text{—R}'\text{C}(=\text{O})\right]_n\text{Cl} \qquad (23)$$

(R = aryl)

TABLE III
POLYPHENYL ESTERS BY INTERFACIAL POLYCONDENSATION[a]

Bisphenol	Acid chloride[b]	η_{inh}[c]	PMT (°C)[d]	Film toughness
Hydroquinone	Isophthaloyl	(Insoluble)	(Infusible)	—
	Terephthaloyl	(Insoluble)	(Infusible)	—
	Succinoyl	0.15	230	Very brittle
	Adipoyl	0.15	190	Very brittle
	Sebacoyl	0.17	150	Very brittle
Resorcinol	Isophthaloyl	1.38	245	Very tough
	Terephthaloyl	(Insoluble)	(Infusible)	—
	I/T 50/50	1.01	220	Very tough
4-Chlororesorcinol	I/T 50/50	1.47	220	Very tough
2,2-Bis(4-hydroxyphenyl)propane	Isophthaloyl	1.86	280	Very tough
	Terephthaloyl	1.59	350	Very tough
	4,4′-Sulfonylbibenzoyl	0.45	230	Brittle
	Hydroquinone-diacetyl	0.85	165	Tough
	Fumaroyl	0.30	240	Very brittle
	Oxaloyl	0.12	155	Very brittle
	Adipoyl	0.08	80	—
	Sebacoyl	0.10	(Gum)	—
1,5-Dihydroxynaphthalene	I/T 50/50	1.00	290	Brittle

[a] Reprinted from Eareckson [50]. Copyright 1959 by the *Journal of Polymer Science*. Reprinted by permission of the copyright owner.

[b] I/T 50/50 = mixture of equal parts of isophthaloyl and terephthaloyl chlorides.

[c] Measured at 0.5% concentration in *sym*-tetrachloroethane/phenol, 40/60 (by weight), at 30°C.

[d] Polymer melt temperature.

To a blender is added 0.05 mole of a given bisphenol or dihydroxybenzene and 0.1 mole of sodium hydroxide in 300 ml of water. A solution of 3.0 gm of sodium lauryl sulfate in 30 ml of water is also added. The speed is regulated so that at low speed a second solution of 0.05 mole of the acid chloride of a dicarboxylic acid in 150 ml of a nonreactive solvent is quickly added.

The emulsified reaction mixture is stirred at high speed for 5–10 min and then poured into acetone to coagulate the polymer. The polymer is filtered, washed with water, and dried to afford an 80–100% yield.

Some typical polymers prepared by this process are shown in Table III.

3. MISCELLANEOUS METHODS

1. Polyesters from *N*-substituted bis(β-hydroxyethyl)amine and maleic–phthalic anhydride [56].
2. Preparation of nitrogen-containing unsaturated polyesters by the polycondensation of unsaturated discarboxylic anhydrides, glycols, and primary or secondary amines [57].
3. Preparation of polyesters with tertiary nitrogen atoms [58].
4. Preparation of unsaturated nitrogen-containing polyesters [59].
5. Preparation of unsaturated polyesters [60].
6. Synthesis of cyclic di(ethylene terephthalate) [61].
7. Reaction of dimethylketene and acetone to form a polyester [62].
8. Reaction of dinitriles with glycols in the presence of hydrochloric acid [63].
9. Reaction of epoxides with anhydrides to give polyesters [64].
10. Reaction of trimesic acid with polyhydric alcohols in the presence of 0.1 to 5% tetrabutyl titanate to give polyesters [65].
11. Preparation of polyamide-esters [66].
12. Preparation of soluble polyesters from poly(vinyl alcohol) and high molecular weight acid chlorides [67].
13. Preparation of a polyester by the self-condensation of 12-hydroxymethyltetrahydroabietic acid and 12-hydroxymethyltetrahydroabietanol [68].
14. Preparation of polyesters from bicyclic alcohols or bicyclic carboxylic acids [69].
15. Catalysts for the production of polyesters [70].
16. Preparation of halogenated esters having flame-retardant properties [71].
17. Aromatic polyesters for filaments or fibers [72].
18. Polyester-amides [73].
19. Polycondensation of ω-amino alcohol with some aliphatic dicarboxylic acids [74].

20. Lactone polyesters [75].
21. Photosensitive polyesters [76].
22. Polyesters from benzenedicarboxylic acids and an alkylene oxide [77].
23. Poly(β-hydroxybutyrate) [78].
24. Preparation of crystalline thermoplastic poly(ethylene terephthalate) [79].
25. Preparation of aromatic polyesters with large cross-planar substituents [80].

REFERENCES

1. J. M. Hawthorne and C. J. Heffelfinger, *Encycl. Polym. Sci. Technol.* **11**, 1 (1969).
2. I. Goodman, *Encycl. Polym. Sci. Technol.* **11**, 62 (1969).
3. H. V. Boening, *Encycl. Polym. Sci. Technol.* **11**, 729 (1969).
4. V. V. Korshak and S. V. Vinogradova, "Polyesters." Elsevier, Amsterdam, 1953.
5. V. V. Korshak and S. V. Vinogradova, "Polyesters." Pergamon, Oxford, 1969.
6. R. Hill and E. E. Walker, *J. Polym. Sci.* **3**, 609 (1948).
7. H. J. Hagemeyer, U.S. Patent 3,043,808 (1962).
8. H. Batzer and F. Wiloth, *Makromol. Chem.* **8**, 41 (1952).
9. B. M. Grievson, *Polymer* **1**, 499 (1960).
10. M. J. Hurwitz and E. W. Miller, French Patent 1,457,711 (1966).
11. Borg-Warner Corp., British Patent 1,034,194 (1966).
12. J. Berzelius, *Rapp. Annu.* **26**, 1 (1847); *Jahresbericht* **12**, 63 (1833).
13. A. V. Lorenzo, *Ann. Chim. Phys.* [2] **67**, 293 (1863).
14. J. R. Whinfield, *Nature (London)* **158**, 930 (1946); S. K. Agarawal, *Indian Chem. J.* **5**, 25 (1970).
15. O. Bayer, *Justus Liebigs Ann. Chem.* **549**, 286 (1941).
16. W. H. Carothers and E. J. Van Natta, *J. Amer. Chem. Soc.* **55**, 4714 (1933).
17. R. Gilkey and J. R. Caldwell, *J. Appl. Polym. Sci.* **2**, 198 (1959).
18. E. R. Walsgrove and F. Reeder, British Patent 636,429 (1950).
19. R. E. Wilfong, *J. Polym. Sci.* **54**, 385 (1961).
20. C. S. Marvel and J. H. Johnson, *J. Amer. Chem. Soc.* **72**, 1674 (1950).
20a. J. T. Dickson, H. P. W. Huggill, and J. C. Welch, British Patent 590,451 (1947).
21. P. J. Flory and F. S. Leuther, U.S. Patents 2,623,034 and 2,589,688 (1952).
21a. K. Yamaguchi, M. Takayanagi, and S. Kuriyama, *J. Chem. Soc. Jap. Ind. Chem. Sect.* **58**, 358 (1955).
22. J. T. Dickson, H. P. W. Huggill, and J. C. Welch, British Patent 590,451 (1947).
23. R. H. Kienle and H. G. Hovey, *J. Amer. Chem. Soc.* **52**, 3636 (1930); W. H. Carothers and J. W. Hill, *ibid.* **54**, 1577 (1932).
24. C. A. Bishoff and A. von Hendenström, *Chem. Ber.* **35**, 3431 (1902).
25. W. H. Carothers and J. A. Arvin, *J. Amer. Chem. Soc.* **51**, 2560 (1929); D. Vorländer, *Justus Liebigs Ann. Chem.* **280**, 167 (1894).
25a. F. E. Critchfield and R. D. Lundberg, French Patent Appl. 2,026,274 (1970).
26. M. Callahan, U.S. Patents 1,191,732 and 1,108,329–39 (1914).
27. I. Goodman, *Encycl. Polym. Sci. Technol.* **11**, 62 (1969).
27a. J. M. Hawthorne and C. J. Heffelfinger, *Encycl. Polym. Sci. Technol.* **11**, 1 (1969).
28. H. von Boenig, *Encycl. Polym. Sci. Technol.* **11**, 129 (1969).
29. R. G. Mraz and R. P. Silver, *Encycl. Polym. Sci. Technol.* **1**, 663 (1964).

30. S. R. Sandler and W. Karo, "Organic Functional Group Preparations," Vol. 1, Chapter 10. Academic Press, New York, 1968.
31. C. A. Bischoff and P. Walden, *Chem. Ber.* **26**, 262 (1893); *Justus Liebigs Ann. Chem.* **279**, 45 (1894).
32. W. H. Carothers, G. L. Dorough, and F. J. Van Natta, *J. Amer. Chem. Soc.* **54**, 761 (1932).
33. E. E. Blaise and L. Marcilly, *Bull. Soc. Chim. Fr.* [3] **31**, 308 (1904).
34. F. Fichter and A. Beisswenger, *Chem. Ber.* **36**, 1200 (1903).
35. P. Chuit and J. Hausser, *Helv. Chim. Acta* **12**, 463 (1929); W. H. Lycan and R. Adams, *J. Amer. Chem. Soc.* **51**, 625 and 3450 (1929).
36. R. H. Kienle and H. G. Hovey, *J. Amer. Chem. Soc.* **51**, 509 (1929).
37. H. Batzer, H. Holtschmidt, F. Wiloth, and B. Mohr, *Macromol. Chem.* **7**, 82 (1951); M. J. Hurwitz and E. W. Miller, French Patent 1,457,711 (1966).
38. V. V. Korshak and S. V. Vinogradova, "Polyesters." Pergamon, Oxford, 1965; M. Davies and D. R. J. Hill, *Trans. Faraday Soc.* **49**, 395 (1953); M. T. Pope and R. J. P. Williams, *J. Chem. Soc., London* p. 3579 (1959); C. Y. Huang, Y. Simono, and T. Unizaka, *Chem. High Polym.* **23**, 408 (1966); P. J. Flory, "Principles of Polymer Chemistry," pp. 79–83. Cornell Univ. Press, Ithaca, New York, 1953.
39. F. G. Baddar, M. H. Nosseir, G. G. Gabra, and N. E. Ikladious, *J. Polym. Sci.*, A–1 **9**, 1947 (1971).
40. J. R. Whinfield and J. T. Dickson, British Patent 578,079 (1946); J. R. Whinfield, *Nature (London)* **158**, 930 (1946).
41. C. J. Kibler, A. Bell, and J. G. Smith, U.S. Patent 2,901,466 (1959).
42. W. Briehl and G. Schnock, *J. Polym. Sci.* **40**, 413 (1958).
43. E. W. Wallsgrove and R. Reeder, British Patent 636,429 (1950); M. Levine and S. C. Temin, *J. Polym. Sci.* **28**, 179 (1958); Inventa A.G. für Forschung und Patenverwertung, French Patent 1,220,725 (1960); S. C. Temin, *J. Org. Chem.* **26**, 2518 (1961); I. Goodman, J. E. McIntyre, and J. W. Stimpson, British Patent 989,552 (1965); J. G. N. Drewilt and J. Lincoln, U.S. Patent 2,595,343 (1952).
44. A. J. Conix, *Ind. Chem. Belge* [2] **22**, 1457 (1957); *Ind. Eng. Chem.* **51**, 147 (1959); F. F. Holub and J. W. Kantor, U.S. Patents 3,036,990–3,036,992 (1962) and 3,160,604 (1964); A. J. Conix, British Patent 883,312 (1961); A. R. Macon, U.S. Patent 3,225,003 (1965); A. A. D'Onofrio, U.S. Patent 3,219,627 (1965).
45. I. Vancso-Szmercsanyi, K. Maros-Greger, and E. Makay-Bodi, *J. Polym. Sci.* **53**, 251 (1961).

45a. W. H. Carothers, *J. Amer. Chem. Soc.* **51**, 2560 (1929).

46. G. S. Papava, N. A. Maisuradze, S. V. Vinogradova, V. V. Korshak, and P. D. Tsiskarishvili, *Soobshch. Akad. Nauk Gruz. SSR* **62**, 581 (1971); *Chem. Abstr.* **75**, 77340s (1971).
47. A. Conix and R. V. Kerpel, *J. Polym. Sci.* **50**, 521 (1959); M. Levine and S. C. Temin, *ibid.* **28**, 179 (1958); C. W. Bunn, *ibid.* **26**, 323 (1955); K. W. Doak and H. N. Campbell, *ibid.* **28**, 215 (1955); E. F. Izard, *ibid.* **8**, 503 (1952); O. B. Edgar and R. Hill, *ibid.* p. 1.
48. C. D. Hurd and C. H. Blunck, *J. Amer. Chem. Soc.* **60**, 2419 (1938); J. Hine, "Physical Organic Chemistry," p. 453. McGraw-Hill, New York, 1956.
49. F. D. Trischler and J. Hollander, *J. Polym. Sci., Part A-1* **7**, 971 (1969).
50. W. M. Eareckson, III, *J. Polym. Sci.* **40**, 399 (1959); V. V. Korshak, S. V. Vinogradova, and J. P. Antanova-Antanova, *Vysokomol. Soedin.* **1**, 1543 (1965).
51. F. D. Trischler and J. Hollander, *J. Polym. Sci., Part A-1* **7**, 971 (1969).
52. W. H. Carothers, *J. Amer. Chem. Soc.* **51**, 2560 (1929); R. H. Kienle and F. E. Petke, *ibid.* **62**, 1083 (1940).

53. L. H. Flett and W. H. Gardner, "Maleic Anhydride Derivatives," p. 176. Wiley, New York, 1952.
54. W. M. Smith, Jr., K. C. Eberly, E. E. Hanson, and J. L. Binder, *J. Amer. Chem. Soc.* **78**, 626 (1956).
55. C. S. Marvel and C. H. Young, *J. Amer. Chem. Soc.* **73**, 1066 (1951).
56. A. D. Valgin, *Mater. Plast. Elastomeri* **37**, 750 (1971).
57. V. K. Skubin, A. D. Valgin, D. F. Kutepow, and V. V. Korshak, U.S.S.R. Patent 283,571 (1970).
58. E. Eimers, H. Rudolp, and W. Kloeker, German Patent 1,943,954 (1971).
59. A. D. Valgin, B. P. Vorob'ev, V. V. Korshak, and D. F. Kutepow, U.S.S.R. Patent 294,840 (1971).
60. V. K. Skubin, D. F. Kutepow, V. V. Korshak, and A. D. Valgin, U.S.S.R. Patent 299,519 (1971).
61. H. Repin and E. Papanikolau, *J. Polym. Sci., Part A-1* **7**, 3426 (1969).
62. G. Natta, G. Mazzanti, G. F. Pregaglia, and G. Pozzi, *J. Polym. Sci.* **58**, 1201 (1962.
63. E. N. Zilberman, A. E. Kulikova, and N. M. Teplyakov, *J. Polym. Sci.* **56**, 417 (1962).
64. R. F. Fischer, *J. Polym. Sci.* **54**, 155 (1960).
65. R. E. Dunbar and E. C. Hutchins, *J. Polym. Sci.* **21**, 550 (1956).
66. J. Preston, *J. Polym. Sci., Part A–1*, **8**, 3135 (1970).
67. A. G. Reynolds, U.S. Patent 3,560,465 (1971).
68. J. A. Osman and C. S. Marvel, *J. Polym. Sci., Part A–1*, **9**, 1213 (1971).
69. L. Taimr and J. G. Smith, *J. Polym. Sci., Part A–1*, **9**, 1203 (1971).
70. F. Kobayashi, K. Takakura, K. Tomita, and R. Yamaguchi, Japanese Patent 71/06,596 (1971); Asahi Chem. Industry Co. Ltd., French Patent Appl. 2,029,486 (1970); Chem. Abstr. **75**, 64639–66644 (1971).
71. P. L. Smith and L. R. Comstock, German Patent 1,933,064 (1971).
72. F. B. Cramer, British Patent 1,228,007 (1971); K. Nishikawa and N. Ito, Japanese Patent 70/32,995 (1970); Asahi Chem. Industry Co. Ltd., French Patent Appl. 2,033,421 (1970).
73. K. Fukui, T. Kagiyer, S. Narusawa, T. Maeda, and M. Obata, Japanese Patent 70/39,830 (1970).
74. B. Jasse, *Bull. Soc. Chim. Fr.* [6] p. 2264 (1971).
75. A. A. L. Schoen, German Patent 2,104,246 (1971).
76. Eastman Kodak Co., French Patent 2,041,411 (1971).
77. N. Izawa, Y. Iizuka, and Y. Kubota, German Patent 2,060,421 (1971).
78. J. R. Shelton, J. B. Lando, and D. E. Agostini, *J. Polym. Sci.* B. **9**(3) 173 (1971).
79. W. Herwig and G. Freund, German Patent 1,945,967 (1971).
80. P. W. Morgan, *Macromolecules* **3**, 536 (1970).

Chapter 3

POLYCARBONATES

I. INTRODUCTION

Structurally, polycarbonates are polyesters derived from the reaction of carbonic acid or its derivatives with dihydroxy compounds (aliphatic, aromatic, or mixed type compounds).

$$\mathrm{H}\left[\text{—OROC—}\atop{}\right]_n\mathrm{OROH} \qquad \text{(C=O above the carbonyl carbon)}$$

Polycarbonates are almost unaffected by water and many inorganic and organic solvents. These properties allow these resins to fill many applications, especially in view of their attractive mechanical properties. The stability of polycarbonates is good below 250°C and this property has been reviewed [1].

Einhorn [2] first prepared polycarbonates by the Schotten-Baumann reaction of phosgene with either hydroquinone or resorcinol in pyridine

(Eq. 1). Bischoff and von Hedenström [3] later reported that the same polycarbonates can be prepared from the ester-interchange reaction of hydroquinone or resorcinol with diphenyl carbonate (Eq. 1).

$$n\mathrm{HO{-}R{-}OH} + n\mathrm{COCl_2} \xrightarrow{-2n\mathrm{HCl}} \left[\mathrm{{-}OR{-}O\overset{\overset{\displaystyle O}{\|}}{C}{-}} \right]_n$$

$$n\mathrm{HO{-}R{-}OH} + n\mathrm{C_6H_5{-}O\overset{\overset{\displaystyle O}{\|}}{C}{-}OC_6H_5} \xrightarrow{-2n\mathrm{C_6H_5OH}} \left[\mathrm{{-}OR{-}O\overset{\overset{\displaystyle O}{\|}}{C}{-}} \right]_n \qquad (1)$$

Carothers and Hill [4] were the first to systematically study methods to prepare high molecular weight polymers. Later Peterson [5] improved the procedure and developed a way to prepare high molecular weight polycarbonates having useful film and fiber properties.

The subject of polycarbonates has been reviewed earlier and many references to their applications are given [6–8].

Polycarbonates of 2,2-bis(4-hydroxyphenyl) propane (Bisphenol A) are commercially the most important polycarbonates [6a,b]. Polycarbonates are also prepared by the reaction of bischloroformates with dihydroxy compounds. Some typical examples of the preparation of polycarbonates are shown in Table I

2. CONDENSATION REACTIONS

A. Aromatic Polycarbonates [3, 9–37]

The Schotten-Baumann reaction of phosgene with aromatic dihydroxy compounds is carried out in either an organic or an aqueous alkaline system. In the aqueous system a water-insoluble organic solvent is used (interfacial polycondensation reaction) to obtain high molecular weight products. For solution polymerization the use of pyridine or triethylamine in a chlorinated hydrocarbon solvent (chloroform or chlorobenzene) gives good results. Both reactions are carried out at room temperature. The reactions are exothermic and catalyzed by pyridine or triethylamine. The final polymer is soluble in the tertiary amine and is precipitated out by the addition of methanol. For the aqueous system quaternary ammonium halides are effective catalysts and vigorous stirring is required. These two methods are illustrated in Preparations 2-1 and 2-2.

The two major methods for preparing aromatic polycarbonates are summarized in Eq. (2) and examples are described below in more detail.

TABLE I

PREPARATION OF POLYCARBONATES BY CONDENSATION REACTIONS

Hydroxy compound	Reactant	Catalyst	Reaction conditions Temp. (°C)	Time (hr)	M.p. (°C)	Mol. wt.	Ref.
$(HO-C_6H_4-)_2CH_2$	$COCl_2$	R_3N or $R_3N^+X^-$	20–30	1–2	300	20,000–80,000	*a*
$(HO-C_6H_4-)_2C(CH_3)_2$	$(C_6H_5O)_2CO$	—	150–300	1–2	220–230	20,000–80,000	*a*
$HO-C_4(CH_3)_4-OH$ *cis* or *trans*	$(C_2H_5O)_2CO$	—	150–300	—	—	—	*b*
$C_2H_5OC(=O)-O(CH_2)_2O-C_6H_4-O-(CH_2)_2-OC(=O)-OC_2H_5$		Sodium hydrogen titanium butoxide (1 ml)	225–250	3–5	100	—	*c*
HO—Ar—OH	$COCl_2$	$AlCl_3$ in chlorobenzene	133	6–12	90–99	—	*d*

[a] H. Schnell, *Ind. Eng. Chem.* **51**, 157 (1959).

[b] H. Gawlak, R. P. Palmer, J. B. Rose, D. J. Sandiford, and A. Turner-Jones, *Chem. Ind.* (*London*) **25**, 1148 (1962); E. U. Elam, J. C. Martin, and R. Gilkey, British Patent 962,913 (1964); Union Carbide Corp. British Patent 1,011,283 (1965).

[c] D. D. Reynolds and J. Van Den Berghe, U.S. Patents 2,789,964 and 2,789,965 (1957).

[d] M. Matzner, R. P. Kurkjy, and R. J. Cotter, *J. Appl. Polym. Sci.* **9**, 3295 (1965).

The latter procedure may involve an interfacial polymerization [9–16] of an aromatic dihydroxy compound with phosgene or bis(chloroformate) in

$$\text{HO—Ar—OH} + \text{COCl}_2 \xrightarrow{2\text{R}_3\text{N}} \left[\text{—O—Ar—O}\overset{\overset{\text{O}}{\|}}{\text{C}}\text{—} \right]_n + 2\text{R}_3\text{NHCl} \qquad (2)$$

$$\text{HO—Ar—OH} + 3\text{COCl}_2 \longrightarrow \text{Cl—}\overset{\overset{\text{O}}{\|}}{\text{C}}\text{—OAr—O}\overset{\overset{\text{O}}{\|}}{\text{C}}\text{—Cl} \xrightarrow{\text{HO—Ar—OH},\ 2\text{R}_3\text{N}} \left[\text{—O—Ar—O}\overset{\overset{\text{O}}{\|}}{\text{C}}\text{—} \right]_n$$

the presence of an inert solvent. This reaction can also be carried out in a mixed aqueous–organic immiscible layer system as described below.

Another method which is useful under certain conditions is the transesterification of dialkyl and more preferably diaryl carbonates with aryl dihydroxy compounds (Eq. 3). The monohydroxy compound should be removed from the reaction to shift the equilibrium toward product formation.

$$\text{X—HO—Ar—OH} + \text{ArO}\overset{\overset{\text{O}}{\|}}{\text{C}}\text{—O—Ar} \longrightarrow \text{Ar—O}\overset{\overset{\text{O}}{\|}}{\text{C}}\text{—}\left[\text{—O—Ar—O}\overset{\overset{\text{O}}{\|}}{\text{C}}\text{—} \right]_x\text{—OAr} + 2\text{XArOH} \qquad (3)$$

Catalysts [18,22] are effective in speeding up the rate of polycarbonate formation. Catalysts such as alkali and alkaline earth metal oxides, hydroxides, and amides and acidic catalysts (H_3PO_4 and salts) are effective. Large amounts of alkaline catalysts tend to give discolored products and partially insoluble polycarbonates. The reaction is usually carried out at 260°–300°C to give products of molecular weights of 100,000. An advantage of the transesterification process is that the product need not be contaminated with amine and subjected to extraction. However, the major disadvantages are that the process requires high temperatures and vacuum and must be carried out at the melt temperature of the polycarbonate. The condensation reactions for preparation of polycarbonates are illustrated in Table I.

2-1. *Preparation of Poly[2,2-propanebis(4-phenyl carbonate)]. Interfacial Polycondensation with Phosgene Gas* [35,36,37]

CAUTION: Use an explosion-proof blender and a well-ventilated hood. Use great care in handling phosgene gas.

To a blender jar containing 30 ml of 1,2-dichloroethane is added 2.85 gm (0.0125 mole) of 2,2-bis(4-hydroxyphenyl)propane, 1.0 gm (0.025 mole) of

$$NaO-C_6H_4-C(CH_3)_2-C_6H_4-ONa + COCl_2 \longrightarrow \left[-C_6H_4-C(CH_3)_2-C_6H_4-OC(=O)-O- \right]_n + 2NaCl \quad (4)$$

sodium hydroxide, and 1.0 gm of tetraethylammonium chloride. While the mixture is vigorously stirred, phosgene gas is passed in ($\frac{1}{2}$–1 hr) until the pH of the reaction mixture drops from 12 to 7. The polymer is isolated by precipitation with acetone, filtration, and drying to afford 2.96 gm (93%), η_{inh} = 0.5146, m.p. 240°C.

A similar experiment was carried out with phenolphthalein and the phosgene added until the color faded. The color is restored with several drops of 20% aqueous sodium hydroxide and the phosgene addition repeated. This process was repeated two more times to afford the polycarbonate polymer in 93% yield, η_{inh} = 2.20.

Poly[2,2-propanebis(4-phenyl carbonate)] can also be prepared by the reaction of phosgene with 2,2-bis(4-hydroxyphenyl)propane dissolved in excess pyridine at 25°–30°C [19,36]. The stoichiometric amounts of reactants are used and the pyridine hydrochloride precipitates during the reaction. The polymer is isolated by pouring the reaction mixture into water to give a polymer, m.p. 240°C, η_{inh} = 0.6–0.8 [0.5% conc. at 25°C in *sym*-tetrachloroethane/phenol (40/60)].

2-2. Preparation of Poly[2,2-propane-bis(4-phenyl carbonate)] [35,36]

$$HO-C_6H_4-C(CH_3)_2-C_6H_4-OH + COCl_2 \xrightarrow[\text{pyridine}]{25^\circ-30^\circ C} \left[-O-C_6H_4-C(CH_3)_2-C_6H_4-O-C(=O)- \right]_n \quad (5)$$

CAUTION: Use extreme care when working with phosgene. Use a well-ventilated hood.

To a 500 ml resin kettle equipped with a mechanical stirrer, gas inlet tube,

TABLE II

PHYSICAL PROPERTIES OF SLIGHTLY CRYSTALLINE POLYCARBONATE FILMS CAST FROM SOLUTION[d]

Polycarbonate from	Melting range[a] (°C)	Second-order transition point (°C)	Density (gm/cm³)	Refractive index
4,4′-Dihydroxydiphenyl-				
methane	>300	—	—	—
1,2-ethane	290–300	—	—	—
1,1-ethane	185–195	130[b]	1.22	1.5937
1,1-butane	150–170	123[b]	1.17	1.5792
1,1-isobutane	170–180	149[b]	1.18	1.5702
2,2-propane	220–230	149[b]	1.20	1.5850
2,2-butane	205–222	134[b]	1.18	1.5827
2,2-pentane	200–220	137[b]	1.13	1.5745
2,2(4-methylpentane)	200–220	—	1.14	1.5671
2,2-hexane	180–200	—	—	—
4,4-heptane	190–200	148[c]	1.16	1.5602
2,2-nonane	170–190	—	—	—
phenylmethylmethane	210–230	176[c]	1.21	1.6130
diphenylmethane	210–230	121[c]	1.27	1.6539
1,1-cyclopentane	240–250	167[c]	1.21	1.5993
1,1-cyclohexane	250–260	175[c]	1.20	1.5900
ether	230–235	—	—	—
sulfide	220–240	—	—	—
sulfoxide	230–250	—		—
sulfone	200–210	—	—	—
4,4′-Dihydroxy-				
3,3′-dichlorodiphenyl-2,2-propane	190–210	147[c]	1.32	1.5900
3,3′,5,5′-tetrachlorodiphenyl-2,2-propane	250–260	180[c]	1.42	1.6056
3,3′,5,5′-tetrachlorodiphenyl-1,1-cyclohexane	260–270	163[c]	1.38	1.5858
3,3′,5,5′-tetrabromodiphenyl-2,2-propane	240–260	157[c]	1.91	1.6147
3,3′-dimethyldiphenyl-2,2-propane	150–170	95[b]	1.22	1.5783

[a] In contrast to crystalline polyesters and polyamides, polycarbonates have no sharply defined melting point.

[b] Determined by refractometer.

[c] Determined by dilatometer.

[d] Reprinted from Schnell [36]. Copyright 1959 by the American Chemical Society. Reprinted by permission of the copyright owner.

reflux condenser with calcium chloride drying tube, thermometer, and an aqueous sodium hydroxide–alcohol bath is added 49 gm (0.22 mole) of bisphenol A and 350 ml of pyridine. The phosgene gas is bubbled (75 ml/min) into the reaction mixture at room temperature and the mixture cooled if the temperature exceeds 30°C. At the midpoint of the reaction pyridine hydrochloride begins to precipitate and gradually thickens to a heavy slurry. The end point (approx. 2½ hr) of the reaction is sometimes accompanied by a yellow–red color change. The polymer is precipitated in 1600 ml of water, washed with 500 ml of hot water, the solid suspended in 1 liter of water, stirred for 10 min at 80°C, filtered, washed with 500 ml of cold water, and dried under reduced pressure to afford 54.9 gm (95%), softening point 215°–230°C. The infrared spectrum of this polymer compares well with the published spectrum. The polymer is also soluble in methylene chloride and films can easily be cast from solution.

Some physical properties of polycarbonates prepared in a similar process are shown in Table II.

The use of thiophosgene has been reported to also give polythiocarbonates on reaction with dihydroxy compounds [37].

$$\text{HOR—OH} + \text{CSCl}_2 \longrightarrow \text{H}\left[\text{—O—RO}\overset{\overset{\displaystyle S}{\|}}{\text{C}}\text{—}\right]_n\text{Cl} \tag{6}$$

2-3. *Preparation of Poly[2,5-bis(3′-propyl carbonate)-1,4-benzoquinone]* [36]

$$\text{HOCH}_2\text{CH}_2\text{CH}_2\text{—}(\text{1,4-benzoquinone-2,5-diyl})\text{—CH}_2\text{CH}_2\text{CH}_2\text{OH} + \text{COCl}_2 \xrightarrow[\substack{\text{THF} \\ -20^\circ\text{C} \\ -2\text{HCl}}]{\text{pyridine}}$$

$$\left[\text{—CH}_2\text{CH}_2\text{CH}_2\text{—}(\text{1,4-benzoquinone-2,5-diyl})\text{—CH}_2\text{CH}_2\text{CH}_2\text{O}\overset{\overset{\displaystyle O}{\|}}{\text{C}}\text{—O—}\right]_n \tag{7}$$

CAUTION: Use a well-ventilated hood only. Great care should be exercised when working with phosgene.

To a flask equipped with a mechanical stirrer, dropping funnel, and Dry Ice condenser is added 1.79 gm (0.008 mole) of 2,5-bis(3′-hydroxypropyl)-1,4-

benzoquinone dissolved in 25 ml of tetrahydrofuran (THF) and 5.0 ml of pyridine. The solution is cooled to −20°C, and 1.0 ml (0.01 mole) of phosgene is slowly added with vigorous stirring. The mixture is stirred for 1 hr at 0°C, the excess phosgene is removed by displacement with a nitrogen gas stream, and the solid product filtered. The filtrate is poured into 300 ml of *n*-pentane to afford a dark yellow polymeric oil. The latter oil is separated and redissolved in THF and again precipitated by *n*-pentane to afford brown-yellow flakes weighing 2.0 gm (80%). The polycarbonate is soluble in acetone, chloroform, THF, and dimethylformamide. The polymer is insoluble in water, acetic acid, or aliphatic hydrocarbons.

The polycarbonate possesses the following properties: $\eta_{sp}/c = 9.3\ cm^3/gm$ (in 95% THF); UV, $\lambda_{max} = 256\ m\mu$, $\varepsilon_{max} = 16.5 \times 10^3$ (in methylene chloride); IR, 1655 cm^{-1} (quinone group), 1745 cm^{-1} (carbonate group) (recorded as a film on a sodium chloride plate).

Polycarbonates can also be prepared by the reaction of bischloroformates with dihydroxy aliphatic or aromatic compounds as shown in Preparation 2-4 [38,39].

2-4. *Preparation of the Polycarbonate Formed by the Reaction of 2,5-Bis(3′-hydroxypropyl)-1,4-benzoquinone and Diethylene Glycol Bis(chloroformate)* [38]

$$HOCH_2CH_2CH_2\text{–}C_6H_2O_2\text{–}CH_2CH_2CH_2OH + ClC(=O)OCH_2CH_2OCH_2CH_2OC(=O)Cl \xrightarrow[0^\circ\text{–}5^\circ C,\ -2HCl]{\text{pyridine}}$$

$$\left[\text{–}CH_2CH_2CH_2\text{–}C_6H_2O_2\text{–}CH_2CH_2CH_2\text{–}O\text{–}C(=O)OCH_2CH_2OCH_2CH_2O\text{–}C(=O)\text{–}O\text{–}\right]_n \quad (8)$$

CAUTION: This reaction should be carried out in a well-ventilated hood and great care should be exercised in handling phosgene.

(*a*) *Preparation of diethylene glycol bis(chloroformate)* [38]

$$HOCH_2CH_2OCH_2CH_2OH + 2COCl_2 \longrightarrow Cl\text{–}C(=O)OCH_2CH_2\text{–}O\text{–}CH_2CH_2OC(=O)Cl + 2HCl \quad (9)$$

To a flask equipped as in Preparation 2-3 is added 5.6 ml (0.08 mole) of phosgene, and a solution of 4.24 gm (0.04 mole) of diethylene glycol in 50 ml of benzene is slowly added with vigorous stirring. The phosgene refluxes from the Dry Ice condenser and after 2 hr the reaction temperature is allowed to rise to 20°–25°C. The reaction mixture is heated for 3 hr at 50°C and then nitrogen gas sweeps the excess phosgene into an alcoholic potassium hydroxide solution. The benzene is removed by distillation under reduced pressure at 50°C to give the bischloroformate, which is then used in the next step of this preparation.

(*b*) *Polycarbonate formation.* To a dry flask equipped with a magnetic stirrer, dropping funnel, and condenser with drying tube is added 1.79 gm (0.008 mole) of 2,5-bis(3′-hydroxypropyl)-1,4-benzoquinone dissolved in 30 ml of THF. Then 1.85 gm (0.008 mole) of diethylene glycol bis(chloroformate) is quickly added at 0°–5°C. Pyridine (4.0 ml; 0.005 mole) is added with vigorous stirring. As the pyridine hydrochloride precipitates the solution becomes viscous. The mixture is stirred for 2 hr at room temperature, filtered, and the polymer isolated as in Preparation 2-2 to afford 2.14 gm (70%), η_{sp}/c = 10 cm³/gm (in 95% THF); UV, 255 mμ, $\varepsilon_{max} = 16.5 \times 10^3$ (in methylene chloride); IR, 1659 (quinone group), 1745, 1260 cm^{-1} (carbonate group) (recorded as a film on a sodium chloride plate).

B. Aliphatic Polycarbonates

Aliphatic polycarbonates are prepared by similar processes. The bischloroformate process yields low molecular weight polycarbonates but the transesterification method gives polycarbonates of molecular weights of about 3000 or more [23–25]. The latter procedure can be modified using vacuum to give high molecular weight aliphatic polycarbonates suitable for fiber and film-forming applications [27, 28].

$$n\text{HOROH} + n(\text{C}_2\text{H}_5\text{O})_2\text{CO} \longrightarrow \text{H}\left[-\text{OR}\text{O}\overset{\text{O}}{\overset{\|}{\text{C}}}-\right]_n\text{OC}_2\text{H}_5 + (2n-1)\text{C}_2\text{H}_5\text{OH} \qquad (10)$$

Carothers [23] also showed that aliphatic diols form polycarbonates but where in the structure unit —O—$(CH_2)_n$—CO, n is 5 or 6, then a cyclic monomer is formed. These classical results are given in Table III.

2-5. Preparation of Poly(hexamethylene carbonate) [23]

To a polymer tube is added 12 gm (0.10 mole) of hexamethylene glycol and 12 gm (0.10 mole) of ethyl carbonate. A small piece of sodium metal is added

TABLE III

Names of carbonates	Formula of structural unit	No. of atoms in chain of structural unit	Anal. calcd. for structural unit
Ethylene	$-O(CH_2)_2-O-\underset{\underset{\displaystyle O}{\Vert}}{C}-$	5	C 40.91 H 4.54
Trimethylene (monomeric)	$-O(CH_2)_3-O-\underset{\underset{\displaystyle O}{\Vert}}{C}-$	6	C 47.06 H 5.92
Trimethylene (polymeric)	$-O(CH_2)_3-O-\underset{\underset{\displaystyle O}{\Vert}}{C}-$	6	C 47.06 H 5.92
Tetramethylene	$-O(CH_2)_4-O-\underset{\underset{\displaystyle O}{\Vert}}{C}-$	7	C 51.71 H 6.94
Pentamethylene	$-O(CH_2)_5-O-\underset{\underset{\displaystyle O}{\Vert}}{C}-$	8	C 55.96 H 7.75
Hexamethylene	$-O(CH_2)_6-O-\underset{\underset{\displaystyle O}{\Vert}}{C}-$	9	C 58.33 H 8.33
Decamethylene	$-O(CH_2)_{10}-O-\underset{\underset{\displaystyle O}{\Vert}}{C}-$	13	C 65.96 H 10.07
Diethylene	$-O(CH_2)_2-O-(CH_2)_2-O-\underset{\underset{\displaystyle O}{\Vert}}{C}-$	8	C 45.44 H 6.11
p-Xylylene (soluble)	$-O-CH_2C_6H_4CH_2-O-\underset{\underset{\displaystyle O}{\Vert}}{C}-$	9	C 65.84 H 4.91
p-Xylylene (insoluble)	$-O-CH_2C_6H_4CH_2-O-\underset{\underset{\displaystyle O}{\Vert}}{C}-$	9	C 65.84 H 4.91

[a] The analyses were carried out by the Pregl micro method.
[b] D. Vorländer, *Justus Liebigs Ann. Chem.* **280**, 187 (1894).

GLYCOL ESTERS OF CARBONIC ACID

Analysis found[a]	Mol. wt. obs.	Method	Av. no. of structural units per molecule	Physical properties
41.07[b]	90 93[b]	F.p. C_6H_5OH[b]	1	M.p. 39°C[b]
4.88	76	Vapor density		B.p. 238°C
46.97 47.14	105 114	F.p. C_6H_6	1	Colorless needles
5.96 5.98	112 114	B.p. C_6H_6		M.p. 47°–48°C
				B.p. 133°C at 4 mm
47.06 47.08	4670 3880	B.p. C_6H_6	38–45	Glass gradually changes to powder
6.01 6.19				
52.18 52.32	1450 1400	B.p. C_6H_6		
7.05 7.06	1350 1310		11–12	Powder
	1290 1370	F.p. C_6H_6		M.p. 59°C
55.29 55.47	2840 2830	B.p. C_6H_6	20–22	Powder
7.90 7.83	2550			M.p. 44°–46°C
58.58 58.54	2740 2970	B.p. C_6H_6	18–21	Horny and elastic
8.46 8.21	2610			M.p. 55°–60°C
65.80 66.04	1880 1800			
10.18 10.10	1810 1640	B.p. C_6H_6	8–10	Powder
	1770 1830	F.p. C_6H_6		M.p. 55°C
45.53 45.33	1540 1550	B.p. C_6H_6	12	Sirup
6.30 6.38				
65.81 66.00	840 780 810	B.p. $C_2H_4Cl_2$	5	Powder
5.60 5.63				M.p. 137–138°C
65.02 65.03	1010 1030	M.p. camphor	6	Powder
5.29 5.19				M.p. 177°–185°C

and then the mixture is heated from 130° to 170°C over a 2 hr period. Approximately 86% of the ethanol is collected. After removal of sodium the residue is subjected to distillation under reduced pressure. A small amount of starting diol is collected. The residue, 10 gm (67%), is a light-colored, horny,

$$HOCH_2(CH_2)_4CH_2OH + (C_2H_5O)_2CO \xrightarrow{Na} \left[-O-CH_2(CH_2)_4CH_2O\overset{O}{\overset{\|}{C}}- \right]_n + 2C_2H_5OH \quad (11)$$

tough mass, m.p. 55°–60°C. The average number of structural units per molecule is 18–21 because the determined molecular weight (boiling method of benzene) is 2740, 2970, and 2610. The X-ray diffraction pattern shows that it is crystalline. The polymer is soluble in benzene, acetone, and chloroform but insoluble in ether and alcohol.

More recently Sarel and Pohoryles [26] described the effect of alkyl substituents at carbons 1 and 2 in 1,3-propanediol (more specifically substituted neopentyl glycols).

$$(C_2H_5CO)_2O + HOCH_2-\underset{R^2}{\overset{R^1}{C}}-CH_2OH \longrightarrow \left[-CH_2-\underset{R^2}{\overset{R^1}{C}}-CH_2OCOO- \right]_n \quad (12)$$

where $R^1 = R^2 = H$ $n = 38–45$
$R^1 = R^2 = CH_3$ $n = 20–22$
$R^1 = CH_3, R^2 = n\text{-}C_3H_7$ $n = 14–15$

2-6. Preparation of Poly(2,2-dimethyl-1,3-propanediol carbonate) [poly(neopentylene carbonate)] [26]

$$(C_2H_5O)_2O + HOCH_2-\underset{CH_3}{\overset{CH_3}{C}}-CH_2OH \longrightarrow \left[-CH_2-\underset{CH_3}{\overset{CH_3}{C}}-CH_2OCOO- \right]_{20-22} + 2C_2H_5OH \quad (13)$$

To a flask equipped with a 3 ft Vigreaux column is added 20.8 gm (0.2 mole) of neopentyl glycol, 24 gm (0.2 mole) of diethyl carbonate, and 0.05 gm of dry sodium methoxide. The mixture is gradually heated and the ethanol that is formed is distilled off at 77°–84°C. After the ethanol stops distilling the pressure is reduced and the mixture heated to 200°C to complete the reaction.

The residue (25 gm) on cooling is an opaque, glassy, and tough polymer. The polymer is purified from ligroin–benzene to afford 23.5 gm (90%) of a white powder, m.p. 107°–109°C. Intrinsic viscosity of the polymer is 0.035; mol. wt., 2500. The polymer is soluble in benzene but insoluble in most other organic solvents. The polymer is stable to 1 *N* aq. ethanolic NaOH on refluxing for 2–3 hr.

Cyclic carbonates can also be polymerized to give high molecular weight polycarbonates [23, 24, 30].

$$n(CH_2)_x\langle O, O \rangle C{=}O \xrightarrow[\text{OH}^-,\ 200^\circ\text{C}]{\text{H}_2\text{O or ROH}} \text{RO}\overset{\text{O}}{\overset{\|}{\text{C}}}\text{—}\left[\text{—O(CH}_2)_x\text{—O}\overset{\text{O}}{\overset{\|}{\text{C}}}\text{—}\right]_n\text{—OH} \quad (14)$$

$$(x = >5)$$

Aromatic–aliphatic polycarbonates can also be prepared. For example, *p*-xylylene glycol can be reacted with bisallyl carbonates using catalytic amounts of titanium tetrabutoxide or titanium tetrachloride [31]. The reaction of bisaryl or bisalkyl carbonates with bishydroxyethyl ethers of aromatic compounds also gives high molecular weight polycarbonates [32–34].

3. MISCELLANEOUS METHODS

1. Reduction of polycarbonates containing quinone groups with sodium hydrosulfite in a THF–water mixture (saturated with NH_4Cl) or with hydrogen using Pd/C catalyst [38].
2. Polycarbonates derived from bisphenol A structures containing unsaturated substituents [40].
3. A new synthesis of carbonates. The reaction of carbon monoxide with alkoxides in the presence of selenium [41].
4. Polymerization of cyclic carbonates [42–44].
5. Polycarbonates by the reaction of *N,N'*-carbonyldiimidazoles with bisphenol A in the melt and elimination of imidazoles [45–47].
6. Polycarbonates by the reaction of diesters of bisphenol with diphenyl carbonate and elimination of phenyl acetate [48].
7. Preparation of copolycarbonates [36,49,50].
8. Preparation of copoly(ester-carbonates) [51–54].
9. Preparation of poly(carbonate-urethanes) [55–57].
10. Preparation of cycloaliphatic polycarbonates [58].
11. Recovering polycarbonates from emulsions using inorganic salts [59].
12. Accelerated preparation of polycarbonates by a transesterification [60].

13. Condensation of organic dihalides with bifunctional carbonate esters using $AlCl_3$ catalyst to give polycarbonates [61].

14. Preparation of bisphenol di- and polycarbonates [62].

15. Preparation of polyamide–polycarbonate resin [63].

16. Preparation of bisphenol A–tetrachlorobisphenol A–polycarbonate copolymers [64].

REFERENCES

1. A. Davis and J. H. Golden, *J. Macromol. Sci., Rev. Macromol. Chem.* **3**, 49 (1969).
2. A. Einhorn, *Justus Liebigs Ann. Chem.* **300**, 135 (1890).
3. C. A. Bischoff and A. von Hedenström, *Chem. Ber.* **35**, 3431 (1902).
4. W. H. Carothers and J. W. Hill, *J. Amer. Chem. Soc.* **54**, 1559, 1566, and 1579 (1932).
5. W. R. Peterson, U.S. Patent 2,210,817 (1940).

6a. H. Schnell, *Angew. Chem.* **68**, 633 (1956).

6b. H. Schnell, "Chemistry and Physics of Polycarbonates." Wiley (Interscience), New York, 1964.

7. L. Bottenbruch, *Encycl. Polym. Sci. Technol.* **10**, 710 (1969).
8. W. F. Christopher and D. W. Fox, "Polycarbonates." Van Nostrand-Reinhold, Princeton, New Jersey, 1962.
9. H. Schnell, L. Bottenbruch, and H. Krimm, Belgian Patent 905,141 (1945).
10. H. Schnell, L. Bottenbruch, and H. Krimm, Belgian Patent 532,543 (1954).
11. L. H. Lee and H. Keskkula, U.S. Patent 3,177,179 (1965).
12. H. Willersinn, German Patent 1,100,952 (1961).
13. N. Melnikov, *J. Prakt. Chem.* [N.S.] **128**, 233 (1931).
14. A. J. Conix, Belgian Patent 601,392 (1961).
15. L. H. Lee and H. Keskkula, U.S. Patent 3,085,992 (1963).
16. H. Schnell, L. Bottenbruch, H. Schwartz, and H. G. Lotter, Belgian Patent 603,106 (1961).
17. General Electric Co., British Patent 839,858 (1960).
18. Kanoshima Kagaku Kogyo K.K., British Patent 883,540 (1961).
19. H. Schnell, *Angew. Chem.* **68**, 633 (1956).
20. H. Vernaleken, C. Wulff, L. Bottenbruch, and H. Schnell, Belgian Patent 691,898 (1967).
21. H. Schnell and G. Fritz, German Patent 1,024,710 (1958).
22. H. Vernaleken, L. Bottenbruch, and H. Schnell, Belgian Patent 677,424 (1966).
23. W. H. Carothers and F. J. Van Natta, *J. Amer. Chem. Soc.* **52**, 314 (1930).
24. J. W. Hill and W. H. Carothers, *J. Amer. Chem. Soc.* **55**, 5031 (1933).
25. S. Sarel, L. A. Pohoryles, and R. Ben-Shoshan, *J. Org. Chem.* **24**, 1873 (1959).
26. S. Sarel and L. A. Pohoryles, *J. Amer. Chem. Soc.* **80**, 4596 (1958).
27. W. R. Peterson, U.S. Patent 2,210,817 (1940).
28. H. Gawlak, R. P. Palmer, J. B. Rose, D. J. Sandiford, and A. Turner-Jones, *Chem. Ind. (London)* **25**, 1148 (1962).
29. Union Carbide Corp., British Patent 1,011,283 (1965).
30. H. C. Stevens, British Patent 872,983 (1961).
31. J. L. R. Williams and K. R. Dunham, U.S. Patent 2,843,567 (1958).
32. D. D. Reynolds and J. Van Den Berghe, U.S. Patent 2,789,964 (1957).

33. D. D. Reynolds and J. Van Den Berghe, U.S. Patents 2,789,509 and 2,789,965 (1957).
34. H. Rinke, W. Lehmann, and H. Schnell, German Patent 1,050,544 (1959).
35. P. W. Morgan, *J. Polym. Sci., Part A* **2**, 437 (1964).
36. H. Schnell, *Ind. Eng. Chem.* **51**, 157 (1959).
37. N. S. McPherson, M. L. Clachan, and K. R. Tatchell, British Patent 927,178 (1963).
38. G. Wegner, N. Nakabayashi, S. Duncan, and H. G. Cassidy, *J. Polym. Sci., Part A-1* **6**, 3395 (1968).
39. A. Y. Yakubovich, G. Y. Gordon, L. I. Maslennikova, E. M. Grohman, K. I. Tretyakova, and N. I. Kokoreva, *J. Polym. Sci.* **55**, 251 (1961).
40. J. R. Caldwell and W. J. Jackson, French Patent 1,333,561 (1963).
41. K. Kondo, N. Sonoda, and S. Tsutsumi, *Tetrahedron Lett.* No. 51, p. 4485 (1971).
42. H. Schnell and L. Bottenbruch, *Makromol. Chem.* **57**, 1 (1962).
43. R. J. Prochaska, Belgian Patent 630,530 (1963).
44. B. C. Oxenrider and R. M. Hetterly, French Patent 1,381,791 (1964).
45. H. Staab, *Justus Liebigs Ann. Chem.* **609**, 75 (1957).
46. H. Staab and F. Ebel, French Patent 1,208,196 (1960).
47. P. Beicrsdorf and Co. A.-G., French Patent 1,202,213 (1960).
48. M. Sander, German Patent 1,067,213 (1960).
49. D. W. Fox, Belgian Patent 570,529 (1958).
50. L. Bottenbruch, G. Fritz and H. Schnell, Belgian Patent 568,545 (1958).
51. E. P. Goldberg, Belgian Patent 570,531 (1958).
52. E. P. Goldberg, U.S. Patent 3,030,331 (1962).
53. H. Schnell, V. Boellert, and G. Fritz, Belgian Patent 597,654 (1960).
54. M. L. Clachan, N. S. McPherson, K. R. Tatchell, and T. A. Abott, U.S. Patent 3,000,849 (1961)
55. Bexfort, Ltd., Belgian Patent 675,389 (1959).
56. V. S. Foldi and T. W. Campbell, *J. Polym. Sci.* **56**, 1 (1962).
57. W. E. Bissinger, F. Strain, H. C. Stevens, W. R. Dial, and R. S. Chisholm, U.S. Patent 3,215,668 (1965).
58. Chemische Werke Huels A.-G. French Patent 1,410,431 (1965).
59. A. V. Pinter, U.S. Patent 3,251,802 (1960).
60. V. Boellert, U. Curtius, G. Fritz, and H. Schnell, French Patent 1,403,441 (1965).
61. R. F. J. Gattys, Italian Patent 643,921 (1962).
62. Allied Chem. Corp., Netherland Patent Appl. 6,400,872 (1964).
63. M. Goulay and E. Marechal, *Bull. Soc. Chim. Fr.* [6] No. 3, p. 854 (1971).
64. K. Ikeda and Y. Sekine, *Ind. Eng. Chem. Prod. Res. Develop.* **12**, 212 (1973).

Chapter 4

POLYAMIDES

I. INTRODUCTION

Polymeric amides are found in nature in the many polypeptides (proteins) which constitute a variety of animal organisms and the composition of silk and wool. It was a search to prepare substitutes for the latter which led to the commercial development of the synthetic polyamides known as nylons.

Several early investigators reported polyamide type materials. Balbiano and Trasciatti [1] heated a mixture of glycine in glycerol to give a yellow amorphous glycine polymer [2]. Manasse [3] obtained nylon 7 by heating 7-aminoheptanoic acid to the melting point. Curtius [4] found that ethyl glycinate

$$n\mathrm{NH_2(CH_2)_6COOH} \longrightarrow [\mathrm{-NH(CH_2)_6CO-}]_n + n\mathrm{H_2O} \qquad (1)$$

polymerizes on standing in the presence of moisture or in the dry state in ethyl ether to afford polyglycine [5].

$$n\mathrm{NH_2CH_2COOC_2H_5} \longrightarrow [\mathrm{-NHCH_2CO-}]_n + n\mathrm{C_2H_5OH} \qquad (2)$$

The β-amino acids on heating afford ammonia but no polymers. In addition, the γ- and δ-amino acids on heating afford stable lactams and no polymers. However, the ϵ-, ζ-, and η-amino acids give polyamides.

The reaction of dibasic acids with diamines was reported in the early literature [6a-h] to give low molecular weight cyclic amides as infusible and insoluble products. It was Carothers [7a–e] who first recognized that polymeric amides were formed by the reaction of diamines with dibasic acids. Many of the polyamides were able to be spun into fibers and the fibers were called "nylon" by du Pont [7a–g].

The major methods of preparing polyamides are summarized in Scheme 1. Other methods of less importance are summarized in Scheme 2.

Polyamides with molecular weights above 7000 are useful since they possess properties which allow them to be spun into fibers [8a].

Cross-linked polyamides may be prepared from polyfunctional amino acids, triamines, or tricarboxylic acids, etc. The chapter will describe only linear polyamides.

The preparation of polyamides has recently been reviewed and this review is worthwhile consulting [8b].

Some of the main uses [8a,b] of polyamides or nylons are for synthetic fibers for the tire, carpet, stocking, and upholstery industries. Use of polyamides as molding and extrusion resins for the plastics industry is also of increasing importance [8d].

SCHEME 1

The Major Preparative Methods for the Synthesis of Polyamides

$$H_2N(CH_2)_xCOOH \xrightarrow{-H_2O} [\text{—}NH(CH_2)_xCO\text{—}]_n \xleftarrow{\text{catalyst}} (CH_2)_x\,\overset{C=O}{\underset{NH}{|}}$$

$$\left[\text{—}O\overset{O}{\overset{\|}{C}}(CH_2)_x\overset{O}{\overset{\|}{C}}O\text{—}\right]^{2-}\ H_3\overset{+}{N}(CH_2)_x\overset{+}{N}H_3 \xrightarrow{-2H_2O} [\text{—}NH(CH_2)_xCO\text{—}]_n$$

$$R(COZ)_2 + R'(NH_2)_2 \xrightarrow{-HZ} \left[\text{—}\overset{O}{\overset{\|}{C}}\text{—}R\text{—}\overset{O}{\overset{\|}{C}}NHR'NH\text{—}\right]_n$$

$$(Z = Cl, OR'', \text{ or } NH_2)$$

$$R(COOH)_2 + R'(NHCOR'')_2 \longrightarrow RCO\left[\text{—}HNR'NH\overset{O}{\overset{\|}{C}}R\overset{O}{\overset{\|}{C}}\text{—}\right]_nOH + R''COOH$$

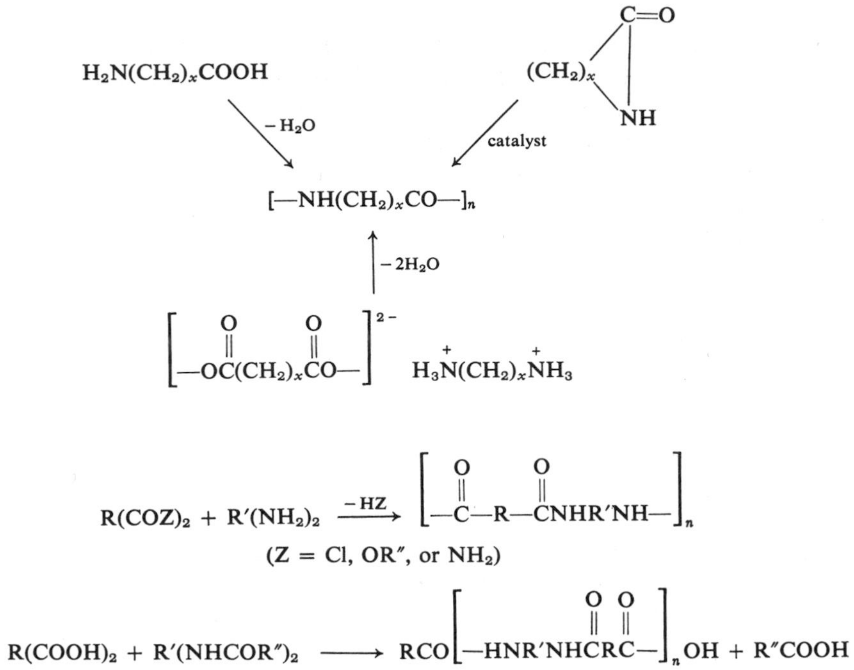

SCHEME 2

Miscellaneous Methods for the Synthesis of Polyamides from Nitriles

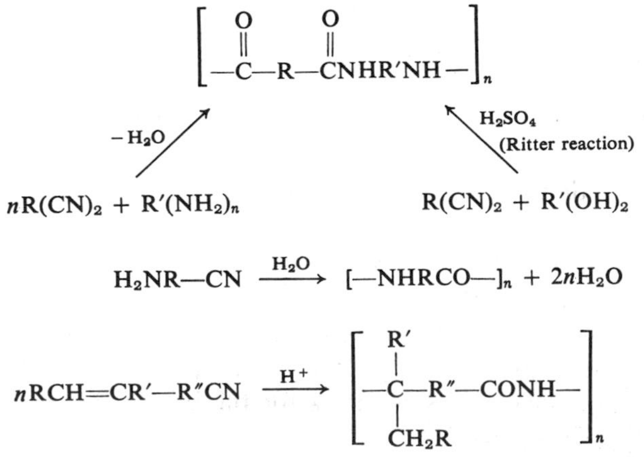

$$nR(CN)_2 + R'(NH_2)_n \xrightarrow{-H_2O} \left[\text{—}\overset{O}{\overset{\|}{C}}\text{—}R\text{—}\overset{O}{\overset{\|}{C}}NHR'NH\text{—}\right]_n \xleftarrow[\text{(Ritter reaction)}]{H_2SO_4} R(CN)_2 + R'(OH)_2$$

$$H_2NR\text{—}CN \xrightarrow{H_2O} [\text{—}NHRCO\text{—}]_n + 2nH_2O$$

$$nRCH{=}CR'\text{—}R''CN \xrightarrow{H^+} \left[\text{—}\overset{R'}{\underset{CH_2R}{\overset{|}{\underset{|}{C}}}}\text{—}R''\text{—}CONH\text{—}\right]_n$$

2. CONDENSATION REACTIONS

A. Amidation of Dibasic Acids with Diamines

In order to achieve a good reactant balance the diammonium salt is first prepared by the reaction of the diamine with the dibasic acid, as shown in Eq. (3). The salt is then heated to eliminate water and form the polyamide.

$$nR(NH_2)_2 + nR'(COOH)_2 \longrightarrow n[R'(COO^-)_2][R(\overset{+}{N}H_3)_2] \xrightarrow[\text{heat}]{-2nH_2O} \left[-\overset{\overset{O}{\|}}{C}-R'-\overset{\overset{O}{\|}}{C}NHRNH- \right]_n \quad (3)$$

The salt is conveniently formed by bringing together alcoholic solutions of equivalent amounts of the reactants and following the reaction electrometrically. The white salt that separates from the cooled solution is dried and then used directly in the polymerization reaction [7e]. Recrystallization of the salts may be effected to give pure crystalline compounds of melting points as shown in Table I.

TABLE I

MELTING POINTS OF POLYMETHYLENE DIAMMONIUM DICARBOXYLATES AND THEIR RESULTING POLYAMIDES[a]

Diamine $(CH_2)_n(NH_2)_2$ $n=$	Dicarboxylic acid $(CH_2)_n(COOH)_2$ $n=$	Melting point (°C)	
		Salt	Polyamide
4	4	204	278
4	5	138	233
4	7	175	223
5	8	129	195
6	4	183	250
6	8	170	209
8	4	153	235
8	8	164	197
9	4	125	204–205
9	8	159	174–176
10	4	142	230
10	8	178	194
11	8	153	168–169
12	4	144	208–210
12	8	157	171–173

[a] Data taken from D. D. Coffman, G. J. Berchet, W. R. Peterson, and E. W. Spanagel, *J. Polym. Sci.* **2**, 306 (1947).

The salts are converted to polyamides (see Table I) either by fusing them under reduced pressure or by heating them in an inert solvent such as cresol or xylenol.

The polyamides derived from polymethylene diamines and polymethylene dicarboxylic acids can be spun into fiber filaments by drawing a cold rod from the molten polymer.

2-1. Preparation of Poly(pentamethylenesebacamide) [8c]

$$H_2N—(CH_2)_5—NH_2 + (CH_2)_8(COOH)_2 \longrightarrow [—NH(CH_2)_5—NHCO(CH_2)_8CO—]_n + 2H_2O \quad (4)$$

(*a*) *Preparation of pentamethylenediammonium sebacate.* To a flask containing 510 gm (2.51 moles) of sebacic acid dissolved in 2350 ml of *n*-butanol is added 272 gm (2.67 moles) of pentamethylenediamine. The mixture is heated at the reflux temperature to effect the solution and then cooled. The crystalline product is filtered, washed with acetone, and dried to afford 729 gm (94%), m.p. 129°C (soluble in water, ethanol, and hot butanol but insoluble in acetone, ethers, and hydrocarbons).

(*b*) *Polyamide formation.* To a flask equipped with a condenser, nitrogen inlet, thermometer, and a viscometer and containing 150 gm (0.50 mole) of pentamethylenediammonium sebacate is added 2.5 gm of sebacic acid as a stabilizer. Then 152 gm of xylenol (b.p. 218°–222°C) is added as a reaction solvent. The mixture is refluxed for 6½ hr, cooled, and then poured into a large volume of alcohol. The precipitated polymer is filtered and dried to afford a white, powdery polymer, m.p. 190°C; intrinsic viscosity, 0.65 in *m*-cresol.

The polyamide can also be formed by placing 30 gm (0.1 mole) of pentamethylenediammonium sebacate together with 0.5 gm of sebacic acid in a Pyrex polymer tube (2.5 cm diameter × 40 cm long), evacuating, filling with nitrogen, reevacuating, sealing, and heating for 2–3 hr at 150°–200°C. The tube is cooled, opened, and then heated under reduced pressure at 195°–225°C for an additional 2 hr to complete the polymerization.

Polyamides prepared from diamino ethers and polymethylene dicarboxylic acids, such as $[—NH(CH_2)_2O(CH_2)_2NHCO(CH_2)\ CO—]_n$, have been reported to be soluble in hot water [9]. However, the introduction of ether groups as in the latter polyamide lowers the polymer melt temperature to about 150°–160°C compared to poly(octamethylene adipamide) which melts at 230°C. The greater flexibility of the ether groups accounts for the lowering of the melting point. However, if piperazine is used as the diamine and is condensed with diglycolic acid to give poly(piperazine diglycolamide) the melting point is raised to 258°C but the polyamide is soluble in hot water [10].

B. Amidation of Dicarboxylic Acid Esters with Diamines

The preparation of polyamides by the aminolysis of reactive esters proceeds under mild conditions and without racemization [11]. Speck [12] reported that the phenyl esters of some substituted malonic acid derivatives condense with aliphatic diamines to produce heat-stable high molecular weight polyamides. Heating the free malonic acid with the same diamines caused only decarboxylation to take place at 115°–185°C. The polyamides are prepared as follows: Equivalent amounts of diamine and the phenyl ester of malonic acid are weighed into a glass tube, and the tube is evacuated, sealed, and heated at 200°–210°C. The tube is cooled, opened, and heated to 260°–280°C in the presence of nitrogen. The polymerization is completed by heating under reduced pressure for several hours as described in Table II.

This method has also been used to prepare long-chain peptides [13] and for the polymerization of tripeptides [14]. Recently Overberger and Sebenda reported the rate of polymerization of several esters of bicyclo[2.2.2]octane-*trans*-dicarboxylic acid with piperazine [15].

2-2. *Polycondensation of Piperazine with Bis(2,4-dinitrophenyl) (±)-Bicyclo[2.2.2]octane-trans-2,3-dicarboxylate* [15]

COOR, COOR + HN NH $\xrightarrow{-2ROH}$ [CON N— CO—]$_n$ (5)

($R = 2,4\text{-}(NO_2)_2C_6H_3$)

To an ice-cooled flask containing 48 gm (0.56 mole) of piperazine is added 296.3 gm (0.56 mole) of bis(2,4-dinitrophenyl) (±)-bicyclo[2.2.2]octane-*trans*-2,3-dicarboxylate (see Note), 110 ml of cold triethylamine, and 5 liters of cold chloroform. The reaction mixture is stirred for 1 hr at 0°C and 48 hr at room temperature. The reaction mixture is added to an ethanol–ether mixture to precipitate the polymer. The filtered polymer is boiled with $1\frac{1}{2}$ liters of 0.1 *N* sodium hydroxide solution (50% excess ethanol), filtered, washed with ethanol, and then with acetone. The polyamide is dried under reduced pressure to afford 111 gm (80%), $[\eta]$ 25°C = 0.295 ($CHCl_3$).

TABLE II

PREPARATION AND PROPERTIES OF POLYAMIDES FROM PHENYL DISUBSTITUTED MALONATES[a]

Reactants (equiv. amts.)	Polymerization			Polymer		Remarks
	Time (hr)	Temp. (°C)	Press. (mm)	Stick temp.[b] (°C)	η_{inh}[c]	
Decamethylenediamine, ethyl malonate	4	200	Sld.[d]	185–190	0.93	Orange-red, brittle, partially gelled
	½	283	Atm.			
	1¼	283	3			
Decamethylenediamine, ethyl ethylmalonate	20	200	Sld.	95–100	0.26	Clear, light yellow, brittle, not spinnable
	1	283	Atm.			
	2¼	283	3			
Decamethylenediamine, ethyl diethylmalonate	20	200	Sld.	120–125	0.72	Pale yellow
	1	283	Atm.			
	7	283	3			
Decamethylenediamine, phenyl dimethylmalonate	6	200	Sld.	120–125	0.93	Clear, colorless, tough. Mechanically spun into tough drawable fibers
	1	259	Atm.			
	2	273	14			
Hexamethylenediamine, phenyl dimethylmalonate	6	200	Sld.	115–120	0.93	Clear, colorless, tough, spinnable
	1	260	Atm.			
	2	273	14			
	6	273	3			
Bis(*p*-aminomethyl)benzene, phenyl dimethylmalonate	5	210	Sld.	110–115	0.51	Manually spinnable
	1	273	Atm.			
	½	273	14			
	5	273	3			
p-Xylylenediamine, phenyl dimethylmalonate	5	210	Sld.	260	0.35	White, opaque, brittle
	1	283	Atm.			
	3	283	3			
Tetramethylenediamine, phenyl diethylmalonate	4	210	Sld.	100–105	0.30	Clear, colorless, brittle
	2	260	Atm.			
	6	273	3			

Hexamethylenediamine,	5	210	Sld.	95–100	0.69	Clear, tough, spinnable.
phenyl diethylmalonate	1	273	Atm.			Sol. in ethanol, $CHCl_3$-methanol (88/12)
p-Xylylenediamine,	5	210	Sld.	165–170	0.38	Clear, brittle
phenyl diethylmalonate	1	273	Atm.			
	3	273	3			
Hexamethylenediamine,	20	200	Sld.	—	—	Complete decarboxylation at 283°C
n-butylethylmalonic acid		283	Atm.			
Hexamethylenediamine,	2	210	Sld.	115–120	0.41	Clear, little color. Manually spinnable and drawable
phenyl *n*-butylethylmalonate	1	260	Atm.			
	9	260	3			
p-Xylylenediamine,	5	210	Sld.	170–175	0.48	Clear, colorless, brittle
phenyl *n*-butylethylmalonate	1	273	Atm.			
	2½	273	3			
Hexamethylenediamine,	14	210	Sld.	140–145	0.66	Tough and strong
phenyl *n*-dibutylmalonate	1	265	Atm.			
	5	265	3			
Hexamethylenediamine,	14	210	Sld.	193–197	0.63	Semi-opaque. Hard and tough. Mechanically spun into strong, drawable resilient fibers
phenyl 1,1-cyclohexanedecarboxylate	1	273	Atm.			
	5	273	3			
Bis(*p*-aminomethyl)benzene,	1¼	260	Atm.	185–190	0.52	Clear, pale yellow. Spinnable
phenyl 1,1-cyclohexanedecarboxylate	3¾	273	3			
Hexamethylenediamine,	14	210	Sld.	—	0.17	Soft and sticky at room temp.
phenyl phenylethylmalonate	1	273	Atm.			
	5	273	1			

[a] Reprinted from S. B. Speck, *J. Amer. Chem. Soc.* **74**, 2876 (1952). Copyright 1952 by the American Chemical Society. Reprinted by permission of the copyright owner.

[b] Temp. at which polymer begins to stick when a fresh surface of polymer is touched to a polished copper block.

[c] η_{inh} = inherent viscosity = $(2.3 \log \eta_{rel})/C$. Determined in *m*-cresol at 0.5% concentration.

[d] Sld. = reagents heated in an evacuated sealed tube.

NOTE: Bis(2,4-dinitrophenyl) (±)-bicyclo[2.2.2]octane-*trans*-2,3-dicarboxylate is prepared by the condensation of 2,4-dinitrophenol (21.1 mmole) in 15 ml of DMF with 10.4 moles of bicyclo[2.2.2]octane-*trans*-2,3-dicarboxylic acid in the presence (added slowly at 0°C) of 24.9 mmole of *N,N'*-dicyclohexylcarbodiimide in 10 ml of DMF at 0°C. The mixture is allowed to stand in the refrigerator several hours and then is filtered to remove dicyclohexylurea (theoretical). The residue is concentrated under reduced pressure and the solid after recrystallization from toluene is obtained in 33.5% yield, m.p. 200°–201°C.

C. Condensation of Amino Carboxylic Acids and Their Derivatives

Amino carboxylic acids condense in the absence of catalysts at approx. 260°C to liberate water. The reaction may be accelerated by use of reduced pressure. In the case of nylon 7 and nylon 11 the probability of occurrence of intramolecular dehydration to from 8- or 12-membered ring lactams is very remote, but nevertheless approx. 0.5% do form in both cases [16].

In a continuous industrial process utilizing 11-aminoundecanoic acid, the monomer dissolved in water is dried by being sprayed on the hot walls of the reactor and the resulting liquid monomer is then polymerized [16].

The physical properties of nylons 6–11 are shown in Table III. Note that the even-numbered polymethylene polyamides melt higher than the odd-numbered ones.

TABLE III
PHYSICAL PROPERTIES OF POLYAMIDES PREPARED FROM ω-AMINO ACIDS[a]

Polyamide	M.p. (°C)	Density of crystalline polymer
—HN—$(CH_2)_5$—CO—	214	1.14
—HN—$(CH_2)_6$—CO—	225	1.10
—HN—$(CH_2)_7$—CO—	185	1.08
—HN—$(CH_2)_8$—CO—	194	1.06
—HN—$(CH_2)_9$—CO—	177	1.04
—HN—$(CH_2)_{10}$—CO—	182	1.04

[a] Data taken from [17].

2-3. *Preparation of Nylon 7 [Poly(7-heptanamide)]* [17]

$$n\mathrm{NH_2(CH_2)_6COOCH_3} \longrightarrow [\mathrm{—NH(CH_2)_6CO—}]_n + n\mathrm{CH_3OH} \quad (6)$$

To a Pyrex test tube equipped with a nitrogen capillary is added 20.0 gm (0.12 mole) of methyl 7-aminoheptanoate. The amino acid ester is heated at 270°C for 5 hr while nitrogen is passing through the molten material. After this time the tube is cooled to afford 15.2 gm (100%) with reduced viscosity of 0.89 [0.2 gm/100 ml phenol tetrachloroethane (3:2) at 30°C]. In a similar manner 7-aminoheptanoic acid was heated for 1½ hr to give the same polyamide of reduced viscosity of 1.64. Nylon 7 is a white semicrystalline polyamide and is stable in the molten state at 250°C for over 24 hr. Products of molecular weight 15,200–30,000 give useful fiber, film, and molding resins of properties similar to nylon 66.

Recently it has been reported that acrylates add to amino alcohols and the resulting *N*-(hydroxyalkyl)-β-alanine esters give polyamides by undergoing a base-catalyzed condensation at room temperature (Eq. 7) [18].

$$\mathrm{CH_2{=}CH—COOR} + \mathrm{H_2N—R'—OH} \longrightarrow \mathrm{HOR'—NHCH_2CH_2COOR}$$

$$\xrightarrow{-\mathrm{ROH}} \left[\begin{array}{c}\mathrm{—N—CH_2CH_2CO—} \\ | \\ \mathrm{R'OH}\end{array}\right]_n \qquad \xrightarrow{-\mathrm{ROH}} \left[\mathrm{—O—R'NHCH_2CH_2CO—}\right]_n \quad (7)$$

D. Condensation of Lactams to Give Polyamides (Addition Type Reaction)

Lactams may be polymerized [19] under acidic or basic conditions to give polyamides by an addition type reaction.

The nomenclature in Table IV is used for the more common lactams.

Schlack [20] reported that the polymerization of caprolactam to nylon 6 can be carried out in the absence of air in a sealed tube at 220°–250°C for a prolonged period. Hanford and Joyce [21] confirmed these results and also reported two rapid methods (hydrolytic and catalytic) for converting caprolactam to nylon 6. It is interesting to note that at first Carothers [22] and later Hermans [23] reported that caprolactam under rigorously anhydrous conditions is hardly affected by heating at 250°C for 600 hr in a sealed evacuated tube.

TABLE IV
NOMENCLATURE OF LACTAMS

Ring size	Lactam name	Other name
4	β-Propiolactam	2-Azetidinone
5	γ-Butyrolactam	2-Pyrrolidone, pyrrolidinone
6	δ-Valerolactam	2-Piperidone
7	Caprolactam	6-Hexanolactam or 2-oxohexamethylenimine
8	Enantholactam	7-Heptanolactam or 2-oxoheptamethylenimine
9	Capryllactam	8-Octanolactam or 2-oxooctamethylenimine

a. Base-Catalyzed Polymerization [24]

Hanford and Joyce [21] reported that caprolactam can be polymerized smoothly and rapidly by heating at 200°–280°C in a closed vessel with an inert atmosphere in the presence of small amounts of alkali or alkaline earth metals [25]. The alkali metals such as sodium or lithium are more effective than calcium and magnesium. In practice the metal (0.5 to 1% of the lactam) is reacted with the lactam at about 100°C and then the temperature raised to 200°–280°C. The melt becomes viscous after 10 min heating at 250°C and the reaction after 2 hr still contains 3–7.5% residual caprolactam. Heating for prolonged periods at 285°C causes a decrease in molecular weight. The reaction is thought to go through the steps shown in Eq. (8).

The following other catalysts have been reported as effective catalysts for the polymerization of ε-caprolactam: alkali metal salts of carboxylic acids [26], alkali metal cyanides or azides [26], alkali metal hydroxides [27,28], Grignard reagents [29], or aluminum trialkyls [30]. Acylating agents (car-

$$(CH_2)_5\left\langle\begin{matrix} NH \\ C{=}O \end{matrix}\right. + Na \longrightarrow (CH_2)_5\left\langle\begin{matrix} \overset{-}{N}\text{—}\overset{+}{Na} \\ C{=}O \end{matrix}\right. \xrightarrow{(CH_2)_5\left\langle\begin{matrix} NH \\ C{=}O \end{matrix}\right.} (CH_2)_5\left\langle\begin{matrix} N\text{—}CO(CH_2)_5\overset{-}{N}H\overset{+}{Na} \\ C{=}O \end{matrix}\right.$$

$$\xrightarrow{(CH_2)_5\left\langle\begin{matrix} NH \\ C{=}O \end{matrix}\right.} (CH_2)_5\left\langle\begin{matrix} N\text{—}[CO(CH_2)_5NH]_n^{-}\overset{+}{Na} \\ C{=}O \end{matrix}\right. \tag{8}$$

boxylic acid chlorides or anhydrides, isocyanates, inorganic anhydrides such as P_2O_5) act as very effective activators [31,31a] in the anionic polymerization of lactams [31,32,32a,32b]. The ease of polymerization of lactams varies with ring size as follows: caprolactam ≥ pyrrolidone > piperidone [32b]. Polymers are often produced having a broad molecular weight distribution because of the proton-transfer reaction with amide groups on a polymer instead of with monomer [32a].

Some typical catalysts which are effective for the polymerization of α-piperidone are sodium hydride, sodium borohydride, lithium hydride, lithium aluminum hydride, and triethylaluminum–ethylmagnesium bromide [32b]. The following showed no catalytic effect: titanium tetrachloride and triethylaluminum, triethylboron and diethylcadmium [32b].

Recently Konomi and Tani reported that five, six-, and seven-membered ring lactams can be polymerized at low temperature by using Na piperidonate-AlEt (piperidone)$_2$ system or the salt of $MAlEt_4$ (M = Li, Na, K) [33]. Other catalysts such as $MAlEt_4$ or $MOAlEt_2$–$AlEt_3$ were also reported to be effective for the polymerization of ε-caprolactam [34].

Hall [35] reported that water and alkali metals were equally potent catalysts at temperatures above 200°C. In addition alkyl or aryl groups on the ring caused a decrease in polymerizability, as is shown in Table V.

b. Acid-Catalyzed Polymerization

Caprolactam is also polymerized to nylon 6 in the presence of catalytic amounts of hydrogen chloride [36,37]. Lactams polymerize at rates corresponding to the following order of lactam ring size: 8 > 7 > 1 > 5 and 6 [3f]. These results are related to the relative basicities of the ring lactams toward hydrogen chloride [38].

Water can also initiate the polymerization of lactams at elevated temperature as shown in Eq. (9) [23,39].

Water reacts with the lactam to form amino acid, which then initiates the

$$\overbrace{(CH_2)_5\begin{matrix}C{=}O\\ |\\ NH\end{matrix}} + H_2O \longrightarrow HOOC(CH_2)_5NH_2 \xrightarrow{(CH_2)_5\,C{=}O\,/\,NH}$$

$$HOOC(CH_2)_5NH\overset{\overset{\displaystyle O}{\|}}{C}\underset{\underset{\displaystyle NH_2}{|}}{\underset{(CH_2)_5}{|}} \xrightarrow{-H_2O} \left[-CO(CH_2)_5-NHCO(CH_2)_5NH-\right]_n \qquad (9)$$

TABLE V
POLYMERIZATION OF LACTAMS[l]

Polymerized	Did not polymerize
Four-Membered Rings	
2-Azetidinone[b]	
Five-Membered Rings	
2-Pyrrolidinone[c]	1-Methyl-2-pyrrolidinone[a]
	5-Methyl-2-pyrrolidinone[a]
	5,5-Dimethyl-2-pyrrolidinone[a]
	5,5-Pentamethylene-2-pyrrolidinone[a]
	5-Carbethoxy-2-pyrrolidinone[a]
	2-Pyrrolidinethione (d.)[a]
	4-Thiazolidinone[a]
	2-Phenyl-4-thiazolidinone[a]
	2-Phenyl-4-thiazolidinone-1-dioxide[a]
Six-Membered Rings	
2,5-Piperazinedione[d]	1,4-Dimethyl-2,5-piperazinedione[a]
	1-Methyl-2,5-piperazinedione[d]
	3,6-Dimethyl-2,5-piperazinedione[d]
	1,4-Diphenyl-2,5-piperazinedione[d]
	2-Piperidone[a]
	3-Morpholone (d.)[a,e]
	3-Thiamorpholone (d.)[a]
	1-Isopropyl-5,5-dimethyl-2,3-piperazinedione[a]
	Benzo-2*H*-1,4-oxazin-3(4*H*)-one[a]
	Benzo-2*H*-1,4-thiazin-3(4*H*)-one[a]
Seven-Membered Rings	
2-Oxohexamethylenimine[f] (caprolactam)	1-Methyl-2-oxohexamethylenimine[a]
3-Methyl-2-oxohexamethylenimine[g,h]	1-Phenyl-2-oxohexamethylenimine[a]
4-Methyl-2-oxohexamethylenimine[a,g,h]	1-Methylol-2-oxohexamethylenimine[a]
5-Methyl-2-oxohexamethylenimine[a,g,h]	1-Ethylthiomethyl-2-oxohexamethylenimine[a]
6-Methyl-2-oxohexamethylenimine[h]	5-Isopropyl-2-oxohexamethylenimine[h]
7-Methyl-2-oxohexamethylenimine[h]	7-Isopropyl-2-oxohexamethylenimine[h]
5-Ethyl-2-oxohexamethylenimine[h]	5-*n*-Propyl-2-oxohexamethylenimine[h]
2-Hexamethyleneiminethione[i,j]	5-*t*-Butyl-2-oxohexamethylenimine[h]
Endomethylene-2-oxohexamethylenimine[k]	5-Phenyl-2-oxohexamethylenimine[a]
Endoethylene-2-oxohexamethylenimine[k]	5-Cyclohexylmethyl-2-oxohexamethylenimine[h]
	4-Methyl-7-isopropyl-2-oxohexamethylenimine[a]
	5-Methyl-2-hexamethyleneiminethione (d.)[a]
	4,6-Dimethyl-2-hexamethyleneiminethione (d.)[a]
	2-Oxo-5-thiahexamethyleneimine (d.)[a]

TABLE V (*cont.*)

Polymerized	Did not polymerize
	2-Oxo-5-thiahexamethyleneimine-5-dioxide (d.)[a]
	1-Benzenesulfonyl-1,4-diazepin-5-one[a]
	2,3-Benzo-1,4,5,6-tetrahydroazepin-7-one[h]
	Perhydro-3,4-benzazepin-7-one[a]
	2,3-Benzotetrahydro-1,4-thiazepin-5-one (d.)[a]
	2,3-Benzotetrahydro-1,4-thiazepin-5-one-1-monoxide (d.)[a]
	2,3-Benzotetrahydro-1,4-thiazepin-5-one-1-dioxide (d.)[a]
Eight-Membered Rings	
2-Oxoheptamethylenimine[b]	
2-Heptamethyleneiminethione[a,i]	
1,5-Diazacyclooctane-2,6-dione[a]	
Nine-Membered Rings	
2-Oxooctamethylenimine[f]	
2-Octamethyleneiminethione[a,i]	

[a] Hall (1958), see below. The lactams were treated with water at 200°C or higher, and with sodium or sodium hydride at several temperatures between the melting point of the monomer and 250°C. *N*-Acetylcaprolactam was added in many of the alkali-catalyzed experiments; (d.) signifies serious decomposition.

[b] R. W. Holley and A. D. Holley, *J. Amer. Chem. Soc.* **71**, 2129 (1949).

[c] W. O. Ney, Jr., W. R. Nummy, and C. E. Barnes, U.S. Patent 2,638,463 (1953); W. O. Ney, Jr. and M. Crowther, U.S. Patent 2,739,959 (1956).

[d] A. B. Meggy, *J. Chem. Soc., London*, p. 1444 (1956).

[e] R. Leimu and J. I. Jansson, *Suom. Kemistilehti B* **18**, 40 (1945); *Chem. Abstr.* **41**, 769 (1947).

[f] D. D. Coffman, N. L. Cox, E. L. Martin, W. E. Mochel, and F. J. van Natta, *J. Polym. Sci.* **3**, 85 (1948); W. E. Hanford and R. M. Joyce, *ibid.*, p. 167; H. R. Mighton, U.S. Patent 2,647,105 (1953).

[g] Z. A. Rogovin, E. Khait, I. L. Knunyants, and Y. Rymashevskaya, *J. Gen. Chem. USSR* **17**, 1316 (1947); *Chem. Abstr.* **42**, 4939 (1948).

[h] A. Schäffler and W. Ziegenbein, *Chem. Ber.* **88**, 1374 (1955); W. Ziegenbein, A. Schäffler, and R. Kaufhold, *ibid.*, p. 1906.

[i] N. L. Cox and W. E. Hanford, U.S. Patent 2,276,164 (1942).

[j] Swiss Patents 270,546 and 276,924 (1951); 280,367 (1952).

[k] But see C. E. Barnes, W. R. Nummy, and W. O. Ney, Jr., U.S. Patent 2,806,841 (1957). Repetition of this work gave only material which dissolved instantly in acetone to form nonviscous solutions and therefore was not polymeric.

[l] Reprinted from H. K. Hall, Jr., *J. Amer. Chem. Soc.* **80**, 6404 (1958). Copyright 1958 by the American Chemical Society. Reprinted by permission of the copyright owner.

polymerization by the proton transfer from the carboxylic acid. This type of initiation is effective with readily polymerized lactams, for example, caprolactam > enantholactam ≃ capryllactam. The latter is said to be less reactive because the amide function may exist in the *trans* or *anti* conformation compared to the *cis* or *syn* conformation of caprolactam and enantholactam [39].

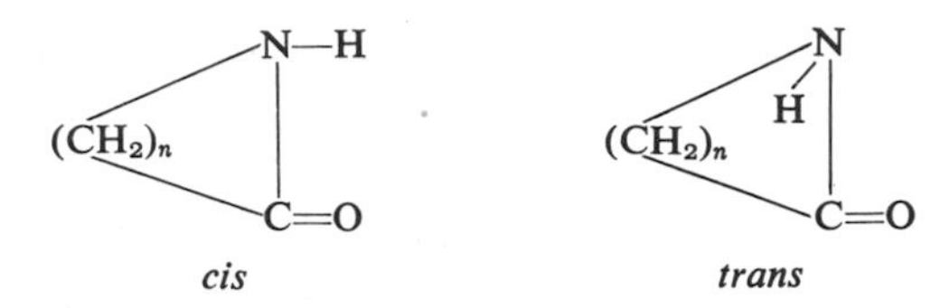

Van der Want found that carboxylic acids such as trifluoroacetic acid are not as effective as amino acids as an initiator. The effectiveness of HCl, water, and other initiators is described in Table VI.

TABLE VI
POLYMERIZATION OF ε-CAPROLACTAM AT 254°C USING VARIOUS INITIATORS[a]

Initiator	Moles initiator per mole lactam	Reaction time (hr)	% Conversion
H_2O	0.025	2	5
		4	56
HCl	0.025	2	52
$HClO_4$	0.025	1	Very rapid (100%)
NH_2SO_3H	0.025	1	Very rapid (100%)
p-CH_3—C_6H_4—SO_3H	0.025	2	31
CF_3COOH	0.05	2	0
		4	6
$H_2N(CH_2)_5COOH$	0.025	2	65
		3	76
$(CH_2CH_2CH_2NH_3^+Cl^-)_2$	0.0125	1	39
NH_4Cl	0.05	1	34
NH_4BF_4	0.025	1	34
$(NH_4)_2SO_4$	0.0125	4	69

[a] Data taken from G. M. van der Want and C. H. Kruissink, *J. Polym. Sci.* **35**, 119 (1959).

2-4. *Preparation of Poly(γ-butyramide) from 2-Pyrrolidinone Using Anionic Catalysis* [35]

(*a*) *Cocatalytic polymerization.* To a long glass test tube containing 6.5 gm (0.076 mole) of anhydrous 2-pyrrolidinone is added 0.13 gm (0.006 mole) of sodium hydride. After the evolution of hydrogen ceases 0.10 gm of acetic anhydride is added as a cocatalyst. An exothermic reaction takes place,

$$\text{2-pyrrolidinone (cyclic NH–C=O)} \longrightarrow \left[-HN-CH_2CH_2CH_2\overset{O}{\overset{\|}{C}}- \right]_n \qquad (10)$$

leading to a hard white polymer. The tube is broken and the polymer broken up into small pieces. The polymer is extracted with water and acetone and then dried to afford 5.7 gm (88%) of poly(γ-butyramide), $\eta_{inh} = 0.88$ (obtained in *m*-cresol).

Other effective cocatalysts are *N*-acyl lactams, acyl halides, isocyanates, isothiocyanates, esters or dimethylcyanamide, nitriles, aromatic nitro compounds, fluorene, dialkyl amides, and polyhalo aliphatics.

(*b*) *Non-cocatalytic polymerization.* To a long glass test tube containing 10.0 gm (0.11 mole) of anhydrous 2-pyrrolidinone is added 0.10 gm of sodium hydride. The resulting turbid solution is kept for approx. 18 hr at room temperature and then 0.8 gm (8%) of poly(γ-butyramide) is isolated, $\eta_{inh} =$ 0.20 (*m*-cresol).

2-5. *Preparation of Nylon 6 by the Anionic Polymerization of ε-Caprolactam* [21]

$$(CH_2)_5\!\left\langle\begin{matrix} C{=}O \\ | \\ NH \end{matrix}\right. \xrightarrow{Na} \left[-NH-(CH_2)_5-\overset{O}{\overset{\|}{C}}- \right]_n \qquad (11)$$

To a 2.2 liter stainless steel autoclave containing 1385 gm (12 moles) of anhydrous caprolactam (2-oxohexamethylenimine) is added 1.77 gm (0.77 gm atom) of sodium metal. The autoclave is sealed, purged with nitrogen, and heated at 90°–124°C for 15 min to form sodiocaprolactam. The temperature of the autoclave is raised to 250°–270°C and kept there for $1\frac{1}{2}$ hr. The resulting polymer, nylon 6 [poly(ε-caproamide)], is hard and light colored with an intrinsic viscosity of 0.78 (0.5% solution in *m*-cresol at 25°C).

2-6. *General Procedure for the Preparation of Nylon 6 by the Anionic Polymerization of Caprolactam in the Presence of N-Acylcaprolactam Cocatalysts* [31a]

A test tube containing 11.3 gm (0.1 mole) of caprolactam is placed in an oil bath at 150°–160°C and nitrogen is bubbled through the melt for $\frac{1}{2}$ hr to remove traces of water. Then 0.225 gm (0.0042 mole) of sodium methoxide is added and the nitrogen capillary is placed in the melt again for $\frac{1}{2}$ hr to remove the resulting methanol. Next 0.1–1.5% of the *N*-acylcaprolactam cocatalyst is added and the mixture kept in the heating bath until it solidifies. The

$$\underset{\text{NH}}{(\text{CH}_2)_5\overset{\text{C=O}}{|}} \xrightarrow[N\text{-acylcaprolactam}]{\text{NaOMe}} \left[-\text{HN}-(\text{CH}_2)_5-\overset{\text{O}}{\overset{\|}{\text{C}}}- \right]_n \qquad (12)$$

nylon 6 sample is ground up and extracted with chloroform to remove unreacted caprolactam. The vacuum-dried polymer is dissolved in sulfuric acid. Table VII gives the results of a series of preparations with various *N*-acylcaprolactam cocatalysts. The inherent viscosities were determined in a No. 75 viscometer.

2-7. *Preparation of Nylon 6 from Caprolactam Using Sodium Hydride and N-Acetylcaprolactam* [35]

$$\underset{\text{NH}}{(\text{CH}_2)_5\overset{\text{C=O}}{|}} \xrightarrow[N\text{-acetylcaprolactam}]{\text{NaH}} \left[-\text{HN}-(\text{CH}_2)_5-\overset{\text{O}}{\overset{\|}{\text{C}}}- \right]_n \qquad (13)$$

To a polymer tube containing 25.0 gm (0.22 mole) of caprolactam is added 0.60 gm (0.025 mole) of sodium hydride. After the tube is evacuated and filled with nitrogen several times the lactam is melted to allow the sodium hydride to react. When the hydrogen evolution ceases 0.33 gm (0.002 mole) of *N*-acetylcaprolactam is added. The tube is shaken and then placed in a heating bath set at 139°C. The contents quickly solidify, and after $\frac{1}{2}$ hr the tube is cooled, broken, and the polymer ground up. The polymer is extracted with hot water and dried to afford 18.9 gm (74.7%), $\eta_{inh} = 1.0$ (*m*-cresol) and $\eta_{rel} = 26.77$ (in formic acid). Experiments carried out in the absence of *N*-acetylcaprolactam gave no polymer below 150°C.

The experimental procedure for the cationic polymerization of caprolactam is similar except that *N*-acyl derivatives are not necessary. The use of HCl requires a polymerization temperature of 170°C and H_3PO_4 requires 185°C [37].

2-8. *Preparation of Nylon 6 by the Hydrolytic Polymerization of ε-Caprolactam* [21]

To a 2.2 liter stainless steel autoclave is added 937 gm (8.3 mole) of caprolactam, 600 gm (33.3 moles) of water, and 2.0 gm (0.03 mole) of glacial acetic

TABLE VII

PREPARATION OF NYLON 6 USING *N*-ACYLCAPROLACTAM COCATALYSTS IN THE ANIONIC POLYMERIZATION OF CAPROLACTAM IN PROCEDURE 2-6[a]

N-Acyl substituent (mole %)	Time for polymerization (sec)	% Yield	Inherent viscosity
$CH_3\overset{O}{\overset{\|}{C}}-$[b]			
(1.0)	138	100	0.74
(1.5)	70	85.5	0.69
$C_3H_7\overset{O}{\overset{\|}{C}}-$[c]			
(1.0)	338	90.0	0.61
(1.5)	161	97.4	0.84
$C_{17}H_{35}\overset{O}{\overset{\|}{C}}-$[d]			
(1.0)	—	4.5	0.25
(1.5)	—	4.3	0.32
$C_6H_5\overset{O}{\overset{\|}{C}}-$[e]			
(1.0)	625	81.6	0.65
(1.5)	144	100	0.65
$p\text{-}CH_3O-C_6H_4\overset{O}{\overset{\|}{C}}-$[f]			
(1.0)	1520	48.3	0.52
(1.5)	325	87.9	0.68

[a] Data taken from R. P. Scellia, S. E. Schonfeld, and L. G. Donaruma, *J. Polym. Sci.* **8**, 1363 (1964).

[b] B.p. 126°C (14 mm Hg).

[c] B.p. 100°C (0.1 mm Hg).

[d] B.p. 156°C (0.1 mm Hg), m.p. 37°–39°C.

[e] M.p. 69°C.

[f] M.p. 95°C.

acid. The reactor is purged with nitrogen and heated at 250°C under 250 psi pressure. The pressure is slowly released over a 30 min period and the polymerization is completed for 1 hr at atmospheric pressure to give 881 gm (94%).

$$(CH_2)_5\!\begin{array}{c}C{=}O\\ |\\ NH\end{array} \xrightarrow{H_2O,\ HOAc} \left[-NH-(CH_2)_5-\overset{O}{\overset{\|}{C}}-\right]_n \qquad (14)$$

Approx. 6% of the caprolactam is probably in the water phase. The polymer has an intrinsic viscosity of 0.74 (as determined in 0.5% solution in *m*-cresol at 25°C).

E. Low-Temperature Polycondensation of Diacid Chlorides with Diamines

The Schotten-Baumann [40] reaction (described in Volume 1 of *Organic Functional Group Preparations* [41]) can be applied to the preparation of polyamides using bifunctional reagents. Since the reaction is very rapid at room temperature it can be carried out at low temperatures (a) in solution or (b) by an interfacial polycondensation technique.

In the solution polycondensation method the reaction is carried out in a single inert liquid in the presence of an acid acceptor. The polymer may precipitate out of the solution or it may be soluble.

In the interfacial polycondensation method the reaction is carried out at the interface of two immiscible solvents. The amine is dissolved in water and the acid chloride is dissolved in the hydrocarbon layer. Earlier (1938) Carothers [7c] reported on the potential use of this reaction to prepare polyamides. However, it was not until the work of Magat and Strachan [42] and later Morgan and co-workers [43,43a–43c] that this reaction was exploited for the preparation of various polyamides.

The use of the low-temperature interfacial condensation technique to prepare polyamides and various other polymers has recently been reviewed in a monograph on this subject [44,44a].

Some of the important variables involved in interfacial polymerization are (a) organic solvent; (b) reactant concentration; and (c) use of added detergents [43].

The organic solvent is the most important variable since it controls partition and diffusion of the reactants between the two immiscible phases, the reaction rate, solubility, and swelling of permeability of the growing polymer. The solvent should be of such composition as to prevent precipitation of the polymer before a high molecular weight has been attained. The final polymer does not have to dissolve in the solvent. The type of solvent will also influence the characteristics of the physical state of the final polymer. Solvents such as chlorinated or aromatic hydrocarbons make useful solvents in this system.

Concentrations in the range of approximately 5% polymer based on the combined weights of water and organic solvent usually are optimum. Too low concentrations may lead to hydrolysis of the acid halide and too high concentrations may cause excessive swelling of the solvent in the polymer.

In some cases the addition of 0.2 to 1% of sodium lauryl sulfate has been found to give satisfactory results. In many cases it may not be necessary.

The reactants should be pure but need not be distilled prior to use. An exact reactant balance is not essential as in the melt polymerization process since the reactions are extremely rapid. A slight excess (5–10%) of the diamine usually helps to produce higher molecular weights.

The advantage of the interfacial polymerization process is that it is a low-temperature process requiring ordinary equipment. It also allows one to prepare those polyamides that are unstable in the melt polymerization process. Random or block polymers can easily be prepared depending on the reactivity of the reactants and their mixing (consecutively versus all at once.)

2-9. *Preparation of Poly(hexamethylenesebacamide) (Nylon 6–10) by the Interfacial Polymerization Technique* [43b]

$$(CH_2)_6(NH_2)_2 + (CH_2)_8(COCl)_2 \longrightarrow \left[-HN-(CH_2)_6-NH\overset{\overset{\displaystyle O}{\|}}{C}-(CH_2)_8-\overset{\overset{\displaystyle O}{\|}}{C}- \right]_n \quad (15)$$

To a tall-form beaker is added a solution of 3.0 ml (0.014 mole) of sebacoyl chloride dissolved in 100 ml of distilled tetrachloroethylene. Over this acid chloride solution is carefully poured a solution of 4.4 gm (0.038 mole) of hexamethylenediamine (see Note) dissolved in 50 ml of water. The polyamide film which begins to form at the interface of these two solutions is grasped with tweezers or a glass rod and slowly pulled out of the beaker in a continuous fashion. The process stops when one of the reactants becomes depleted. The resulting "rope"-like polymer is washed with 50% aqueous ethanol or acetone, dried, and weighed to afford 3.16–3.56 gm (80–90%) yields of polyamide, $\eta_{inh} = 0.4$ to 1.8 (*m*-cresol, 0.5% conc. at 25°C), m.p. 215°C (soluble in formic acid).

NOTE: In this experiment excess diamine is used to act as an acid acceptor.

2-10. *Preparation of Poly(hexamethyleneadipamide) (Nylon 6–6) by the Interfacial Polymerization Technique* [43c]

$$(CH_2)_6(NH_2)_2 + (CH_2)_4(COCl)_2 \longrightarrow \left[-HN-(CH_2)_6-NH-\overset{\overset{\displaystyle O}{\|}}{C}-(CH_2)_4-\overset{\overset{\displaystyle O}{\|}}{C}- \right]_n \quad (16)$$

In an ice-cooled blender jar containing a solution of 3.95 gm (0.034 mole) of hexamethylenediamine dissolved in 200 ml of water containing 3.93 gm (0.070 mole) of potassium hydroxide is added with agitation a solution of 6.22 gm (0.034 mole) of adipoyl chloride dissolved in 200 ml of xylene. The addition takes about 5 min and the speed of the blender agitation is slow at first and then speeded up toward the end of the addition period. The product

TABLE VIII

PREPARATION OF VARIOUS POLYAMIDES BY THE INTERFACIAL POLYMERIZATION TECHNIQUE

Diamine (moles) $R(NH_2)_2$ R =	or (R′ NH R′ NH ring) R′ =	Water (ml)	Diacid chloride $R''(COCl)_2$ (moles) R″ =	Organic solvent (ml)	Base (moles)	Temp. (°C)	Time (hr)	% Yield	η_{inh}	Ref.
$(CH_2)_6$ (0.038)	—	50	$—(CH_2)_8—$ (0.014)	Tetrachloro-ethylene (100)	—	25	½	80–90	0.4–1.8	*a*
(0.0374)	—	374	(0.0374)	CCl_4 (632)	NaOH (0.075)	25	½	81–90	1.8	*b*
(0.019)	—	188	(0.019)	CCl_4 (316)	NaOH (0.038)	25	½	80–90	2.06	*c*
(0.034)	—	200	$—(CH_2)_4—$ (0.034)	Xylene (200)	KOH (0.070)	0–10	¼	73	1.16	*c*
(0.05)	—	70	$COCl_2$ (0.05)	CCl_4 (200)	NaOH (0.1)	25	½	70	0.93	*c*
(0.05)	—	225(d)	*m*-C_6H_4— (0.05)	Toluene (50)	NaOH (0.105)	25	½	55	0.81	*c*
(0.0275)	—	210	—(S)— (0.025)	CH_2Cl_2 (125)	Na_2CO_3 (0.05)	25	½	78	1.05	*c*
(0.05)	—	150	$—CH_2—CH{=}CHCH_2—$ (0.05)	C_6H_6 (150)	Na_2CO_3 (0.1)	25	½	—	1.20	*c*

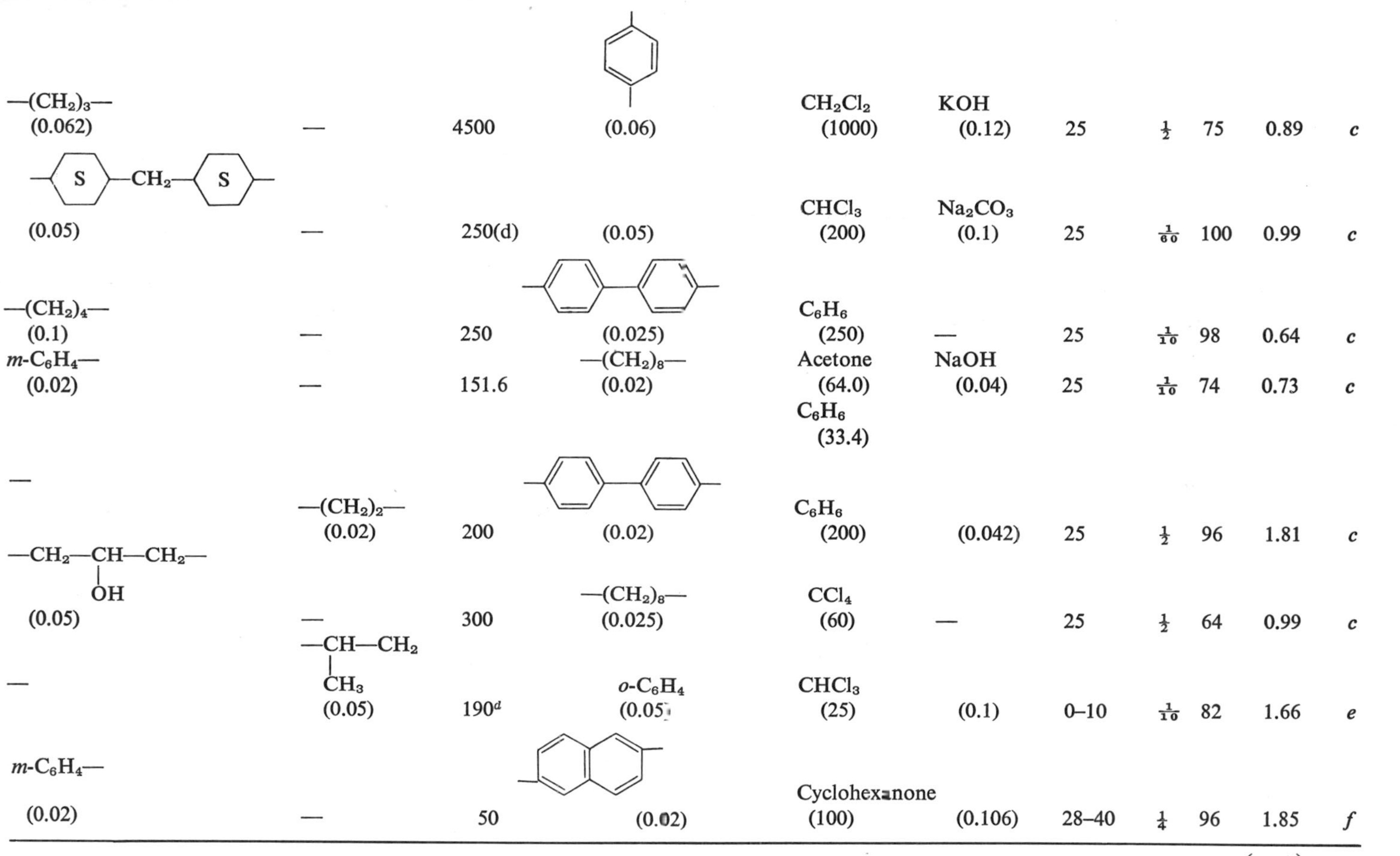

$—(CH_2)_3—$ (0.062)	—	4500	(0.06)	CH_2Cl_2 (1000)	KOH (0.12)	25	$\frac{1}{2}$	75	0.89	*c*
S —CH_2— S (0.05)	—	250(d)	(0.05)	$CHCl_3$ (200)	Na_2CO_3 (0.1)	25	$\frac{1}{60}$	100	0.99	*c*
$—(CH_2)_4—$ (0.1)	—	250	(0.025)	C_6H_6 (250)	—	25	$\frac{1}{10}$	98	0.64	*c*
m-C_6H_4— (0.02)	—	151.6	$—(CH_2)_8—$ (0.02)	Acetone (64.0) C_6H_6 (33.4)	NaOH (0.04)	25	$\frac{1}{10}$	74	0.73	*c*
—	$—(CH_2)_2—$ (0.02)	200	(0.02)	C_6H_6 (200)	(0.042)	25	$\frac{1}{2}$	96	1.81	*c*
$—CH_2—CH(OH)—CH_2—$ (0.05)	—	300	$—(CH_2)_8—$ (0.025)	CCl_4 (60)	—	25	$\frac{1}{2}$	64	0.99	*c*
—	$—CH(CH_3)—CH_2$ (0.05)	190[d]	*o*-C_6H_4 (0.05)	$CHCl_3$ (25)	(0.1)	0–10	$\frac{1}{10}$	82	1.66	*e*
m-C_6H_4— (0.02)	—	50	(0.02)	Cyclohexanone (100)	(0.106)	28–40	$\frac{1}{4}$	96	1.85	*f*

(*cont.*)

TABLE VIII (*cont.*)

Diamine (moles)											
$R(NH_2)_2$ R =	or	R′(NH)₂R′ ring, R′ =	Water (ml)	Diacid chloride $R''(COCl)_2$ (moles) R″ =	Organic solvent (ml)	Base (moles)	Temp. (°C)	Time (hr)	% Yield	η_{inh}	Ref.
(0.04)		—	100	$-C_{10}H_6-$ (0.032) + m-C_6H_4 (0.008)	Cyclohexanone (300)	(0.22)	25–40	$\frac{1}{4}$	92	1.40	*f*
—		$-CH(CH_3)-CH_2-$ (0.0129)	200	$-C_6H_4-SO_2-C_6H_4-$ (0.0125)	CH_2Cl_2 (50)	Na_2CO_3 (0.025)	25	$\frac{1}{10}$	98	3.13	*g*

[a] P. W. Morgan and S. L. Kwolek, *J. Chem. Educ.* **36**, 182 (1959).
[b] P. W. Morgan and S. L. Kwolek, *J. Polym. Sci.* **40**, 299 (1959).
[c] R. G. Beaman, P. W. Morgan, C. R. Koller, and E. L. Wittbecker, *J. Polym. Sci.* **40**, 329 (1959).
[d] Contains 0.5 gm sodium lauryl sulfate.
[e] M. Katz, *J. Polym. Sci.* **40**, 337 (1959).
[f] L. Starr, *J. Polym. Sci.* **4**, 3041 (1966).
[g] C. W. Stephens, *J. Polym. Sci.* **40**, 359 (1959).

is filtered, washed with water, and dried to afford 5.6 gm (73%), $\eta_{inh} = 1.16$.

In Table VIII the preparations of a number of different polyamides by the interfacial polymerization technique are summarized.

2-11. Preparation of Poly(N,N′-terephthaloyl-trans-2,5-dimethylpiperazine) by the Solution Polymerization Technique [44]

$$\text{HN}(\text{CH}_3)\text{C}_4\text{H}_6(\text{CH}_3)\text{NH} + \text{ClC(=O)}-\text{C}_6\text{H}_4-\text{C(=O)Cl} \xrightarrow{2\text{Et}_3\text{N}} \left[-\text{N}(\text{CH}_3)\text{C}_4\text{H}_6(\text{CH}_3)\text{N}-\text{C(=O)}-\text{C}_6\text{H}_4-\text{C(=O)}- \right]_n \tag{17}$$

To a 500 ml Erlenmeyer flask equipped with a magnetic stirrer is added 2.28 gm (0.019 mole) of *trans*-2,5-dimethylpiperazine and 5.6 ml (0.04 mole) of triethylamine dissolved in 100 ml of chloroform. To this cooled (10°C) solution is added rapidly with stirring a solution of 4.06 gm (0.02 mole) of terephthaloyl chloride in 80 ml of chloroform. Approx. 20 ml of additional chloroform is used to immediately rinse the residues of terephthaloyl chloride. The mixture remains clear and the temperature rises from 10° to 25°C. At this point the solution increases in viscosity and after 5 min is coagulated by pouring the solution with stirring into hexane. The fibrous precipitate and triethylamine hydrochloride are filtered, washed with water and acetone, and then dried at 100°C to afford 4.3 gm (92%), $\eta_{inh} = 3.1$ (in *m*-cresol at 30°C).

3. MISCELLANEOUS METHODS

1. Condensation of tripeptides with tetramethyl pyrophosphite [45].
2. Condensation of tripeptides by means of dicyclohexylcarbodiimide [46].
3. Preparation of ferrocene polyamides by the condensation of 1,1′-dichlorocarbonylferrocene with primary diamines in xylene–pyridine solvent [47].
4. Polyamides via the Ritter reaction [48].
5. Preparation of polyamides containing *N,N′*-dipiperazyl groups [49].
6. Preparation of redox polyamides by condensing a hydroquinone bislactone of 1,4-dihydroxybenzene-2,5-bis(ethyl-2′-carboxylic acid) with diamines [50].
7. Preparation of polyamides by the condensation of bissuccinimides with diamines [51].

8. Preparation of polyamides by the condensation of bisimidazoline with dicaroboxylic acids [52].

9. Preparation of fluorinated *N*-alkyl-substituted polyamides [53].

10. Preparation of aromatic polyketones and aromatic polyamides [54].

11. Condensation of aminophenols with diacid chlorides to give polyamide-esters [55].

12. Preparation of poly(amide-acetals) [56].

13. Preparation of polyamides from neocarboranedicarboxylic dichloride [57].

14. Ring-opening polymerization of *N*-carboxyanhydrides [58].

15. Addition of amines to activated double bonds [59].

16. Reaction of formaldehyde with dinitriles [60].

17. Polymerization of isocyanates [61].

18. Reaction between diazlactone with diamines [62].

19. Aromatic polyamides [63].

20. Polyamide-imides [64].

21. Photodegradable polyamides [65].

22. Poly-*p*-benzamide and the polyterephthalamide of *p*-phenylenediamine and other aromatic polyamides [66,67].

23. Polyamide-hydrazides and other aromatic polyamides [67].

24. Preparation and polymerization of aminimides [68].

REFERENCES

1. L. Balbiano and D. Trasciatti, *Ber. Deut. Chem. Ges.* **33**, 2323 (1900).
2. L. C. Maillard, *Ann. Chim. Anal. Chim. Appl.* [9] **1**, 519 (1914); [9] **2**, 210 (1914).
3. A. Manasse, *Ber. Deut. Chem. Ges.* **35**, 1367 (1902).
4. I. Curtius, *Ber. Deut. Chem. Ges.* **37**, 1284 (1904).
5. M. Frankel and A. Katchalsky, *Nature* **144**, 330 (1939).

6a. E. Fischer and H. Koch, *Justus Liebigs Ann. Chem.* **232**, 227 (1886).
6b. A. W. Hofmann, *Ber. Deut. Chem. Ges.* **5**, 247 (1872).
6c. M. Freund, *Ber. Deut. Chem. Ges.* **17**, 137 (1884).
6d. F. Anderlini, *Gazz. Chim. Ital.* **24** (1) 397 (1894).
6e. E. Fischer, *Ber. Deut. Chem. Ges.* **46**, 2504 (1913).
6f. H. Meyer, *Justus Liebigs Ann. Chem.* **347**, 17 (1906).
6g. P. Ruggli, *Justus Liebigs Ann. Chem.* **392**, 92 (1912).
6h. C. L. Butler and R. Adams, *J. Amer. Chem. Soc.* **47**, 2614 (1925).
7a. W. H. Carothers and G. J. Berchet, *J. Amer. Chem. Soc.*, **52**, 5289 (1930).
7b. W. H. Carothers and J. W. Hill, *J. Amer. Chem. Soc.* **54**, 1566 (1932).
7c. W. H. Carothers, U.S. Patent 2,071,250 (1937).
7d. W. H. Carothers, U.S. Patent 2,130,523 (1938).
7e. W. H. Carothers, U.S. Patent 2,130,947 (1938).
7f. W. H. Carothers, U.S. Patent 2,130,948 (1938).
7g. H. K. Livingston, M. S. Sioshansi, and M. D. Glick, *J. Macromol. Sci.—Revs. Macromol. Chem.* **6** (1), **29** (1971).

8a. O. E. Snider and R. J. Richardson, *Encycl. Polym. Sci. Technol.* **10**, 347 (1969).
8b. W. Sweeny and J. Zimmerman, *Encycl. Polym. Sci. Technol.* **10**, 483 (1961).
8c. D. D. Coffman, G. J. Berchet, W. R. Peterson, and E. W. Spanagel, *J. Polym. Sci.* **2**, 306 (1947).
8d. E. C. Schule, *Encycl. Polym. Sci. Technol.* **10**, 460 (1969).
9. Imperial Chemical Industries, Ltd., British Patent 562,370 (1944).
10. W. J. Peppel, *J. Polym. Sci.* **51**, S64 (1961); K. Sastome and K. Sato, *ibid.* **4**, 1313 (1955).
11. G. H. L. Nefkens, G. I. Tesser, and J. R. F. Nivard, *Rec. Trav. Chim. Pays. Bas* **81**, 683 (1962); M. W. Williams and G. T. Young, *J. Chem. Soc., London* p. 881 (1963); G. W. Anderson, J. E. Zimmerman, and F. M. Calahan, *J. Amer. Chem. Soc.* **86**, 1839 (1964); G. W. Anderson, F. M. Callahan, and J. E. Zimmerman, *ibid.* **89**, 178 (1967); F. Weygand, A. Pox, and W. Konig, *Chem. Ber.* **99**, 1451 (1966).
12. S. B. Speck, *J. Amer. Chem. Soc.* **74**, 2876 (1952).
13. M. Bodanszky and V. du Vigneaud, *J. Amer. Chem. Soc.* **81**, 5688 (1959).
14. D. F. DeTar, W. Honsberg, U. Honsberg, A. Wieland, M. Gouge, H. Bach, A. Tahara, W. S. Bringer, and F. F. Rogers, Jr., *J. Amer. Chem. Soc.* **85**, 2873 (1963); S. M. Bloom, S. K. Dasgupta, R. P. Patel, and E. R. Blont, *ibid.* **88**, 2035 (1966); D. F. DeTar and N. F. Estrin, *Tetrahedron Lett.* p. 5985 (1966); D. F. DeTar, M. Gouge, W. Honsberg, and U. Honsberg, *J. Amer. Chem. Soc.* **89**, 988 (1967); D. F. DeTar and T. Vajda, *ibid.*, p. 998; J. Kovacs and A. Kapoor, *ibid.* **87**, 118 (1965); J. Kovacs, R. Gianotti, and A. Kapoor, *ibid.* **88**, 2282 (1966).
15. C. G. Overberger and J. Sebenda, *J. Polym. Sci., Part A-1* **7**, 2875 (1969).
16. R. Aelion, *Ind. Eng. Chem.* **53**, 826 (1961).
17. C. F. Horn, B. T. Freure, H. Vineyard, and H. J. Decker, *J. Polym. Sci.* **7**, 887 (1963).
18. K. Sanui, T. Asahara, and N. Ogata, *J. Polym. Sci.* **6**, 1195 (1968); K. Sanui and N. Ogata, *J. Polym. Sci., Part A-1* **7**, 889 (1969).
19. P. A. Small, *Trans. Faraday Soc.* **51**, 1717 (1955); F. S. Dainton and R. J. Ivin, *Quart. Rev., Chem. Soc.* **12**, 82 (1958).
20. P. Schlack, U.S. Patent 2,241,321 (1941).
21. W. E. Hanford and R. M. Joyce, *J. Polym. Sci.* **3**, 167 (1948).
22. W. H. Carothers, *J. Amer. Chem. Soc.* **52**, 5289 (1938).
23. P. H. Hermans, D. Heikens, and P. F. Van Velden, *J. Polym. Sci.* **30**, 81 (1958).
24. R. M. Joyce and D. M. Ritter, U.S. Patent 2,251,519 (1941).
25. J. Kralicek and J. Sebenda, *J. Polym. Sci.* **30**, 493 (1958).
26. O. Wichterle, J. Kralicek, and J. Sebenda, *Collect. Czech. Chem. Commun.* **24**, 755 (1959).
27. W. Voss, *Chem. Tech. (Leipzig)* **1**, 111 (1949).
28. W. Voss, British Patent 538,619 (1941).
29. W. Griehl, *Faserforsch. Textiltech.* **6**, 260 (1955); **7**, 207 (1956).
30. M. Ito, Japanese Patent 18590 (1969); O. Fujumoto, Japanese Patent 13794 (1962).
31. W. O. Ney, Jr. and W. Crowther, U.S. Patent 2,739,959 (1956); S. Bar-Zakay, M. Levi, and D. Vofsi, *J. Polym. Sci.* **5**, 965 (1967).
31a. R. P. Scelia, S. E. Schonfeld, and L. G. Donaruma, *J. Appl. Polym. Sci.* **8**, 1363 (1964).
32. W. O. Ney, Jr., W. R. Nummy, and C. E. Barnes, U.S. Patent 2,638,463 (1951); R. Graf, G. Lohans, K. Börner, E. Schmidt, and H. Bestian, *Angew. Chem., Int. Ed. Engl.* **1**, 481 (1962).
32a. O. Wichterle, J. Sebenda, and J. Kralicek, *Fortschr. Hochpolym.-Forsch.* **2**, 578 (1961).

32b. N. Yoda and W. Miyake, *J. Polym. Sci.* **43**, 117 (1960).
33. H. Tani and T. Konomi, *J. Polym. Sci., Part A* **4**, 301 (1966); T. Konomi and H. Tani, *ibid.* **6**, 2295 (1968).
34. T. Konomi and H. Tani, *J. Polym. Sci., Part A-1* **7**, 2269 (1969).
35. H. K. Hall, Jr., *J. Amer. Chem. Soc.* **80**, 6404 (1958).
36. G. M. van der Want and C. A. Kruissink, *J. Polym. Sci.* **35**, 119 (1959).
37. M. Rothe, G. Reinisch, W. Jaeger, and I. Schopov, *Makromol. Chem.* **54**, 183 (1962).
38. N. Ogata, *J. Polym. Sci., Part A* **1**, 3151 (1963).
39. R. C. P. Cubbon, *Polymer* **4**, 545 (1963).
40. C. Schotten, *Ber. Deut. Chem. Ges.* **15**, 1947 (1882); C. Schotten and J. Baum, *ibid.* **17**, 2548 (1884); C. Schotten, *ibid.* p. 2545; **21**, 2238 (1888); **23**, 3430 (1890); E. Baumann, *ibid.* **19**, 3218 (1886).
41. S. R. Sandler and W. Karo, "Organic Functional Group Preparations," Vol. 1, pp. 279–281, Academic Press, New York, 1968.
42. E. Magat and D. R. Strachan, U.S. Patent 2,708,617 (1955).
43. E. L. Wittbecker and P. W. Morgan, *J. Polym. Sci.* **40**, 289 (1959).
43a. P. W. Morgan and S. L. Kwolek, *J. Polym. Sci.* **40**, 299 (1959); P. W. Morgan, *SPE* (*Suc. Plast. Eng.*) *J.* **15**, 485 (1959).
43b. P. W. Morgan and S. L. Kwolek, *J. Chem. Educ.* **36**, 182 (1959).
43c. R. G. Beaman, P. W. Morgan, C. R. Koller, and E. L. Wittbecker, *J. Polym. Sci.* **40**, 329 (1959).
44. P. W. Morgan, "Condensation Polymers: By Interfacial and Solution Methods." Wiley (Interscience), New York, 1965.
44a. P. W. Morgan and S. L. Kwolek, *J. Polym. Sci.* **2**, 181 (1964).
45. N. S. Andreeva, V. A. Debabov, M. I. Millionova, V. A. Shibnev, and Y. N. Chirgadze, *Biofizika* **6**, 244 (1961).
46. H. G. Khorana, *Chem. Ind.* (*London*) p. 1087 (1955).
47. C. V. Pittman, Jr., *J. Polym. Sci.* **6**, 1687 (1968).
48. F. L. Ramp, *J. Polym. Sci., Part A* **3**, 1877 (1968).
49. M. Draivert, C. Burba, and E. Griebseh, U.S. Patent 3,565,837 (1971).
50. N. Nakabayashi and H. G. Cassidy, *J. Polym. Sci., Part A-1* **7**, 1275 (1969).
51. T. Kagiya, M. Izu, T. Matsuda, and K. Fukui, *J. Polym. Sci.* **5**, 15 (1967).
52. T. Kagiya, M. Izu, M. Hatta, T. Matsuda, and K. Fukui, *J. Polym. Sci., Part A-1* **5**, 1129 (1967).
53. B. S. Marks and G. C. Schweiker, *J. Polym. Sci.* **42**, 229 (1960).
54. Y. Iwakura, K. Uno, and T. Takiguchi, *J. Polym. Sci., Part A-1* **6**, 3345 (1968).
55. J. Preston, *J. Polym. Sci., Part A-1* **8**, 3135 (1970).
56. E. H. Pryde, D. J. Moore, H. M. Teeter, and J. C. Cowan, *J. Polym. Sci.* **58**, 611 (1962).
57. V. V. Korshak, N. I. Bekasova, and L. G. Komarova, *J. Polym. Sci., Part A-1* **8**, 2351 (1970).
58. D. Coleman and A. C. Farthing, *J. Chem. Soc. London*, p. 3213 (1950).
59. A. S. Matlack, U.S. Patent 2,672,480 (1954).
60. A. Cannepin, G. Champetier, and A. Pansat, *J. Polym. Sci.* **8**, 35 (1952); E. E. Magat, L. B. Chandler, B. F. Faris, J. E. Reith, and L. F. Salisbury, *J. Amer. Chem. Soc.* **73**, 1031 (1951).
61. V. E. Shashoua, W. Sweeny, and R. F. Tietz, *J. Amer. Chem. Soc.* **82**, 866 (1960).
62. C. S. Cleaver and B. C. Pratt, *J. Amer. Chem. Soc.* **77**, 1541 (1955).
63. S. Hara, M. Seo, T. Yoshida, and M. Uchida, Japanese Patent 71/27,819 (1971); N. Yoda, M. Kurihara, I. Kojuro, S. Toyama, and R. Nakanishi, Japanese Patent 68/28,835 (1968).

64. W. M. Alvino and L. W. Frost, *J. Polym. Sci., Part A-1* **9**, 2209 (1971).
65. H. Takahashi, M. Sakuragi, M. Hasegawa, and H. Takahashi, *J. Polym. Sci., Part A-1* **10**, 1399 (1972).
66. H. W. Hill, Jr., S. L. Kwolek, and W. Sweeny, U.S. Patent 3,094,511 (1963); S. L. Kwolek, P. W. Morgan, and W. R. Sorenson, U.S. Patent 3,063,966 (1962); L. F. Beste and C. W. Stephens, U.S. Patent 3,068,188 (1962); T. F. Bair and P. W. Morgan, German Patent 1,816,106 (1969); D. C. Pease, U.S. Patent 3,197,443 (1965); W. Sweeny, U.S. Patent 3,287,324 (1966).
67. *Chem. & Eng. News* **50** (16), 33 (1972); W. B. Black, *Trans. N.Y. Acad. Sci.* [2] **32**, 765 (1970); W. B. Black and J. Preston, *Man-Made Fibers* **2**, 297 (1968).
68. W. J. Mckillip, E. A. Sedor, B. M. Culbertson, and S. Wawzonek, *Chem. Rev.* **73**, 255 (1973).

Chapter 5

POLYMERIZATION OF ALDEHYDES

I. INTRODUCTION

In the polymerization of aldehydes the reaction shown in Eq. (1) occurs to give acetal resins. These polymers may be considered polyethers or polyacetals. This chapter will not consider in detail polyacetals prepared by the reaction of aldehydes with polyols as described in Eq. (2).

$$n\mathrm{RCH{=}O} \longrightarrow \left[\begin{matrix} \mathrm{R} \\ | \\ \mathrm{-CH-O-} \end{matrix} \right]_n \qquad (1)$$

$$n\mathrm{RCH{=}O} + n\mathrm{HOROH} \xrightarrow{n\mathrm{H_2O}} \left[\begin{matrix} \mathrm{R} \\ | \\ \mathrm{-ROCH-O-} \end{matrix} \right]_n \qquad (2)$$

The first polymerization of an aldehyde to be reported was that of formaldehyde in 1859 [1]. Butlerov [1] erroneously designated the structure of this material as a dimer, $(CH_2O)_2$. Its structure was later investigated by Hofmann [2], Tollens and Mayer [3], Lösekann [4], and Delepine [5]. Staudinger [6] confirmed Delepine's suggestion that these materials are high molecular weight substances. Staudinger and co-workers [7] also succeeded in preparing high molecular weight polymers of formaldehyde and also found that end-capping helped to prevent hemiacetal degradation. These polymers are now known as polyoxymethylenes. In 1942 E. I. du Pont de Nemours & Co. [8,8a,8b] was awarded the first of many patents describing high molecular weight polyoxymethylenes derived by the homopolymerization of formaldehyde, polymers which are now known under the trade name Delrin [9]. Celanese [10a–10c] is marketing a polyacetal copolymer under the trade name Celcon; it is prepared by the copolymerization of trioxane [10,10a–10c] and ethylene oxide.

These acetal resins are in the category of engineering plastics since they have excellent mechanical, chemical, and electrical properties [11]. The acetal resins find use in automotive components, plumbing hardware (sheets, rods, and tubes), electrical components, appliances, consumer products (aerosol containers), and machinery parts (glass bearings). The acetal resins compete for end use application with metals, polyamides (nylon), polycarbonates, acrylonitrile butadiene styrene (ABS), polysulfones, polyphenylene oxides, and polyimides.

The resins are used predominantly for injection molding and extrusion operations. The resins are also available in grades useful for preparing glass or fluorocarbon-reinforced fibers.

United States consumption of acetal resins is estimated to reach 98 million pounds by 1975, up from 10 million pounds in 1961 [12].

Other polyaldehydes have also been prepared from acetaldehyde [13,13a], butyraldehyde [14,14a], chloral [15,15a–15c], monochloroacetaldehyde [15c,16], dichloroacetaldehyde [15c,17], and other aldehydes [18,19].

Recently chloral has been reported to be polymerized by a monomer casting method using an anionic initiator to give polychloral [15]. The polymer does not support combustion in oxygen and does not melt or drip but decomposes. Polychloral degrades at 200°C and gives mainly the monomer when this degradation is carried out under reduced pressure.

The use of high-resolution NMR spectroscopy for the analysis of the structure of polyaldehydes has been reported to be able to give a quantitative measure of the tacticity components [20].

For additional background material up to about 1965 one should consult the available reviews [21] and texts [19,22]. This chapter will cover the literature up to 1973.

2. POLYMERIZATION OF FORMALDEHYDE TO POLYOXYMETHYLENE

A. Monomeric Formaldehyde. Properties and Generation

Monomeric formaldehyde is a colorless gas that condenses on cooling to give a liquid with a b.p. −19°C and m.p. −118°C [23]. The density of liquid formaldehyde at −80°C is 0.9151 and at −20°C it is 0.8153 [24].

The liquid and gas polymerize readily and can be kept in the monomeric state for only a limited time. Formaldehyde can be dissolved in water to give a stable 37% solution (stabilized by methanol or other additives) or polymerized to paraformaldehyde prior to shipment.

Formaldehyde gas that is 90–100% pure is stable at 100°–150°C. Formaldehyde gas does not decompose at temperatures below 400°C [25].

CAUTION: Formaldehyde gas is flammable and forms explosive mixtures with air or oxygen. Formaldehyde–air mixtures in the range of 7–72% formaldehyde are potentially explosive [26].

Formaldehyde monomer can be generated either by heating a mineral oil (b.p. 350°C) slurry containing 10–40% dry paraformaldehyde to 115°–140°C or by adding paraformaldehyde–di(2-ethylhexyl) phthalate to a heated generator [27]. Dry monomer is obtained from the pure anhydrous liquid formaldehyde.

Monomeric formaldehyde can also be obtained by the thermal decomposition of cyclohexyl hemiformal [28], α-polyoxymethylene [8,8a], and trioxane [29].

Liquid formaldehyde is readily prepared by vaporizing alkali-precipitated

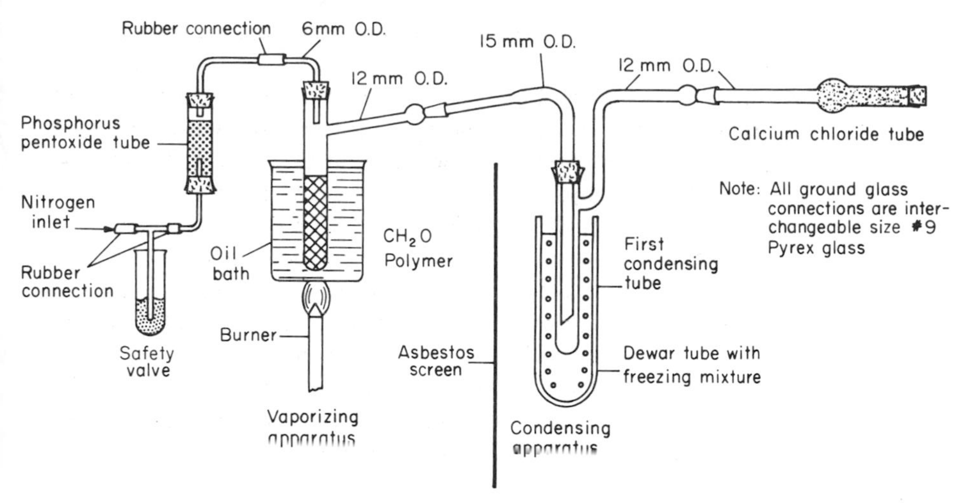

FIG. 1. Apparatus for the preparation of liquid formaldehyde. [Reprinted from Walker [30]. Copyright 1933 by the American Chemical Society. Reprinted by permission of the copyright owner.]

α-polyoxymethylene, condensing the vapors, and then redistilling. A typical set-up is shown in Fig. 1. α-Polyoxymethylene usually contains only 0.1% combined water, whereas paraformaldchyde contains 2–4% water even after drying.

Walker [30] prepared liquid formaldehyde in the apparatus shown in Fig. 1. Alkali-precipitated α-polyoxymethylene [31] was vaporized by heating to 150°–200°C while a slow current of nitrogen was swept over its surface. Thc tubes leading from the vaporizer to the condensing tube were kept hot by the use of resistance wires wound around these areas. Approximately 10 gm of pure liquid formaldehyde is obtained from 20 gm of starting polymer.

The liquid formaldehyde polymerizes in ether at 25°C or at 0°C to give a polymer, m.p. 170°–172°C. The starting α-polyoxymethylene melts at 172°–173°C and paraformaldehyde melts at 121°–123°C.

B. Polymerization of Formaldehyde in Aqueous Solutions

Polyoxymethylene glycols such as paraformaldehyde are obtained from aqueous formaldehyde solutions by either atmospheric [3,32] or vacuum concentration [33,33a]. Several such processes have been patented [33,33a,34].

$$n\mathrm{CH_2{=}O} \xrightarrow{\mathrm{H_2O}} \mathrm{HO(CH_2O)}_n\mathrm{H} \tag{3}$$

where n = less than 8 = lower polyoxymethylene glycols
n = 8–100 = paraformaldehyde

A German process azeotropically removes water from aqueous solutions to give paraformaldehyde [35].

α-Polyoxymethylene is obtained by the addition of acids [36] or bases [31] to aqueous formaldehyde solution. Staudinger, Singer, and Schweitzer [36] obtained this polymer in 58% yield by the addition of 1.0 mole KOH to a methanol-free 40% formaldehyde solution (100 moles $CH_2{=}O$).

Other polyoxymethylenes such as β [37] and γ [38] have been reported to be isolated by the addition of sulfuric acid to formaldehyde solutions containing methanol. The γ structure is thought to be polyoxymethylene dimethyl ether. A δ-polyoxymethylene is reported to be produced by treatment of γ polymer in boiling water [39]. ε-Polyoxymethylene is made by repeated sublimation of trioxane to obtain a small amount of a nonvolatile residue [40]. The structure of all these forms [41] is still in question and further research is required to clarify the present situation.

2-1. Preparation of Paraformaldehyde [42]

$$n\,HOCH_2OH(H_2O) \xrightarrow{(n-1)H_2O} HO(\text{—}CH_2O\text{—})_nH \qquad (4)$$

$$\text{where } n = <200$$

Method A: Evaporation at room temperature of 37% formaldehyde. On a large plate is placed 100 gm of 37% formaldehyde (inhibited with 10–15% methanol) and it is allowed to evaporate slowly at room temperature to afford 14.0 gm (40%), m.p. 94°–95°C.

Method B: Azeotropic removal of water. To a flask equipped with a Dean and Stark trap is added 100 gm of 37% formaldehyde solution (inhibited with 10–15% methanol) and 150 ml of *n*-hexane. The contents are heated to reflux and the water collected in the Dean and Stark trap. The product precipitates out as a white solid, which after drying affords 22 gm (60%), m.p. 125°–128°C. The yield is only 60% because formaldehyde vapors are lost up the condenser. Using heptane gives similar results but benzene gives somewhat lower yields of paraformaldehyde.

C. Polymerization of Anhydrous Gaseous Formaldehyde

Hofmann first reported that gaseous anhydrous formaldehyde polymerizes slowly to a solid polymer [43]. The rate follows a unimolecular rate equation as determined by Trautz and Ufer [44] by studying pressure changes. Spence [45] reported that deviations from unimolecularity occur later in the polymerization due to the fact that a polymer film covers the surface of the reaction. Carrothers and Norrish [46] studied the reaction kinetics in detail using formic acid catalyst at 100°C. Bevington and Norrish [47] reported that

HCl, $SnCl_4$, and BF_3 are more powerful catalysts than formic acid but also give rise to some chain-branching reactions.

D. Polymerization in Bulk and Solution of Monomeric Liquid or Gaseous Formaldehyde

Kekulé [24] in 1892 reported that anhydrous liquid formaldehyde undergoes polymerization. Several years later Staudinger and co-workers [48,49] reported that liquid formaldehyde gels when kept at $-80°C$ for 1 hr and is converted to a solid in 24 hr. This polymer is of a higher molecular weight than the one produced from aqueous solution and possesses film- and fiber-forming properties. The polymer decomposes at 180°–185°C to give free formaldehyde without even melting. Staudinger noted that bulk polymerization is very rapid and almost explosive at elevated temperatures and gives a lower molecular weight polymer. Solution polymerization of anhydrous formaldehyde can be carried out with or without catalysts in an inert solvent at low temperatures.

Acidic (BCl_3) [48,49] and basic $(CH_3)_3N$ [48,49], n-$C_4H_9NH_2$ [50], $C_2H_5—NH_2$ [50] catalysts greatly accelerate the polymerization of gaseous formaldehyde dissolved in inert solvents. For example, 1 part of n-$C_4H_9NH_2$ is effective per 60,000 parts of $CH_2═O$. Many of the catalysts used are described in the patent literature. Some typical ones are triorgano phosphines, stibines, and arsines [51]; sulfur compounds [52]; group VIII metal carbonyls [53]; onium salts such as quaternary phosphonium or tertiary sulfonium halides [54]; tertiary amine polymers [55]; and tertiary amine salts of Friedel-Crafts metal halides [56]. Several other catalysts are described in detail in a recent paper by Brown [57]. Metals, metal oxides, hydroxides, hydrated metal oxides, or derivatives are also effective catalysts for polymerizing monomeric gaseous formaldehyde at $-80°C$ to $+70°C$ [58]. Metals and aluminum alloy catalysts have also been reported to be effective for the polymerization of gaseous formaldehyde monomer [59].

Staudinger was able to prepare films and fibers from these polyoxymethylenes (Eu-polyoxymethylene) but their thermal stability was not good enough for practical applications.

A redox catalyst has recently been reported to be effective in polymerizing anhydrous formaldehyde [60]. The catalyst comprises a redox system consisting of an organic peroxide and a reducing agent such as a metal compound or an amine. The mechanism of polymerization using this catalyst may not be free radical but still can be ionic.

Machaecek and co-workers [61] studied the polymerization of anhydrous formaldehyde with dibutylamine catalyst in diethyl ether at $-58°C$. The rate of polymerization increased with an increase in dibutylamine concentration. Water did not affect the rate but lowered the molecular weight of the polymer.

The end groups and molecular weight of the polymer have been studied by examining an acetylated polymer. The end groups contained acetate and methoxy groups. The latter probably is derived from methanol impurities in the formaldehyde monomer. These results also indicate that water and methanol induce the polymerization and thus give rise to the presence of these groups. Water and methanol also function as chain-transfer agents in the polymerization process. Gander [62] has described methods of controlling molecular weights of the polymers with various chain-transfer agents (aliphatic alcohols, carboxylic acids, esters, amides, acid anhydrides, phenols, and mercaptans).

The blocking or end-capping of terminal groups in polyoxymethylene has been reported to be affected by a variety of techniques [63–67].

Polyoxymethylenes undergo autoxidative degradation at about 160°C even when the ends are capped with methoxy or acetoxy groups. This can be prevented by the addition to the polymer of antioxidants such as secondary or tertiary aryl amines [68], phenols [69], urea [70], benzophenone derivatives [71], thiazoles [72], manganese and copper salts [73], sulfides [74], polyureas [75], *N*-vinyllactam polymers [76].

Polymerization of pure liquid formaldehyde is rarely used commercially because of the problems of temperature control. Usually formaldehyde monomer is polymerized in solution. The monomer which is obtained by pyrolysis of α-polyoxymethylene, a hemiformal or trioxane [77], is passed through several traps at −15°C before being passed into the inert liquid medium containing the catalyst.

2-2. *Preparation of Polyoxymethylene by the Polymerization of Anhydrous Formaldehyde Monomer in an Inert Medium* [8a,8b]

$$\underset{\alpha\text{-Polyoxymethylene}}{(CH_2O)_x} \longrightarrow CH_2{=}O \xrightarrow{-30^\circ C} \underset{\substack{\text{High molecular weight}\\\text{polyoxymethylene}}}{[-CH_2O-]_n} \qquad (5)$$

To a pyrolysis flask containing nitrogen inlet and outlet connections is added 100 gm (3.34 mole equiv.) of anhydrous α-polyoxymethylene (see Note). The polymer is pyrolyzed over a period of 100 min at 150°–190°C and the monomeric formaldehyde that forms is continuously removed with a slow stream of nitrogen. The formaldehyde monomer is passed through a bubbler to two or three traps at −15°C and then into a rapidly agitated mixture of 626 gm of pentane containing 0.15 gm of tri-*n*-butylamine and 0.1 gm of diphenylamine held at 25°C. It is advisable to also have a bubbler after the reaction flask. After 4 hr 12–60 gm (12–60%) of a white granular formaldehyde polymer is obtained, $\eta_{inh} = 2.2$ (0.5% in *p*-chlorophenol containing 2% α-pinene as a stabilizer at 60°C). The polymer degrades at 0.9% by weight per minute at 222°C. The polymer can be formed into a tough, translucent 3–7 mil thick film by pressing at 190°–240°C and 2000 lbs/in^2, m.p. 178°C.

NOTE: α-Polyoxymethylene is obtained by the alkaline precipitation of aqueous formaldehyde solutions, washed with water and acetone, and dried for several hours in a vacuum oven at 160°–180°C [31]. The aqueous formaldehyde can be generated by adding 225 gm of purified paraformaldehyde to 350 ml water at 90°C. A few drops of 20% NaOH is added to the hot (90°C) solution to adjust the pH to 7.0. The fresh formaldehyde solution is filtered and cooled and then 18.6 gm of 50% sodium hydroxide is added dropwise at 40°C over a 2 hr period to precipitate the α-polyoxymethylene. The solution is stirred for 24 hr at 40°C to precipitate more α-polyoxymethylene and is worked up as described above to give about 80 gm of product.

Traces of water should be absent from the α-polyoxymethylene but small amounts will be trapped in the Dry Ice trap prior to entry into the reaction vessel. Traces of water also cause the polymer to clog the connecting tubes and therefore a bubbler before the traps at −15°C and a bubbler after the reaction flask (for N_2) is essential.

Other examples of polyoxymethylenes prepared by a similar procedure are shown in Table I.

The polymerization of gaseous formaldehyde monomer by various onium catalysts is described in Table II.

Recently Goidea described a continuous process for the polymerization of formaldehyde monomer using diethylamine or tributylamine catalysts [78].

2-3. *Preparation of High Molecular Weight Polyoxymethylene from a Methanolic Solution of Formaldehyde* [57,79]

$$n\mathrm{CH_3OH} + n\mathrm{CH_2{=}O} \longrightarrow \mathrm{CH_3O[CH_2O]_nH} \qquad (6)$$

To a flask is added 953 gm of a 72% formaldehyde (see Note) (22.9 moles) in methanol by weight sample, 29.0 gm (0.257 mole) of *N*-ethylpiperidine, and 23.6 gm (0.287 mole) of 85% phosphoric acid. The solution should have pH 5.2 to 5.3. A 20 ml sample of this solution is kept at 60°C for 5 min and then heated for 5 min at 85°C to give a polymeric precipitate. The solution is next heated to 105°C and 400 ml of the remaining clear methanolic formaldehyde solution is added. The reaction mixture is stirred well and kept at 105°C for 3 days while making the following periodic checks and additions to reactants:

1st Day—Mol. wt. of polymer is 21,000. Add 250 ml of methanolic formaldehyde prepared as above but analyzes at 71.1% $CH_2{=}O$ by weight.

TABLE I

PREPARATION OF POLYOXYMETHYLENE BY THE POLYMERIZATION OF GASEOUS FORMALDEHYDE [8a,8b]

Gaseous $CH_2{=}O$ by pyrolysis of α-polyoxymethylene (gm)	Solvent (ml)	Catalyst (gm)	Reaction conditions		% Yield polymer[l]	η_{inh}[b]	% Degrad. at 222°C/min
			Temp. (°C)	Time (hr)			
	Decahydronaphthalene[a]						
69	(523)	—	−30	1–2	87	1.66	—
100	(698)[c]	—	−30	1–2	52	2.11	—
	Propane[d]						
100	(468)	—	−40	2	56	1.94	—
	Pentane[a]						
51	(376)	—	−30	2	78	1.69	—
	Decahydronaphthalene						
100	(528)	—	−30	2	35.5	5.72	—
100	(880)[e]	—	50	3.75	58	2.57	—
100	(704)[e]	—	25	3.5	38	1.6	—
	Pentane	$(n\text{-}C_4H_9)_3N$					
100	(626)	(0.12)	25	4	12	2.2	0.9
	Petroleum fraction[f]						
34	800	g	25–45	1	65	3.4	0.67
	C_6H_6	$(n\text{-}C_4H_9)_3N$					
100	(800)	(0.2)	25	2.5	62	1.36	0.77
		$+(C_6H_5)_2NH$ (0.1)					
	Cyclohexane	$(n\text{-}C_4H_9)_3N$					
100	(956)	(0.16)	25	1.5	25	2.87	0.42
		$+(C_6H_5)_2NH$ (0.10)					

100	(780)	$(C_6H_5)_2NH$ (0.1) +*h*	25	2	41	3.4	1.91
33	(780)	$(C_6H_5)_2NH$ (0.1) +*i*	25	0.66	85	1.32	0.84
100	Propionitrile (468)	—	−80	3	35	1.5	—
75	Carbon tetrachloride (800)	$(C_6H_5)_2NH$ (0.05) +$(C_4H_9)_3N$ (0.078)	25	2	34	1.75	0.67
100	Pentane (626)	*j*	25	4.1	28.5	2.0	0.67
100	(630)	*k*	25	4	13	1.81	—
100	(630)	$(C_6H_5)_3P$ (0.2)	25	1	9	1.9	1.32
100	Cyclohexane (780)	$(C_4H_9)_3N$ (0.078) +phenothiazine (0.1)	25	2.5	35.5	2.4	0.42

[a] 5 gm of a nonionic dispersing agent added (ester of polyethylene glycol of mol. wt. 400 with oleic acid).
[b] *p*-Chlorophenol solvent.
[c] 2 gm nonionic dispersing agent, footnote (a).
[d] 4 gm nonionic dispersing agent, footnote (a).
[e] 3 gm nonionic dispersing agent, footnote (a).
[f] B.p. 212°–284°F.
[g] 1.0 gm of octadecyldimethylamine containing 25% dimethylhexylamine.
[h] 0.1 gm of *N*,*N*-dimethyl-*p*-aminoazobenzene.
[i] 0.09 gm of *N*,*N*-diethylaminoethanol.
[j] 0.33 gm of octadecyldimethylamine.
[k] 0.1 gm of octadecyldimethylamine.
[l] M.p. 180°–220°C.

TABLE II

POLYMERIZATION OF MONOMERIC FORMALDEHYDE (480 GM/HR) BY ONIUM CATALYSTS[a]

Initiator	Initiator concentration (mg/liter reaction medium)	Solvent (ml)	Reaction conditions		Hold-up volume (ml)	Grams polymer per hr	η_{inh}
			Temp. (°C)	Time (min)			
$(n\text{-Bu})_4NI$ + lauric acid	1.87 0.99	Toluene (1000)	65	10	600	396	0.67
$(n\text{-Bu})_4N^+$laurate$^-$	0.5	Heptane (1000)	33	5	600	363	0.82[b]
$(n\text{-Bu})_4N^+$laurate$^-$	1.0	Cyclohexane (1000)	33	14	700	150	1.55
$(C_2H_5)_4PI$	1.0	1:1 C_6H_6–Acetone (1000)	30–35	—	—	43	1.10
$(n\text{-Bu})_3SBr$	5.0	Dioxane (400)	25	—	—	17.5	—

[a] From Goodman and Sherwood [54].

[b] Thermal degradation at 222°C is 0.06% by wt per minute.

2nd Day—Mol. wt. of polymer is 30,000–31,000. Add 400 ml of above methanolic formaldehyde solution of 69.7% $CH_2{=}O$ by weight.

Note that the methanolic formaldehyde solutions also contain *N*-ethylpiperidine and H_3PO_4 in the same ratio found in the original solution used.

After the third day 17.2 gm (2.6%) of polymer is isolated, mol. wt. 34,000. This polymer is acetylated by treatment with acetic anhydride in the presence of sodium acetate as a catalyst to give a thermally stable polymer having a 0.11% wt loss/min at 222°C.

NOTE: The formaldehyde is prepared by bubbling anhydrous formaldehyde into methanol.

Preparation 2-3 is only given for information and is not recommended as a method for preparing high molecular weight polyoxymethylene. As will be noted, the yield is only 2.6% after a considerable reaction time.

A similar process has been reported by the same authors using a completely aqueous system [80] to also give low yields of polymer (mol. wt. approx. 38,000–39,000). Larger yields are said to be obtained from a mixed aqueous dioxane system but the polymer has a lower molecular weight (17,000).

The number average molecular weights were determined both by osmometry and by end group analysis [81].

E. Radiation-Initiated Polymerization of Formaldehyde

X-Rays [82,82a] and γ-rays [83] catalyze the polymerization of monomeric formaldehyde. This γ-ray irradiation polymerization has been shown to proceed by a cationic rather than an anionic or radical mechanism [84]. Ultraviolet light has also been reported to initiate the polymerization of formaldehyde monomer [85].

Magat and co-workers also studied the γ-ray irradiation polymerization of solid formaldehyde monomer at -118°C [82].

3. POLYMERIZATION OF TRIOXANE TO POLYOXYMETHYLENE

Trioxane, which is obtained by distilling an acidic solution of aqueous formaldehyde, has m.p. 60.3°C and boiling point of 114.3°C.

The polymer obtained through the ring-opening polymerization reaction of trioxane has the same chemical structure as that of linear polyoxymethylene, $[CH_2O]_n$ [86]. When the trioxane is free from formaldehyde gas, polymer formation is very slow. Thus a trace of formaldehyde gas tends to induce the polymerization reaction [87,88].

$$BF_3 + CH_2{=}O \longrightarrow F_3^-B{-}O{-}CH_2^+ \xrightarrow{CH_2=O}$$

$$F_3^-B{-}OCH_2O{-}CH_2^+ \xrightarrow{\text{trioxane}} F_3^-B{-}OCH_2OCH_2^+{-}O\langle\text{trioxane ring}\rangle \longrightarrow$$

$$F_3^-B{-}OCH_2(OCH_2)_3OCH_2^+ \xrightarrow{\text{trioxane}} \text{polyoxymethylene} \quad (7)$$

The addition of 60 mmoles/liter of formaldehyde monomer to 5.3 moles/liter of trioxane in methylene chloride and 0.012 mole/liter of BF_3 caused a polymerization without an induction period at 30°C. The rate of sublimation–polymerization of trioxane is slow but the rate of cationic polymerization is rapid and quantitative. The cationic polymerization can be carried out in bulk or in solution (methylene chloride, cyclohexane, nitrobenzene, etc.) [88] using such catalysts as boron trifluoride and its complex compounds [87–90], antimony trifluoride [10], tin tetrachloride, titanium tetrachloride, thionyl chloride, phosphorus trichloride, phosphorus triiodide and other metal salts [10,10a,87], alkyl sulfonic acids [10c], perchloric acid [87,91], acetyl perchlorate [19], and aryldiazonium fluoroborates [19].

The catalyst is thought to coordinate with the oxygen of trioxane to cause cleavage (Eq. 8) [88,92].

In spite of thorough drying, water that is still present acts as a chain terminator to give hydroxyl groups [93]. Price and McAndrew [94] found that H_2O, CH_3OH, and $CH_3OCH_2OCH_3$ serve as chain-transfer agents and supply the ends of the polymer chain as shown in Fig. 2.

TABLE III

POLYMERIZATION OF TRIOXANE[a]

Initiator[a]	Induction period	Reaction time (hrs)	Yield (%)	$\overline{M}_w$
CH_3COClO_4	25 sec	1	47	29,000
$FeCl_3$	40 sec	1	38	27,000
$SnCl_4$	3 min	1	27	16,000
$BF_3 \cdot O(Et)_2$	18 min	1	13	8,500
$SbCl_5$	~30 min	1	5	3,000
H_2SO_4	24 hr	24	0	—

[a] Reprinted from Price and McAndrew [94]. Copyright 1966 by the American Chemical Society. Reprinted by permission of the copyright owner.

[b] Mole initiator/mole trioxane = 4×10^{-5}.

$$\begin{array}{c}\text{O—CH}_2\\ \text{CH}_2\quad\quad\text{O}\\ \text{O—CH}_2\end{array} + \text{BF}_3 \longrightarrow \begin{array}{c}\text{O—CH}_2\\ \text{CH}_2\quad\quad\text{O}^{\delta+}\\ \text{O—CH}_2\end{array} \longrightarrow$$

$$^{\delta-}\text{BF}_3 \longrightarrow [\text{—}\overset{+}{\text{C}}\text{H}_2\text{OCH}_2\text{OCH}_2\overset{-}{\text{O}}\text{BF}_3] \xrightarrow{\text{trioxane}}$$

$$\overset{+}{\text{C}}\text{H}_2\text{O(CH}_2\text{O)}_2[(\text{CH}_2\text{O})_3]_p\text{—CH}_2\text{O}^- \longrightarrow [\text{—CH}_2\text{O—}]_n \quad (8)$$

The effects of initiator on the induction period and other reaction parameters as found by Jaacks and Kern [93] are shown in Table III.

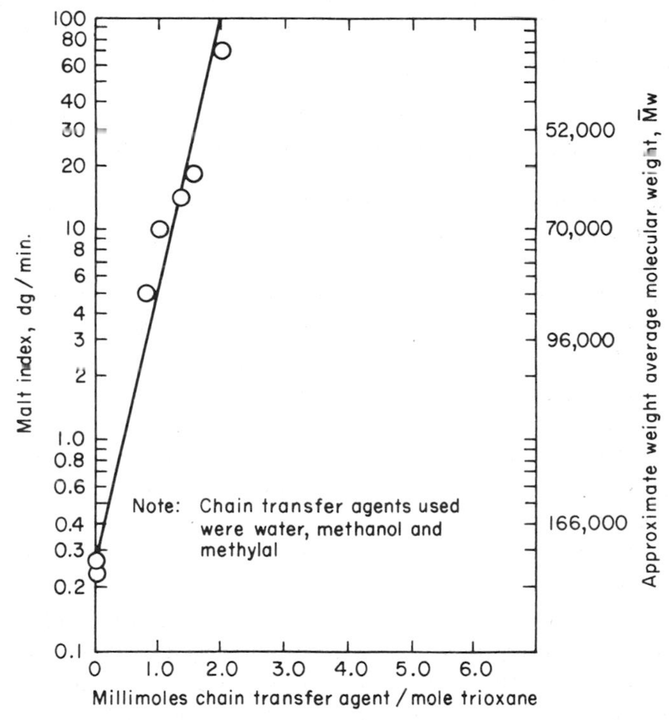

FIG. 2. Effect of chain transfer agents in the acid-catalyzed trioxane copolymerization. (Reprinted from Price and McAndrew [94]. Copyright 1966 by the American Chemical Society. Reprinted by permission of the copyright owner.)

3-1. Polymerization of Trioxane Using Boron Trifluoride Etherate Catalyst [95]

To a 23 cm long test tube equipped with a ground glass joint is added 25.0 gm (0.278 mole) of trioxane (see Note) and 25 gm of dry cyclohexane. The tube is stoppered and placed in a Dry Ice–acetone bath and cooled to

$-80°$ to $-60°C$ until solidification occurs. Then 0.1 ml of BF_3 etherate is added and the sealed tube is placed in a rotor which is immersed in a water bath at 66°–68°C for 4 hr. The tube is opened and the polymer isolated by washing successively with hot 5% Na_2CO_3, hot water, acetone, and then ether.

$$\text{trioxane} \xrightarrow{BF_3 \cdot O(C_2H_5)_2} [-OCH_2-OCH_2-O-CH_2-]_n \quad (9)$$

The polymer is dried overnight at 63°C to afford 6.0 gm (24%), m.p. 190°–195°C, 84.1% wt loss at 225° ± 5°C for 2 hr period. The polymer can be stabilized by acetylation in pyridine.

NOTE: Prior to use the trioxane should be distilled (b.p. 115°C) and then recrystallized (m.p. 64°C) twice from methylene chloride.

Recent results have shown that tetraoxane can be detected in large quantities by gas chromatography during the polymerization of solutions of trioxane [96].

The details of the polymerization mechanism can be obtained from the previous references and from two recent reviews [97,98].

The use of trioxane has the advantage over that of monomeric formaldehyde in that it is more easily handled and purified.

Trioxane also copolymerizes with other aldehydes, cyclic ethers, cyclic esters, vinyl compounds, nitriles, and trithiane. The copolymerization with ethylene oxide or 1,3-dioxolane in the amounts of 0.1 to 15 mole % are the most important commercially [99].

Studies conducted at Celanese [94] and also by Farbwerke Hoechst on the copolymerization of ethylene oxide with trioxane indicates that an induction period occurs which is much longer than for the homopolymer. It was established that during this induction period ethylene oxide is converted to 1,3-dioxolane, 1,3,5-trioxepanes, and to low molecular weight linear copolymers. It is not until the ethylene oxide has been consumed that the copolymerization gives solid polymer at a rate as shown in Fig. 3.

Using 1,3-dioxolane in place of ethylene oxide gives an induction period comparable to the homopolymerization reaction time. When the reaction is stopped at the early stages no 1,3-dioxolane is found but it is in the polymer as monoethyleneoxy groups.

The ethyleneoxy units become randomly distributed through the chain via transacetylization reactions [94]. For example, the reaction with BF_3 of poly-1,3-dioxolane (mol. wt. approx. 1000) dissolved in trioxane leads to a copolymer which has ethyleneoxy units randomly distributed through the polymer [94].

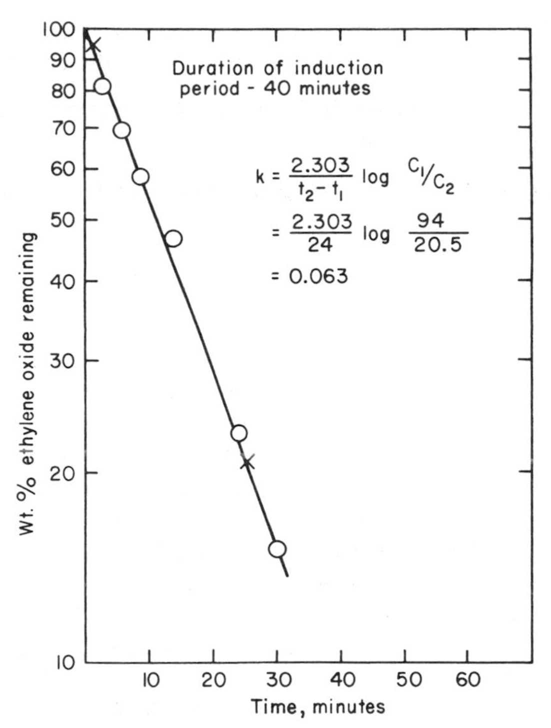

FIG. 3. Rate of disappearance of ethylene oxide during the induction period of a trioxane (98 wt%) ethylene oxide (2 wt%) copolymerization reaction. Reprinted from [94]. Copyright 1966 by the American Chemical Society; reprinted by permission of the copyright owner.

3-2. Copolymerization of Trioxane with Ethylene Oxide [95]

$$x\ \text{(trioxane)} + y\,\underset{\text{O}}{CH_2\text{—}CH} \longrightarrow (\text{—}OCH_2\text{—}OCH_2\text{—}OCH_2\text{—})_x(\text{—}CH_2CH_2O\text{—})_y \quad (10)$$

In a 23 cm long test tube is placed 23.75 gm (0.264 mole) of trioxane, and 25.0 gm of cyclohexane, and the mixture cooled to $-80°$ to $-60°C$. Then 1.25 ml (0.057 mole) of ethylene oxide (b.p. 10.7°C) is added followed by 0.10 ml of $BF_3 \cdot O(C_2H_5)_2$. The tube is sealed and the reaction and workup carried out as described in Preparation 3-1. The polymer yield is 4.5 gm (19%), m.p. 173°C, and the polymer has a 32.4% weight loss after 2 hr heating at 225° ± 5°C. The polymers prepared without ethylene oxide have a 84.1% weight loss. Using 0.03 ml (0.006 mmole) of ethylene oxide gave a polymer with m.p. 189°C and a 28.9% weight loss upon heating at 225° ± 5°C for

2 hr. The ethylene oxide content of the samples was determined by titration with HBr–glacial acetic acid (0.25 *N*) to a methyl violet end point.

NOTE: Trioxane should be distilled and then recrystallized twice from methylene chloride prior to use.

3-3. Polymerization of Trioxane Using Antimony Trifluoride Catalyst [10]

$$\text{(trioxane)} \xrightarrow{SbF_3} [-CH_2O-]_n \tag{11}$$

A Pyrex Carius tube with a ground glass joint is washed with 10% $NaHCO_3$, rinsed very well several times with distilled water, dried in a 100°C vacuum oven, and then cooled down under nitrogen.

Using completely anhydrous conditions 10 gm (0.111 mole) of trioxane and 0.1 gm (0.00056 mole) of antimony trifluoride are added and the tube is attached to a high-vacuum system to cause some evaporation of trioxane. The tube is then sealed under vacuum and placed in an oven or oil bath at 120°C for 48 hr. The tube is cooled, cut in the middle, the polymer removed from each end and dried separately of residual trioxane. The top section of the tube affords 6.0 gm of polymer and the bottom section affords 3.0 gm for a total of 90% yield. The polymer isolated from the top of the tube is tougher than that obtained from the bottom of the tube. This may be due to the fact that the trioxane vaporizing toward the cooler top of the tube gives a purer polyoxymethylene polymer.

Trioxane is also polymerized by irradiation with γ-rays [100–103] and ultraviolet light. Benzoyl chloride [100] also initiates the polymerization even under reduced pressure or at above 50°C.

Recently an announcement has been made that tetraoxocane, the tetramer of formaldehyde, will be commercially produced [104]. This material may also find use as a starting material for the preparation of polyoxymethylene.

4. POLYMERIZATION OF ACETALDEHYDE

Polyacetaldehyde is formed by condensing acetaldehyde vapors in a liquid oxygen trap [13,13a].

Furukawa [105,106] and Letort [107] described the cationic polymerization of neat acetaldehyde or acetaldehyde dissolved in inert solvents (toluene, propylene, or ethylene). Vogl [108,108a] reported that Lewis acids such as BF_3, $AlCl_3$, $ZnCl_2$, SbF_3, AsF_3, $SbCl_3$, and $AsCl_3$ are effective, as well as H_3PO_4, CF_3COOH, HNO_3, HCl, and adipoyl chloride.

Ineffective catalysts were CCl_3COOH, $ZrCl_4$, $TiCl_3$, $FeCl_3$/pyridine, phosphotungstic acid, H_2SO_4, P_2O_5, AlF_3, $BeCl_3$, SbOCl, BiOCl, BiF_3, $CdCl_2$, $B(OAc)_3$, $TiCl_2$, TiF_3, VCl_3, COF_2, CrF_3, PbF_2, ZrF_4, NH_4Cl, NH_4HF_2, $(C_6H_5)_3P$, $(BuO)_3P$, silica, and LiOAc.

The cationic polymerization can be carried out at −40°C or lower to give a rubbery polyacetaldehyde similar to that prepared by Letort [109,110]. Cationic catalysis by BF_3 etherate is also known to give crystalline polyacetaldehyde [111]. The freezing point is not the ceiling temperature of this polymerization process if a catalyst is present. Letort found that in the triethylaluminum-catalyzed polymerization, −40°C is sufficient to cause polymerization [112].

The mechanism of the cation-catalyzed polymerization of acetaldehyde has been suggested to involve generation of a carbonium ion and addition to the carbonyl oxygen atom of each aldehyde group [113].

$$CH_3CH{=}O + BF_3 \longrightarrow CH_3\overset{H}{C}{=}O{:}BF_3 \longrightarrow {}^{+}\underset{H}{\overset{CH_3}{C}}{-}O\bar{B}F_3 \xrightarrow{CH_3CH=O}$$

$$BF_3^{-}O{-}\underset{H}{\overset{CH_3}{C}}{-}O{-}\underset{H}{\overset{CH_3}{C}}{}^{+} \xrightarrow{nCH_3CH=O} BF_3^{-}O{-}\underset{H}{\overset{CH_3}{C}}{-}\left[{-}O{-}\underset{H}{\overset{CH_3}{C}}{-}\right]_n{-}O{-}\underset{H}{\overset{CH_3}{C}}{}^{+} \qquad (12)$$

Polyacetaldehyde degrades readily to acetaldehyde unless the polymer is treated with acetic anhydride and pyridine [18]. Etherification with ortho esters is also effective. The addition of antioxidants and the addition of thermal stabilizers of the amide type give a further improvement in stability [18]. These polymers to date have not found any commercial use because they still do not have enough thermal stability.

4-1. *Preparation of Polyacetaldehyde by Cationic Polymerization of Acetaldehyde Using BF_3 Catalysis* [108a]

$$nCH_3CHO \xrightarrow[\text{ethylene}\ -130°C]{BF_3\ \text{etherate}} {-}O{-}\underset{H}{\overset{CH_3}{C}}{-}O{-}\underset{H}{\overset{CH_3}{C}}{-}O{-}\underset{H}{\overset{CH_3}{C}}{-}O{-}\underset{H}{\overset{CH_3}{C}}{-}O{-}\underset{H}{\overset{CH_3}{C}}{-}O{-}\underset{H}{\overset{CH_3}{C}}{-} \qquad (13)$$

CAUTION: Use a well-ventilated hood.

Into a 1 liter round-bottomed flask equipped with a mechanical stirrer, nitrogen inlet and outlet, and a −200°C thermometer is condensed by means of liquid nitrogen 250 ml of ethylene (b.p. −104°C). Then 48 gm (1.2 moles) of acetaldehyde (see Note) is slowly injected through a serum cap using a hypodermic syringe, and stirring is begun. The temperature is kept at −130°C and then 5 drops of BF_3 etherate is injected. After 15–20 min the reaction mixture increases in viscosity. It is allowed to react for 1 hr. Next 100 ml of pyridine is slowly added at −104°C to deactivate the catalyst and dissolve the polymer. The pyridine solidifies and the reaction is kept at −78°C overnight in the hood to allow the ethylene to evaporate slowly. Then a solution of 20 ml of pyridine and 200 ml of acetic anhydride is added to the pyridine solution of the polymer and the mixture stirred for 2 hr. The polymer is isolated by precipitating it in ice–water. The polymer is kneaded by hand in water using rubber gloves, to destroy all the excess acetic anhydride. The polymer is dissolved in ether, extracted several times with water, dried, filtered, and concentrated under reduced pressure to afford 44 gm (92%) of polyacetaldehyde, $\eta_{inh} = 3.67$ (in 0.1% butanone solution at 25°C).

NOTE: The acetaldehyde is twice distilled in a low-temperature still over an antioxidant (0.1% β,β′-dinaphthyl-*p*-phenylenediamine).

Polyacetaldehyde has also been prepared by the depolymerization of paraldehyde by phosphoric acid and contacting the distilled acetaldehyde with Fluorosil–phosphoric acid (Floridin Co., Tallahassee, Florida; activated by heating for 24 hr at 450°–500°C) to give a rubbery amorphous polymer in 40–70% yield, $\eta_{rel} = 0.16$ as a 1% solution in DMF [114]. The polymer degrades to acetaldehyde unless it is kept in a cool place.

$$\text{(paraldehyde: 2,4,6-trimethyl-1,3,5-trioxane, } (CH_3CHO)_3) \xrightarrow{H_3PO_4} 3CH_3CH{=}O \xrightarrow[H_3PO_4]{\text{Fluorosil}} \left[-\underset{\underset{}{}}{\overset{CH_3}{\overset{|}{C}H}}-O- \right]_n \qquad (14)$$

Vogl [108,111,115,115a] reported that acetaldehyde polymerizes to give a stereoregular crystalline polymer (isotactic) using anionic catalysts such as alkali alkoxides (lithium *sec*-butoxide or potassium triphenylmethoxide), $LiAlH_4$, and alkali metal alkyls (*n*-BuLi). The polymerization is carried out in such solvents as toluene, propylene, and isobutylene at −40° to −80°C. The acetaldehyde was converted to polymer in about 77% yield and 50% of this was crystalline polymer, m.p. 165°C. The crystalline polymer differs in solubility from the elastomeric polyacetaldehyde obtained by freezing acetaldehyde (Letort procedure) since it is not soluble in common solvents at room temperature. The yield of polyacetaldehyde increases by the addition of $TiCl_4$ or $BF_3 \cdot OEt_2$ to an organo–alkali metal or an organo–alkaline earth metal complex or aluminum isopropoxide [116]. The reaction of triethylaluminum with lithium hydroxide also affords a catalyst which yields crystalline polyacetaldehyde [117].

Furukawa and co-workers reported that alumina can also act as a catalyst at $-70°C$ to give a rubbery polyacetaldehyde [106,118]. A relationship between the absorptive power and catalytic activity for polymerization exists with the various types of alumina [119]. Other metal oxides such as CrO_3 and MoO_3 are also effective catalysts. Less effective catalysts are B_2O_3, P_2O_5, MgO, ThO_2, SiO_2, CuO, TiO_2, ZrO_2, V_2O_5, ZnO, MnO_2, Fe_2O_3, BaO, PtO_2, PdO, Ni_2O_3, CaO, and silica-alumina (SiO_2 87%, Al_2O_3 13%). The silica gel used in chromatography is effective below $-30°C$ to give a spongelike polyacetaldehyde [120].

4-2. *Polymerization of Acetaldehyde Using Potassium Triphenylmethoxide Catalysis* [115a]

$$n\mathrm{CH_3CH{=}O} \xrightarrow{(\mathrm{C_6H_5})_3\mathrm{COK}} \left[\begin{array}{c}\mathrm{CH_3} \\ | \\ -\mathrm{CH}-\mathrm{O}-\end{array}\right]_n \tag{13}$$

A 500 ml three-necked, round-bottomed flask is equipped with a thermometer, stirrer, nitrogen inlet and outlet, and serum stopper. Nitrogen gas is added to flush the flask out and then reduced to such a rate as to keep a static nitrogen atmosphere. The flask is immersed in a Dry Ice–acetone bath and 200 ml of propylene is condensed in. When the temperature in the flask reaches $-75°C$ 5 ml of 0.125 *N* potassium triphenylmethoxide (see Note) in benzene–heptane is added using a hyperdermic syringe. While the reaction mixture is being vigorously stirred, 44 gm (1.0 mole) of acetaldehyde is added with a hyperdermic syringe. The polymer starts to precipitate as a gelatinous polymer and after 1 hr 200 ml of cold acetone containing 0.5 ml of acetic acid is added. The polymer is filtered after the reaction mixture warms to room temperature and 31.7 gm (72%) is obtained. Approximately 45% of this material is crystalline. The polymer may be stabilized by end-capping with pyridine and acetic anhydride as described previously.

NOTE: Potassium triphenylmethoxide is prepared as follows: To a 500 ml dry three-necked flask equipped with a dropping funnel, condenser and drying tube, and magnetic stirrer is added 2.2 gm of potassium sand in 10 ml of heptane. Then 50 ml of dry benzene is added followed by the dropwise addition (2 hr) of 13 gm (0.050 mole) of triphenylmethanol dissolved in 250 ml of benzene. Next the reaction mixture is boiled for $\frac{1}{2}$ hr while the original blue color changes to tan. The mixture is filtered under nitrogen into a dry flask and is ready for use.

Vogl [115,121], Natta [122], and Furukawa [123,124] and co-workers reported independently that crystalline polyacetaldehyde is obtained by polymerizing acetaldehyde at low temperature in the presence of organometallic compounds or metal alkoxides. The infrared spectrum of the crystalline polymer gives sharper absorption bands with more distinct relative intensities. The X-ray diffraction diagrams indicate that the polymer is definitely crystalline.

4-3. Preparation of Polyacetaldehyde Using Triethylaluminum Catalyst [125]

To a flask equipped as in Preparation 4-2 and containing 100 gm of ether and 200 gm (5.0 moles) of acetaldehyde at $-30°C$ is added 5.0 gm of triethylaluminum. The reaction mixture is stirred for 1 hr, cooled to $-78°C$, and then allowed to react for 1 hr to give 146.6 gm (73.3%) of polymer. The polymer is isolated as described previously.

In contrast to the results described in Preparation 4-3, Furukawa [124] reported only a 25% yield using $Al(C_2H_5)_3$ in hexane at the same temperature, as is shown in Table IV. The results with other organometallic catalysts are also described in Table IV.

The γ-ray radiation of acetaldehyde in the liquid and solid phase has also been reported to give rubbery solids with molecular weights in the range of $1.4–1.5 \times 10^5$ (liquid phase) and 1.5×10^6 (solid phase) [126].

Stereo-block polyacetaldehyde may be prepared using a combination of two catalysts such as alumina and diethylzinc [127], partially hydrolyzed aluminum alkoxide [128], and $AlEt_3–H_2O$ $AlEt_3$–alkali metal hydroxide, $AlEt_3$–ketone–H_2O [129], or $AlEt_3$–alcohol [130].

It is interesting to note that the polyaldolization of acetaldehyde has been reported to give either low molecular weight poly(vinyl alcohol) [131,131a]

TABLE IV

POLYMERIZATION OF ACETALDEHYDE (0.25 MOLE) IN 30 ML OF *n*-HEXANE AT $-78°C$ USING VARIOUS ORGANOMETALLIC CATALYSTS[a]

Organometallic catalyst (0.0125 mole)	Percent Polyacetaldehyde
$Al(C_2H_5)_2$	25
$B(C_4H_9)_3$	38
$Sn(C_2H_5)_4$	0
LiC_4H_9	5
$Zn(C_2H_5)_2$	45
$Cd(C_2H_5)_2$	0
C_2H_5MgBr	26
$Zn(C_2H_5)_2 + H_2O$	32
$LiAlH_4$	2

[a] From Furukawa *et al.* [124].

or cyclic structures [132,133]. NMR data suggest that branched structures are obtained rather than linear polyvinyl alcohol structures [134].

$$CH_3{-}CHO + B^- \xrightarrow{-BH} \bar{C}H_2CHO \xrightarrow{nCH_3CHO} CH_3\left[{-}\underset{|}{\overset{OH}{|}}\!\!\!\!CH{-}CH_2{-}\right]_n{-}CHO \quad [131, 131a] \tag{16}$$

$$CH_3{-}CHO + B^- \xrightarrow[1000\ atm]{triethylamine} CH_3{-}(CH_2{-}\underset{OH}{\underset{|}{C}}H)_m(CH{=}CH)_nCH{=}O \quad [131]$$

$$CH_3{-}CHO + B^- \xrightarrow{pyridine} \text{2,4-Dimethyl-6-hydroxy-1,3-dioxane (ring: } CH_3\text{, O, O, HO, } CH_3\text{)} \quad [132]$$

2,4-Dimethyl-6-hydroxy-1,3-dioxane

[132]

5. POLYMERIZATION OF HALOGENATED ALDEHYDES

The polymerization of trichloroacetaldehyde, or chloral, was first reported by Liebig in 1832 [15a]. Chloral can be polymerized by either acid or base catalysis [15a,15b,15e,135]. One can also obtain a cyclic trimer of chloral in some cases. Novak and Whalley [136,137] reported that infrared spectral analysis indicates that polychloral is a linear polyacetal structure $HO[CH(CCl_3)O]_nH$.

Monochloroacetaldehyde [16] and dichlorometaldehyde [138] were also earlier reported to be polymerized.

The preparation of polymeric fluoroaldehydes has been described by Husted and Ahlbrecht [139].

A. Polymerization of Chloral

Polychloral has been polymerized best (45% yield) at −78°C by Furukawa [124] using diethylzinc. Other catalysts ($AlEt_3$, $SnEt_4$, $LiAlH_4$, LiBu) gave low yields of polymer. The polychloral prepared by Furukawa was insoluble in most organic solvents and was not readily decomposed by hot pyridine.

Rosen [15c,140] reported that $AlBr_3$ was more effective for the cationic polymerization of chloral because of its good solubility in organic solvents. In addition, $AlCl_3$, and $ZrCl_4$ were more effective than $SnCl_4$. The best temperature to get a degree of polymerization of 190 was reported to be 30°C for a 6 mmole $AlBr_3$-catalyzed chloral (1 mole)–methylene chloride (6 moles) solution. Temperature had the greatest effect on degree of polymerization

(DP). For example, −75°C, DP = 60; −40°C, DP = 100; and 10°C, DP = 70.

The noncationic catalyst which gave a DP = 220 and a 78% conversion was $(C_4H_9)_3N$. Other catalysts that gave even higher DP values but lower conversions were $(C_4H_9)_3CH_3NI$ in acetonitrile, Et_2Zn, and 2,6-dimethoxyphenyllithium + Et_2Zn. Both the DP and the percent yield decreased as the water content of the system was increased.

Polychloral trimer could not be polymerized like trioxane with cationic catalysts, probably because of steric hindrance.

Polychloral of DP approx. 200 produced by either cationic or anionic catalysts had the same structure and physical properties. Both were highly intractable under conventional molding conditions.

Polychloral can be stabilized by end-capping using acid anhydrides or acid chlorides [141]. End-capping cannot be done using alkaline materials. The end-capped polychloral is stable up to 255°C (1.08% weight loss/min). Polychloral is estimated from copolymer data to have a melting point of approx. 460°C.

Recently Vogl, Miller, and Sharkey have found that LiO*t*-Bu, Bu_4NCl, and R_3SOCl also initiate polymerization of chloral [142,142a]. Triphenylphosphine rapidly reacted with chloral to give triphenyl(2,2-dichlorovinyloxy)-phosphonium chloride, which also acts as an initiator for polymerization. The polymers are obtained in high yield and of good thermal stability. Other catalysts such as $AlCl_3$, and $SbCl_5$ are also effective, but R_3N compounds such as pyridine gave polymers of lower thermal stability.

Since polychloral cannot be molded or processed it is polymerized in molds below the ceiling temperature (58°C) by first pouring in the warm (60°C) liquid monomer mixture and anionic catalyst and polymerizing by cooling to room temperature or below. The ceiling temperature can be varied by using inert solvents as diluents. A 95% conversion of chloral to polymer can be achieved by using 20% inert diluent. Otherwise 75–80% conversion is obtained and then baking at 100°–120°C for 1–4 hr removes the unreacted monomer. The polychloral thus produced can resist impact better than poly(methyl methacrylate) and does not support combustion in oxygen even at 1.1 atm. The polymer does not melt in the flame but decomposes to gaseous products. Polychloral is stable towards acids such as fuming nitric acid but is not stable towards bases such as amines. The polychloral can also be reacted further with isocyanates to end-cap it and thus further improve its heat stability and solvent resistance [142a].

5-1. Preparation of Polychloral [140]

A resin kettle is dried in an oven at 110°C, cooled in a nitrogen atmosphere, and equipped with a magnetic stirrer, nitrogen inlet–outlet, serum cap, and

$$nCCl_3CH{=}O \longrightarrow \left[\begin{array}{c} CCl_3 \\ | \\ -CH-O- \end{array} \right]_n \tag{17}$$

thermometer. Then 147.5 gm (1.0 mole) of chloral (see Note) dissolved in 510 gm (6.0 moles) of methylene chloride is added under a nitrogen atmosphere. The solution is cooled to $-30°C$ and 1.6 gm (6.0 mmole) of aluminum bromide in 5 ml of methylene chloride is injected through the serum cap. After 5–10 hr the polymer is filtered, washed with acidified methanol, then methanol, filtered, and dried to afford 118 gm (8%), DP = 190.

NOTE: The chloral used was 99.9% pure as seen by vapor-phase chromatographic analysis and had <0.1% dichloroacetaldehyde as the major impurity. The water content should not be higher than 20 ppm.

B. Polymerization of Dichloroacetaldehyde

Rosen and Sturm [143] have recently reported that dichloroacetaldehyde polymerizes readily in the presence of Lewis acid catalysts [$AlCl_3$, $AlBr_3$, $SnCl_4$, $NbCl_5$, $BF_3 \cdot O(C_2H_5)_2$] but not with basic or neutral catalysts such as tertiary amines, tertiary phosphines, and quaternary ammonium salts. Using the latter catalysts gave practically no polymer even after prolonged standing at 0°–25°C.

The thermal stability of the capped polymer (using acetic anhydride) is less than that of polychloral. Polydichloroacetaldehyde gives a 3.4% weight loss per minute at 220°C, whereas chloral has only a weight loss of 0.84% per minute at 220°C. The polymers are self-extinguishing when burned.

5-2. Preparation of Polydichloroacetaldehyde [143]

$$nCCl_2H-CH{=}O \longrightarrow \left[\begin{array}{c} CHCl_2 \\ | \\ -CH-O- \end{array} \right]_n \tag{18}$$

To a resin kettle that has been oven-dried at 120°C and cooled in a nitrogen atmosphere is attached a magnetic stirrer, nitrogen inlet-outlet, thermometer, and serum cap. Then 109 gm (1.0 mole) of dichloroacetaldehyde is added with a hypodermic syringe, followed by 0.036 gm (0.74 mole) of $BF_3 \cdot O(C_2H_5)_2$. After 5 hr reaction at 0°C, 76.3 gm (70%) of polymer is isolated. The polymer is ground in a blender in the presence of acidic methanol to remove the catalyst and then dried. The reduced viscosity (η_{sp}/C) of a polymer prepared in this manner was obtained by measuring the flow time of 0.1 gm of polymer in THF at 25°C and in this case was 3.8 (uncapped polymer). The estimated DP = 440. The polymer did not melt up to 240°C.

Commercial dichloroacetaldehyde can be used, but it should be dried over P_2O_5 and then distilled to give at least 95% purity. In most cases the final impurity is chloral.

C. Polymerization of Monochloroacetaldehyde

$$n\mathrm{ClCH_2CH{=}O} \longrightarrow \left[\begin{array}{c} \mathrm{CH_2Cl} \\ | \\ -\mathrm{CH}-\mathrm{O}- \end{array} \right]_n \tag{19}$$

Monochloroacetaldehyde (MCA) has not been extensively studied for its polymerizability. The results of a paper by Furukawa [144] and co-workers are summarized in Table V.

Sedlmeier [145] reported that BF_3 catalyzed the polymerization of monochloroacetaldehyde at $-50°$ to $-10°C$ to give an amorphous elastic thermoplastic.

5-3. Preparation of Polymonochloroacetaldehyde [144]

$$n\mathrm{ClCH_2CH{=}O} \longrightarrow \left[\begin{array}{c} \mathrm{CH_2Cl} \\ | \\ -\mathrm{CH}-\mathrm{O}- \end{array} \right]_n \tag{20}$$

To a glass polymer tube which has been flushed with nitrogen is added 40 ml of dry toluene, 5.8 gm (78 mmole) of monochloroacetaldehyde (see Note), and 0.0889 gm (0.78 mmole) of triethylaluminum. The tube is immersed in a cold bath and kept at $-78°C$ for 43 hr. Then methanol is added and the precipitate filtered to afford 3.0 gm (52%). The latter product consists of 31% chloroform-soluble material (amorphous polymer) and 69% chloroform-insoluble material (crystalline polymer). Polymonochloroacetaldehyde (uncapped) gave a 20% wt loss after heating for 4 hr at 100°C.

TABLE V
MONOCHLOROACETALDEHYDE POLYMERIZATION[a]

Catalyst	Yield (%)	% Amorphous	% Crystalline
$ZnEt_2$	19	0	100
$Al(OEt)_3$	28	0	100
$AlEt_2Cl$	31	47	53
$AlEtCl_2$	28	43	57
$BF_3 \cdot Et_2O$	38	100	0

[a] Reprinted from Rosen [15c]. Copyright 1966 by the American Chemical Society. Reprinted by permission of the copyright owners.

NOTE: Monochloroacetaldehyde monomer is prepared by the pyrolysis of paramonochloroacetaldehyde (m.p. 86°C, prepared by H_2SO_4 reaction of a 40% monochloroacetaldehyde aqueous solution at −10° to −20°C, washed with water, and recrystallized three times with ethanol) under a nitrogen atmosphere using a $CuSO_4$–$CaCl_2$ catalyst mixture. The aldehyde distills at 86°C and the vapors are passed through a $CaCl_2$ drying tube kept at 100°C. The liquid aldehyde is relatively stable and does not polymerize for a long period without a catalyst.

The thermal stabilities of the substituted acetaldehydes may be ranked in the following order: $CCl_3CH{=}O > Cl_2CHCH{=}O > ClCH_2CHO > CH_3CH{=}O$.

D. Polymerization of Fluoro-Substituted Acetaldehydes

Polyfluoral was first prepared in 1950 and was recently reinvestigated by Temple and Thornton [146]. Some data on the fluoro-substituted acetaldehydes are summarized in Table VI.

TABLE VI
POLYMERS OF FLUOROALDEHYDES[g]

Monomer	Catalysts	Ref.
CF_3CHO	Spontaneous, anionic, cationic, and free radical	*a,b,c,h*
C_2F_5CHO	Spontaneous, peroxide, acid	*a,b,c,h*
C_3F_7CHO	Acid	*a*
$CClF_2CHO$	Spontaneous	*d*
CF_2HCHO	$P(OEt)_3$ at −80°C	*e*
$CH_2FCH{=}O$	Spontaneous	*a,f*

[a] D. R. Husted and A. H. Ahlbrecht, British Patent 719,877 (1954).
[b] S. Temple and R. L. Thornton, *J. Polym. Sci., Part A-1* **10**, 709 (1972).
[c] W. K. Busfield and E. O'Malley, *Can. J. Chem.* **43**, 2289 (1965).
[d] G. Woolf, U.S. Patent 2,870,213 (1959).
[e] A. L. Barney, U.S. Patent 3,067,173 (1963).
[f] H. Schechter and F. Conrad, *J. Amer. Chem. Soc.* **72**, 3371 (1950).
[g] Reprinted in part from Rosen [140]. Copyright 1966 by the American Chemical Society. Reprinted by permission of the copyright owner.
[h] Nonflammable, chemically resistant, insoluble, high decomposition point.

Temple and Thornton [146] found that fluoral (trifluoroacetaldehyde) can be polymerized at low temperatures using ionic initiators in inert diluents to give high molecular weight polyfluoral. The polyfluoral is amorphous and tractable and can be end-capped by the usual techniques to give a polymer which is thermally stable at 350°–400°C. The stabilized polymer can be fabricated from solution or melt-pressed at about 125°C to give clear films and coatings. The films and coatings are oil and water repellent but have a high vapor permeability toward oxygen and moisture.

Table VII gives a more detailed summary of the catalysts and conditions used to polymerize fluoral. Preparation 5-4 gives a typical example of the polymerization conditions used for fluoral.

5-4. Preparation of Polyfluoral [146,147]

$$CF_3CH{=}O \xrightarrow[-78°C]{CsF} \left[\begin{array}{c} CF_3 \\ | \\ -CH-O \end{array} \right]_n \qquad (21)$$

To a 250 ml three-necked flask (see Note a) equipped with a mechanical stirrer, nitrogen inlet and outlet tubes, and a serum-capped joint is added 60 ml of trichloropentafluoropropane. The flask is back-pressured with dry nitrogen and cooled to $-78°C$. Then 45 gm (0.46 mole) of fluoral (Note b) is distilled into the flask from a calibrated trap. After gently stirring the clear solution at $-78°C$ the CsF initiator is added (20 μl) (Note c) to give 3.8 ppm CsF/fluoral monomer. A clear gelatinous solid is formed after stirring for 2 hr at $-78°C$. The mixture is additionally stirred for 2 hr at $-78°C$, 1 hr at $-60°C$, and 1 hr at $-40°C$ before gently warming to room temperature. Then 100 ml of 2,2-di(trifluoromethyl)-5-methyl-1,3-dioxalane (Note d) is added and stirred to give a clear solution. Then 10 ml of acetyl chloride and 4 drops of pyridine are added and stirred at reflux for 48 hr. The solution is cooled and poured into hexane to give a white rubbery solid. The solid is cooled with Dry Ice and then quickly powdered by stirring in a blender. The crumb solid is washed with hexane and dried under reduced pressure to afford 34.3 gm (76%), $\eta_{inh} = 3.13$ [0.5% in 2,2-di(trifluoromethyl)-1,3-dioxolane (Note d)].

NOTES: (a) All glassware is acid-washed, rinsed with distilled water, and dried for 10–20 hr at 125°C and cooled under a nitrogen atmosphere. (b) Fluoral (b.p. $-19°C$) is prepared by the vapor-phase fluorination of chloral with HF over a chromium oxide gel catalyst [147]. (c) The soluble CsF initiator is prepared by passing fluoral vapor into an equimolar solution of dry CsF and tetraglyme (tetraethylene glycol dimethyl ether) in dry ether to give the initiator $F{-}CH(CF_3)_3{-}O^-{-}Cs^+$. (d) Prepared from hexafluoroacetone and ethylene or propylene oxide as described by D. G. Coe and D. B. Patteson, U.S. Patent 3,324,144 (1967).

TABLE VII

POLYMERIZATION OF FLUORAL[a]

Fluoral (moles)	Solvent (ml)	Catalyst (ppm)	Reaction conditions: Temp. (°C)	Reaction conditions: Time (hr)	% Yield	η_{inh}[b]	Thermal decomp.[c]
0.53	Hexane (88)	$F—OCH(CF_3)—\bar{O}\ Cs^+$ (0.02)	25–40	—	80	Insol.	105,260
0.46	$CCl_3CF_2CF_3$ (67)	CsF (3.8)	−78	4	76	3.13	373
0.38	Toluene (58)	CsF (3.5)	−78	4	49	0.99	157,395
0.30	$CCl_3CF_2CF_3$ (50)	PF_5 (not reported)	−78	3	100	0.83	399
0.34	$CCl_3CF_2CF_3$ (67)	Electron beam (4 Mrad)	−78	94	65	0.21	130,136

[a] Data taken from Temple and Thorton [146].

[b] 0.5% in 2,2-di(trifluoromethyl)-1,3-dioxolane at 30°C.

[c] Decomposition of acetate-capped polymer under nitrogen (first derivative TGA plot).

6. POLYMERIZATION OF OTHER ALDEHYDES

Upon polymerization of aldehydes other than formaldehyde one obtains an asymmetric center. Isotactic, syndiotactic, and atactic polyaldehydes have been synthesized and characterized [130].

nR—C(H)=O →

—O—C(R)(H)—O—C(R)(H)—O—C(R)(H)—O—C(R)(H)— Isotactic

—O—C(R)(H)—O—C(H)(R)—O—C(R)(H)—O—C(H)(R)— Syndiotactic (22)

—O—C(R)(H)—O—C(H)(R)—O—C(R)(H)—O—C(R)(H)— Atactic

Bridgman and Conant [148–150] in 1929 reported the high-pressure polymerization of n- and isobutyraldehydes, n-valeraldehyde, and n-heptaldehyde to give atactic polymers [14a].

Letort [13a] and Travers [13] reported in 1936 that freezing acetaldehyde at -123°C and then melting afforded atactic polyacetaldehyde.

Starting in about 1960 Vogel [108,121,151], Furakawa [123], and Natta [122] reported on a series of investigations of stereoregular polyaldehydes. These polymers have a tendency to depolymerize unless stabilized by reacting the terminal hydroxyl groups [151].

Polyaldehydes are generally obtained by using either anionic or cationic catalyst systems as described below.

See the Introduction section of this chapter for references to additional background material.

Vogl [108] and Natta [122] reported preparation of the following crystalline polyaldehydes in 80–95% yields: polypropionaldehyde, poly-n-butyraldehyde, polyisobutyraldehyde, poly-n-valeraldehyde and poly-n-heptaldehyde [113]. The noncrystalline polymers of the following aldehydes have been prepared: n-octaldehyde, phenylacetaldehyde, and 3-ethoxypropionaldehyde.

Glutaraldehyde is polymerized with the aid of aluminum triisopropoxide at about -55°C in toluene, or with $BF_3 \cdot O(C_2H_5)_2$ in THF at -70°C. Suberaldehyde polymerizes with similar catalysts at 20°C [152].

Terephthaldehyde has been polymerized with the aid of aluminum alkyl compounds. However, benzaldehyde has not yielded to polymerization [153].

Unsaturated aldehydes such as acrolein, crotonaldehyde, and 3-cyclohexen-1-carboxaldehyde also can be polymerized with polyoxymethylene derivatives using triethylaluminum initiator at -78°C [154].

In the case of the polymerization of *n*-butyraldehyde isotactic polymer is formed at about the same polymerization rate regardless of the counter ion such as Li, Na, K, Cs, Mg, or transition metals [18].

Vogl [111] reported on a series of low-temperature polymerizations of various aldehydes in solvents of low dielectric constants. Alkali metal alkoxides, alkali metal alkyls, aluminum organic compounds, and cationic catalysts such as $SnBr_4$ and $BF_3 \cdot O(C_2H_5)_2$ were effective in giving isotactic polyaldehydes as described in Table VIII.

6-1. Preparation of Polypropionaldehyde Using Triethylaluminum Catalyst [124]

$$n\,C_2H_5CHO \longrightarrow \left[\begin{array}{c} C_2H_5 \\ | \\ \text{—CHO—} \end{array} \right]_n \qquad (23)$$

The apparatus consists of two tubes connected by a series of stopcocks. In one tube is placed 13.2 gm (0.25 mole) of propionaldehyde (b.p. 48.5°–49°C) and in the other 30 ml of dry *n*-hexane and 1.43 gm (0.0125 mole) (commercial) of triethylaluminum. The two tubes are cooled in a Dry Ice–acetone bath and the system evacuated to 3–5 mm Hg. The vacuum stopcock is closed and the stopcock connecting each tube is opened. The Dry Ice is removed from the propionaldehyde and this is distilled over a $1\frac{1}{2}$ hr period through a capillary tube into the hexane $AlEt_3$-containing tube reactor. After the monomer has been completely added the stopcock is closed and the reactor tube is kept at −78°C for 2 hr. Then 50 ml of methanol is slowly added, and the insoluble polymer is filtered and dried to afford 5.7 gm (43%). The polymer is insoluble in chloroform.

Recently Tani *et al.* [155] reported that an aldehyde aluminum complex $Me_2AlOCMeNPh \cdot RCHO$ can be used to polymerize acetaldehyde, propionaldehyde, and *n*-butyraldehyde. This appears to be the first time that such a complex has been isolated.

7. MISCELLANEOUS METHODS

1. Preparation of polyacetaldehyde with improved mechanical properties by use of a graphite-Li catalyst [156].
2. Preparation of a hydrogenated polyacetaldehyde produced by the polymerization of acetaldehyde and Li-naphthalene [157].
3. Preparation of polyoxymethylene-acrylamide copolymers of high thermal stability (decomp. 0.2%/min at 220°C) [158].
4. Preparation and polymerization of MeO-, EtO-, PrO-, and BuO-substituted butyraldehydes, propionaldehydes, and acetaldehydes [159].
5. Preparation of acetaldehyde–formaldehyde copolymers [160].

TABLE VIII

POLYMERIZATION OF VARIOUS ALDEHYDES BY ANIONIC AND CATIONIC CATALYSTS TO GIVE ISOTACTIC POLYMERS

Aldehyde	Solvent	$(C_6H_5)_3COK$	BuLi	$ZnBr_4$	$Al(C_2H_5)_3$	$Zn(Et)_2$	$(C_2H_5)_3Al-TiCl_4$	Ref.
Acetaldehyde	Propylene	75+	75	—	—	—	—	*a*
	Hexane	—	5	—	25	45	—	*b*
Propionaldehyde	Propylene	75+	—	—	—	—	—	*a*
	Hexane	—	0	—	43	31	—	*b*
n-Butyraldehyde	Propylene	88.5	—	—	—	—	—	*a*
	Pentane	79	—	83	20–30	—	75+	*a*
	DMF	0	—	—	—	—	—	*a*
i-Butyraldehyde	Pentane	75+	—	—	—	—	—	*a*
	Propylene	75+	—	—	—	—	—	*a*
n-Valeraldehyde	Propylene	75+	—	—	—	—	—	*a*
n-Heptaldehyde	Propylene	75+	—	—	—	—	—	*a*
n-Octaldehyde	Propylene	20–50	—	—	—	—	—	*a*
Chloral	Propylene	75+	—	—	—	—	—	*a*
3-*H*-Perfluoropropionaldehyde	Propylene	75+	—	—	—	—	—	*a*
Phenylacetaldehyde	Propylene	75+	—	—	—	—	—	*a*
3-Methoxypropionaldehyde	Propylene	75+	—	—	—	—	—	*a*
Cyclohexaldehyde	Propylene	Trace	—	—	—	—	—	*a*

[a] From Vogl [111].
[b] From Furukawa *et al.* [124].

6. Polymerization of formaldehyde under high pressure [161].
7. Copolymerization of formaldehyde with chloral [162].
8. Copolymerization of acetaldehyde and propionaldehyde [163].
9. Copolymerization of dimethylketene with formaldehyde [164].
10. Copolymerization of monomeric formaldehyde with fluoroaldehyde [165].
11. Copolymerization of formaldehyde with isocyanic acid [166].
12. Copolymerization of trioxane with lactones [167].
13. Copolymerization of acetaldehyde with acrylonitrile [168].
14. Copolymerization of trioxane with styrene [169].
15. Polymerization of cyclic ethers in the presence of maleic anhydride [170].
16. Copolymerization of trioxane with dicyclopentadiene [171].
17. Preparation of graft copolymers of polyoxymethylene with ester groups [172].
18. Copolymers of trichloroacetaldehyde and dichloroacetaldehyde [173].
19. Copolymer of trichloroacetaldehyde with formaldehyde [174,175].
20. Copolymer of monochloroacetaldehyde with acetaldehyde [176].
21. Copolymer of trifluoroacetaldehyde or high fluoroaldehydes with formaldehyde [165].
22. Cationic copolymerization of trioxane and 1,3-dioxolane [177].
23. Preparation of optically active aldehyde polymers derived from citronellal [178].
24. Preparation of polyoxymethylene polymer with epoxy side groups by the polymerization of glycidaldehyde [179].
25. Polymerization of β-cyanopropionaldehyde [180].
26. Polymerization of suberaldehyde [153].
27. Preparation of poly(vinyl)oxymethylene by polymerization of acrolein with cyanide ion at low temperatures [181].
28. Preparation of polyoxymethylene polymers from glyoxal [182], glutaraldehyde [152,183], 3-phenylglutaraldehyde [184], 3-methylglutaraldehyde [185,186], and related succinaldehydes [152].
29. Preparation of polyoxymethylene copolymers by the copolymerization of chloral with 1,3,6-trioxocane (formal of diethylene glycol) [187].
30. Polyoxymethylenes from 1,3,6-trioxocane [188].
31. Preparation of tetraoxane, a source of pure formaldehyde [189].
32. Polymerization of trioxane to high molecular weight polyoxymethylene using $C_6H_5OPCl_2$ catalyst [190].
33. Preparation of α-polyoxymethylenes [191].
34. Polymerization of propiolaldehyde to [poly(ethynyloxymethylene)] [192].
35. Polymerization of dialdehydes [193].

REFERENCES

1. A. M. Butlerov, *Ann. Chem. Pharm.* **111**, 242 (1859).
2. A. W. Hofmann, *Ann. Chem. Pharm.* **145**, 357 (1868); *Ber. Deut. Chem. Ges.* **2**, 156 (1869).
3. B. Tollens and F. Mayer, *Ber. Deut. Chem. Ges.* **21**, 1566, 2026, and 3503 (1888).
4. G. Lösekann, *Chem.-Ztg.* **14**, 1408 (1890).
5. M. Delepine, *C. R. Acad. Sci.* **124**, 1528 (1897); *Ann. Chim. Phys.* [4] **15**, 530 (1898).
6. H. Staudinger, "Die Hochmolekularen Organischen Verbindungen," pp. 224–287. Springer-Verlag, Berlin and New York, 1960 (new ed.).
7. H. Staudinger and A. Gaule, *Ber. Deut. Chem. Ges.* **49**, 1897 (1916); H. Staudinger and R. Signer, *Helv. Chim. Acta* **11**, 1847 (1958); H. Staudinger, R. Signer, H. Johner, O. Schweitzer, M. Lüthy, W. Kern, and D. Fussidis, *Justus Liebigs Ann. Chem.* **474**, 145 (1929).
8. P. R. Austin and C. E. Frank, U.S. Patent 2,296,249 (1942); R. N. MacDonald, U.S. Patent 2,828,286 (1958); H. H. Goodman, Jr., U.S. Patent 2,994,687 (1961).

8a. R. N. MacDonald, U.S. Patent 2,768,994 (1956).

8b. C. E. Schweitzer, R. N. MacDonald, and J. O. Punderson, *J. Appl. Polym. Sci.* **1**, 158 (1959).

9. J. C. Bevington and H. May, *Encycl. Polym. Sci. Technol.* **1**, 609 (1964).
10. A. K. Schneider, U.S. Patent 2,795,571 (1957).

10a. K. W. Bartz, U.S. Patent 2,947,728 (1960).

10b. O. H. Axtell and C. M. Clarke, U.S. Patent 2,951,059 (1960); C. J. Bruni, U.S. Patent 2,989,510 (1961); C. L. Michaud, U.S. Patent 2,982,758 (1961); Farbwerke Hoechst, A.G., Australian Patent Appl. 55,706/59 (1960); British Industries Plastics Ltd., Belgian Patent 592,599 (1961); D. E. Hudgin and F. M. Berardinelli, French Patent 1,216,327 (1960); G. J. Bruni, French Patent 1,226,988 (1960).

10c. D. E. Hudgin and F. M. Berardinelli, U.S. Patents 2,989,505–2,989,509 (1961).

11. D. Oosterhof, "Chemical Economics Handbook," Acetal Resins, No. 580,0121A and 580,0122A, Dec. 1971.
12. D. Oosterhof, "Chemical Economics Handbook," Acetal Resins No. 580,0121 C, Dec. 1971.
13. M. S. Travers, *Trans. Faraday Soc.* **32**, 246 (1936).

13a. M. Letort, *C. R. Acad. Sci.* **202**, 767 (1936).

14. J. B. Conant and W. R. Peterson, *J. Amer. Chem. Soc.* **52**, 1659 (1930).

14a. A. Novak and E. Whalley, *Can. J. Chem.* **37**, 1710 and 1718 (1959).

15. Chemical & Engineering News, *Chem. Eng. News* **52** (12), 41 (1972).

15a. J. Böeseken and A. Schimmel, *Rec. Trav. Chim. Pays-Bas* **32**, 112 (1913).

15b. S. Gaertner, *C. R. Acad. Sci.* **1**, 513 (1906).

15c. I. Rosen, *Polym. Prepr., Amer. Chem. Soc., Div. Polym. Chem.* **7**, 221 (1966).

15d. J. Liebig, *Ann. Pharm.* **1**, 194 and 209 (1832).

15e. A. W. H. Kolbe, *Ann. Chem. Pharm.* **54**, 183 (1854).

16. K. Natterer, *Monatsh, Chem.* **3**, 442 (1882).
17. O. Jacobsen, *Ber. Deut. Chem. Ges.* **8**, 87 (1875).
18. O. Vogl, *Polym. Prepr., Amer. Chem. Soc., Div. Polym. Chem.* **7**, 216 (1966).
19. J. Furukawa and T. Saegusa, "Polymerization of Aldehydes and Oxides." Wiley (Interscience), New York, 1963.
20. E. G. Brame, Jr. and O. Vogl, *Polym. Prepr., Amer. Chem. Soc., Div. Polym. Chem.* **7**, 227 (1966).

21. J. Furukawa and T. Saegusa, *Kogyo Kagaku Zasshi* **65**, 649 (1962); T. Saegusa and H. Fujii, *Shokubai* **7**, 395 (1965); O. Vogl, *High Polym.* **23 (pt 1)**, 419 (1968).
22. O. Vogl, ed., "Polyaldehydes." Dekker, New York, 1967.
23. R. Spence and W. Wild, *J. Chem. Soc., London* p. 506 (1935).
24. A. Kekule, *Ber. Deut. Chem. Ges.* **25**, 2435 (1892).
25. W. A. Bone and H. L. Smith, *J. Chem. Soc., London* **87**, 910 (1905).
26. H. L. Callendar, *Engineering (London)* **123**, 147 (1927); R. Spence, *J. Chem. Soc., London* p. 649 (1936).
27. A. R. Miller, Jr., U.S. Patent 2,460,592 (1949); F. Walker, *J. Amer. Chem. Soc.* **55**, 2821 (1933).
28. H. H. Goodman, Jr. and L. T. Sherwood, Jr., U.S. Patent 2,994,687 (1961).
29. E. I. du Pont de Nemours & Co., British Patent 770,717 (1957).
30. F. Walker, *J. Amer. Chem. Soc.* **55**, 2821 (1933).
31. H. Staudinger, R. Singer, and O. Schweitzer, *Ber. Deut. Chem. Ges. B* **64**, 398 (1931).
32. A. Butlerov, *Ann. Chem. Pharm.* **111**, 245 (1859).
33. J. Nasch, British Patent 420,993 (1934).

33a. B. W. Greenwald and R. K. Cohen, U.S. Patent 2,490,206 (1950).

34. H. Finkenbeiner and W. Schmäche, U.S. Patent 2,116,783 (1938); E. Naujoks, U.S. Patent 2,093,422 (1937); C. Pyle and J. A. Lane, U.S. Patent 2,527,654 (1950); U.S. Patent 2,581,881 (1952).
35. Deutsche Gold-und Silber-Scheideanstalt. vorm. Roseelen, British Patent 479,255 (1938); German Patent 588,470 (1932).
36. H. Staudinger, R. Singer, H. Johner, M. Lüthy, W. Kern, D. Russidis, and O. Schweitzer, *Justus Liebigs Ann. Chem.* **474**, 251 (1929).
37. F. Auerbach and H. Barschall, "Studien über Formaldehyde—Die Festen Polymeren des Formaldehyde," pp. 20 and 26. Springer-Verlag, Berlin and New York, 1907.
38. T. E. Londergain, U.S. Patent 2,512,950 (1950).
39. F. Auerbach and H. Barschall, "Studien über Formaldehyde—Die Festen Polymeren der Formaldehyde," p. 34. Springer-Verlag, Berlin and New York, 1907.
40. D. L. Hammick and A. R. Boeree, *J. Chem. Soc., London* **121**, 2738 (1922).
41. H. Staudinger and M. Lüthy, *Helv. Chim. Acta* **8**, 41 (1925).
42. Author's Laboratory (SRS).
43. A. W. Hofmann, *Ber. Deut. Chem. Ges.* **1**, 200 (1868); **2**, 152 (1869).
44. M. Trautz and E. Ufer, *J. Prakt. Chem.* [N.S.] **113**, 105 (1926).
45. R. Spence, *J. Chem. Soc., London* p. 1193 (1933).
46. J. E. Carrothers and R. G. W. Norrish, *Proc. Roy. Soc., Ser. A* **205**, 217 (1951).
47. J. C. Bevington and R. G. W. Norrish, *Proc. Roy. Soc., Ser. A* **205**, 517 (1951).
48. H. Staudinger, "Die Hochmoleckularen Organischen Verbindungen." Springer-Verlag, Berlin and New York, 1960 (new ed.).
49. W. Fukuda and H. Kakiuchi, *Bull. Fac. Eng., Yokohama Nat. Univ.* **10**, 51 (1961)
50. R. N. MacDonald, U.S. Patent 2,768,994 (1956).
51. R. N. MacDonald, U.S. Patent 2,828,286 (1958).
52. J. Behrends, O. Schweitzer, W. Kern, and H. Hopff, U.S. Patent 3,020,264 (1962).
53. F. C. Stair, Jr., U.S. Patent 2,734,886 (1956).
54. H. H. Goodman, Jr. and L. T. Sherwood, Jr., U.S. Patent 2,994,687 (1961).
55. M. G. Bechtold and R. N. MacDonald, U.S. Patent 2,844,561 (1958).
56. T. E. O'Connor, U.S. Patent 3,002,952 (1961).
57. N. Brown, *Polym. Prepr., Amer. Chem. Soc., Div. Polym. Chem.* **7**, 199 (1966).
58. Farbenfabriken Bayer Akt.-Ges., British Patent 857,321 (1960).

59. K. Wagner, U.S. Patent 3,005,799 (1961).
60. Deutsche Gold- und Silber-Scheideanstadt Vormals Roessler, British Patent 836,288 (1960); W. Querfurth and O. Schweitzer, U.S. Patent 2,985,623 (1961).
61. Z. Machaecek, J. Mejzlik, and I. Pac, *J. Polym. Sci.* **52**, 309 (1961).
62. F. W. Gander, U.S. Patent 2,780,652 (1957); E. I. du Pont de Nemours & Co., British Patent 796,863 (1958).
63. F. Szilagyi, French Patent 1,348,553 (1964).
64. L. S. Shpichinetskaya, U.S.S.R. Patent 159,983 (1964).
65. J. Mejzlik and J. Pac, *Makromol. Chem.* **82**, 226 (1965).
66. J. Mejzlik and L. Janeckova, *Makromol. Chem.* **82**, 238 (1965).
67. J. Mejzlik, M. Lesna, P. Osecky, I. Srackova, and E. Sazovska, *Makromol. Chem.* **82**, 253 (1965).
68. R. N. MacDonald and M. J. Roedel, U.S. Patent 2,920,059 (1960).
69. R. N. MacDonald, Canadian Patent 565,816 (1952).
70. M. A. Kubico and R. N. MacDonald, U.S. Patent 2,893,972 (1960).
71. L. T. Sherwood, Jr., Canadian Patent 603,284 (1960).
72. G. Louis, H. Penning, H. Pohlemann, and H. Wilhelm, German Patent 1,076,363 (1960).
73. W. Runkel and E. Becker, German Patent 1,098,713 (1961).
74. H. D. Hermann, C. Heuck, K. Kullman, O. Mauz, M. Reiber, and J. Winter, German Patent 1,117,868 (1961).
75. Kurashiki Rayon Co., Belgian Patent 608,917 (1962).
76. Farbwerke Hoechst A.G. Vorm. Meister Lucius & Brüning, Belgian Patent 607,874 (1962).
77. J. F. Walker, U.S. Patent 2,304,431 (1942).
78. D. Goidea, *Polym. Prepr. Div. Polym. Chem. Amer. Chem. Soc.*, **13**, 409 (1972).
79. N. Brown, D. L. Funck, and C. E. Schweitzer, U.S. Patent 3,000,860 (1961).
80. N. Brown, D. L. Funck, and C. E. Schweitzer, U.S. Patent 3,000,861 (1961); W. M. D. Bryant and J. B. Thompson, *J. Polym. Sci., Part A-1* **9**, 2523 (1971).
81. T. A. Koch and P. Lindvig, *J. Appl. Polym. Sci.* **1**, 164 (1959).
82. C. Chachaty, M. Magat, and L. TerMinassian, *J. Polym. Sci.* **48**, 139 (1960).
82a. R. Fourcade, U.S. Patent 3,093,560 (1963).
83. S. Okamura, S. Nakashio, K. Hayashi, and I. Sakurada, *Isotop. Radiat.* **3**, 242 (1959).
84. S. Okamura, K. Hayashi, H. Yamoaka, and Y. Nakamura, *14th Annu. Meet. Jap. Chem. Soc., Tokyo, 1961*, oral paper (1961).
85. S. Okamura, S. Nakashio, K. Hayashi, and I. Sakurada, *Isotop. Radiat.* **3**, 242 (1960); Sumitomo Chemical Co. Ltd. and Sumitomo Atomic Energy Industries, Ltd., British Patent 939,477 (1963).
86. D. L. Hammick and A. R. Boeree, *J. Chem. Soc., London* **121**, 2738 (1922); H. W. Kohlschütter, *Justus Liebigs Ann. Chem.* **482**, 75 (1930); H. W. Kohlschütter and L. Sprenger, *Z. Phys. Chem., Abt. B* **16**, 284 (1932).
87. W. Kern, H. Cherdron, and V. Jacobs, *Angew. Chem.* **73**, 177 (1961).
88. W. Kern and V. Jacobs, *J. Polym. Sci.* **48**, 399 (1960).
89. A. W. Schnizer, U.S. Patent 2,989,511 (1961).
90. T. Higashimura and S. Okamura, *J. Polym. Sci., Part A-1* **5**, 95 (1967).
91. W. Kern, *Chem.-Ztg.* **88**, 623 (1964).
92. L. Leese and M. Baumber, *Polymer* **6**, 269 (1965).
93. V. Jaacks and W. Kern, *Makromol. Chem.* **62**, 1 (1963).
94. M. B. Price and F. B. McAndrew, *Polym. Prepr., Amer. Chem. Soc., Div. Polym. Chem.* **7**, 207 (1966).

95. C. T. Walling, F. Brown, and K. W. Bartz, U.S. Patent 3,027,352 (1962).
96. T. Miki, T. Higashimura, and S. Okamura, *J. Polym. Sci., Part A-1* **5**, 95, 2977, 2989, and 2997 (1967).
97. M. Baccaredda, *Chem. Ind. Genie Chim.* **97**, 95 (1967).
98. K. Weissermel, E. Fischer, K. Gutweiler, H. D. Hermann, and H. Cherdron, *Angew. Chem., Int. Ed. Engl.* **6**, 526 (1967).
99. G. W. Polly, Jr. and W. E. Heiz, U.S. Patent 3,210,322 (1965); C. S. H. Chen and A. D. Edwardo, *Advan. Chem. Ser.* **91**, 359 (1969).
100. K. Takakura, K. Hayashi, and S. Okamura, *J. Polym. Sci.* **4**, 1731 (1966).
101. S. Rosinger, H. Hermann, and K. Weissermel, *J. Polym. Sci., Part A-1* **5**, 183 (1967).
102. H. Rao and D. S. Ballantine, *J. Polym. Sci., Part A* **3**, 2579 (1965).
103. C. F. Doyle, French Patent 1,398,493 (1965).
104. Chemical & Engineering News, *Chem. Eng. News* **50** (26), 4 (1972).
105. J. Furukawa, T. Saegusa, T. Tsuruta, H. Fujii, and T. Tatano, *J. Polym. Sci.* **36**, 546 (1959).
106. J. Furukawa, T. Saegusa, T. Tsuruta, H. Fujii, A. Kawasaki, and T. Tatano, *Makromol. Chem.* **33**, 32 (1959).
107. M. Letort and P. Mathis, *C. R. Acad. Sci.* **249**, 274 (1959).
108. O. Vogl, *J. Polym. Sci.* **46**, 261 (1969).
108a. O. Vogl, *J. Polym. Sci., Part A* **2**, 4591 (1964).
109. M. Letort and V. Petry, *J. Chim. Phys. Physiocochim. Biol.* **48**, 594 (1951).
110. M. Letort and P. Mathis, *C. R. Acad. Sci.* **241**, 1765 (1955).
111. O. Vogl, *J. Polym. Sci., Part A* **2**, 4607 (1964).
112. M. Letort and X. Duval, *C. R. Acad. Sci.* **216**, 58 (1943).
113. O. Vogl and W. M. D. Bryant, *J. Polym. Sci., Part A* **2**, 4633 (1964).
114. J. Brandrup and M. Goodman, *Makromol. Syn.* **3**, 74 (1969).
115. O. Vogl, *J. Polym. Sci., Part A* **2**, 4621 (1964).
115a. O. Vogl, German Patent 1,144,921 (1963).
116. H. Takida, *Kobunshi Kagaku* **24**, 68 (1967).
117. H. Tani, N. Araki, and T. Aoyagi, Japanese Patent 14,465 (1967).
118. J. Furukawa, T. Tsuruta, T. Saegusa, H. Fujii, and T. Tatano, *J. Polym. Sci.* **36**, 546 (1959).
119. S. J. Gregg and K. S. Sing, *J. Phys. Chem.* **56**, 388 (1952).
120. K. Weissermel and W. Schmeider, *Makromol. Chem.* **51**, 39 (1962).
121. O. Vogl, Belgian Patent 580,553 (1959).
122. G. Natta, G. Mazzanti, P. Coeradini, and I. W. Bassi, *Makromol. Chem.* **37**, 156 (1960).
123. J. Furukawa, T. Saegusa, H. Fujii, A. Kawasaki, H. Imai, and Y. Fujii, *Makromol. Chem.* **37**, 149 (1960).
124. J. Furukawa, T. Saegusa, and H. Fujii, *Makromol. Chem.* **44–46**, 398 (1960).
125. R. Wakasa, S. Ishida, and H. Fukuda, Japanese Patent 5351 (1967).
126. C. Chachaty, *C. R. Acad. Sci.* **251**, 285 (1960).
127. H. Fujii, J. Furukawa, and T. Saegusa, *Makromol. Chem.* **40**, 226 (1960).
128. H. Fujii, J. Furakawa, and T. Saegusa, *10th Annu. Meet. Soc. High Polym., Tokyo, 1961*, oral paper (1961).
129. H. Tani, T. Aoyagi, and T. Araki, *J. Polym. Sci., Part B* **2**, 921 (1964).
130. G. Natta, P. Corrandini, and I. W. Bassi, *J. Polym. Sci.* **51**, 505 (1961).
131. T. Imoto and T. Matsubara, *J. Polym. Sci.* **56**, 54 (1962).
131a. W. H. Smyers, U.S. Patent 2,274,749 (1942); S. M. Vokanyan, N. G. Karapetyan,

and G. A. Chukhadzhyan, *Arm. Khim. Zh.* **21**, 1019 (1968); T. Marsubara and T. Imoto, *Makromol. Chem.* **117**, 215 (1968); T. Imoto, K. Aotani, and T. Matsubara, Japanese Patent 4991 (1967); *Chem. Abstr.* **67**, 54625 (1951); S. R. Sandler and E. C. Leonard, U.S. Patent 3,422,072 (1969).
132. E. Späth, R. Lorenz, and E. Freund, *Chem. Ber.* **76**, 57, 523, and 725 (1943).
133. E. Späth and H. Schmid, *Chem. Ber.* **74**, 859 (1941).
134. C. M. Modena, G. Carraro, and G. Cossi, *J. Polym. Sci., Part B* 5, 613 (1966).
135. S. Gaertner, German Patent 165,984 (1903).
136. A. Novak and E. Whalley, *Trans. Faraday Soc.* **55**, 1490 (1959).
137. A. Novak and E. Whalley, *Can. J. Chem.* **37**, 1722 (1959).
138. O. Jacobsen, *Ber. Deut. Chem. Ges.* **8**, 87 (1875).
139. D. R. Husted and A. H. Ahlbrecht, British Patent 719,877 (1954).
140. I. Rosen, C. L. Sturm, G. H. McCain, R. M. Wilhjelm, and D. E. Hudgin, *J. Polym. Sci., Part A* **3**, 1545 (1965).
141. I. Rosen, British Patent 1,010,091 (1965).
142. O. Vogl, H. C. Miller, and W. H. Sharkey, *Chem. Eng. News* **50** (12), 41 (1972); O. Vogl, H. C. Miller, and W. H. Sharkey, *Macromolecules* **5**, 658 (1972); O. Vogl, U.S. Patent 3,454,527 (1969).
142a. O. Vogl, French Patent 1,528,327 (1968).
143. I. Rosen and C. L. Sturm, *J. Polym. Sci., Part A* **3**, 3741 (1965).
144. T. Iwata, G. Wasai, T. Saegusa, and J. Furukawa, *Makromol. Chem.* **77**, 229 (1964).
145. J. Sedlmeier, German Patent 1,189,714 (1965).
146. S. Temple and R. L. Thornton, *J. Polym. Sci., Part A-1* **10**, 709 (1972).
147. British Patent 1,036,870 (1966).
148. P. W. Bridgman and J. B. Conant, *Proc. Nat. Acad. Sci. U.S.* **15**, 680 (1929).
149. J. B. Conant and C. O. Tongberg, *J. Amer. Chem. Soc.* **52**, 1659 (1930).
150. J. B. Conant and W. R. Peterson, *J. Amer. Chem. Soc.* **54**, 628 (1932).
151. O. F. Vogl, *Chem. Ind. (London)* p. 748 (1961).
152. W. W. Moyer, Jr. and D. A. Grev, *J. Polym. Sci., Part B* **1**, 29 (1963).
153. J. N. Koral and E. M. Smolin, *J. Polym. Sci., Part A* **1**, 2831 (1963).
154. W. Sweeny, *J. Appl. Polym. Sci.* **7**, 1983 (1963).
155. H. Tani, T. Araki, and H. Yasuda, *J. Polym. Sci., Part B* **6**, 389 (1968).
156. Charbonnages de France, French Patent 1,566,796 (1969).
157. H. Moriguchi, T. Goto, Y. Kishi, and S. Yoshida, Japanese Patent 05,792 (1970).
158. K. D. Kiss, J. W. L. Fordham, and J. T. Reed, German Patent 1,495,536 (1970).
159. R. F. Nichol, *Orbital* **2**, 13 (1969).
160. H. Daimon, K. Kamio, and S. Kojima, Japanese Patent 2835 (1967).
161. T. Imoto and T. Nakajima, *Mem. Fac. Eng., Osaka City Univ.* **6**, 203 (1964).
162. W. Thuemmler, G. Lorenz, and K. Thinius, *Plaste Kaut.* **11**, 386 (1964); C. E. Lorenz, German Patent 1,181,417 (1964).
163. A. Tanaka, Y. Hozumi, S. Endo, and K. Taniguchi, U.S. Patent 3,405,094 (1968).
164. G. Natta, G. F. Pregaglia, G. Mazzanti, V. Zamboni, and M. Biraghi, *Eur. Polym. J.* **1**, 25 (1965).
165. T. L. Cairns, E. T. Cline, and P. J. Graham, U.S. Patent 2,828,287 (1958).
166. C. D. Lewis, R. N. McDonald, and C. E. Schweitzer, U.S. Patent 3,043,803 (1962).
167. R. J. Kray and C. A. Defazio, U.S. Patent 3,026,299 (1962).
168. S. Kambara, N. Yamazake, K. Ikebe, and R. Niiumi, Japanese Patent 25,517 (1968).
169. T. Higashimura, A. Tanaka, T. Miki, and S. Okamura, *J. Polym. Sci., Part A-1* **5**, 1927 (1967).

170. K. Takakura, K. Hayashi, and S. Okamura, *J. Polym. Sci., Part A-1* **4**, 1731 (1966).
171. K. Nakatsuka, F. Ide, and M. Takamura, Japanese Patent 03,106 (1970).
172. Daicell Co., Ltd., French Patent 1,578,678 (1969).
173. G. H. McCain, D. E. Hudgen, and I. Rosen, *J. Polym. Sci., Part A-1* **5**, 975 (1967).
174. E. F. T. White, British Patent 902,602 (1962).
175. C. E. Lorenz, British Patent 995,770 (1965).
176. T. Iwata, G. Wasai, T. Saegusa, and J. Furukawa, *Makromol. Chem.* **77**, 229 (1964).
177. V. Jaacks, *Advan. Chem. Ser.* **91**, 371 (1969).
178. M. Goodman and A. Abe, *J. Polym. Sci.* **59**, S37 (1962).
179. W. J. Sullivan, U.S. Patent 3,067,174 (1962).
180. H. Sumitomo and K. Kobayashi, *J. Polym. Sci., Part A-1* **4**, 907 (1966).
181. R. C. Schulz, G. Wegner, and W. Kern, *Angew. Chem.* **78**, 613 (1966).
182. W. T. Brady and H. R. O'Neal, *J. Polym. Sci., Part B* **2**, 649 (1964).
183. C. G. Overberger, S. Ishida, and H. Ringsdorf, *J. Polym. Sci.* **62**, S1 (1962).
184. K. Meyersen, R. C. Schulz, and W. Kern, *Makromol. Chem.* **58**, 204 (1962).
185. C. Aso and Y. Aito, *Bull. Chem. Soc. Jap.* **35**, 1426 (1962).
186. C. Aso, A. Furuta, and Y. Aito, *Makromol. Chem.* **84**, 126 (1965).
187. Farbwerke Hoechst, A.-G., French Patent 1,391,539 (1965).
188. E. Fischer, K. Kuellmar, and K. Weissermel, U.S. Patent 3,210,297 (1965).
189. Y. Miyake and K. Shoji, Japanese Patent 23,531 (1970).
190. K. Nakatsuka, F. Ide, and M. Takamura, Japanese Patent 26,506 (1970).
191. Borden Co., British Patent 928,118 (1963).
192. K. Kobayashi and H. Sumitomo, *J. Polym. Sci., Part B* **10**, 703 (1972).
193. R. J. Cotter and M. Matzner, "Ring-Forming Polymerizations," Vol. 1, Part B, pp. 291–296. Academic Press, New York, 1972; H. Yasuda and H. Tani, *Makromol.* **6**, 303 (1973).

Chapter 6

POLYMERIZATION OF EPOXIDES AND CYCLIC ETHERS

1. INTRODUCTION

Ethylene oxide, various epoxides, and cyclic ethers can be polymerized with anionic, cationic, and coordination type catalysts.

Wurtz [1] was the first to report the base-initiated polymerization of ethylene oxide. Roithner [2] extended this work and showed that raising the temperature accelerated the polymerization (degree of polymerization approx. 30). In addition he showed that aqueous alkali was also an effective catalyst.

Wurtz [3] also showed that polyoxyethylene acetates could be prepared by heating ethylene oxide with acetic anhydride. Staudinger [4,4a] later re-investigated the polymerization of ethylene oxide and systematically studied the effect of various anionic and cationic catalysts on the polymerization. For example, the most effective catalysts found were $(CH_3)_3N$, Na, K, and $SnCl_4$. Other compounds showing catalytic activity were $NaNH_2$, ZnO, SrO, and CaO. The latter catalysts gave a degree of polymerization varying from 250 to 2500. Activated alumina was also effective but the following were ineffective: ultraviolet rays, Florida earth, ferric oxide, magnesium oxide, lead oxide, silica gel, and activated carbon.

Staudinger [4a] also demonstrated that as the molecular weight of polyoxyethylene increased (to 100,000) the melting points (40°–175°C) and the viscosity of a 1% aqueous solution of the polymer increased but the polymer solubility in water tended to diminish (especially at molecular weight 100,000).

In 1957 Hill [5] and co-workers of Union Carbide [5a] reported on the commercial process of polymerizing ethylene oxide with specially prepared calcium carbonates and calcium amides.

Pruitt and Baggett [6] of the Dow Chemical Company reported that the reaction product of ferric chloride and propylene oxide afforded a coordination catalyst capable of polymerizing propylene oxide to a high molecular weight crystalline or amorphous rubbery polymer.

The polymerization of propylene oxide to a high molecular weight product has also been reported to occur with a variety of metal alkoxides, halides, and alkyls [7,8,8a].

Low molecular weight polyoxypropylene glycols can be obtained by the potassium hydroxide-catalyzed polymerization of propylene oxide [9]. The latter products are suitable intermediates for the preparation of flexible polyurethane foams, elastomers, and adhesives.

The mechanisms of polymerization for epoxides and cyclic ethers may be summarized as in Scheme 1, using ethylene oxide as an example. Routes (A) through (C) illustrate the following mechanisms, respectively: cationic, anionic, and coordinated anionic–cationic. The coordinate mechanism probably involves two metal atoms and will be discussed later in greater detail

Today polyethylene glycols are available in a range from water-soluble viscous liquids (mol. wt. below 700) to grease and waxes (mol. wt. 1000 to 20,000) from a variety of sources [10,10a]. Products of molecular weight of about 100,000 to 5,000,000 are available from Union Carbide under the Polyox trademark [10a]. The latter have melting points of about 65° ± 2°C and a specific gravity of 1.21 gm/cm³.

Very low molecular weight polyoxyethylene glycols, $HO(CH_2—CH_2O)_nH$, where $n = 2–6$ are useful starting materials for the preparation of diesters and polyesters.

SCHEME 1

Outline of the Mechanisms of Polymerization of Alkylene Oxides

$H_2C\overset{O}{—}CH_2 + X^+(Y^-) \xrightarrow{(A)} \left[H_2C\overset{\overset{X}{|}}{\overset{O^+}{—}}CH_2 \right] \xrightarrow{(A)} [+CH_2—CH_2—OX]$

(C) X^+Y^- → $CH_2—CH_2$, O ↓ $Y^-—X^+$ (C) →

(B) $Y^-(X^+)$ ↓

$YCH_2CH_2O^-(X^+)$ —(B), $H_2C\overset{O}{—}CH_2$ → $YCH_2—CH_2O[CH_2—CH_2O]_nX$

$[+CH_2—CH_2—OX]$ —X^+Y^-, $CH_2\overset{O}{—}CH_2$ ↓ $YCH_2—CH_2O[CH_2—CH_2O]_nX$

(C) $H_2C\underset{O}{—}CH_2$ ↓

$CH_2—CH_2$, O ↓ $O^-—X^+$, CH_2, CH_2, Y —(C), $H_2C\underset{O}{—}CH_2$ → $YCH_2—CH_2O[CH_2—CH_2O]_nX$

Low to medium range of molecular weight polyoxyethylene glycols are useful in the pharmaceutical, cosmetic, polyurethane, rubber, and textile fields. Other applications are additives for metal plating baths, lubricants, plasticizers, humectants, spreading agents for sprays, wood preservatives, and antistatic agents.

High molecular weight polyoxylene glycol polymers find use in such areas as warping sizing agents, water-soluble films, flocculants and coagulants, beer foam stabilization, thickeners, and protective colloids binders, and in the areas mentioned above for the low molecular weight polymers.

The polyoxyethylene glycols enjoy a low order of toxicity and specific toxicity data have been published [10a,11].

Several earlier reviews describing the polymerization of epoxides and cyclic ethers have appeared and are worthwhile consulting for further information [12].

2. CONDENSATION METHODS

The condensation methods to be discussed in consecutive order in this section are anionic, cationic, and coordinated anionic–cationic by homogeneous and heterogeneous catalyst systems.

A. Anionic Polymerization

Perry and Hibbert [13,13a] established the stepwise mechanism [13a] of polymerization of ethylene oxide by aqueous alkali. They also showed that the intermediate polyoxyethylene glycols $HO(CH_2CH_2O)_nH$ ($n = 2$ to 18) [13a] are identical in structure to the products and are reactive with ethylene oxide to initiate further polymerization. A lower ratio of polyglycol to ethylene oxide gave a more complete reaction. This was also true for the use of water. Perry and Hibbert [13a] also showed that the percent polymerization increases linearly with time (Fig. 1) and that the degree of polymerization (and molecular weight) increased as the reaction proceeded (Table I), in agreement with the stepwise mechanism (Eq. 1).

$$\underset{\text{O}}{CH_2\text{—}CH_2} + NaOH \longrightarrow HOCH_2CH_2ONa \xrightarrow{\underset{\text{O}}{CH_2\text{—}CH_2}}$$

$$HOCH_2CH_2OCH_2CH_2ONa \xrightarrow{\underset{\text{O}}{CH_2\text{—}CH_2}} HO[\text{—}CH_2CH_2O]_nNa \qquad (1)$$

Perry and Hibbert [13a] further showed that the polyoxyethylene glycols and the products after the complete reaction of ethylene oxide are capable of again reacting ethylene oxide in stages by successive addition (stages 1–5) to increase the degree of polymerization from 6 to 74 as shown in Table II.

This principle of multistage polymerizations has been applied in some patents to prepare polyoxyethylene oxide [14] and block copolymers with propylene oxide [15].

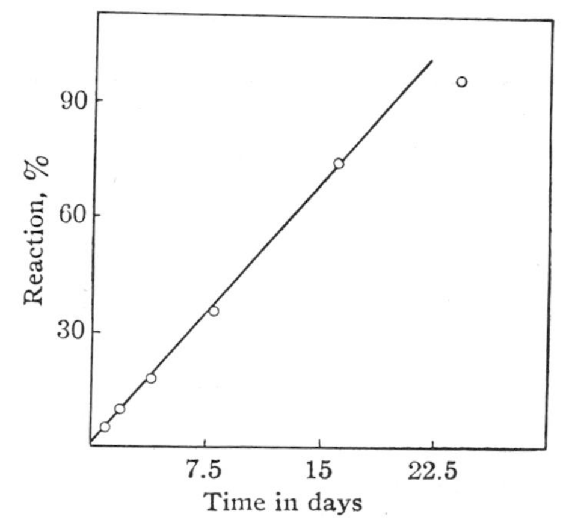

FIG. 1. Polymerization of ethylene oxide in the presence of water and alkali. Reprinted from Perry and Hibbert [13a], copyright 1940 by the American Chemical Society; reprinted by permission of the copyright owner.

$$n\mathrm{CH_3{-}\underset{\diagdown O \diagup}{CH{-}CH_2}} \xrightarrow[\ +\ \mathrm{NaOH}]{\mathrm{CH_3CH(OH){-}CH_2OH}} \left[-\underset{|}{\overset{\mathrm{CH_3}}{}}\mathrm{CH{-}CH_2{-}O} \right]_n \xrightarrow{\mathrm{CH_2{-}CH_2 \ (O)}}$$

$$\mathrm{HO[CH_2CH_2O]_x} \left[\overset{\mathrm{CH_3}}{\mathrm{CH}}\mathrm{{-}CH_2{-}O} \right]_n \mathrm{[CH_2CH_2O]_yH} \qquad (2)$$

2-1. *Preparation of Polyoxyethylene Glycol by Polymerization of Ethylene Oxide with Aqueous Alkali* [13a]

$$\mathrm{\underset{\diagdown O \diagup}{CH_2{-}CH_2}} \xrightarrow{\mathrm{KOH{-}H_2O}} [\mathrm{-CH_2{-}CH_2{-}O-}]_n \qquad (3)$$

To a cooled Hoke pressure cylinder containing 44 gm (1.0 mole) of ethylene oxide is added 2.0 gm (0.0145 mole) of potassium hydroxide and 180 gm (10.0 moles) of water. The pressure cylinder is allowed to stand for 2 weeks at 35°C, cooled to −10°C, and opened. No residual ethylene oxide remains and the brown viscous reaction mixture is neutralized with 5% HCl and concentrated at 50°C at 15 mm Hg. The residue is dissolved in 800 ml of anhydrous ether containing 5% absolute ethanol in order to remove inorganic constituents. The solution is decolorized with activated carbon, filtered, and concentrated under reduced pressure to afford 40.8 gm (92.8%) of liquid polyethylene oxide.

St. Pierre and Price [16] reported that propylene oxide could be polymerized readily by powdered potassium hydoxide at atmospheric pressure

TABLE I

CHANGE IN DEGREE OF POLYMERIZATION WITH PROGRESS OF REACTION[a,b]

Reaction time (days)	Percentage reaction	M.p. of product (°C)	Refractive index of product (60°C)
1	5	5–7	1.4450
2	9.8	28–30.0	1.4550
4	17.7	30–32.0	1.4556
8	35.4	30–33.5	1.4561
16	74.3	45–47.0	1.4563
24	96.1	47–49.5	1.4578

[a] Reaction mixture: 44 gm ethylene oxide, 1.8 gm H_2O, and 0.1 gm KOH at 20°–25°C

[b] Reprinted from Perry and Hibbert [13a]. Copyright 1940 by the American Chemical Society. Reprinted by permission of the copyright owner.

TABLE II
PRODUCTS FROM A STEPWISE SERIES OF REACTIONS[a]

Sample	M.p. of product (°C)	Refractive index of product (60°C)	Av. mol. wt.
Stage 1	−19 to −17	1.4507	287
Stage 2	9 to 11	1.4536	424
Stage 3	20 to 22	1.4545	526
Stage 4	40 to 43	1.4571	1090
Stage 5	54 to 58	1.4579	3270

[a] Reprinted from Perry and Hibbert [13a]. Copyright 1940 by the American Chemical Society. Reprinted by permission of the copyright owner.

at 25°–30°C to give a polymer of molecular weight 3000–5000. The polymer has a high proportion of unsaturated end groups, which are mainly the allyl type with a minor proportion of the propenyl type (Eq. 4). Several other catalysts were studied and of these only KOH gave the best results. For example, $NaOCH_3$, LiOH, NaOH, K_2CO_3, $Ba(OH)_2$, $C_6H_5CH_2N^+(CH_3)_3$-OH^-, or $C_6H_5CH_2N(CH_3)_3OC_2H_5$ gave no polymer and $NaNH_2$ gave only 1% of a low molecular weight polymer after 92 hr at 25°C.

$$\underset{\substack{|\\CH_3}}{CH_2\overset{O}{—}CH} + \bar{O}H \longrightarrow HOCH_2\underset{\substack{|\\CH_3}}{CH}O^- \xrightarrow{\underset{\substack{|\\CH_3}}{CH_2\overset{O}{—}CH}} HOCH_2\underset{\substack{|\\CH_3}}{CH}—\left[OCH_2\underset{\substack{|\\CH_3}}{CH}\right]_n—OCH_2—\underset{\substack{|\\CH_3}}{CH}—O^- \tag{4}$$

$$\longrightarrow HOCH_2\underset{\substack{|\\CH_3}}{CH}—\left[OCH_2\underset{\substack{|\\CH_3}}{CH}\right]_{n+1}OH$$

$$\xrightarrow{\left(—HO\underset{\substack{|\\CH_3}}{CH}—CH_2O—\right)} HOCH_2—\underset{\substack{|\\CH_3}}{CH}—\left[OCH_2—\underset{\substack{|\\CH_3}}{CH}\right]_{n-1}—OCH_2—CH{=}CH_2$$

The specificity of solid potassium hydroxide suggests that the potassium ion plays an important role in the mechanism of polymerization.

2-2. *Preparation of Polyoxypropylene Glycol by the Room-Temperature Polymerization of Propylene Oxide Using KOH* [16]

To a flask equipped with a mechanical stirrer and condenser is added 200 gm (3.45 moles) of dried propylene oxide (dried over KOH and distilled before use) and 40 gm (1.4 mole) of finely powdered technical grade potassium hydroxide. The mixture is stirred for 42 hr at room temperature and then ether is added to dissolve the polymer. Hydrogen chloride is then bubbled into the solution and the precipitated KCl filtered. The base (KOH) can also be removed by shaking the ether solution with Amberlite IR-120 (H) (Rohm & Haas Co.). The filtrate is concentrated under reduced pressure (150°C at 3 mm Hg) to afford 160 gm (80%) of product. This and several other related experiments are summarized in Table III.

TABLE III

ROOM-TEMPERATURE POLYMERIZATION OF PROPYLENE OXIDE WITH POWDERED POTASSIUM HYDROXIDE[e]

Monomer (gm)	KOH (gm)	Time (hr)	Yield (gm)	OH^- (Eq./gm $\times 10^4$)	C=C (Eq./gm $\times 10^4$)	η_{sp}/c[a]	$n^{20}D$
200	40	42	160[b]	3.82	—	—	—
150	20	64	36.5[c]	2.96	—	0.141	—
150	10	48	—	4.88	—	0.138	—
420	10	141	250[c]	—	1.35	0.154[d]	1.4509
406	30	42	209[c]	3.61	—	0.136	1.4509
800	60	220	360[c]	6.02	1.37	0.161	—

[a] 4% solution in benzene; c = gm of polymer in 100 gm of solution.

[b] Isolated by washing product in ether with hydrochloric acid and evaporating to 150°C at 3 mm.

[c] Isolated by shaking an ether solution with Amberlite IR-120(H) to remove base, followed by vacuum evaporation.

[d] Cryoscopic molecular weight (benzene) 3440.

[e] Reprinted from St. Pierre and Price [16]. Copyright 1956 by the American Chemical Society. Reprinted by permission of the copyright owner.

Recently it has been reported that propylene oxide is also polymerized readily with RbOH and CsOH [17]. The rate of reaction can be markedly increased by the addition of small amounts of alcohol to the propylene oxide–solid KOH system. Too much alcohol slows down the rate of polymerization [17].

In the absence of alcohols the KOH probably reacts with propylene oxide to form the monoalkoxide of 1,2-propanediol, which then initiates the polymerization (see Eq. 4) [18].

The polymerization of *l*-propylene oxide by potassium hydroxide affords a low molecular weight, optically active polymer having m.p. 55°–56.5°C. This is in sharp contrast to the liquid polymer obtained from *dl*-propylene oxide under identical conditions [19].

2-3. *Polymerization of l-Propylene Oxide with Solid KOH Catalyst* [19]

$$\overset{-}{OH} + \underset{\diagdown O \diagup}{CH_2{-}\overset{CH_3}{\overset{|}{C}}H_*} \longrightarrow HOCH_2\underset{*}{\overset{CH_3}{\overset{|}{C}H}}{-}O^- \longrightarrow$$

$$HOCH_2\underset{*}{\overset{CH_3}{\overset{|}{C}H}}{-}OCH_2{-}\underset{*}{\overset{CH_3}{\overset{|}{C}HO}} \longrightarrow HO\left[{-}CH_2\underset{*}{\overset{CH_3}{\overset{|}{C}H}}{-}O\right]_n H \quad (5)$$

To a cooled Pyrex tube is added 2.5 gm (0.043 mole) of *l*-propylene oxide and 0.5 gm (0.00895 mole) of powdered potassium hydroxide. The tube is flushed with nitrogen, sealed, and put in a shaker water bath at 25°C for 50 hr. The mixture finally solidifies to a wax after 50 hr and then is dissolved in 250 ml of benzene. The benzene solution is washed successively with water, dilute aqueous sulfuric acid, aqueous sodium bicarbonate solution, and then with distilled water until neutral. The benzene solution is freeze-dried (under reduced pressure) to afford 2.2 gm (88%) of a white crystalline poly(*l*-propylene oxide), m.p. 55.5°–56.5°C.

Using similar conditions *dl*-propylene oxide affords 88% of a liquid poly(*dl*-propylene oxide).

Recently Patat and Wojtech reported that sodium phenoxide does not react with ethylene oxide in the absence of phenol [20]. Phenol alone preferentially gave β-phenoxyethyl alcohol [21]. Ethylene oxide also reacts under other conditions with alkyl phenols to give alkylphenoxy ethanols [22]. Several substituted alkylene oxides have been studied and their rates of polymerization have been reported [23,23a]. The anionic polymerization appears to be favored by electron-withdrawing groups [24].

Glycidol, a substituted propylene oxide, has been shown to polymerize at a greater rate than propylene oxide using such catalysts as triethylamine, pyridine, lithium hydroxide, potassium hydroxide, sodium hydroxide, sodium methoxide, and sodium amide [25].

$$n\overset{\diagup O \diagdown}{CH_2{-}\underset{CH_2OH}{\underset{|}{C}H}} \longrightarrow \left[{-}CH_2{-}\underset{CH_2OH}{\underset{|}{C}H}{-}O{-}\right]_n \quad (6)$$

2-4. *Preparation of Polyglycidol Using Several Anionic Catalysts* [25]

To an Erlenmeyer flask containing 74 gm (1.0 mole) of glycidol is added 0.74 gm (0.013 mole) of KOH and the flask is placed in a water bath kept at 23°C for 27 hr. After this time the mixture is dissolved in methanol and then shaken with 3.0 gm of Amberlite ion-exchange resin IR-120-H. The solution is filtered, and the solvent and monomer are removed by gradually heating under reduced pressure to 140°C at 0.5 mm Hg to afford 88% of polyglycidol

TABLE IV
ROOM-TEMPERATURE POLYMERIZATION OF GLYCIDOL AND PROPYLENE OXIDE USING INORGANIC AND ORGANIC BASES[a]

Catalyst	Catalyst (%)	Time (hr)	Conversion (%)	
			Glycidol	Propylene oxide[b]
KOH	1	27	88	
	10	51		33
$NaOCH_3$	1	20	87	
	10	72		0
LiOH	1	72	84	
	10	72		0
NaOH	1	72	81	
	10	72		0
$NaNH_2$	1	72	82	
	10	92		1
Et_3N	1	19.5	73	—
Pyridine	1	25	78	—
$CaCl_2$	1	25	38	—

[a] Reprinted from Sandler and Berg [25]. Copyright 1966 by the *Journal of Polymer Science*. Reprinted by permission of the copyright owners.

[b] A zero signifies an experiment has been performed and no polymer was obtained; however, a dash indicates that an experiment was not performed. Data from St. Pierre and Price [16].

(mol. wt. 440), $\eta_{inh} = 0.070$ (in DMSO at 32.5°C). Some additional data on useful catalysts for glycidol as compared to propylene oxide are shown in Table IV.

B. Cationic Polymerization

Cationic polymerization [26] affords relatively higher molecular weight polymers as compared to anionic catalysis. Furthermore, cationic catalysis is effective for cyclic ethers with three-, four-, or five-membered rings [27,27a]. Six-membered cyclic ethers are rather stable and do not tend to polymerize [27,27a]. However, there is some tendency for the seven- and eight-membered cyclic ethers to polymerize [28].

Cationic polymerization of cyclic ethers is thought to occur via initial formation of the oxonium ion or salt. Several stable trialkyl oxonium salts have been isolated as crystalline solids [28,29] [$(C_2H_5)_3^+OBF_3^-$, m.p. 92°C] [29]. An intermediate carbonium ion has also been proposed.

Recently it was suggested that the boron trifluoride-catalyzed rearrangement of 1,1-disubstituted ethylene oxide may give a carbonium ion intermediate [30].

$$\text{(}t\text{-Bu)(H}_3\text{C)C(-O-)CD} + BF_3 \longrightarrow \text{(}t\text{-Bu)(H}_3\text{C)C}^+\text{—C(OBF}_3^-\text{)D} \longrightarrow \text{(}t\text{-Bu)(H}_3\text{C)CD—CH=O} \quad (7)$$

In addition to the polyether products, many 12-membered ring polyethers have been obtained in 30–40% yields from the $(C_2H_5)_3^+OBF_3^-$ and BF_3-catalyzed polymerization of propylene, 1,2-butylene oxide, and epichlorohydrin [31].

Ethylene oxide reacts with sulfuric acid to give a series of low molecular weight oligomers, $HO(CH_2CH_2O)_nH$, where $n = 1\text{–}4$ [32]. Furthermore, using boron trifluoride in a homogeneous system affords polymers of 500–900 molecular weight [33,33a,b]. Stannic chloride-initiated polymerization of ethylene oxide affords molecular weights up to 20,000 [33a,b,34,34a,b]. Antimony pentachloride [35] affords polymers with molecular weights in the range of 10,000, whereas aluminum chloride is mildly active at 80°C [33]. Other effective catalysts are $FeCl_3$, $FeBr_2$, BCl_3, $TiCl_4$, and $ZnCl_2$. Aluminum, titanium, and germanium chlorides are ineffective for the low-temperature polymerization of ethylene oxide [34].

Typical data using the stannic chloride and boron trifluoride catalyst are shown in Tables V and VI. It is noted that the molecular weights obtained

TABLE V

FORMATION OF THE POLYMER OVER THE COURSE OF THE REACTION[a,b]

Time (min)	C_2H_4O consumed (moles/liter)	Polymer formed (base moles/liter)	Mol. wt.	Mole ratio polymer/catalyst
13	0.20	0.2	*ca.* 600	2.0
29.5	0.40	0.35	1010	2.1
51	0.60	0.51	1530	2.0
81	0.80	0.60	2090	1.7
120	1.00	0.73	2560	1.7
190	1.20	0.81	3000	1.6
315	1.42	0.91	3630	1.5
—	1.67	1.01	3830	1.6

[a] (C_2H_4O), 2 M; $(SnCl_4)$, 0.0072 M; temp. 20°C.

[b] Reprinted from Worsfold and Eastham [34]. Copyright 1957 by the American Chemical Society. Reprinted by permission of the copyright owner.

TABLE VI
EFFECT OF REACTANT CONCENTRATION ON POLYMER FORMATION AT 20°C[a]

BF_3 (mole/liter)	C_2H_4O (moles/liter)	H_2O (mole/liter)	Wt. polymer/liter (gm)	Mol. wt.	Polymer (mole/liter)
0.02	2.0	0.100	60.5	730	0.083
0.02	2.0	0.080	57.1	—	—
0.02	2.0	0.074	54.9	760	0.072
0.02	2.0	0.050	36.0	770	0.047
0.02	2.0	0.039	32.2	—	—
0.02	2.0	0.026	21.8	—	—
0.02	2.0	0.020	20.0	770	0.026
0.03	2.0	0.030	22.4	730	0.031
0.04	2.0	0.040	30.6	760	0.041
0.05	2.0	0.050	32.0	670	0.048
0.02	1.0	0.050	27.6	560	0.049
0.02	1.0	0.100	35.3	—	—
0.02	0.5	0.050	13.1	—	—
0.02	0.5	0.100	21.0	—	—

[a] Reprinted from Worsfold and Eastham [34a]. Copyright 1957 by the American Chemical Society. Reprinted by permission of the copyright owner.

with the $SnCl_4$ catalyst are much higher than with the BF_3 catalyst regardless of the amount of water present.

2-5. *Preparation of Polyethylene Oxide Using $SnCl_4$ (or BF_3) Catalyst* [34]

*Materials.** Ethylene oxide is purified by low-temperature distillation, by distillation at −78°C from dried sodium hydroxide pellets, and finally by drying over barium oxide, all under vacuum.

Stannic chloride is fractionally distilled from tin metal, then twice distilled under vacuum.

Solvents are fractionated from phosphorus pentoxide. Ethylene chloride is treated with aluminum chloride, washed and dried before fractionating.

*Method.** Polymerizations are carried out in an all-glass vacuum system and the rate of reaction in ethylene chloride solution followed by the decrease in vapor pressure of the reaction mixture. The relationship between oxide concentration and vapor pressure is linear and hence readily determined from known mixtures.

Stannic chloride and water are measured out by allowing the liquid, at controlled temperature, to evaporate into bulbs of known volume until

* From Worsfold and Eastham, [34]. Copyright 1957 by the American Chemical Society. Reprinted by permission of the copyright owner.

equilibrium is established. The quantity is then determined from the vapor pressure of the liquid.

Ethylene oxide and ethyl chloride are measured in the gas phase, but the less volatile solvents are pipetted into a trap containing phosphorus pentoxide and degassed before distilling into the reaction vessel.

Catalyst, water if any, and solvent in that order are condensed into the reaction vessel and brought to reaction temperature. Ethylene oxide is then rapidly distilled in from a small side arm on the main reaction vessel. Magnetic stirring establishes equilibrium within 3 minutes.

Ethyl chloride is employed as solvent when reaction products are to be determined because it can be separated readily by distillation. Ethylene chloride is used for most of the kinetic studies because of its lower vapor pressure.

Reaction products are isolated by treating a 10% solution of ethylene oxide with 1 mole % stannic chloride at 20°C. Solvent, excess oxide and volatile products are removed by distillation, leaving a white wax having an infrared spectrum identical with that of a polyethylene glycol of similar melting point. The volatile product is found, by mass spectrometer analysis, to consist of 92% dioxane and 8% 2-methyl-1,3-dioxolane.

*Molecular weights.** Molecular weights are determined viscometrically in ethylene chloride, using Ostwald viscometers with flow times of about 200 sec at 20°C. For molecular weights less than 1000 the relationship

$$\eta_{sp}/c = 0.048 + 0.000204\,M$$

is used where c is in base moles per liter. This expression is obtained using commercial polyglycols whose molecular weights are determined by end-group analysis and by the depression of the benzene freezing point.

Molecular weights above 1000 are given by the expression

$$\eta_{sp}/c = 0.160 + 0.000089\,M$$

which is obtained from commercial polyglycols whose molecular weights are first determined in carbon tetrachloride using the viscometric data of Fordyce and Hibbert [34c].

Polymer samples are obtained from the reaction mixture by quenching aliquots in aqueous alcohol, evaporating on a steam-bath, then drying for 2 or 3 hr under vacuum at 70°C.

As mentioned in the preparation heading, the same technique is also used for the polymerizations with BF_3 [34]. Decreasing the temperature in the

* From Worsfold and Eastham, [34]. Copyright 1957 by the American Chemical Society. Reprinted by permission of the copyright owner.

latter case raised the molecular weight from 760 at 20°C to 880 at 0°C and 925 at −20°C.

Earlier Staudinger [4,4a] polymerized propylene oxide with homogeneous Lewis acid catalysts to give viscous liquid (atactic) polymers of high molecular weight. Typical Lewis acids which are active initiators and which are soluble in propylene are $AlCl_3$, $InCl_3$, $SbCl_3$, $BiCl_2$, $FeCl_3$, $FeBr_3$, $SnCl_4$, $TiCl_4$, $ZnCl_2$, and $ZrCl_4$ [35]. Inactive catalysts include $SbCl_3$, $NiCl_2$, CuCl, $CuCl_2$, $CrCl_3$, $CoCl_2$, $HgCl_2$, and $MgCl_2$. Sulfuric acid affords low molecular weight oligomers [36].

The use of organometallic zinc and aluminum catalysts will be discussed in the next section.

Other epoxide monomers which are also polymerized by Lewis acid catalysts are styrene oxide [35], phenyl glycidyl ether [37], epichlorohydrin [37], 2-butene oxide [38], and alicyclic [39] oxides. The reactivity of the alicyclic oxide follows the order: $6 > 5 > 7 \gg 8$. A variety of 3,3-*sym*-disubstituted oxacyclobutanes polymerize with the use of homogeneous cationic catalysts to give high molecular weight polymers [40,40a].

2-6. *Preparation of Polyoxacyclobutane Using Boron Trifluoride Dihydrate Catalyst* [40]

$$\begin{matrix} CH_2-CH_2 \\ | \quad\quad | \\ CH_2-O \end{matrix} \xrightarrow{BF_3\cdot 2H_2O} [-O-CH_2CH_2CH_2-]_n \tag{8}$$

Into a flask cooled in liquid nitrogen and equipped with a mechanical stirrer and Dry Ice condenser is distilled in under reduced pressure 195 gm (3.36 mole) of oxacyclobutane and 490 ml of methyl chloride. The mixture is warmed to −60°C and 22.1 ml of $BF_3 \cdot 2H_2O$ is added. The polymerization starts and methyl chloride refluxes. After 18 hr the mixture is poured into 2 liters of ether (*in a hood*) and the methyl chloride boils off. The solution is treated with aqueous sodium hydroxide and then with water. The ether solution is dried over K_2CO_3 and concentrated to afford 171 gm (88%) of the polymer, m.p. 25°C, viscosity (approx.) 0.05 dl/gm. In addition 1 gm (0.5%) of tetramer is isolated.

2-7. *Preparation of Poly(phenyl glycidyl ether) Using Dibutylzinc Catalyst* [37]

$$C_6H_5OCH_2-\underset{\diagdown O \diagup}{CH-CH_2} \xrightarrow{(C_4H_9)_2Zn} \left[\begin{matrix} -CH_2-CH-O- \\ | \\ CH_2 \\ | \\ OC_6H_5 \end{matrix}\right]_n \tag{9}$$

To a glass vial is added by means of a syringe 10 gm (0.0745 mole) of phenyl glycidyl ether, 15 gm of toluene, 0.15 gm (0.08 mole) of dibutylzinc, and

1.34 gm (0.0745 mole) of water. The vial is cooled, sealed under nitrogen, and placed in a circulating air oven at 90°C for 24 hr. The polymer is isolated by opening the tube, dissolving out the material with acetone, and adding it to 200 ml of toluene in a Waring Blender. After blending, the polymer is obtained as a finely divided swollen product which is precipitated in 3 liters of ethanol, filtered, washed with ethanol, and dried under reduced pressure for 15 hr at 40°–60°C at 10–20 mm Hg. The polymer is finally obtained in the amount of 9.7 gm (97%) with a reduced viscosity of 10.2 (in *p*-chlorophenol 0.05 gm/25 ml containing 2% α-pinene at 47°C).

2-8. *Preparation of Poly(cyclohexene oxide) Using Triethylaluminum Catalyst* [41]

$$\text{cyclohexene oxide} \xrightarrow{Al(C_2H_5)_3} [\text{-}C_6H_{10}\text{-}O\text{-}]_n \tag{10}$$

To a dry nitrogen-filled flask equipped with a stirrer and condenser and cooled to −78°C is added 100 ml of *n*-hexane and 1.0 ml (0.01 mole) of triethylaluminum. Cyclohexene oxide is slowly added; the reaction is exothermic and solid polymer precipitates. Addition of the monomer is continued until 20.5 gm (0.21 mole) has been added, and the temperature throughout the reaction is kept at −70°C. After 25 min reaction time the unreacted catalyst is decomposed with methanol and concentrated HCl and then filtered. The polymer is washed several times with methanol and then dried under reduced pressure. The polymer is purified by dissolving in benzene, filtering, precipitating in methanol, filtering, and drying to afford 16.8 gm (82%). $[\eta] = 2.28$ (in C_6H_6 at 30°C), softening temperature 70°C. Several other alicyclic epoxides were polymerized by a similar technique and some of them are shown in Table VII [41].

Tetrahydrofuran also polymerizes well with cationic catalysts [27a]. Effective catalysts are $BF_3(O(C_2H_5)_2$–epichlorohydrin [27a,42], $C_6H_5N_2PF_6$ [43], PF_5 [44], and metal halides ($FeCl_3$, $AlCl_3$, $SnCl_4$, $SbCl_5$) in reactions with labile halogen compounds (chloromethyl acid chloride, etc.), ortho esters, and acetic anhydride [45]. Trialkyl oxonium salts, $(C_2H_5)_3O^+BF_3^-$ [46], and other inorganic halides are also effective ($POCl_3$, PCl_3, $SOCl_2$, $FeCl_3$–$AlCl_3$, $FeCl_3$–$TiCl_4$, $ClSO_3H$, FSO_3H) [47]. Several other cyclic ethers and epoxides have been polymerized and are mentioned in Table VII and in the Miscellaneous Methods section.

It should be noted that the polymerization of epoxides using typical Lewis acids such as $FeCl_3$ does not always mean that a cationic mechanism is followed. In fact it may be that a ferric alkoxide structure really does the catalysis and that a coordinate anionic–cationic mechanism is involved, as described in the next section [35–48].

C. Coordination-Catalyzed Polymerizations

The polymerizations induced by several Lewis acid type catalysts, organometallics, and alkaline earth oxides appear to fit a coordinate mechanism. In order to explain steric inversion, the transition state seems to require at least two separate metal sites, as shown below:

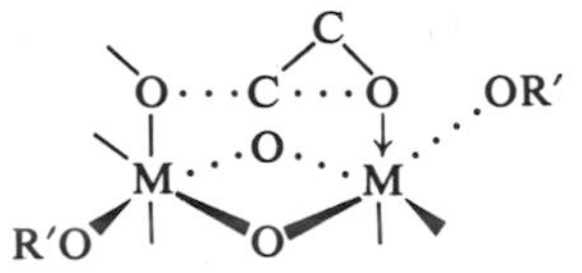

Vandenberg's [49,49a] work supports the following conclusions:

1. Epoxides polymerize with inversion of configuration at the carbon atom undergoing ring opening.
2. Coordination catalysts cause attack at the primary carbon atom of monosubstituted epoxides.
3. The coordination polymerization of epoxides involves two or more metal atoms [49,49a,50].

For example, using aluminum Eq. (11) illustrates some of these conclusions:

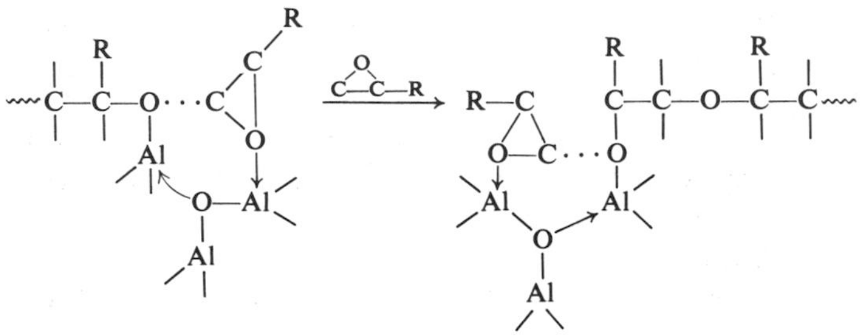

(11)

Propagation occurs largely by rearward displacement at the primary carbon atom of the epoxide. This is in line with the retention of configuration of the asymmetric carbon atom of *l*-propylene oxide which was earlier observed by Price and Osgan [19], and with Parker's results that nucleophilic attack on epoxides occurs more readily on primary rather than on secondary carbon atoms [51].

In order to clarify the stereochemistry and mechanism of epoxide polymerization Price prepared *cis*- and *trans*-1,2-dideuterioethylene oxides and polymerized them with cationic, anionic, and coordination catalysts [50].

TABLE VII

POLYMERIZATION OF ALICYCLIC EPOXIDES USING ALUMINUM ALKYL CATALYSTS[a]

Epoxide	Wt. of monomer	*n*-Hexane (ml)	$Al(C_2H_5)_3$ (ml)	Reaction conditions: Temp. (°C)	Reaction conditions: Time (min)	% Polymer	S.P. (°C)	$[\eta_i]$
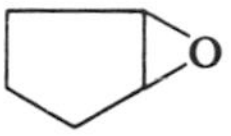	1.0	6	0.1	−60	10	29	—	1.27
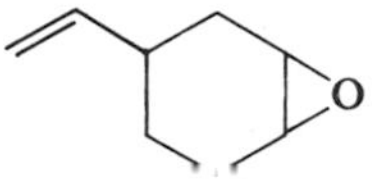	1.92	2	0.1	−78 to +30	5	93.7	65	0.83
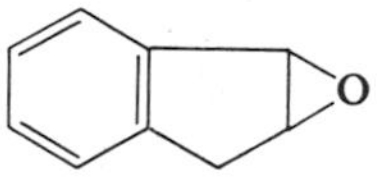	1.0	(1 ml *m*-xylene)	0.1	−20 to +30	5	25	—	
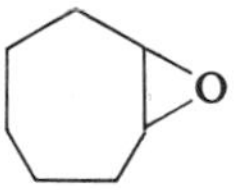	1.0	6	0.1	−60	10	32	60	0.7
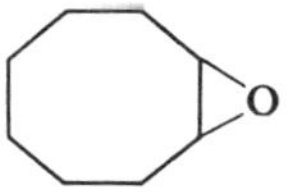 *cis*	0.95	—	0.1	+125	960	47	65	0.04
trans	1.0	1	0.1	−78		No reaction		
	1.0	1	0.1	25		Violent reaction		

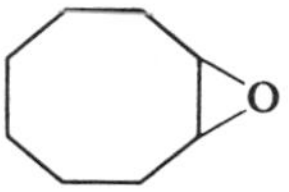

[a] Data taken from Bacskai [41].

Regardless of the catalyst used each polymerization proceeded with inversion of configuration at the carbon atom undergoing ring opening [50]. Thus, for example, *cis*(meso)-dideuterioethylene oxide monomer will give *dl* units in the polymer and the *trans*(*dl*)-dideuterioethylene oxide monomer will give meso polymer units as shown in Eqs. (12) and (13). The data for these polymerizations is shown in Table VIII.

Price's data show that a similar steric course is followed whether the carbon undergoing attack is primary (as in ethylene oxide) or secondary (as in *cis*-2-butene oxide [49]) for coordination and anionic catalysts.

For descriptive purposes the polymerization of epoxides by coordination

TABLE VIII

BULK POLYMERIZATION OF DEUTERATED ETHYLENE OXIDES[a]

Monomer (1.0 gm)	Catalyst	Yield (gm)	Intrinsic viscosity (dl/gm)	Av. mol. wt.[b]
cis-DEO	$AlEt_3$–H_2O	0.30	0.39	27,000
	$ZnEt_2$–H_2O	0.75	0.65	60,000
	$FeCl_3$–PO	0.33	0.65	60,000
trans-DEO	$AlEt_3$–H_2O	0.60	0.30	16,000
	$ZnEt_2$–H_2O	0.68	0.61	56,000
	$FeCl_3$–PO	0.60	1.34	214,000
	KOH	0.45	0.25	13,000
	K-*t*-BuO in DMSO	0.63	0.43	32,000

[a] Reprinted from Price and Spector [50]. Copyright 1966 by the American Chemical Society. Reprinted by permission of the copyright owner.

[b] $[\eta] = 9.8 \times 10^{-4} M^{0.59}$.

catalysts will be discussed according to the two main catalyst types: (1) metal–oxygen-bonded catalysts and (2) alkaline earth catalysts.

$$RO^{\ominus} + \text{cis (meso) CHD–CHD epoxide} \xrightarrow{O} RO–CHD–CHD–O^{\ominus}\ (dl) \tag{12}$$

cis (meso) → (dl)

$$trans(dl) \xrightarrow{O} (meso) \tag{13}$$

a. Metal–Oxygen-Bonded Catalysts

Pruitt and Baggett [6] reported that a ferric chloride–propylene oxide catalyst reacts with propylene oxide and various olefin oxides to give polyethers when heated for 100–200 hr at 80°C in a closed vessel [6]. Price reported similar findings with propylene oxide [19].

2-9. Polymerization of l-Propylene Oxide with Ferric Chloride–Propylene Oxide Catalyst [19]

$$CH_3–\overset{*}{C}H–CH_2 (–O–) \longrightarrow \left[–O\overset{*}{C}H(CH_3)–CH_2– \right]_n \tag{14}$$

(*a*) *Preparation of ferric chloride–propylene oxide catalyst* [6,19]. To a cooled flask containing 1.0 gm (0.0062 mole) of anhydrous ferric chloride and

18.0 ml of dry ethyl ether under a nitrogen blanket is added dropwise with stirring 6.0 ml (0.086 mole) of propylene oxide at such a rate to keep the temperature at 25°C. The mixture instantly becomes cloudy and a dark brown oil separates before all the propylene oxide is added. After the addition of the propylene oxide the reaction mixture is allowed to come to 28°–35°C and stirred for an additional 10 min. The volatiles are removed by evaporation under reduced pressure at 25°C to afford approx. 4.0 ml of an oily red-brown material. The residue is left for 1 hr in a high-vacuum system at room temperature and then stored under nitrogen at −15°C.

(*b*) *Polymerization of l-propylene oxide.* To a Pyrex polymer tube containing 10 ml (8.5 gm or 0.147 mole) of *l*-propylene oxide is added 0.1 gm of the above ferric chloride–propylene oxide catalyst in 10 ml of ether. The tube is covered, flushed with nitrogen, sealed, and placed in an 80°C constant temperature bath for 265 hr. The tube is then removed from the bath, cooled in a Dry Ice–acetone bath, opened, and the volatiles removed under reduced pressure at room temperature to afford 4.5 gm (52%) of a white snappy rubbery polymer. The polymer is dissolved in 45 gm of hot acetone and cooled to −30°C. The precipitated crystalline polymer is filtered to afford 2.1 gm (24.8%), and then 2.4 gm (28.3%) of noncrystalline polymer is obtained by concentration of the acetone. The physical properties of the *l*-polymer are outlined in Table IX. In addition the *dl*-polymer properties are also shown for comparison in Table IX.

The noncrystalline polymer obtained from concentration of the acetone has been recently shown to originate from head-to-head and tail-to-tail units in the polymer chain as shown below.

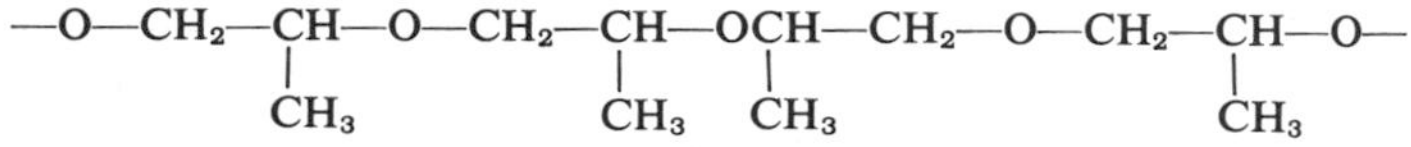

$$-O-CH_2-\underset{\substack{|\\CH_3}}{CH}-O-CH_2-\underset{\substack{|\\CH_3}}{CH}-O\underset{\substack{|\\CH_3}}{CH}-CH_2-O-CH_2-\underset{\substack{|\\CH_3}}{CH}-O-$$

Ozonization and lithium aluminum hydride reduction [49] or *n*-butyllithium reaction [52] have been shown to give predominantly polymer (A) from head-to-tail monomer units in crystalline poly(propylene oxide); (B) and (C) from head-to-head and tail-to-tail polymerization occur predominantly in the

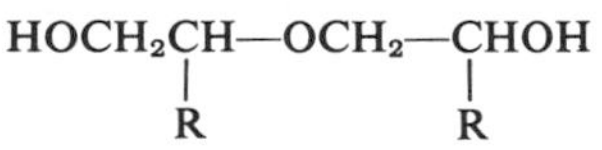

$$HOCH_2\underset{\substack{|\\R}}{CH}-OCH_2-\underset{\substack{|\\R}}{CH}OH \qquad \text{(A)}$$

$$HOCH_2\underset{\substack{|\\R}}{CH}-O\underset{\substack{|\\R}}{CH}CH_2OH \qquad \text{(B)}$$

$$HO\underset{\substack{|\\R}}{CH}-CH_2-OCH_2-\underset{\substack{|\\R}}{CH}OH \qquad \text{(C)}$$

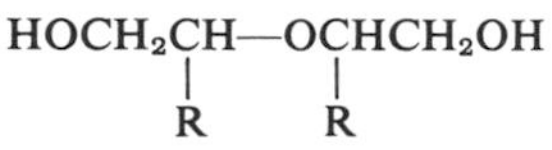

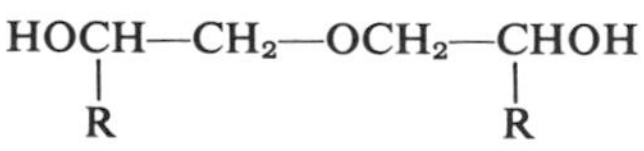

$$[R=CH_3 \text{ or } C(CH_3)_3]$$

TABLE IX
COMPARATIVE DATA FOR POLYMERS FROM FERRIC CHLORIDE (1%)-CATALYZED POLYMERIZATION OF *l*- AND *dl*-PROPYLENE OXIDE IN ETHER AT 80°C[a]

	From *l*-	From *dl*-
Polymer $[\alpha]^{20}D$ unfractd.	$+17 \pm 5°$ ($CHCl_3$)	—
$[\alpha]^{20}D$ cryst.	$+25 \pm 5°$ ($CHCl_3$)	—
	$-20 \pm 5°$ (C_6H_6)	—
$[\alpha]^{20}D$ amorph.	$+3 \pm 5°$ ($CHCl_3$)	—
	$-6 \pm 5°$ (C_6H_6)[b]	—
Softening point (cryst.) (°C)	72–74	72–74
Freezing point (cryst.) (°C)	75–72	75–72
Wt. cryst. fraction (gm)	2.1	2.5
$(\ln \eta_{rel})/c$, cryst. fraction	3.11	3.12
$(\ln \eta_{rel})/c$, amorphous	1.37	1.18
$(\ln \eta_{rel})/c$, crude polym.	3.08	2.28
Total yield (gm)	4.5	7.8

[a] Reprinted from Price and Osgan [19]. Copyright 1956 by the American Chemical Society. Reprinted by permission of the copyright owner.
[b] ($c = 2.6\%$).

noncrystalline polymer. The work has recently [53] been extended to (*d*-)*t*-butylethylene oxide and shown to give similar results.

Using a similar ferric chloride–propylene oxide catalyst system, copolymers of propylene oxide with varying proportions by weight of other olefin oxides were prepared as shown in Table X. The catalyst concentration was kept at 2.0% by weight and the polymerization temperature was 80°C for the reaction times indicated. Each of the copolymers has a softening temperature above 50°C and was able to be made into films which were oriented on stretching. It is interesting to note that *cis*- and *trans*-2,3-epoxybutanes copolymerize about equally as well with propylene oxide.

Other monomers copolymerizing satisfactorily with propylene oxide were reported to be ethylene oxide, epichlorohydrin, and styrene oxide.

Pruitt and Baggett also reported that solid polypropylene oxides are also made from other iron compounds, preferably the hydrates or hydroxides such as $Fe(OH)_3$, $Fe(OH)(C_2H_3O_2)_2$, $FeCl_3 \cdot 6H_2O$, and $FeCl_2 \cdot 4H_2O$ [34]. Some iron compounds are ineffective, such as Fe powder, FeF_3, $FeSO_4 \cdot H_2O$, and $Fe(NO_3)_3 \cdot 9H_2O$ [54]. Anhydrous $FeCl_3$ is too violent and affords only liquid polymers [54]. The preferred catalyst is the $FeCl_3$–propylene oxide pre-condensate [54]. Recently Osgan [55] reported that hydrated ferric oxides prepared in the presence of acetate ions [approx. composition, $2Fe_2O_3 \cdot$

TABLE X

COPOLYMERIZATION OF OLEFIN OXIDES WITH PROPYLENE OXIDE AT 80°C, USING 2% BY WT. FERRIC CHLORIDE–PROPYLENE OXIDE CATALYST[a]

Co-monomer	% Used	Reaction time (hr)	% Yield
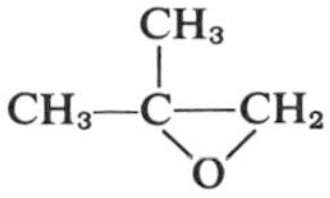 CH_3—C(CH_3)—CH_2 (epoxide, —O— bridge)	10.3 50	112 596	29.8 11.4
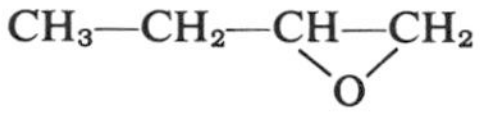 CH_3—CH_2—CH—CH_2 (epoxide, —O— bridge)	10.7 50	113 496	25.5 6.7
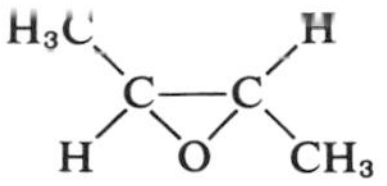 H_3C(H)C—C(H)(CH_3) (epoxide, —O— bridge; trans)	10.4 52	112 496	28.9 11.8
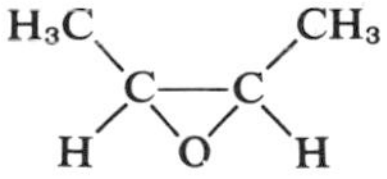 H_3C(H)C—C(H)(CH_3) (epoxide, —O— bridge; cis)	10.3	112	20.1

[a] Data taken from Pruitt and Baggett [54].

Fe(OH) · FeO · $OCOCH_3$] [56] show higher catalytic activity for the polymerization of propylene oxide than ordinary hydrated ferric oxide.

The latter reaction of $FeCl_3$ with propylene oxide has been studied in detail by Colclough [35] and the reaction was suggested to occur by the path shown in Eq. (15).

Other ferric chloride–epoxy complexes [57] have been reported which are also effective in the polymerization of propylene oxide [58]. In addition, ferric trialkoxide, $Fe(OR)_3$, is an active catalyst [59].

Price [8,60,61] extended the applicability of the Pruitt and Baggett catalyst system by showing that a wide variety of Lewis acid–metal alkoxide or halides serve as catalysts for the stereospecific polymerization of propylene oxide as detailed in Table XI.

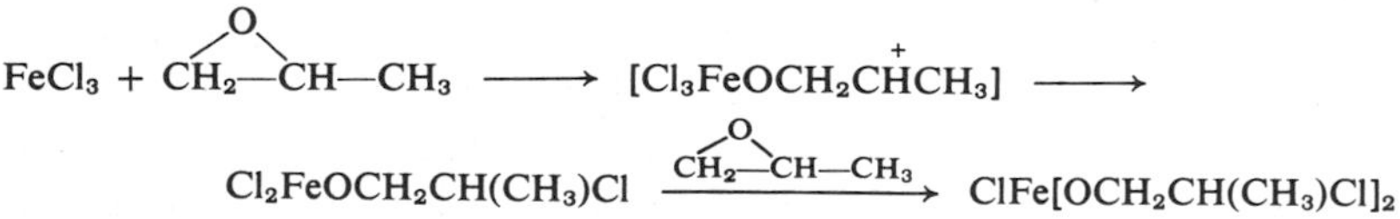

$$FeCl_3 + \overset{O}{CH_2{-}CH}{-}CH_3 \longrightarrow [Cl_3FeOCH_2\overset{+}{C}HCH_3] \longrightarrow$$

$$Cl_2FeOCH_2CH(CH_3)Cl \xrightarrow{\overset{O}{CH_2-CH}-CH_3} ClFe[OCH_2CH(CH_3)Cl]_2 \quad (15)$$

2-10. *Preparation of Polypropylene Oxide Using Aluminum Isopropoxide–Zinc Chloride Catalyst* [62]

To a glass-lined stainless steel pressure vessel wrapped with heating wire and containing a thermocouple going well to the center of the reactor is

TABLE XI

POLYMERIZATION OF PROPYLENE OXIDE AT 80°C[a]

Catalyst[b]	Wt. %	Days[f]	% Atactic	% Isotactic	ln η_{rel}/c
$FeCl_3$ × propylene oxide[c]	1	10	63	29	3–4
$FeCl_2OR$[d]	1	1	40	5	3–4
$Al(O\text{-}i\text{-}C_3H_7)_3$ (AIP)	1	2	11[e]	2[e]	—
AIP–$ZnCl_2$					
95:5	2	1	43	16	2.6
80:20	2	1	58	14	2.6
50:50	2	1	84	14	3.7
25:75	2	1	67	9	3.8
5:95	2	1	19	6	2.7
AIP–$FeCl_2$	1	10	20	8	—
$Ti(O\text{-}i\text{-}C_3H_7)_4$–$ZnCl_2$	1	10	70	5.4	—
$Mg(OCH_3)_2$	2	10	25	2.5	—
$MgCl_2 \cdot OEt_2$	1	10	35	8.2	12.0

[a] Reprinted from Osgan and Price [8]. Copyright 1959 by the *J. Polymer Sci.* Reprinted by permission of the copyright owner.

[b] Very slight polymerization was observed for $FeCl_2$, $ZnCl_2$, $CoCl_2$, $Ti(OR)_4$, or $NiCl_2$ alone. No polymer was obtained with Al_2O_3, $Al(OH)_3$–$CrOCl$, $Mo(OH)_3$, $CdCl_2$, $CuCl_2$, or CuCl.

[c] Prepared according to Pruitt and Baggett [54].

[d] Prepared by suspending $FeCl_2 \cdot 4H_2O$ in ether, adding one mole of propylene oxide and then one equivalent of chlorine.

[e] No increase at 10 days.

[f] Polymerizations were carried out in sealed ampoules under nitrogen at 80°C. The crude reaction mixture was dissolved in acetone (10 ml/gm) with 2% water added, refluxed for ½ hr, cooled, and filtered. Isotactic and atactic materials were obtained by chilling to −35°C to −30°C and filtering. The acetone-solubles were mainly amorphous polymers.

added 300 gm (5.17 moles) of propylene oxide (distilled after refluxing over CaH_2), 200 ml of anhydrous heptane, 1.54 gm (1.88 mole) of aluminum isopropoxide (distilled at 138°–138°C/8.0 mm Hg), and 1.5 gm (11.0 mole) of zinc chloride (reagent grade and fused before use). The reactor is sealed and heated for 90 hr at 80°–90°C. After this time the reactor is cooled and the volatiles distilled off. The remaining product is dissolved in 3.65 liters of acetone containing 0.1 gm of hydroquinone by refluxing for 10 hr. Then 75 ml of water is added and refluxing continued for another hour. The solution is cooled to −30°C and centrifuged to afford approx. 1200 ml of a thick gel.

$$\underset{\text{(propylene oxide: } CH_3\text{–}CH\text{–}CH_2\text{, O bridging)}}{CH_3CH\overset{O}{—}CH_2} \xrightarrow{Al(O\text{—}i\text{-}Pr)_3\text{—}ZnCl_2} \left[—CH_2—\underset{}{\overset{CH_3}{\overset{|}{C}H}}—O— \right]_n \qquad (16)$$

The latter is redissolved in 2.5 liters of acetone and cooled again to $-30°C$ to give a gel precipitate.

The gel is dissolved in 1500 ml of benzene containing 0.5 gm of hydroquinone, and the benzene solution washed successively with 500 ml of dilute hydrochloride acid, 500 ml of water, 500 ml of 5% aqueous sodium bicarbonate, and 3 times with 500 ml of water. The benzene solution is freeze-dried to afford 56.6 gm (23%) of crystalline polymer, m.p. 68.5°C; intrinsic viscosity, 2.53 (mol. wt. 215,000, $[\eta] = 1.4 \times 10^{-4} M^{0.8}$).

Price [63] recently reported that the copolymerization of phenyl glycidyl ether by triethylaluminum–water is promoted by electron-donating groups. Diethylzinc–water and ferric chloride–propylene oxide catalyst do not show as strong a Lewis acid character as the triethylaluminum–water catalyst.

Water acts as an important activator for many Lewis acid catalyst systems (see also Preparations 2-6 and 2-7 in the previous section). For example, Coclough reported that trimethylaluminum must be reacted with water to give an active catalyst for obtaining high molecular weight polypropylene oxide in good yields [35,64]. In contrast, anhydrous aluminum, zinc, or magnesium alkyls polymerize ethylene oxide and propylene oxide in low yields to high molecular weight liquid or waxy polymers [8a,61,65].

Furakawa reported similar findings in the case of diethylzinc [66]. Vanderberg [67] later reported that magnesium alkyls also must be reacted with water to give active catalysts. These catalysts still contain alkyl–metal bonds. Price has reported some additional examples for the polymerization of propylene oxide as shown in Table XI [67a].

The metal alkyl–oxygen (polymeric) catalysts listed in Table XI polymerize ethylene oxide and monosubstituted derivatives in good yields to very high molecular weight polymers (100,000 millions) by a coordination mechanism. Vanderberg [67] also reported the polymerization of styrene oxide, butadiene monoxide, and epibromohydrin to crystalline polymers with an aluminum alkyl–water catalyst.

Combination catalysts such as aluminum isopropoxide–zinc chloride also actively polymerize ethylene oxide [60]. Furthermore acidic oxide–organometallic compounds are more effective than the acidic oxides alone [68].

Diethylzinc and metal oxides or fluorides give active catalysts for the polymerization of propylene oxide [69].

Triethylaluminum–metal acetylacetonate complexes [70] with and without water [71] are active catalysts for the polymerization of propylene oxide and epichlorohydrin [71].

b. Alkaline Earth Catalysts

Staudinger and Lehmann [4a] earlier reported that alkaline earth oxides are effective for the polymerization of ethylene oxide. In addition, the related

TABLE XII
POLYMERIZATION OF PROPYLENE OXIDE[a]

Monomer	Catalyst	Yield[b] (%)	[α]D (deg)	$\overline{M}_n$[c]
RS	Et_2Zn–H_2O	80[d]	0.0	
RS	Et_2Zn–H_2O	50[e]	0.0	2800
RS	$FeCl_3$–PO	82[d]	0.0	5000
R	Et_2Zn–H_2O	85[e]	−9.2	3000
R	Et_2Zn–H_2O	60[e]	−5.9	2500
S	Et_2Zn–H_2O	75[d]	−13.9	6500
S	$FeCl_3$–PO	87[d]	+13.0	6000
S	$FeCl_3$–PO	69[e]	+4.0	3500
RS	Et_2Zn–H_2O	*f*	0.0	

[a] Reprinted in part from Price *et al.* [67a]. Copyright 1972 by the American Chemical Society. Reprinted by permission of the copyright owner.
[b] Percent yield after separation of crystalline fraction.
[c] Vapor pressure osmometer.
[d] Soluble in acetone at −30°C.
[e] Soluble in *n*-hexane at −78°C.
[f] Crystalline polymer, m.p. 58°C (capillary); $\overline{M}_v$ = 110,000.

alkoxides [72], amides [73], and amide alkoxides [74] are also effective. The alkaline earth oxides are not as effective as the hydroxides [75].

Hill and co-workers [5] reported that strontium, calcium, and barium carbonates polymerize ethylene oxide giving high molecular weight polymers. It was shown that water (up to 0.4%) adsorbed on the surface is essential [5,7]. The water may be needed to form either hydroxide or bicarbonate ions [76] as required in the anionic polymerization mechanism.

Hill [5] reported that strontium carbonates were severalfold more active as catalysts than calcium or barium carbonate. The latter carbonates must not contain nitrate, chlorate, thiosulfate, or tetraborate ion contamination or else they will be inactive. The presence of acetate or chloride ion had no effect on activity.

Strontium carbonate samples obtained from various sources were examined and the infrared spectra are shown in Figs. 2–6. Drying the active strontium carbonate catalyst at 350°C for 24–48 hr caused the catalyst to become inactive. The major contaminant is probably nitrate ion (see 7.2–7.5 and 11.8–12.3 μ). The best way to avoid this is to prepare fresh strontium carbonate by reaction of carbon dioxide with strontium hydroxide.

The surfaces of the carbonate or sulfate probably do not exert any steric control during the polymerization since only liquid polymers are produced.

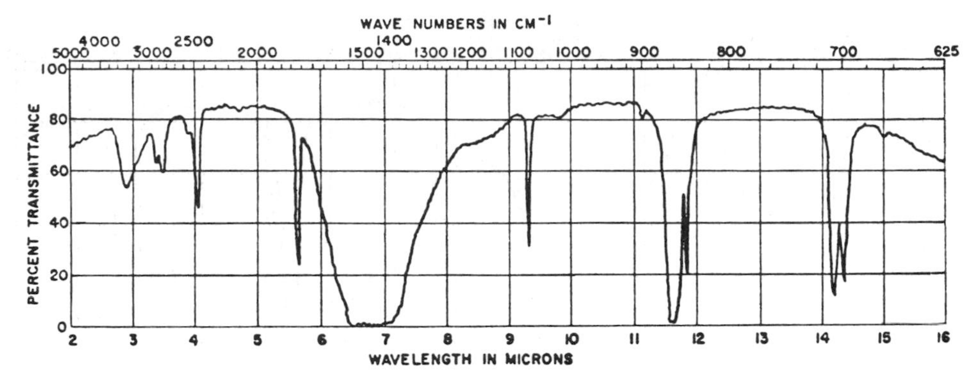

FIG. 2. Infrared spectrum of strontium carbonate. Reprinted from Hill *et al.* [5], copyright 1958 by the American Chemical Society; reprinted by permission of the copyright owner.

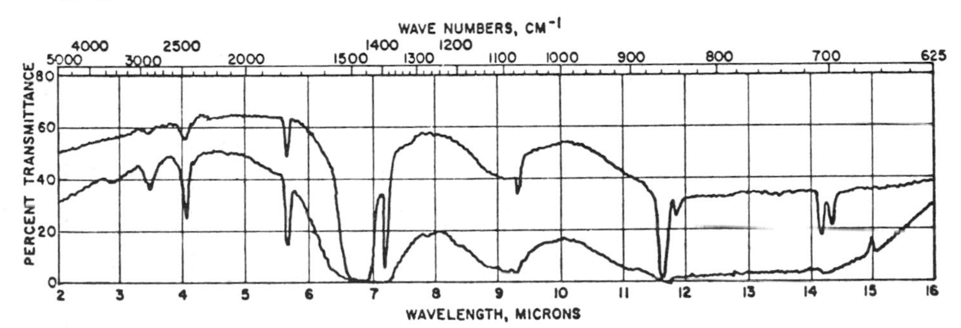

FIG. 3. Infrared spectrum of commercial carbonate (active) after drying at 350°C for 48 hr. Reprinted from Hill *et al.* [5], copyright 1958 by the American Chemical Society; reprinted by permission of the copyright owner.

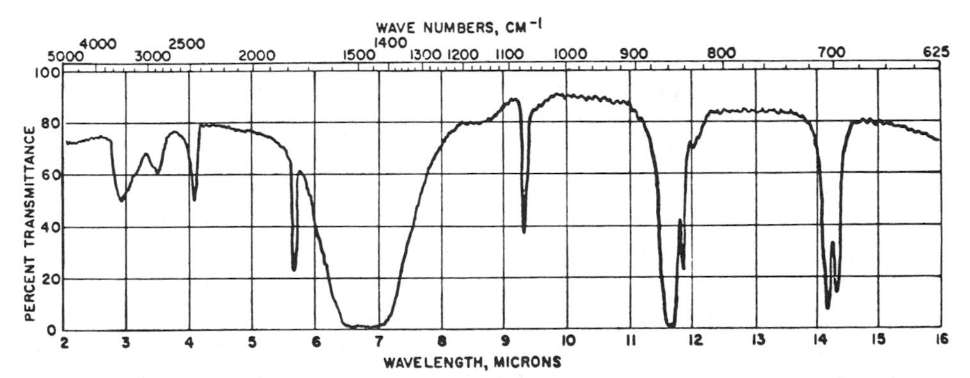

FIG. 4. Infrared spectrum of commercial strontium (inactive). Reprinted from Hill *et al.* [5], copyright 1958 by the American Chemical Society; reprinted by permission of the copyright owner.

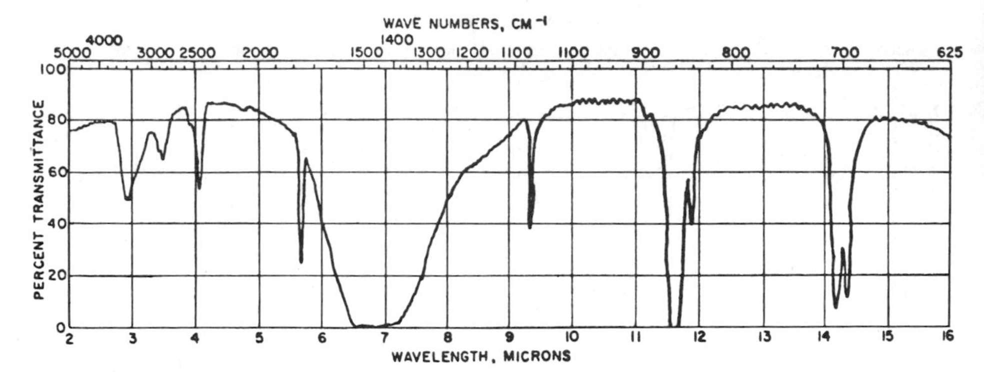

FIG. 5. Infrared spectrum of commercial strontium carbonate (originally inactive) after digestion. Reprinted from Hill *et al.* [5], copyright 1958 by the American Chemical Society; reprinted by permission of the copyright owner.

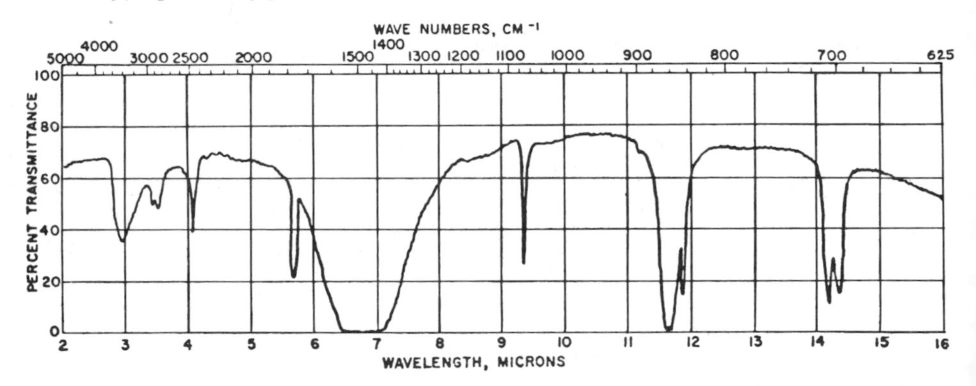

FIG. 6. Infrared spectrum of laboratory-prepared strontium carbonate (highly active). Reprinted from Hill *et al.* [5], copyright 1958 by the American Chemical Society; reprinted by permission of the copyright owner.

Calcium sulfate containing some water is not as active as the carbonate but can polymerize propylene oxide to afford liquid polymers [77].

Other active alkaline earth metal catalysts are the alkoxides [78], amides [79,79a–d], amide–alkoxides [79], and alkaline earth chelated compounds [79a]. The oxides or hexammoniates of Ca, Sr, or Ba polymerize ethylene oxide to high molecular weight poly(ethylene oxide) in an inert organic diluent (heptane) at 25°–32°C [79c]. The monomer is soluble and the polymer precipitates as it is formed [79b]. The addition of fluorene or triphenylmethane to calcium hexammoniate in liquid ammonia also gives a very active catalyst for the polymerization of ethylene oxide [79d]. In contrast, the alkaline earth carboxylates are only fair catalysts for the polymerization of ethylene oxide [80].

2-11. Polymerization of Ethylene Oxide Using Strontium Carbonate Catalyst [5]

$$\underset{\text{(epoxide, O bridging)}}{CH_2\text{—}CH_2} \xrightarrow{SrCO_3} [\text{—}CH_2CH_2\text{—}O\text{—}]_n \qquad (17)$$

(*a*) *Catalyst preparation.* To a 22% aqueous (distilled water) $Sr(OH)_2$ solution at 90°C is added a stream of carbon dioxide until the precipitation of $SrCO_3$ is complete. The solid is filtered, washed with distilled water, and dried to a water content of not less than 0.5 to 1.0% by weight.

(*b*) *Polymerization.* To a cooled polymer tube is added 50 gm of re-distilled ethylene oxide containing less than 50 ppm of aldehyde and 0.2 gm of the above strontium carbonate catalyst. The tube is sealed and placed in a wire mesh shield and metal protective tube. The tube is placed in an 80°C bath and rocked gently back and forth, and after a $1\frac{1}{2}$ hr induction period, the polymerization commences. The polymerization is extremely rapid (CAUTION) and may cause the polymer tube to explode if there is a 50°C or more rise in reaction temperature. The polymerization is complete in an additional 2 hr.

The polymer has interesting solubility properties [81]. For example, the polymer is soluble in water and gives an elastic, nontacky gel. Dissolving 20% in water gives a material which can bounce like a ball even though it contains 80% water. The polymer is also soluble in chloroform, ethylene dichloride, acetonitrile, and anisole. At elevated temperatures it is also soluble in benzene and toluene.

The strontium carbonate catalyst from Preparation 2-11 is also useful in copolymerizing ethylene oxide with other epoxides such as butadiene monoepoxide [81a], chloroprene oxide [81a], isobutylene oxide [81b], and propylene oxide [81b].

3. MISCELLANEOUS METHODS

1. Polymerization of tetrahydrofuran with niobium pentachloride and tantalum pentachloride [82].
2. Polymerization of 2-methyl-2-butene oxide by means of $Al(i\text{-}Bu)_3\text{–}x\text{-}H_2O$ [83].
3. Polymerization of aryl glycidyl ethers by means of ferric chloride–propylene oxide catalyst [84].
4. Polymerization of aryl glycidyl ethers by means of $Al(i\text{-}OPr)_3$, $Al(i\text{-}OPr)_3\text{–}ZnCl_2$ or $AlEt_3$ catalysts [85].
5. Polymerization of propylene oxide with Al—O—Al or Zn—N—Zn catalyst [86].

6. Preparation of soluble polymers from stoichiometric amounts of epichlorohydrin and a bisphenol or from a bisepoxide and a bisphenol [87].

7. Polymerization of tetrahydrofuran with trityl hexachloroantimonate initiator [88].

8. Polymerization of tetrahydrofuran and other cyclic ethers by *p*-chlorophenyldiazonium hexafluorophosphate [89].

9. Radiation-induced polymerization of isobutylene oxide [90].

10. Use of dialkylaluminum acetylacetonate for the polymerization of epoxide [91].

11. Preparation of tetrahydrofuran–propylene oxide copolymers [92].

12. Preparation of polyepichlorohydrin using Grignard reagent catalysts [93].

13. Polymerization of epiperchloroatohydrin [94].

14. Preparation of crystalline epihalohydrin polymers and copolymers [95].

15. Polymerization of olefin oxides and sulfides with S–Et_2Zn and 98% H_2O_2–Et_2Zn [96].

16. Poly(ethylene oxide) grafted on to poly(methyl methacrylate) by transesterification [97].

17. Dialkylzinc–hydrazine catalyst for the polymerization of propylene oxide [98].

18. Polymerization of 1,4-epoxycyclohexane (7-oxabicyclo[2.2.1]heptane) with cationic initiators [99].

19. Polymerization of cyclic ethers with cationic catalysts [100].

20. Polymerization of cyclic formals [101].

21. Polymerization of epoxide and episulfide catalyzed by metal perchlorates [102].

22. Base-catalyzed polymerization of epoxides in dimethyl sulfoxide and hexamethylphosphoric triamide [103].

23. Anionic polymerization of 1,2-butylene oxide [104].

24. Ring-opening polymerization of episulfides [105].

25. Cationic polymerization of cyclic ethers initiated by macromolecular dioxolenium salts [106].

26. Hexacyanometallate salt complexes as catalysts for epoxide polymerization [107].

27. Polymerization of cyclic ethers and sulfides [108].

REFERENCES

1. A. Wurtz, *C. R. Acad. Sci.* **49**, 813 (1859); *Chem. Ber.* **10**, 90 (1877); *Bull. Soc. Chim. Fr.* [2] **29**, 530 (1878).
2. E. Roithner, *Monatsh. Chem.* **15**, 679 (1894); *J. Chem. Soc., London* **68**, 319 (1895).

3. A. Wurtz, *Ann. Chim. Phys.* [3] **69**, 334 (1863).
4. H. Staudinger and O. Schweitzer, *Chem. Ber.* **63**, 2395 (1929).
4a. H. Staudinger and H. Lehmann, *Justus Liebigs Ann. Chem.* **505**, 41 (1933).
5. F. N. Hill, F. E. Bailey, Jr., and J. T. Fitzpatrick, *Ind. Eng. Chem.* **50**, 5 (1958).
5a. Union Carbide and Carbon Corp., Australian Patent Appl. 27792 (1957).
6. M. E. Pruitt and J. M. Baggett, U.S. Patent 2,706,181 (1955).
7. A. B. Borkovec, U.S. Patent 2,873,258 (1959).
8. M. Osgan and C. C. Price, *J. Polym. Sci.* **34**, 153 (1959); R. O. Colclough, G. Geoffrey, and A. H. Jagger, *ibid.* **48**, 273 (1960).
8a. P. E. Ebert and C. C. Price, *J. Polym. Sci.* **34**, 157 (1959).
9. D. M. Simons and J. J. Verbanc, *J. Polym. Sci.* **44**, 303 (1960).
10. Carbowax Polyethylene Glycol, Tech. Bull. F, 4772F. Chem. & Plastics Develop. Div., Union Carbide Corp., New York, 1970; Dow Polyethylene Glycols, Tech. Bull. 125–230–59. Dow Chem. Co., Midland, Michigan, 1959; Polyethylene Glycols, Tech. Bull. Jefferson Chem. Co., Houston, Texas, 1961; Polyethylene Glycols, Tech. Bull. OC–107–1061 and OC–108–1061. Organic Chem. Div., Olin Mathieson Chem. Corp., New York, 1961; Pluracol E. Tech. Data. Wyandotte Chemical Corp., Wyandotte, Michigan, 1960.
10a. Polyox Water Soluble Resins, Tech. Bull. F–40246E. Chem. & Plastics Develop. Div., Union Carbide Corp., New York, 1968.
11. H. F. Smyth, Jr., C. P. Carpenter, and C. S. Weil, *J. Amer. Pharm. Ass., Sci. Ed.* **39**, 349 (1950); **44**, 27 (1955).
12. J. Furukawa and T. Saegusa, "Polymerization of Aldehydes and Oxides," pp. 125–208. Wiley (Interscience), New York, 1963; A. E. Gurgiolo, *Rev. Macromol. Chem.* **1**, 76 (1966); H. Staudinger, "Die Hochmolekularen Organischen Verbindungen," p. 224 ff. Springer-Verlag, Berlin and New York, 1960; E. J. Vandenberg, *J. Polym. Sci., Part A-1* **7**, 525 (1969); L. E. St. Pierre, *in* "Polyethers" (N. G. Gaylord, ed.). Part 1. Wiley (Interscience), New York, 1963.
13. H. Hibbert and S. Perry, *Can. J. Res.* **8**, 102 (1933).
13a. S. Perry and H. Hibbert, *J. Amer. Chem. Soc.* **62**, 2597 (1940).
14. S. D. Holland, British Patent 821,203 (1959).
15. Wyandotte Chem. Co., Tech. Inform. Sheets and Booklets, Form 437–9–52; T. H. Vaughan, D. R. Jackson, and L. G. Lunsted, *J. Amer. Oil Chem. Soc.* **29**, 240 (1952); L. G. Lunsted, U.S. Patent 2,674,619 (1954).
16. L. E. St. Pierre and C. C. Price, *J. Amer. Chem. Soc.* **78**, 3432 (1956).
17. E. C. Steiner, R. R. Pelletier, and R. O. Trucks, *J. Amer. Chem. Soc.* **86**, 4682 (1964); W. H. Snyder, J., Ph.D. Thesis, University of Pennsylvania, Philadelphia, 1961.
18. E. C. Steiner, R. R. Pelletier, and R. O. Trucks, *J. Amer. Chem. Soc.* **86**, 4682 (1964).
19. C. C. Price and M. Osgan, *J. Amer. Chem. Soc.* **78**, 4787 (1956).
20. F. Patat, *Kunst.-Plast.* (*Solothurn*) 1 (1958).
21. R. A. Smith, *J. Amer. Chem. Soc.* **62**, 994 (1940).
22. H. A. Bruson and O. Stein, U.S. Patent 2,075,018 (1937).
23. R. M. Laird and R. E. Parker, *J. Amer. Chem. Soc.* **83**, 4277 (1961).
23a. N. B. Chapman, N. S. Isaacs, and R. E. Parker, *J. Chem. Soc., London* p. 1925 (1959); F. D. Trischler and J. Hollander, *J. Polym. Sci., Part A-1* **5**, 2343 (1967).
24. C. C. Price, Y. Atarashi, and R. Yamamoto, *J. Polym. Sci., Part A-1* **7**, 569 (1969).
25. S. R. Sandler and F. R. Berg, *J. Polym. Sci., Part A-1* **4**, 1253 (1966).
26. P. H. Plesch, ed., "The Chemistry of Cationic Polymerization." Pergamon, Oxford, 1963.

27. F. S. Dainton and K. J. Ivin, *Quart. Rev., Chem. Soc.* **12**, 61 (1958).
27a. J. M. Andrews and J. A. Semlyen, *Polymer* **12**, 642 (1971).
28. A. A. Skuratov, A. A. Strepikheev, A. M. Shtekher, and A. V. Volokhina, *Dokl. Akad. Nauk SSSR* **117**, 263 (1957).
29. H. G. Heal, *J. Chem. Educ.* **35**, 192 (1958).
30. B. N. Blackett, J. M. Coxon, M. P. Hartshorn, and K. E. Richards, *J. Amer. Chem. Soc.* **92**, 2574 (1970).
31. R. J. Kern, *J. Org. Chem.* **33**, 388 (1968); R. J. Katnik and J. Schaefer, *ibid.*, p. 384.
32. J. N. Bronsted, Mary Kilpatrick, and Martin Kilpatrick, *J. Amer. Chem. Soc.* **51**, 428 (1929); C. Matignon, H. Mourea, and M. Dode, *Bull. Soc. Chim. Fr.* [5] **1**, 1308 (1934).
33. H. Meerwein, E. Battenburg, H. Gold, E. Pfeil, and G. Willfang, *J. Prakt. Chem.* [3] **154**, 83 (1939); D. J. Worsfold and A. M. Eastham, *J. Amer. Chem. Soc.* **79**, 897 and 900 (1950).
33a. G. A. Latremonille, G. T. Merrall, and A. M. Eastham, *J. Amer. Chem. Soc.* **82**, 120 (1960).
33b. A. M. Eastham, *Fortschr. Hochpolym.-Forsch.* **2**, 18 (1960).
34. D. J. Worsfold and A. M. Eastham, *J. Amer. Chem. Soc.* **79**, 897 (1957).
34a. D. J. Worsfold and A. M. Eastham, *J. Amer. Chem. Soc.* **79**, 900 (1957).
34b. G. T. Merall, G. A. Latremouille, and A. M. Eastham, *Can. J. Chem.* **38**, 1967 (1960).
34c. R. Fordyce and H. Hibbert, *J. Am. Chem. Soc.* **61**, 1912 (1939).
35. R. O. Colclough, G. Gee, W. C. E. Higginson, J. B. Jackson, and M. Litt, *J. Polym. Sci.* **34**, 171 (1959).
36. J. G. Pritchard and F. A. Long, *J. Amer. Chem. Soc.* **78**, 2667 (1956).
37. K. T. Garty, T. B. Gibb, Jr., and R. A. Clendinning, *J. Polym. Sci., Part A* **1**, 85 (1963).
38. E. J. Vandenberg, *J. Polym. Sci.* **47**, 489 (1960); U.S. Patent 3,065,187 (1962).
39. R. Bacskai, *J. Polym. Sci., Part A* **1**, 2777 (1963).
40. J. B. Rose, *J. Chem. Soc., London* p. 542 (1956).
40a. J. B. Rose, *J. Chem. Soc., London* p. 546 (1956); J. G. Pritchard and F. A. Long, *J. Amer. Chem. Soc.* **80**, 4162 (1958).
41. R. Bacskai, *J. Polym. Sci., Part A* **1**, 2777 (1963).
42. H. Meerwein and E. Kroning, *J. Prakt. Chem.* [3] **147**, 257 (1951); H. Meerwein, German Patents 741,478 and 766,208 (1939).
43. M. P. Dreyfuss and P. Dreyfuss, *Polymer* **6**, 13 (1965).
44. E. L. Muetterties, U.S. Patent 2,856,370 (1958).
45. H. Meerwein, D. Delfs, and H. Morschel, *Angew. Chem.* **72**, 927 (1960).
46. D. Vofsi and A. V. Tobolsky, *J. Polym. Sci., Part A* **3**, 3261 (1965).
47. A. M. Eastman, *Fortschr. Hochpolym.-Forsch.* **2**, 18 (1960).
48. A. B. Borkovec, U.S. Patent 2,873,258 (1959).
49. E. J. Vandenberg, *Polym. Lett.* **2**, 1085 (1964).
49a. E. J. Vandenberg, *J. Polym. Sci., Part A-1* **7**, 525 (1969).
50. C. C. Price and R. Spector, *J. Amer. Chem. Soc.* **88**, 4171 (1966).
51. R. E. Parker and N. S. Isaacs, *Chem. Rev.* **59**, 758 (1959).
52. C. C. Price, R. Spector, and A. L. Tumolo, *J. Polym. Sci.* **5**, 407 (1967); E. J. Vanderberg, *Polym. Lett.* **2**, 1085 (1964).
53. C. C. Price, B. T. DeBona, and B. C. Furie, *Polym. Sect. 7th Mid. Atlantic Reg. Meet., Amer. Chem. Soc., 1972* Abstracts, Paper No. 1, p. 161 (1972).
54. M. E. Pruitt and J. M. Baggett, U.S. Patent 2,706,181 (1955).

55. M. Osgan, *J. Polym. Sci., Part A-1* **6**, 1249 (1968).
56. E. Chiellini, P. Salavadori, M. Osgan, and P. Pino, *J. Polym. Sci., Part A-1* **8**, 1589 (1970).
57. S. Oshida and S. Murahashi, *J. Polym. Sci.* **40**, 571 (1959); S. Ishida, *Bull. Chem. Soc. Jap.* **33**, 726 (1960).
58. M. E. Pruitt and J. M. Baggett, U.S. Patent 2,811,491 (1957).
59. A. B. Borkovec, U.S. Patent 2,844,545 (1958).
60. P. E. Ebert and C. C. Price, *J. Polym. Sci.* **46**, 455 (1960).
61. R. A. Miller and C. C. Price, *J. Polym. Sci.* **34**, 161 (1959).
62. C. C. Price and A. L. Tumolo, *J. Polym. Sci., Part A-1* **5**, 175 (1967).
63. C. C. Price and L. R. Brecker, *J. Polym. Sci., Part A-1* **7**, 575 (1969).
64. R. O. Colclough, G. Gee, and A. H. Jagger, *J. Polym. Sci.* **43**, 273 (1960).
65. D. G. Stewart, D. Y. Waddan, and E. T. Borrows, U.S. Patent 2,870,100 (1959); S. Kambara and M. Hatano, *J. Polym. Sci.* **27**, 584 (1958).
66. F. Furakawa, T. Tsuruta, R. Sakata, T. Saegusa, and A. Kawasaki, *Makromol. Chem.* **32**, 90 (1959).
67. E. J. Vandenberg, *J. Polym. Sci.* **47**, 486 (1960).
67a. C. C. Price, M. K. Akkapeddi, B. T. DeBona, and B. C. Furic, *J. Amer. Chem. Soc.* **94**, 3964 (1972).
68. J. Furukawa, T. Saegusa, T. Tsuruta, and G. Kakogawa, *Makromol. Chem.* **36**, 25 (1959); J. Furukawa, T. Saegusa, T. Tsuruta, R. Sakata, and G. Kakogawa, *J. Polym. Sci.* **36**, 541 (1959); *Bull. Jap. Petrol. Inst.* **3**, 39 (1961).
69. K. Okazaki, *Makromol. Chem.* **43**, 84 (1961).
70. E. J. Vanderberg, *J. Polym. Sci., Part B* **2**, 1085 (1964); F. N. Hill, F. E. Bailey, Jr., and J. T. Fitzpatrick, U.S. Patent 2,971,988 (1961).
71. S. Kambara, M. Hatano, and K. Sakaguchi, *J. Polym. Sci.* **51**, 57 (1961).
72. W. L. Bressler and A. E. Gurgiolo, U.S. Patent 2,917,470 (1959).
73. F. E. Bailey, Jr. and J. T. Fitzpatrick, U.S. Patent 2,941,963 (1960).
74. F. N. Hill, F. E. Bailey, Jr., and J. T. Fitzpatrick, U.S. Patent 2,971,988 (1961).
75. O. V. Krylov and Y. E. Sinyak, *Vysokomol. Soedin.* **3**, 898 (1961); *Polym. Sci. USSR* **3**, 719 (1962).
76. K. S. Kazanskii, G. V. Korovina, B. J. Vainshtok, and S. G. Entelis, *Izv. Akad. Nauk SSSR., Ser. Khim.* p. 759 (1964).
77. O. V. Krylov and Y. E. Sinyak, *Polym. Sci. USSR* **3**, 719 (1962).
78. W. L. Bressler and A. E. Gurgiolo, U.S. Patent 2,917,470 (1959).
79. F. N. Hill, J. T. Fitzpatrick, and F. E. Bailey, Jr., Japanese Patent Appl. Pub. 10148 (1960).
79a. F. N. Hill and J. T. Fitzpatrick, U.S. Patent 2,866,761 (1958).
79b. F. N. Hill and J. T. Fitzpatrick, Japanese Patent Appl. Pub. 2197 (1960).
79c. F. N. Hill, J. T. Fitzpatrick, and F. E. Bailey, Jr., British Patent 869,116 (1961).
79d. F. E. Bailey, Jr. and H. G. Frame, British Patent 926,860 (1963).
80. A. E. Gurgiolo, U.S. Patent 2,934,505 (1960).
81. F. E. Bailey, Jr., G. M. Powell, and K. L. Smith, *Ind. Eng. Chem.* **50**, 8–11 (1958).
81a. F. E. Bailey, Jr., British Patent 869,112 (1957).
81b. F. E. Bailey, Jr. and F. N. Hill, British Patent 869,113 (1957).
82. Y. Takegami, T. Ueno, and R. Hirai, *J. Polym. Sci., Part A-1* **4**, 973 (1966).
83. N. D. Field, J. A. Kieras, and A. E. Borchert, *J. Polym. Sci., Part A-1* **5**, 2179 (1967).
84. T. B. Gibb, Jr., R. A. Clendinning, and W. D. Niegisch, *J. Polym. Sci., Part A-1* **4**, 917 (1966).
85. A. Noshay and C. C. Price, *J. Polym. Sci.* **34**, 165 (1959).

86. H. Tani, T. Araki, N. Ogrini, and N. Ureyama, *J. Amer. Chem. Soc.* **89**, 173 (1967).
87. A. S. Carpenter, E. R. Sallsgrove, and F. Reeder, British Patent 652,030 (1951); N. H. Reinking, A. E. Barnabes, and W. F. Hale, *J. Appl. Polym. Sci.* **7**, 2135, 2145, and 2153 (1963); G. E. Myers and J. R. Dagon, *J. Polym. Sci., Part A* **2**, 2631 (1964).
88. I. Kuntz and M. T. Melchior, *J. Polym. Sci.* **7**, 1959 (1969).
89. M. P. Dreyfuss and P. Dreyfuss, *J. Polym. Sci., Part A-1* **4**, 2179 (1966).
90. R. S. Bauer and W. W. Spooncer, *J. Polym. Sci., Part A-1* **8**, 2971 (1970).
91. I. Kuntz and W. R. Kroll, *J. Polym. Sci., Part A-1* **8**, 1601 (1970).
92. L. P. Blanchard and M. D. Baijal, *J. Polym. Sci., Part A-1* **5**, 2045 (1967); I. A. Dickinson, *J. Polym. Sci.* **58**, 857 (1962).
93. Y. Minoura, H. Hironaka, T. Kasabo, and Y. Ueno, *J. Polym. Sci.* **7**, 2505 (1969).
94. J. Radell and J. W. Connolly, *J. Polym. Sci.* **48**, 343 (1960).
95. S. Ishida and J. Murashashi, *J. Polym. Sci.* **40**, 571 (1959).
96. J. Lal, *J. Polym. Sci., Part A-1* **4**, 1163 (1966).
97. M. A. Twaik, M. Tahan, and A. Zilkha, *J. Polym. Sci., Part A-1* **7**, 2469 (1969).
98. N. Calderon and K. W. Scott, *J. Polym. Sci., Part A-1* **5**, 917 (1967).
99. E. L. Wittbecker, H. K. Hall, Jr., and T. W. Campbell, *J. Amer. Chem. Soc.* **82**, 1218 (1960).
100. W. Kern, H. Chedron, and V. Jaacks, *Angew. Chem.* **73**, 177 (1961); S. Okamura, E. Kobayashi, M. Takeda, K. Tomikawa, and T. Higashimura, *J. Polym. Sci., Part C* **4**, 827 (1964).
101. J. W. Hill and W. H. Carothers, *J. Amer. Chem. Soc.* **57**, 925 (1935); M. Okada, Y. Yamashita, and Y. Ishi, *Makromol. Chem.* **80**, 196 (1964).
102. V. Fiala, H. Yamaoka, and S. Okamura, *Polym. Lett.* **9**, 225 (1971).
103. C. C. Price and M. K. Akkapeddi, *J. Amer. Chem. Soc.* **94**, 3972 (1972).
104. L. P. Blanchard, K. T. Dinh, J. Moinard, and F. Tahiani, *J. Polym. Sci., Part A-1* **10**, 1353 (1972).
105. E. J. Vandenberg, *J. Polym. Sci., Part A-1* **10**, 329 (1972).
106. Y. Yamashita, *Polym. Prepr., Amer. Chem. Soc., Div. Polym. Chem.* **13**, 539 (1972).
107. R. J. Herold and R. A. Livigni, *Polym. Prepr., Amer. Chem. Soc., Div. Polym. Chem.* **13**, 545 (1972).
108. P. Dreyfuss, *Chem. Tech.*, 356 (1973)

Chapter 7

POLYUREAS

I. INTRODUCTION

The basic methods involved in the preparation of ureas have been outlined in Chapter 6 of Volume II of *Organic Functional Group Preparations* [1] and will not be repeated here. Essentially the reactions which were earlier described as giving monomeric ureas can be applied to giving polyureas by using di- and polyfunctional starting materials, as shown in Eq. (1).

Polyureas like polyamides have many sites available for hydrogen bonding. The polyureas generally are higher melting and less soluble than the related polyamides. Polyureas are rarely made by melt polymerization techniques because they are thermally unstable at temperatures above 200°C and must be made by solution polymerization techniques.

Polyureas are useful starting materials to react with aldehydes [1a], ethylene oxide [2], reactive halogen compounds [3], and chloromethylphosphonic acid

$$\left[-R'-NH\overset{O}{\overset{\|}{C}}NH- \right]_n \xleftarrow[X = Cl_2S]{COX} H_2NR'NH_2 \xrightarrow{NH_2-\overset{O}{\overset{\|}{C}}-NH_2} \left[-R'-NH\overset{O}{\overset{\|}{C}}NH- \right]_n$$

$$\downarrow R(NCX)_2,\ X = O, S \text{ or } R(NHCOOC_2H_5)_2 \tag{1}$$

$$\left[-R'-NH\overset{O}{\overset{\|}{C}}NHRNH\overset{O}{\overset{\|}{C}}-NH- \right]_n$$

for use in flameproofing [4], cyclic diketones [5], elastomers [6], fibers [7], adhesives [8], and durable press applications [9], to mention a few. In Japan the fiber Urylon [poly(nonamethyleneurea)] has been used in place of nylon because it is hydrolytically more stable [10].

Polyureas may also be named in a manner similar to polyamides. For example, polyamides with n equal to the number of carbons in the diamine and 1 equal to carbonic acid are named $n - 1$ polyamides.

$$\left[-(CH_2)_{10}NH\overset{O}{\overset{\|}{C}}NH- \right]_n$$

10 − 1 polyamide or
poly(decamethyleneurea)

The polymer melt temperature of some of the major polyureas is shown in Table I.

2. CONDENSATION METHODS

A. Reaction of Diamines with Phosgene

Polyureas can be prepared by the interfacial polycondensation of phosgene in an organic solvent with an aqueous solution of a diamine and alkali as described in Preparation 2-1 [11,12].

2-1. Preparation of Poly(hexamethyleneurea) [11]

$$H_2N(CH_2)_6NH_2 + COCl_2 \longrightarrow \left[-(CH_2)_6NH\overset{O}{\overset{\|}{C}}NH- \right]_n + 2HCl \tag{2}$$

CAUTION: This reaction should be run in a well-ventilated hood. (We recommend the presence of small amounts of ammonia vapors in the air to reduce the danger of escaping phosgene.)

To a three-necked, round-bottomed flask equipped with a mechanical stirrer, dropping funnel, and condenser is added a solution of 4.95 gm (0.05 mole) of phosgene in 200 ml of dry carbon tetrachloride. The solution is vigorously stirred while the rapid addition of 5.8 gm (0.05 mole) of hexamethylenediamine and 4.0 gm (0.10 mole) of sodium hydroxide in 70 ml of water takes place. The reaction is exothermic while the polyurea forms.

TABLE I

MELTING POINTS OF SOME REPRESENTATIVE POLYUREAS

$[\text{—CO—NH—R—NH—}]_n$

R	Polymer melt temperature (°C)[a]	Ref.
$\text{—(CH}_2)_2\text{—}$	> 300	*b*
$\text{—(CH}_2)_4\text{—}$	> 320	*c*
$\text{—(CH}_2)_6\text{—}$	270–300	*d*
$\text{—(CH}_2)_8\text{—}$	212–260	*d*
$\text{—(CH}_2)_{10}\text{—}$	209–250	*d*
$\text{—(CH}_2)_{12}\text{—}$	203	*d*
$\text{—C}_6\text{H}_5\text{—}$	260–300	*e*
— naphthalene	360	*e*

[a] Ranges in melting point indicate the differences obtained using various preparative methods.

[b] W. H. Libby, U.S. Patent 3,042,658 (1962).

[c] E. L. Wittbecker, U.S. Patent 2,816,879 (1957); E. Fischer, *Chem. Ber.* **46**, 2504 (1913).

[d] H. V. Boenig, N. Walker, and E. H. Myers, *J. Appl. Polym. Sci.* **5**, 384 (1961).

[e] O. Ya. Fedotova, M. I. Shtilman, and G. S. Kolesinkov, *Vysokomol. Soedin., Ser. B.* **9**, 242 (1967).

After 10 min, the carbon tetrachloride is evaporated off on a steam bath or with the aid of a water aspirator. The polyurea is washed several times in a blender and air-dried overnight to obtain 5.0 gm (70%), inherent viscosity 0.90 (in *m*-cresol, 0.5% conc. at 30°C), polymer melt temperature approximately 295°C.

The addition of phosgene gas rather than the use of a preformed phosgene solution did not give as high molecular weight polyureas [11].

Polyureas can also be prepared in the absence of organic solvents by directly adding phosgene into a water solution of the diamine and sodium hydroxide until pH 7.0 is reached [11].

B. Reaction of Diamines with Biscarbamyl Halides

The reaction of secondary diamines with phosgene at low temperatures affords biscarbamyl chlorides which then can be used to react with diamines to give polyureas which are either homopolymers or copolymers.

$$\mathrm{H{-}N(R''){-}R{-}NH(R') + 2COCl_2 \longrightarrow ClC({=}O){-}N(R''){-}R{-}N(R')C({=}O){-}Cl \xrightarrow{R'''(NH_2)_2} \left[{-}N(R''){-}R{-}N(R'){-}C({=}O){-}NH{-}R'''{-}NHC({=}O){-}\right]_n} \quad (3)$$

A typical example of a secondary amine that undergoes this reaction is piperazine [13–15] and the reaction is illustrated in Preparation 2-2.

2-2. Preparation of an Alternating Copolymer from Piperazine and 1,6-Hexamethylenediamine [13]

(*a*) *Preparation of 1,4-piperazinedicarbonyl chloride*

$$\mathrm{2COCl_2 + HN(CH_2CH_2)_2NH \xrightarrow{(C_2H_5)_3N} ClC({=}O){-}N(CH_2CH_2)_2N{-}C({=}O)Cl} \quad (4)$$

CAUTION: Use a well-ventilated hood and take all due precautions in handling highly toxic phosgene.

To a flask containing 450 gm (4.5 mole) of phosgene at 0° to −10°C is added dropwise a solution of 86 gm (1.0 mole) of anhydrous piperazine and 203 gm (2.0 moles) of triethylamine dissolved in 250 ml of chloroform. After the addition the ice bath is removed and the excess phosgene is allowed to evaporate and caught in a caustic trap. The triethylamine hydrochloride precipitate is filtered and washed with 1 liter of dioxane. Concentration of the combined filtrates affords 97 gm (46%) of 1,4-piperazinedicarbonyl chloride, m.p. 153°–155°C (from chloroform–benzene).

(*b*) *Copolymer preparation*

$$\mathrm{ClC({=}O){-}N(CH_2CH_2)_2N{-}C({=}O)Cl + H_2N(CH_2)_6NH_2 \xrightarrow{NaOH} \left[{-}C({=}O){-}N(CH_2CH_2)_2N{-}C({=}O){-}NH(CH_2)_6{-}NH{-}\right]_n} \quad (5)$$

To a solution of 100 ml of water, 2.91 gm (0.025 mole) of hexamethylenediamine, 5.30 gm (0.05 mole) of sodium carbonate, and 1.0 gm of sodium lauryl sulfate in a blender is added with rapid stirring a solution of 5.27 gm (0.025 mole) of 1,4-piperazinedicarbonyl chloride in 100 ml of chloroform. The white polymer precipitate which forms is then filtered, washed with acetone, and dried under reduced pressure at 80°C (vacuum oven) to give a polymer of inherent viscosity 0.82 in *m*-cresol (0.5 gm/100 ml) at 30°C.

It is interesting to note that the alternating copolyurea from Preparation 2-2 is crystalline and has a polymer melt temperature of 265°C. On the other hand, the random copolyurea prepared by melt polymerization methods is amorphous and has a polymer melt temperature of only 194°C. The latter is also more soluble in dimethylformamide.

Several other alternating copolyureas have been prepared by a similar method [13–18].

C. Reaction of Diamines with Diisocyanates and Derivatives

Diamines are frequently used to cure diisocyanates, especially isocyanate prepolymers, to give polyureas. The reaction has the advantages that no by-products are produced and that it can be carried out at low temperatures. This process is related to the diamine–phosgene reaction in that here the diisocyanate must first be isolated and reacted further with diamines. Copolymers are easily formed in this latter type of reaction when the reaction components have different carbon backbones.

Aromatic diisocyanates are much more reactive (reaction time: approx. 3–10 min) [19,20] than aliphatic diisocyanates (reaction time: approx. 1–5 hr) [21–24] with amines to give polyureas. In addition aromatic diamines are more reactive than aliphatic diamines with diisocyanates.

The use of tertiary amines [25], organotin compounds [26], stannic chloride [19], and lithium chloride [19] has been found to improve the reactivity of *N*-alkylamines which otherwise react very sluggishly with diisocyanates.

The reactions are usually not carried out in bulk unless infusible, insoluble, cross-linked polyureas are desired, as in the case of adhesives [26]. This is due to the fact that the urea groups also can react with isocyanate groups to give biuret groups at the branching site (Eq. 6).

Branching and cross-linking are avoided by carrying out the reaction in a polar solvent in which the polymer is soluble [27].

Polyureas can also be prepared by the interfacial polymerization technique wherein the diisocyanate in an organic solvent is stirred with an aqueous solution of the diamine bishydrochloride at low temperatures [22]. Alkali is slowly added to produce the free diamine for reaction.

$$[\text{—R—NHCONH—}]_n + \text{R(NCO)}_2 \longrightarrow \text{∼R—NHCON∼} \quad (\text{N—C=O—NH—R—NH—C=O—N}) \quad \text{∼R—NHCON∼} \tag{6}$$

2-3. *Preparation of a Polyurea from Bis(3-aminopropyl) Ether and 1,6-Hexamethylene Diisocyanate* [22]

$$(\text{CH}_2)_6(\text{NCO})_2 + \overset{-}{\text{Cl}}\text{H}_3\overset{+}{\text{N}}\text{—(CH}_2)_3\text{—O—(CH}_2)_3\text{—}\overset{+}{\text{N}}\text{H}_3\text{Cl}^- \xrightarrow[\text{H}_2\text{O}]{\text{NaOH}}$$

$$\left[\text{—(CH}_2)_6\text{—NH}\overset{\text{O}}{\overset{\|}{\text{C}}}\text{NH(CH}_2)_3\text{—O—(CH}_2)_3\text{NH}\overset{\text{O}}{\overset{\|}{\text{C}}}\text{NH—}\right]_n + 2\text{NaCl} \tag{7}$$

To a flask containing a mixture of 168 gm (1.0 mole) of 1,6-hexamethylene diisocyanate is added 132 gm (1.0 mole) of bis(3-aminopropyl) ether dissolved in a mixture of 1.3 liters of water and 2 liters of 1 *N* HCl. Then 18 gm of benzyl *p*-hydroxybiphenylpolyglycol is added as an emulsifier. Next 2 liters of ice cold 1 *N* NaOH is added over a 30 min period with stirring and cooling to form the polyurea. The polyurea is filtered to afford 234 gm (78%), m.p. 225°–227°C. At 26°C the polyurea could be drawn into threads which could be stretched in the cold and easily dyed.

In place of a diisocyanate a bisurethane of a diamine may be used to react with a diamine to give a polyurea as shown in Eq. (8) and Preparation 2-4.

$$\text{H}_2\text{N—R—NH}_2 + \text{C}_2\text{H}_5\text{O}\overset{\text{O}}{\overset{\|}{\text{C}}}\text{NH—R—NH}\overset{\text{O}}{\overset{\|}{\text{C}}}\text{OC}_2\text{H}_5 \longrightarrow \left[\text{—R—NH}\overset{\text{O}}{\overset{\|}{\text{C}}}\text{NH—}\right]_n + 2\text{C}_2\text{H}_5\text{OH} \tag{8}$$

2-4. *Preparation of Poly(hexamethylene-decamethyleneurea) Copolymer* [28]

$$\text{H}_2\text{N(CH}_2)_{10}\text{NH}_2 + \text{C}_2\text{H}_5\text{O}\overset{\text{O}}{\overset{\|}{\text{C}}}\text{—NH(CH}_2)_6\text{NH—}\overset{\text{O}}{\overset{\|}{\text{C}}}\text{OC}_2\text{H}_5 \longrightarrow [\text{—(CH}_2)_{10}\text{NH}\overset{\text{O}}{\overset{\|}{\text{C}}}\text{NH(CH}_2)_6\text{NH}\overset{\text{O}}{\overset{\|}{\text{C}}}\text{NH—}]_n + 2\text{C}_2\text{H}_5\text{OH} \tag{9}$$

(*a*) *Preparation of hexamethylenebis(ethylurethane).* To a three-necked flask equipped with a mechanical stirrer, two dropping funnels, and a condenser is added 58 gm (0.5 mole) of hexamethylenediamine in 200 ml of ether. The flask is cooled to 0°–10°C with an ice-water bath, while simultaneously from separate funnels 130 gm (1.2 moles) of ethyl chlorocarbonate, and 48 gm (1.2 moles) of sodium hydroxide in 400 ml of water are added with vigorous stirring. Fifteen minutes after the addition, the solid is filtered and recrystallized from benzene–petroleum ether, m.p. 84°C (percent yield not reported).

(*b*) *Polymerization of decamethylenediamine and hexamethylenebis(ethylurethane).* To a test tube with a capillary tube for nitrogen gas are added 12.4 gm (0.072 mole) of decamethylenediamine and 18.70 gm (0.072 mole) of hexamethylenebis(ethylurethane). The test tube is heated to 202°C while the nitrogen gas is being bubbled through the contents. After 3 hr the polymer is cooled under nitrogen and then isolated to give an inherent viscosity of 0.2–0.4 in *m*-cresol (0.5% conc., 25°C). The polymer melt temperature is about 170°C.

D. Reaction of Diamines with Urea

Urea reacts with diamines at 130°C to give ammonia and polyureas. The polycondensation can be carried out in the melt or in a solvent such as *m*-cresol [29,30] or *n*-alkylpyrrolidones [31]. In some cases tertiary amine catalysts in dioxane solvent are used to accelerate the polymerization [32].

$$NH_2RNH_2 + NH_2CONH_2 \xrightarrow{-2NH_3} [\text{—}NHCONHR\text{—}]_n \qquad (10)$$

A variety of aliphatic and aromatic primary diamines have been used for the polymerization in addition to *N*-substituted diamines [33], unsaturated diamines [34], and heterocyclic-based diamines [35,36].

The preparation of several polymethylene straight-chain polyurea homopolymers have been reported, for example, $[\text{—}CONH\text{—}(CH_2)_n\text{—}NH\text{—}]_x$ where $n = 3$ [37], 6 [38,39], 7 [40–44], 10 [38,42,43], and 12 [43].

2-5. *Preparation of Poly(4-oxyhexamethyleneurea)* [38]

$$NH_2\overset{\overset{O}{\|}}{C}NH_2 + H_2N(CH_2)_3\text{—}O\text{—}(CH_2)_3NH_2 \xrightarrow[\text{heat}]{N_2}$$

$$[\text{—}(CH_2)_3\text{—}O\text{—}(CH_2)_3\text{—}NH\text{—}\overset{\overset{O}{\|}}{C}\text{—}NH\text{—}]_n + 2NH_3 \qquad (11)$$

To a test tube with a side arm are added 7.5 gm (0.125 mole) of urea and

16.5 gm (0.125 mole) of bis(γ-aminopropyl) ether. A capillary tube attached to a nitrogen gas source is placed on the bottom of the tube and the temperature is raised to 156°C and kept there for 1 hr, during which time ammonia is evolved. The temperature is raised to 231°C for 1 hr and then to 255°C for an additional hour. At the end of this time a vacuum is slowly applied to remove the last traces of ammonia. (CAUTION: Frothing may be a serious problem if the vacuum is applied too rapidly.) The polymer is cooled, the test tube broken, and the polymer isolated. The polymer melt temperature is approximately 190°C, and the inherent viscosity is approximately 0.6 in *m*-cresol (0.5% conc., 25°C).

2-6. Preparation of a Polyurea by the Reaction of N,N-Bis(3-aminopropyl)-piperazine with Urea [44]

$$H_2NCH_2CH_2CH_2{-}N\langle\rangle N{-}CH_2CH_2CH_2NH_2 + H_2N{-}\overset{O}{\overset{\|}{C}}{-}NH_2 \longrightarrow$$

$$\left[{-}CH_2CH_2CH_2{-}N\langle\rangle N{-}CH_2CH_2CH_2NH\overset{O}{\overset{\|}{C}}{-}NH{-}\right]_n + 2NH_3 \quad (12)$$

To a flask containing 200.0 gm (1.0 mole) of *N,N*-bis(3-aminopropyl)piperazine and 60.0 gm (1.0 mole) of urea is added 37.7 gm (0.3 mole) of caprolactam. The flask is heated and vigorously stirred at 130°C for 3 hr in a nitrogen stream. After this time the reaction mixture is heated at 250°C under atmospheric pressure for 4 hr and then *in vacuo* for 2 hr to give white polymer, m.p. 231°C; specific viscosity at 30°C, 1.05 (*m*-cresol).

Molecular weight regulators are often used to give polyureas of specified molecular weight ranges. Some typical molecular weight regulators are *N*-pelargonylnonamethylenediamine or phthalimide [45].

3. MISCELLANEOUS METHODS

1. Reaction of diamines with carbon oxysulfide [46].
2. Preparation of linear polyureas by treating the carbonate of a diamine with a diisocyanate in toluene at 70°–80°C [19,47].
3. Reaction of diamines with the reaction product of imidazole and phosgene [48].
4. Reaction of diamines with *N*-methylnitrosourea [49].
5. Reaction of diamines with carbon monoxide in the presence of catalysts such as mercuric acetate in ether, ester, or hydrocarbon solvents [50].

6. Reaction of a diisocyanate with an equivalent amount of benzoic acid in dimethyl sulfoxide [51].

7. Reaction of an aliphatic dihalide with alkali cyanates in solvents such as water, dimethylformamide, or dimethyl sulfoxide [52].

8. Ring-opening polymerization of cyclic ethylene ureas at 100°–200°C using strongly basic catalysts [53].

9. Thermal reaction of bisurethanes in the presence of titanium butoxide catalyst to give polyreas [54].

10. Reaction of diisocyanates with water to give polyureas [55].
11. Reaction of dicarboxylic acid azides with water [56].
12. Preparation of copolyureas containing amide functional groups [57].
13. Preparation of silyl-substituted polyureas [58].
14. Polyureas containing boron atoms [59].
15. Polyureas containing nitrile [60], carboxyl [60a], or sulfonic groups [60b].
16. Copolymers of urea and thiourea groups [61].
17. Preparation of poly(urethan-ureas) [62].
18. Preparation of poly(sulfonamide-ureas) [63].
19. Reaction of polycarbodiimides with water or H_2S to give polyureas or polythioureas [64].
20. Preparation of polythiosemicarbazides [65].
21. Preparation of polyethyleneglycol-urea adducts [66].
22. Preparation of linear poly(spiro-acetalurea) [67].
23. Polymers from urea and chloroalkylphosphonyl dichlorides [68].
24. Polymers of phosphonic acid diamides with urea [69].

REFERENCES

1. S. R. Sandler and W. Karo, "Organic Functional Group Preparations," Vol. 2, Chapter 6. Academic Press, New York, 1971.
1a. H. G. J. Overmars, *Encycl. Polym. Sci. Technol.* **11**, 464 (1969); O. E. Snider and R. J. Richardson, *ibid.*, p. 495; M. A. Dietrich and H. W. Jacobson, U.S. Patent 2,709,694 (1955).
2. H. C. Haas and S. G. Cohen, U.S. Patent 2,835,653 (1958).
3. E. Klein and J. W. Weaver, U.S. Patent 2,911,322 (1959).
4. R. Schiffner and G. Lange, German (East) Patent 18,253 (1960); A. Berger, U.S. Patent 2,781,281 (1957).
5. D. F. Kutepov, A. A. Potashnik, D. N. Khokhlov, and V. A. Tuzhilkina, *Zh. Obshsch. Khim.* **29**, 855 (1959).
6. R. T. Schilit, German Patent 1,183,196 (1964).
7. W. Hentrich and A. Kirstahler, German Patent 924,240 (1955).
8. W. G. Simons, U.S. Patent 2,518,388 (1950); Kyowa Fermentation Industry Co., Ltd., French Patent 1,478,361 (1967).

9. W. S. Talgyesi, *J. Text. Res.* **38**, 416 (1968).
10. Urylon Bull. Toyo Koatsu Industries, 1959; Daily News Record, Toyo Koatsu Industries, 1957.
11. E. L. Wittbecker, U.S. Patent 2,816,879 (1957).
12. L. Alexandra and L. Dascalu, *J. Polym. Sci.* **52**, 331 (1961).
13. D. J. Lyman and S. L. Jung, *J. Polym. Sci.* **41**, 407 (1959).
14. R. J. Cotter, British Patent 902,134 (1962).
15. R. J. Cotter, British Patent 907,829 (1962).
16. R. J. Cotter, U.S. Patent 3,131,167 (1964).
17. R. J. Cotter, U.S. Patent 3,130,179 (1964).
18. R. J. Cotter, British Patent 915,504 (1963).
19. M. Katz, U.S. Patent 2,975,157 (1961).
20. M. Katz, U.S. Patent 2,888,438 (1959).
21. H. V. Boening, N. Walker, and E. H. Meyers, *J. Appl. Polym. Sci.* **5**, 384 (1961).
22. W. Lehmann and H. Rinke, U.S. Patent 2,852,494 (1958).
23. P. Schlack, German Patent 920,511 (1954).
24. W. Lehmann and H. Rinke, U.S. Patent 2,761,852 (1956).
25. O. Y. Fedotova, A. G. Grozdow, and I. A. Rusinovskaya, *Vysokomol. Soedin.* **7**, 2028 (1965).
26. S. R. Sandler and F. R. Berg, *J. Appl. Polym. Sci.* **9**, 3909 (1965).
27. O. Bayer, *Angew. Chem.* **59**, 263 (1947).
28. E. I. du Pont de Nemours, British Patent 528,437 (1940).
29. W. R. Grace Co., British Patent 863,297 (1961).
30. Y. Inaba and K. Ueno, Japanese Patent 10,092 (1956).
31. R. Gabler and H. Müller, U.S. Patent 3,185,656 (1965).
32. Y. Furuya and K. Itoho, *Chem. Ind.* (*London*) p. 359 (1967).
33. H. Iiyama, M. Asakura, and K. Kimoto, *Kogyo Kagaku Zasshi* **68**, 236 (1966).
34. G. Kimura, S. Kaichi, and S. Fujigake, *Yuki Gosei Kagaku Kyokai Shi* **23**, 241 (1965).
35. Y. Inaba, K. Miyake, and G. Kimura, U.S. Patent 3,054,777 (1962).
36. F. Veatch and J. D. Idol, Belgian Patent 614,386 (1962).
37. Y. Iwakura, *Chem. High Polym.* **4**, 97 (1947).
38. E. I. du Pont de Nemours and Co., Inc., British Patent 530,267 (1940).
39. Y. Inaba, K. Miyake, and G. Kimura, U.S. Patent 3,046,254 (1962).
40. R. Gabler and H. Müller, U.S. Patent 3,185,656 (1965).
41. G. Kimura, S. Kaichi, and S. Fujigake, *Yuki Gosei Kagaku Kyokai Shi* **23**, 241 (1965).
42. Y. Inaba, K. Miyake, K. Kimoto, and K. Kimura, French Patent 1,207,356 (1958).
43. P. Borner, W. Gugel, and R. Pasedag, *Makromol. Chem.* **101**, 1 (1967).
44. K. Ueda, H. Mikami, and T. Okawara, Japanese Patent 13,243 (1960).
45. A. G. Schering, French Patent 1,312,385 (1962).
46. G. J. M. van der Kerk, German Patent 1,164,093 (1964).
47. W. Lehmann and H. Rinke, German Patent 838,217 (1952).
48. W. R. Grace Co., French Patent 1,299,698 (1962).
49. H. A. Walter, U.S. Patent 3,006,898 (1961).
50. D. M. Fenton, U.S. Patent 3,277,061 (1966); N. Sonoda, T. Yasuhara, K. Kondo, T. Ikeda, and S. Tsutsumi, *J. Amer. Chem. Soc.* **93**, 6344 (1971).
51. W. R. Sorenson, *J. Org. Chem.* **24**, 978 (1959).
52. W. Gerhardt, German (East) Patent 42,057 (1965); R. C. Doss and J. T. Edmonds, Jr., U.S. Patent 3,379,687 (1968).
53. W. H. Tibby, U.S. Patent 3,042,658 (1962).
54. T. M. Taakov and D. D. Reynolds, *J. Amer. Chem. Soc.* **79**, 5717 (1957).

55. Y. Iwakura, *Chem. High Polym.* **4**, (1947); J. O'Brochta and S. C. Tenin, U.S. Patent 3,160,648 (1964).
56. T. Lieser and H. Gehlen, *Justus Liebigs Ann. Chem.* **556**, 127 (1944); H. Gehlen and T. Lieser, German Patent 869,865 (1941); T. Curtius and H. Clemm, *J. Prakt. Chem.* [N.S.] **62**, 203 (1900).
57. Tayo Kaotsu Industries, Inc., British Patent 773,964 (1957); G. E. Ham, U.S. Patent 3,053,811 (1962); K. Ureda, B. Fukuda, and H. Mitsukami, Japanese Patent 9441 (1960); K. Hayashi, S. Hany, and Y. Iwakura, *Makromol. Chem.* **86**, 64 (1965).
58. J. F. Klebe, *J. Polym. Sci., Part B* **2**, 1079 (1964); J. F. Klebe, U.S. Patent 3,172,874 (1965).
59. V. V. Korshak, N. I. Bekasova, V. A. Zmiyatina, and J. Na, *Vysokomol. Soedin.* **3**, 521 (1961).
60. O. Y. Fedotova and M. I. Shtilman, *Vysokomol. Soedin.* **7**, 312 (1965).
60a. Merck and Co., Inc., Netherlands Patent Appl. 6,515,245 (1965).
60b. W. Thoma, O. Bayer, and H. Rinke, German Patent 1,067,212 (1959); R. Neher, German Patent 1,046,309 (1958); CIBA A.G., Belgian Patent 531,426 (1955).
61. Y. Inaba, K. Miyake, and G. Kimura, U.S. Patent 3,034,777 (1962); G. E. Ham, U.S. Patent 3,080,343 (1963); Tayo Koatsu Industries, Inc., British Patent 773,965 (1957).
62. W. Lehmann and H. Rinke, U.S. Patent 2,852,494 (1958); Y. Iwakura, *Chem. High Polym.* **2**, 323 (1945); Y. Iwakura, K. Hayashi, and K. Inagaki, *Makromol. Chem.* **104**, 56 (1967); **110**, 84 (1967).
63. Y. Iwakura, K. Hayashi, and K. Inagaki, *Makromol. Chem.* **100**, 22 (1967).
64. D. J. Lyman and N. Sadri, *Makromol. Chem.* **67**, 1 (1963).
65. E. A. Tomic, T. W. Campbell, and V. S. Foldi, *J. Polym. Sci.* **62**, 387 (1962); T. W. Campbell and E. A. Tomic, *ibid.*, p. 379.
66. T. Ito, A. Amaba, and Y. Izumi, Japanese Patent 41,306 (1970).
67. T. Hairida, N. Tobita, and E. Takiyoma, Japanese Patent 42,383 (1971); *Chem. Abstr.* **76**, 141746d (1972).
68. A. C. Haven, Jr., U.S. Patent 2,716,639 (1955).
69. H. W. Coover, Jr., U.S. Patent 2,642,413 (1953).

Chapter 8

POLYURETHANES

1. INTRODUCTION

The reaction of isocyanates with alcohols to give *N*-carbamates has been described in detail in Chapter 10 of Volume II of *Organic Functional Group Preparations* [1]. Although polyurethanes were briefly described in that chapter, the subject will be covered here in greater detail, especially from the preparative viewpoint.

Polyurethanes were first reported by Bayer in 1937 [1a,1b] and Rinke and co-workers in 1939 [2], to result from the reaction of diisocyanates with dihydric alcohols. From 1945 on, many other patents were issued on the preparation of polyurethanes from the reaction of diamines and bischloroformates at low temperatures [3,3a]. Soon after, several reports appeared describing the high-temperature solution and melt polymerization methods which involve the reaction of diisocyanates with diols [1b,4]. Recently another method has been reported in which polyurethanes are prepared by the direct reaction of 1,4-dichloro-2-butene with sodium cyanate and diols [5]. However, this latter method may not be of general utility since further research is required. It is presented at this time for information only. These methods are briefly summarized in Scheme 1. Scheme 1 also indicates that ester exchange reactions can also be used to give polyurethanes. This latter reaction is particularly important in one-package type adhesives and coatings.

SCHEME 1
Preparation of polyurethanes

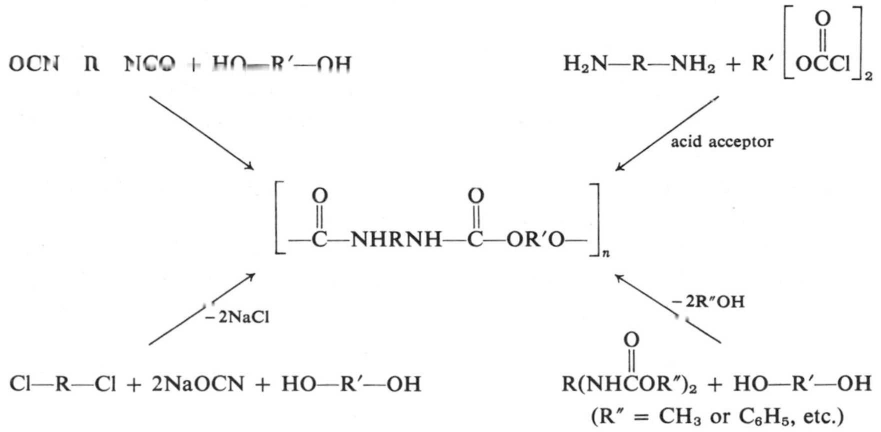

Polyurethanes can be viewed as mixed amide-esters of carbonic acid and thus have properties intermediate between polyesters and polyamides. Polyurethanes usually are lower melting than the related polyamide of the same carbon skeleton. Polyurethanes [6,6a] and urethanes [7] usually decompose at temperatures above 220°C to give either the free isocyanate and alcohol or the free amine, carbon dioxide, and olefin. In the case of phenolic urethanes decomposition starts to take place at temperatures as low as 150°C.

$$RNH_2 + CO_2 + R'CH{=}CH_2 \longleftarrow RNH\overset{\displaystyle O}{\overset{\|}{C}}OCH_2CH_2R' \longrightarrow RNCO + R'CH_2CH_2OH \quad (1)$$

Polyurethanes have found many applications in the areas of adhesives [8], coatings [9], flexible [10] and rigid foams [11], elastomers [12], fibers [13], waterproofing [14], thermoset resins [15], thermoplastic molding compounds [16], rubber vulcanization [17], silicon and boron polymers [18], and the preparation of poly(2-oxazolidone) resins [19]. The commercial applications and forecasts for use of polymethanes for the next three years have recently been reviewed [20].

2. CONDENSATION REACTIONS

The two most important methods of preparing polyurethanes involve the reaction of diisocyanates with dihydric alcohols and the low-temperature reaction of bis(chloroformates) with diamine.

A. Condensation of Diisocyanates with Dihydric Alcohols

The diisocyanate–diol method has the advantage over the bis(chloroformate)–diamine method in that it involves a simple addition reaction without any by-product formation (HCl). Side reactions can be minimized under the appropriate conditions [6].

The reactivity (uncatalyzed) of diisocyanates with a given alcohol depends on the structure of the former as shown in Fig. 1.

Polyurethane resins are usually prepared by the reaction of a long-chain diol with an excess of the diisocyanate to obtain a "prepolymer" with terminal isocyanate groups [21] (Eq. 2). This "prepolymer" can react

$$R(NCO)_2 + HO{\sim\sim\sim}OH \longrightarrow OCN{-}R{-}NHCOO{\sim\sim\sim}OCONH{-}R{-}NCO \quad (2)$$

separately with diols or diamines of low molecular weight to cause further chain extension or polymerization (curing of the prepolymer). On reaction with water, the "prepolymer" can also be used to give foams, amines or tin compounds being used as catalysts in the foaming process. The properties of the polyurethane resin are controlled by the choice of the diisocyanate and polyol. The stiffness of the aromatic portion of the diisocyanate may be increased by using 1,5-naphthalene diisocyanate in place of 2,4-toluene diisocyanate. In addition, the flexibility may be increased by using a polyether polyol derived from propylene oxide or tetrahydrofuran. Using aromatic polyols and triols would lead to chain stiffness and crosslinking in the case of the triols. The use of aromatic diamines leads to a polyurea of greater rigidity than a polyether polyurethane. Some examples of the reactivity of diols with a representative diisocyanate are shown in Table I [22].

2-1. Preparation of a Polyurethane Prepolymer [8]

To a 500 ml, dry, nitrogen-flushed resin kettle equipped with a mechanical stirrer, thermometer, condenser with drying tube, dropping funnel, and

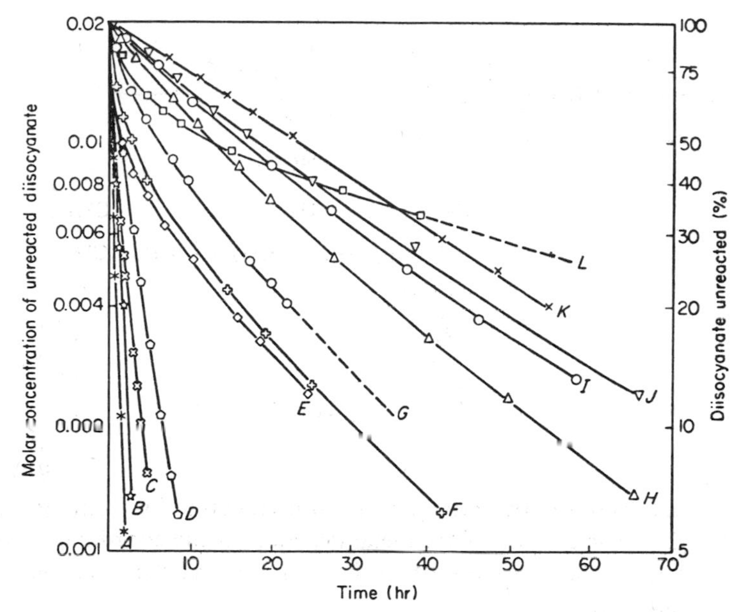

FIG. 1. Reactivity of aromatic diisocyanates 0.02 *M* with 2-ethylhexanol 0.4 *M* and diethylene glycol adipate polyester in benzene at 28°C. A. 1-Chloro-2,4-phenylene diisocyanate. B. *m*-Phenylene diisocyanate. C. *p*-Phenylene diisocyanate. D. 4,4'-Methylene bis(phenyl isocyanate). E. 2,4-Tolylene diisocyanate. F. Tolylene diisocyanate (60% 2,4 isomer, 40% 2,6 isomer). G. 2,6-Tolylene diisocyanate. H. 3,3'-Dimethyl-4,4'-biphenylene diisocyanate (0.002 *M*) in 0.04 *M* 2-ethylhexanol. I. 4,4'-Methylene bis(methylphenyl isocyanate). J. 3,3'-Dimethoxy-4,4'-biphenylene diisocyanate. K. 2,2',5,5'-Tetramethyl-4,4'-biphenylene diisocyanate. L. 80% 2,4 and 20% 2,6 isomer of tolylene diisocyanate with diethylene glycol adipate polyester (hydroxyl No. 57, acid No. 1.6, and average molecular weight 1900). Reprinted from M. E. Bailey, V. Kirss, and K. G. Spaunburgh, *Ind. Eng. Chem.* **48**, 794 (1956). Copyright 1956 by the American Chemical Society; reprinted by permission of the copyright owner.

heating mantle are added 34.8 gm (0.20 mole) of 2,4-toluene diisocyanate and 0.12 gm (0.000685 mole) of *o*-chlorobenzoyl chloride. Polyoxypropylene glycol (Dow Chemical Co. P-1000, mol. wt. 1000) is added dropwise over a 2 hr period at 65° ± 3°C until 100.0 gm (0.1 mole) has been added. Stirring at 65°C is maintained for an additional 2 hr and then the reaction is cooled. The final product is a clear viscous liquid; NCO calc. 6.23%; found 6.27%. Several other preparations of polyurethane prepolymers have recently been reported [23].

Various catalysts help to speed the reaction of isocyanate groups with alcohols.

Tertiary amines activate the isocyanate functional groups for reaction with an active hydrogen compound [24].

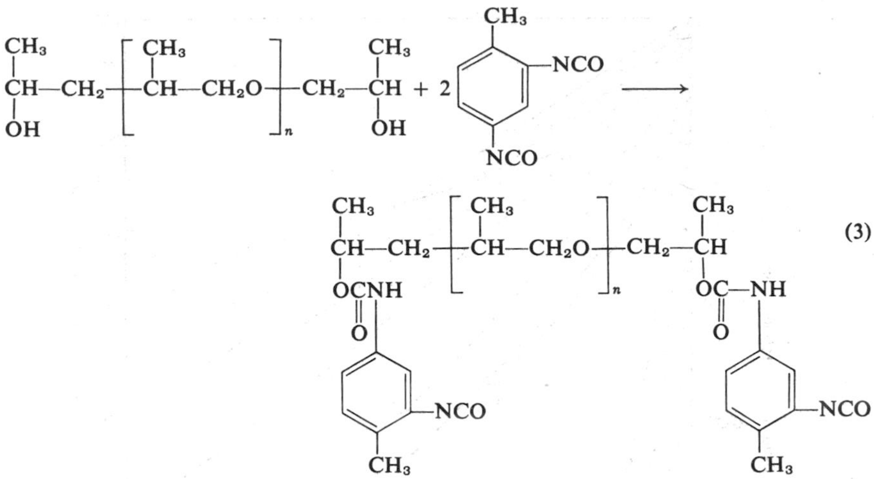

$$CH_3CH(OH)CH_2-[CH(CH_3)CH_2O]_n-CH_2CH(CH_3)OH + 2\ CH_3C_6H_3(NCO)_2 \longrightarrow$$

$$OCN(CH_3)C_6H_3NHC(=O)O-CH(CH_3)CH_2-[CH(CH_3)CH_2O]_n-CH_2CH(CH_3)-OC(=O)NH-C_6H_3(CH_3)NCO \quad (3)$$

Metal salts catalyze by activation of both reactants by forming a ternary complex [25]. The activity of the metal depends on the coordination number of the metal ion, the configuration of the complex, and the ionic radius of the ion in question [25].

Some basic catalysts such as sodium phenates, amines [26], and tri-*n*-butylantimony oxide promote trimerization of isocyanates to isocyanurates and polyisocyanurates [23,27]. The trimerization of phenyl isocyanate with various catalysts has recently been reported as shown in Table II [23,27].

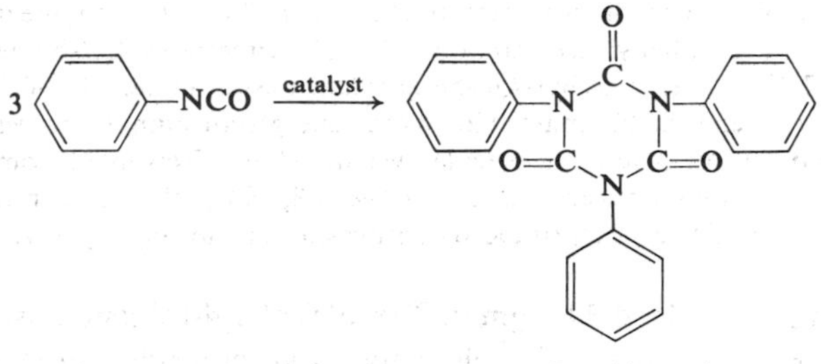

$$3\ C_6H_5-NCO \xrightarrow{\text{catalyst}} (C_6H_5NCO)_3 \text{ (1,3,5-triphenyl isocyanurate)} \quad (4)$$

Some typical examples of the preparation of polyurethanes, with some of the experimental details, are shown in Table III.

2-2. Preparation of Poly[ethylene methylene bis(4-phenylcarbamate)] [6,28]

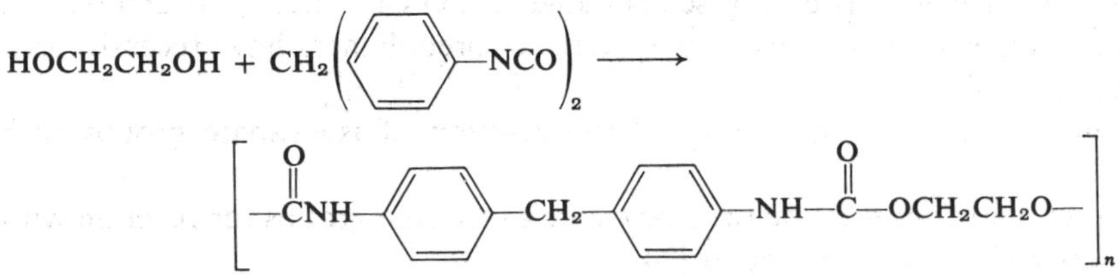

$$HOCH_2CH_2OH + CH_2(C_6H_4-NCO)_2 \longrightarrow$$

$$\left[-\overset{O}{\overset{\|}{C}}NH-C_6H_4-CH_2-C_6H_4-NH-\overset{O}{\overset{\|}{C}}-OCH_2CH_2O-\right]_n \quad (5)$$

TABLE I
REACTIVITY OF DIOLS WITH *p*-PHENYLENE DIISOCYANATE AT 100°C [22]

Diol type	Relative order of reactivity
Polyethylene adipate	100
Polytetrahydrofuran	17–54
1,4-Butanediol	15
1,4-*cis*-Butenediol	7
1,5-Bis(β-hydroxyethoxy)naphthalene	4
1,4-Butynediol	1[a]

[a] Rate of reaction = 0.6×10^{-4} liter/mole sec.

TABLE II
TRIMERIZATION OF PHENYL ISOCYANATE (0.005 *M*) BY VARIOUS CATALYSTS (0.0001 *M*)[a]

Catalyst	Trimer observed[b]		M.p. (°C)[d]
	23°C	50°C	
Triphenylarsine	N.R.[c]	N.R.	—
Bis(tri-*n*-butyltin) oxide	24 hr	7 hr	275–281
Dibutyltin oxide	24 hr	—	—
Tricresyl phosphate	N.R.	N.R.	—
Antimony chloride	N.R.	N.R.	—
Bismuth trioxide	N.R.	N.R.	—
Bismuth oxychloride	N.R.	N.R.	—
Triphenyltin hydroxide	N.R.	N.R.	—
Tributyltin chloride	N.R.	N.R.	—
Triethylamine	N.R.	N.R.	—
Calcium naphthenate	N.R.	24 hr	270–275
Dibutyltin dilaurate	N.R.	7 hr	—
Stannous octoate	24 hr	5 hr	275–281
Aluminum isopropylate	N.R.	24 hr	275–281
Sodium methoxide	48 hr	—	272–278
Calcium oxide	N.R.	N.R.	—
Tri-*n*-butylantimony oxide	15 min	—	265–275

[a] Reprinted from Sandler [23]. Copyright 1967 by the *Journal of Applied Polymer Science*. Reprinted by permission of the copyright owner.
[b] The times listed are the times for complete solidification of the triphenyl isocyanurate.
[c] No reaction.
[d] Literature value for the m.p. of triphenyl isocyanurate is 280°–281°C.

TABLE III

PREPARATION OF POLYURETHANES BY THE REACTION OF DIISOCYANATES WITH DIOLS

Diol (moles)	Diisocyanate (moles)	Solvent (ml)	Catalyst (gm)	Reaction conditions: Temp. (°C)	Reaction conditions: Time (hr)	Yield (%)	Physical properties: Polymer m.p. (°C)	Physical properties: η	Ref.
Butane-1,4-diol (0.10)	$(CH_2)_6(NCO)_2$ (0.10)	C_6H_5Cl (200)	—	132	4–5	87	176–177	0.25[a]	*b*
Benzoquinone-2,5-diol (0.005)	(0.005)	THF (25)	$(C_2H_5)_3N$ (0.05) or $(Bu)_2Sn(OAc)_2$ (0.05)	30	1–2	90–100	—	—	*c*
Hexafluoropentanediol (0.0137)	$(CF_2)_3(NCO)_2$ (0.0137)	$CH_3COOC_2H_5$ (100)	$(C_2H_5)_3N$ (0.05)	87	1–2	90–100	> 300	—	*d*
(0.0142)	$OCN{-}C_6F_4{-}NCO$ (0.0142)	(22)	$(C_2H_5)_3N$ (0.005)	80–82	25	62	> 340	—	*d*
1,5-Pentanediol (0.02)	(0.020)	(25)	—	25–30 77	$1\frac{1}{2}$ $2\frac{1}{2}$	91	280–281	—	*d*

Hexafluoro-pentanediol (0.0125)	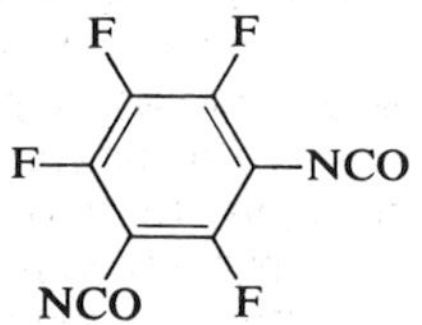 F F F NCO NCO F (0.0125)	—	—	40 90	$\frac{1}{4}$ 1	90–100	—	—	*d*
Tetrafluoro-2-hydroquinone (0.0125)	(0.010)	Xylene (20)	—	140	30	100	225 (dec.)	—	*d*
Hexafluoro-pentanediol (0.05)	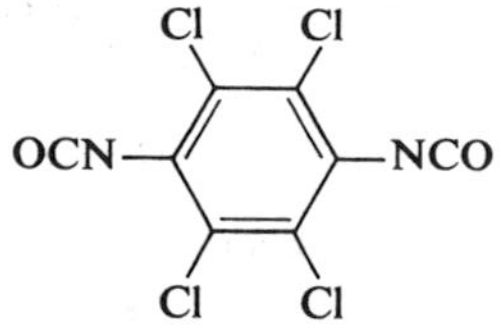 Cl Cl OCN NCO Cl Cl (0.050)	$CH_3COOC_2H_5$ (60)	$(C_2H_5)_3N$ (0.25)	30–35 82	$\frac{1}{2}$ $1\frac{1}{2}$	62	190–196	—	*d*
(0.0267)	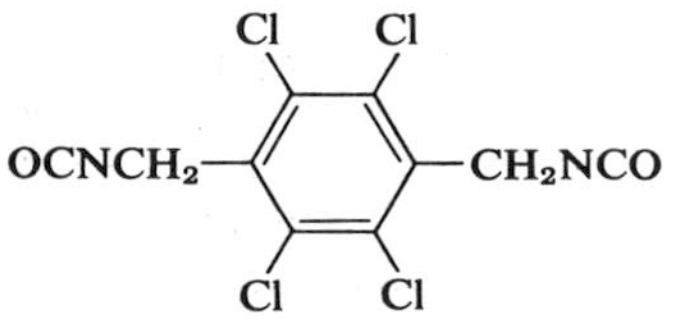 Cl Cl $OCNCH_2$ CH_2NCO Cl Cl (0.0267)	—	—	130–140	1	100	—	—	*d*
Ethylene glycol (0.10)	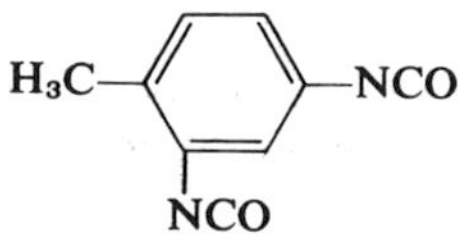 H_3C NCO NCO (0.10)	4-Methyl-pentanone-2 (40) + DMSO (40)	—	115	$1\frac{1}{2}$	95–100	180	0.45[e]	*f*

(*cont.*)

TABLE III (*cont.*)

Diol (moles)	Diisocyanate (moles)	Solvent (ml)	Catalyst (gm)	Reaction conditions		Physical properties			Ref.
				Temp. (°C)	Time (hr)	Yield (%)	Polymer m.p. (°C)	η	
(0.10)	$CH_2(C_6H_4\text{–}NCO)_2$ (0.10)	Same as above	—	115	$1\frac{1}{2}$	95–100	255	1.01[e]	*f,g*
trans-1,4-Cyclohexanediol (0.10)	(0.10)	DMSO (90)	—	105	$1\frac{1}{2}$	90–100	—	0.66[e]	*h*
Butane-1,4-diol (6.7)	$[\text{–}C_6H_3(OCH_3)\text{–}NCO\text{–}]$ (0.67)	$CH_3COC_2H_5$ (1000)	—	80	3	50	133.5	—	*i*
Polyoxy-propyleneglycol (mol. wt. 1000) (0.10)	$H_3C\text{–}C_6H_3(NCO)\text{–}NCO$ (0.20)	—	—	65–68	4	100	—	—	*j*

[a] Measured in *m*-cresol (0.5% conc. at 25°C). [b] C. S. Marvel and J. H. Johnson, *J. Amer. Chem. Soc.* **72**, 1674 (1950).
[c] G. Wegner, N. Nakabayashi, and H. G. Cassidy, *J. Polym. Sci. Part A-1* **6**, 3151 (1968).
[d] J. Hollander, F. D. Trischler, and R. B. Gosnell, *J. Polym. Sci., Part A-1* **5**, 2757 (1967).
[e] 0.05% solution in DMF at 30°C. [f] D. J. Lyman, *J. Polym. Sci.* **45**, 49 (1960).
[g] H. C. Beachell and J. C. Peterson, *J. Polym. Sci., Part A-1* **7**, 2021 (1969). [h] D. J. Lyman, *J. Polym. Sci.* **55**, 507 (1961).
[i] W. Kern, H. Kalsch, K. J. Ranterkus, and H. Sutter, *Makromol. Chem.* **44/46**, 78 (1961).
[j] S. R. Sandler and F. R. Berg, *J. Appl. Polym. Sci.* **9**, 3909 (1965); S. R. Sandler, *ibid.*, **11**, 811 (1967).

To a flask equipped with a mechanical stirrer, condenser, and dropping funnel and containing 40 ml of 4-methylpentanone-2 and 25.02 gm (0.10 mole) of methylene bis(4-phenyl isocyanate) (MDI) is added all at once 6.2 gm (0.10 mole) of ethylene glycol in 40 ml of dimethyl sulfoxide. The reaction mixture is heated at 115°C for $1\frac{1}{2}$ hr, cooled, poured into water, and filtered. The white polymer is chopped up in a blender, washed with water, and dried under reduced pressure at 90°C to afford 29.6–31.2 gm (95–100%), $\eta_i = 1.01$ (0.05% solution in DMF at 30°C), polymer melt temperature, 240°C.

2-3. *Preparation of Poly(hexafluoropentamethylene tetrafluoro-p-phenylenedicarbamate) by Solution Polymerization* [29]

$$HOCH_2(CF_2)_3CH_2OH + OCN-C_6F_4-NCO \longrightarrow \left[-\overset{O}{\overset{\|}{C}}-NH-C_6F_4-NH-\overset{O}{\overset{\|}{C}}-OCH_2(CF_2)_3CH_2-O- \right]_n \tag{6}$$

To a three-necked flask equipped with a stirrer, condenser, dropping funnel, thermometer, and nitrogen inlet, is added 3.02 gm (0.0142 mole) of hexafluoropentanediol dissolved in 10 ml of ethyl acetate. From the dropping funnel is added over a 5 min period 3.30 gm (0.0142 mole) of tetrafluoro-*p*-phenylene diisocyanate dissolved in 12 ml of dry ethyl acetate. The resulting solution is stirred for 24 hr at 80°–82°C under a nitrogen atmosphere. One drop of triethylamine is added and the solution is stirred for an additional hour at 80°–82°C. The ethyl acetate is removed under reduced pressure to afford 4.0 gm (62%) of a light tan waxy polymer, m.p. >342°C, IR (Nujol), (μ), 3.1 (s), 5.75 (s), 6.4 (s), 6.6 (m), 6.75 (s), 7.65 (m), 7.8 (s), 8.2 (m), 8.6 (s), 9.2 (m), 9.75 (m), 10.5 (m), 12.3 (w), 13.2 (m), and 14.75 (w).

2-4. *Preparation of Poly(hexafluoropentamethylene tetrafluoro-p-phenylenedicarbamate) by Melt Polymerization* [29]

To a long Pyrex test tube is added 16.2 gm (0.07 mole) of tetrafluoro-*p*-phenylene diisocyanate and 14.8 gm (0.07 mole) of hexafluoropentanediol. The mixture is heated to 140°C and then to 205°C, at which point the mixture solidifies. The polymer is isolated by cracking open the cooled test tube. In contrast to Preparation 2-3, this polymer is a strong and brittle solid, m.p. (crystalline) 262°–265°C.

B. Condensation of Bischloroformates with Diamines

The reaction of bischloroformates with diamines to give polyurethanes as shown in Eq. (7) has the advantage over the diisocyanate–diol reaction in that it can be carried out at room temperature or lower. The lower temperature reaction thus avoids cross-linking and consequently tends to give linear high molecular weight polymers.

$$H_2NRNH_2 + ClC(=O)OR'OC(=O)Cl \longrightarrow \left[-HNRNHC(=O)OR'OC(=O)- \right]_n + 2HCl \quad (7)$$

Furthermore it is possible to prepare polyurethanes, such as those from piperazine and other nitrogen heterocycles, which cannot be prepared by the diisocyanate–diol process.

The two methods used to prepare polyurethanes by this process are the interfacial and solution processes.

a. Interfacial Polycondensation Method

Factors [3a] important in determining the molecular weight of the product from interfacial polycondensation are (a) type of acid acceptor used; (b) organic solvent type; (c) use and type of dispersing agent; (d) ratio of organic solvent to water; (e) nature and purity of the reactants.

The effect of pH on percent yield and molecular weight of the formation of polyurethanes by this process has also been studied [30].

In the preparation of poly(ethylene *N,N'*-piperazinedicarboxylate) from piperazine and ethylene bis(chloroformate) it was found that the efficiency of bases to give high molecular weight products was in the following order [30]:

$$\text{Piperazine} > Na_2CO_3 > NaOH > NaHCO_3 > NaOAc$$

Piperazine is probably best because it may facilitate the transfer of diamine to the organic phase to give a closer balance of reactants in the organic phase reaction zone [30]. In addition, the inorganic acceptor may cause some hydrolysis of the bischloroformate and may also be less efficient in regenerating the piperazine from its hydrochloride salt [30].

Sodium lauryl sulfate is an efficient emulsifier and tends to aid in obtaining higher inherent viscosities [30].

The most common organic solvents are benzene or methylene chloride and their concentration is not critical.

The purity of the reactants should be high to obtain high molecular weight products. In a few cases this may not be necessary if the other variables mentioned above can be optimized.

b. Solution Polymerization Process

Solution polymerization is usually carried out in a polar solvent such as ethyl acetate with pyridine as the base. The tertiary amines are also used as acid acceptors. The reaction temperatures most commonly used are 22°–50°C.

Side reactions which decrease the availability of chloroformates are the reaction with tertiary amines to give alkyl halide, etc., as shown in Eq. (8) [31].

$$\mathrm{RO\overset{O}{\overset{\|}{C}}Cl} \xrightarrow{R'_3N} \mathrm{RCl + CO_2}$$
$$\mathrm{RO\overset{O}{\overset{\|}{C}}Cl} \xrightarrow{R'_3N} \mathrm{R'Cl + RO\overset{O}{\overset{\|}{C}}NR'_2} \tag{8}$$

Some typical examples of polyurethanes prepared by the bischloroformate–diamine process are described in Table IV.

Polyurethanes can also be prepared by using monoamino chloroformate salts as shown in Eq. (9) [32,33]. The starting materials for this reaction are hydroxy amines.

$$\mathrm{Cl\overset{O}{\overset{\|}{C}}OR{-}NH_3^+X^-} \xrightarrow{K_2CO_3} \left[\mathrm{-\overset{O}{\overset{\|}{C}}{-}OR{-}NH-}\right]_n \tag{9}$$

2-5. *Preparation of Poly(ethylene hexamethylenedicarbamate) by Interfacial Polymerization* [34]

$$\mathrm{Cl\overset{O}{\overset{\|}{C}}OCH_2CH_2O\overset{O}{\overset{\|}{C}}Cl + H_2N(CH_2)_6NH_2} \xrightarrow{-2HCl} \left[\mathrm{-\overset{O}{\overset{\|}{C}}OCH_2CH_2O\overset{O}{\overset{\|}{C}}NH(CH_2)_6NH-}\right] \tag{10}$$

(*a*) *Preparation of ethylene bis*(*chloroformate*) [35]

$$\mathrm{HOCH_2CH_2OH + 2COCl_2} \xrightarrow{-2HCl} \mathrm{Cl\overset{O}{\overset{\|}{C}}OCH_2CH_2O\overset{O}{\overset{\|}{C}}Cl} \tag{11}$$

CAUTION: This preparation should be carried out in a well-ventilated hood and all due precautions in the use of phosgene (toxic) should be strictly observed.

Into an ice-cooled 250 ml three-necked flask equipped with a mechanical stirrer, Dry Ice condenser, dropping funnel, and caustic trap (see Note a) is condensed 89.1 gm (0.90 mole) of phosgene. Then 12.5 gm (0.20 mole) of reagent grade ethylene glycol is added dropwise at 0°–5°C and after the addition the reaction mixture is stirred for 3–4 hr. The excess phosgene is allowed to evaporate and passed through an aqueous methanolic sodium hydroxide trap. The residue is heated at 40°–50°C at 20 mm Hg for a few

TABLE IV

PREPARATION OF POLYURETHANES BY THE REACTION OF BISCHLOROFORMATES WITH DIAMINES

Bischloroformate (mole)	Diamine (mole)	$C_{12}H_{15}ONa$ Surfactant (gm)	Base (gm)	Solvent (ml)	Water (ml)	Reaction conditions Temp. (°C)	Time (hr)	Yield (%)	M.p. (°C)	η	Ref.
Interfacial Polymerization Process											
$Cl—C(=O)—O—CH_2—CH_2—O—C(=O)—Cl$ (0.05)	$(CH_2)_6(NH_2)_2$ (0.05)	1.5	Na_2CO_3 (10.6)	C_6H_6 (125)	150	10	$\frac{1}{6}$	72	180	1.10[a]	b
$Cl—C(=O)—O—CH_2CH_2OC(=O)Cl$ (0.05)	$HN(CH_2CH_2)_2NH$ (0.05)	1.5	(10.6)	(125)	150	10	$\frac{1}{16}$	85	245	1.6[a]	b
$ClC(=O)O—C_6H_{10}—OC(=O)Cl$ (0.05)	$(CH_2)_2(NH_2)_2$ (0.05)	1–2	(10.6)	(125)	150–200	10	$\frac{1}{16}$	93	245	0.57	b

Solution Polymerization Process

$Cl{-}C(=O){-}OCH_2(CF_2)_3CH_2OC(=O)Cl$ (0.07)	$H_2NCH_2(CF_2)_3CH_2NH_2$ (0.07)	—	(14.9)	CH_2Cl_2 (100)	250	0–5	$\frac{1}{6}$	64	—	—	*c*
(0.03)	(0.03)	—	Pyridine (5.2)	$CH_3COOC_2H_5$ (50)	—	22–50	1	84	> 300	—	*c*
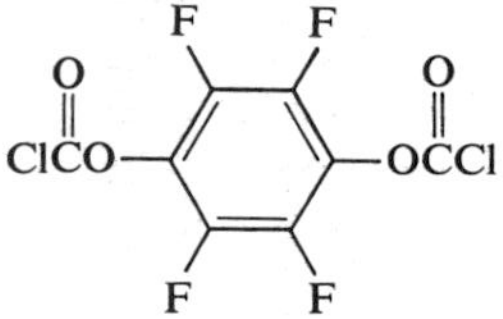 $ClC(=O)O{-}C_6F_4{-}OC(=O)Cl$ (tetrafluoro-p-phenylene) (0.016)	(0.016)	—	THF (2.7)	(100)	—	25 60	1 1	69	175–196	—	*c*
$Cl{-}C(=O)OCH_2CH_2OC(=O){-}Cl$ (0.1)	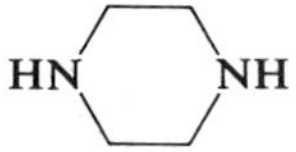 Piperazine (HN⟨⟩NH) (0.1)	—	$(C_2H_5)_3N$ (0.2)	$CHCl_3$ (100)	—	25	$\frac{1}{6}$	72	—	0.63	*d*

[a] η_{inh} determined at 30°C at a concentration of 0.5 gm of polymer/100 ml *m*-cresol.
[b] E. L. Wittbecker and M. Katz, *J. Polym. Sci.* **40**, 367 (1959).
[c] J. Hollender, F. D. Trischler, and R. B. Gosnell, *J. Polym. Sci. Part A-1* **5**, 2757 (1967).
[d] S. L. Kwolek and P. W. Morgan, *J. Polym. Sci., Part A* **2**, 2693 (1964).

minutes to remove volatiles (Note b) and then fractionally distilled through a 10 inch glass helices-packed column to afford 34.0 gm (91%), b.p. 71°–72°C (2.2 mm Hg).

NOTES: (a) A methanolic sodium hydroxide trap is used to trap any uncondensed phosgene. It is advisable to use a small cylinder of phosgene which is kept in the hood and also to have a slight amount of ammonia vapors passing into the hood to react with any free or escaping phosgene. (Use a small ammonia cylinder.) (b) The crude product at this point may be able to be used directly in the next step provided it is first tested on a small scale for its reactivity.

(*b*) *Polyurethane preparation* [34]. To a blender is added a cold (5°C) solution of 5.8 gm (0.05 mole) of hexamethylenediamine, 10.6 gm (0.1 mole) of sodium carbonate, and 1.5 gm of sodium lauryl sulfate in 150 ml of water. While stirring rapidly 9.35 gm (0.05 mole) of ethylene bis(chloroformate) in 125 ml of benzene at 10°C is quickly added. After 5 min the polymer is filtered and dried to afford 8.3 gm (72%) $\eta_{inh} = 1.19$ (0.5 gm polymer/100 ml *m*-cresol at 30°C).

2-6. *Preparation of Poly(hexafluoropentamethylene hexafluoropentamethylenedicarbamate) by the Interfacial Polymerization Process* [29]

$$ClC(=O)OCH_2(CF_2)_3CH_2OC(=O)Cl + H_2NCH_2(CF_2)_3CH_2NH_2 \xrightarrow{-2HCl} \left[-C(=O)OCH_2(CF_2)_3CH_2OC(=O)NHCH_2(CF_2)_3CH_2NH- \right]_n \quad (12)$$

To a three-necked flask equipped with a mechanical stirrer, dropping funnel, condenser, nitrogen inlet, and thermometer is added 14.7 gm (0.07 mole) of hexafluoropentanediamine and 14.9 gm (0.14 mole) of sodium carbonate dissolved in 250 ml of water. Then 23.6 gm (0.07 mole) of hexafluoropentamethylene bis(chloroformate) in 100 ml of methylene chloride is added rapidly with vigorous stirring at 0°–5°C. The slurry is then stirred for 2½ hr at 0°–5°C. The methylene chloride and water are decanted and the oily polymer dried under reduced pressure at 80°–100°C for 24 hr to afford 21.4 gm (64%) of an opaque elastomeric polyurethane product, IR (μ) of polymer film: 3.1 (m), 5.8 (s), 6.6 (s), 7.0 (w), 7.85 (m), 8.75 (s), 10.4 (m), 11.25 (m), and 12.85 (w).

2-7. *Preparation of Poly(hexafluoropentamethylene hexafluoropentamethylenedicarbamate) by the Solution Polymerization Process* [29]

To a flask equipped as in Preparation 2-6 is added 6.3 gm (0.03 mole) of hexafluoropentanediamine, 5.2 gm (0.066 mole) of pyridine, and 50 ml of ethyl acetate. Then 10.1 gm (0.03 mole) of hexafluoropentamethylene bis-(chloroformate) is slowly added at 22°–50°C. The red slurry is stirred for $1\frac{3}{4}$ hr at 30°–50°C and then for 4 hr at 80°C to give a dark green slurry. The reaction mixture is precipitated in 1 liter of rapidly stirred cold water to give a black oily polymer. The water is decanted and the black oil dried under reduced pressure at 90°C to afford 11.9 gm (84%) of a brown, brittle solid.

3. MISCELLANEOUS METHODS

1. Reaction of bis cyclic carbonates with diamines [36].
2. Reaction of bis trichloroacetates with piperazine [34].
3. Reaction of bis(imino cyclic carbonate) with diacids or dihydroxy compounds [37].
4. Reaction of bis *N*,*N*-ethyleneurethanes with dithiols [38].
5. Thermal dissociation of blocked toluene diisocyanates [39].
6. Preparation of grafted copolymers [40].
7. Preparation of flame-retardant polyurethanes [41].
8. Preparation of highly fluorinated polyurethanes [42].
9. Preparation of block copolymers from styrene, methyl methacrylate, or acrylonitrile with polyurethanes containing reactive disulfides [43].
10. Preparation of vinyl copolymers with macro bis peroxicarbamates [44].
11. Reaction of 4,4′-diphenylmethane diisocyanate with alcohols in DMF catalyzed by dibutyltindilaurate [45].
12. Imidazole catalysis of polyurethene production [46].
13. Preparation of blocked isocyanates useful for polyurethane production [47].

REFERENCES

1. S. R. Sandler and W. Karo, "Organic Functional Group Preparations," Vol. 2. Academic Press, New York, 1971.

1a. O. Bayer, *Angew. Chem.* **59**, 275 (1947).

1b. O. Bayer, H. Rinke, W. Siefken, L. Orthner, and H. Schild, German Patent 728,981 (1942).

2. H. Rinke, H. Schild, and W. Siefken, French Patent 845,917 (1939); U.S. Patent 2,511,544 (1950).

3. E. L. Wittbecker and P. W. Morgan, *J. Polym. Sci.* **40**, 289 (1959).

3a. E. L. Wittbecker and M. Katz, *J. Polym. Sci.* **40**, 367 (1959).

4. W. E. Catalin, U.S. Patent 2,284,637 (1942); W. E. Hanford and D. F. Holmes, U.S. Patent 2,284,896 (1942); R. E. Christ and W. E. Hanford, U.S. Patent 2,333,639 (1943); F. B. Hill, C. A. Young, J. A. Nelson, and R. G. Arnold, *Ind. Eng. Chem.*

48, 927 (1956); H. Rinke, *Chimia* **16**, 93 (1962); *Angew. Chem., Int. Ed. Engl.* **1**, 419 (1962); O. Bayer, E. Müller, S. Petersen, H. F. Piepenbrink, and E. Windemuth, *Angew. Chem.* **62**, 57 (1950); C. S. Marvel and J. H. Johnson, *J. Amer. Chem. Soc.* **72**, 1674 (1950); E. Dyer and G. W. Bartels, *ibid.* **76**, 591 (1954).

5. Y. Miyake, S. Ozaki, and Y. Hirata, *J. Polym. Sci., Part A-1* **7**, 899 (1969).
6. D. J. Lyman, *J. Polym. Sci.* **45**, 49 (1960).

6a. V. V. Korshak, Y. A. Strepikheev, and A. F. Moiseev, *Sov. Plast.* **7**, 12 (1961).

7. H. V. Blohm and E. I. Becker, *Chem. Rev.* **51**, 471 (1954); N. Bortnick, L. S. Luskin, M. D. Hurwitz, and A. W. Rytina, *J. Amer. Chem. Soc.* **78**, 4358 (1956); E. Dyer and G. E. Newborn, Jr., *ibid.* **80**, 5495 (1958); E. Dyer and G. C. Wright, *ibid.* **81**, 2138 (1959); H. Larka and F. B. Davis, *ibid.* **51**, 2220 (1929).
8. S. R. Sandler and F. R. Berg, *J. Appl. Polym. Sci.* **9**, 3909 (1965).
9. T. C. Patton, *Off. Dig., Fed. Paint Varn. Prod. Clubs* **34**, 342 (1962); S. N. Glasbrenner, B. Golding, and L. C. Case, *ibid.* **32**, 203 (1960); E. R. Wells and J. C. Hixenbaugh, *Amer. Paint J.* **46**, No. 47, 88 (1962); Wyandotte Chem. Corp. Bulletin, "Urethane Coatings Lab Report C-4, Two Package Polyether Coatings for Concrete." Wyandotte Chem. Corp., 1963.
10. A. Farkas, G. A. Mills, W. E. Erner, and J. B. Maerker, *Ind. Eng. Chem.* **51**, 1299 (1959); F. Hostettler and E. F. Cox, *ibid.* **52**, 609 (1960); H. L. Heiss, Canadian Patent 626,565 (1961).
11. V. V. D'Ancicco, *SPE (Soc. Plast. Eng.) J.* **14**, 34 (1958); W. C. Darr, P. G. Gemeinhardt, and J. H. Saunders, *Ind. Eng. Chem., Prod. Res. Develop.* **2**, 194 (1963).
12. S. L. Axelrood, C. W. Hamilton, and K. C. Frisch, *Ind. Eng. Chem.* **53**, 889 (1961); R. J. Athey, *ibid.* **52**, 611 (1960); A. Damusis, J. M. McClellan, H. G. Wissman, C. W. Hamilton, and K. C. Frisch, *Ind. Eng. Chem., Prod. Res. Develop.* **1**, 269 (1962); O. Bayer and E. Müller, *Angew. Chem.* **72**, 934 (1960).
13. O. Bayer, *Angew. Chem.* **59**, 257 (1947); M. Coenen and P. Schlack, German Patent 915,868 (1954); V. V. Korshak, Y. A. Strepikheev, and A. F. Moiseev, *Sov. Plast.* **7** 12 (1961).
14. W. E. Hanford and D. F. Holmes, U.S. Patent 2,284,896 (1942); British Patent 571,975 (1945).
15. W. Kern, G. Dall'Asta, R. Dieck, and H. Kämmerer, *Makromol. Chem.* **6**, 206 (1951).
16. H. L. Heiss, *Rubber Age (New York)* **88**, 89 (1960).
17. J. S. Jorczak and E. M. Fetters, *Ind. Eng. Chem.* **43**, 324 (1951); G. E. Serniuk and F. P. Baldwin, U.S. Patent 2,918,446 (1959).
18. R. W. Upson, U.S. Patent 2,511,310 (1950).
19. S. R. Sandler, F. Berg, and G. Kitazawa, *J. Appl. Polym. Sci.* **9**, 1944 (1965); S. R. Sandler, *J. Polym. Sci., Part A-1* **5**, 1481 (1967); *J. Appl. Polym. Sci.* **13**, 2699 (1969).
20. Chemical & Engineering News, *Chem. Eng. News* **50** (8), 10–11 (1972).
21. J. H. Saunders and K. C. Frisch, "Polyurethanes—Chemistry and Technology." Wiley (Interscience), New York, 1964.
22. W. Cooper, R. W. Pearson, and J. Drake, *Ind. Chem.* **36**, 121 (1960).
23. S. R. Sandler, *J. Appl. Polym. Sci.* **11**, 811 (1967).
24. J. W. Baker and J. B. Holdsworth, *J. Chem. Soc., London* p. 713 (1947); J. W. Baker, J. Gaunt, and M. M. Davies, *ibid.*, p. 9 (1949); A. Farkas and K. G. Flynn, *J. Amer. Chem. Soc.* **82**, 642 (1950); K. G. Flynn and D. R. Nenortas, *J. Org. Chem.* **28**, 3527 (1963).
25. H. A. Smith, *J. Polym. Sci., Part A* **7**, 85 (1963); A. A. Blabonrevova, G. A. Levkovich, and I. A. Pronina, *J. Polym. Sci.* **52**, 303 (1961).

26. D. A. Chadwick and T. C. Allen, U.S. Patent 2,733,254 (1961); J. M. Lyons and R. H. Thompson, *J. Chem. Soc., London* p. 1971 (1950).
27. J. W. Britain, *Ind. Eng. Chem., Prod. Res. Develop.* **1**, 261 (1962).
28. H. C. Beachell and J. C. Peterson, *J. Polym. Sci., Part A-1* **7**, 2021 (1969).
29. J. Hollander, F. D. Trischler, and R. B. Gosnell, *J. Polym. Sci., Part A-1* **5**, 2757 (1967).
30. P. W. Morgan, "Condensation Polymers By Interfacial and Solution Methods," pp. 272–274. Wiley (Interscience), New York, 1965.
31. T. Hopkins, *J. Chem. Soc., London* **117**, 278 (1920); A. R. Choppin and J. W. Rogers, *J. Amer. Chem. Soc.* **70**, 2967 (1948); R. Delaby, R. Damiens, and G. D'Huyteza, *C. R. Acad. Sci.* **236**, 2076 (1953); W. Gerrard and F. Schild, *Chem. Ind. (London)* **73**, 1232 (1954); J. A. Campbell, *J. Org. Chem.* **22**, 1259 (1957); W. B. Wright and H. J. Brabander, *ibid.* **26**, 4057 (1961).
32. J. R. Schaefgen, F. H. Koontz, and R. F. Tietz, *J. Polym. Sci.* **40**, 377 (1959).
33. I. G. Farbenindustrie A.-G., French Patent 961,754 (1950).
34. W. F. Tousignant and H. Pledger, U.S. Patent 2,894,935 (1959).
35. N. Rabjohn, *J. Amer. Chem. Soc.* **70**, 1181 (1948).
36. J. M. Whelan and R. J. Cotter, U.S. Patent 3,072,613 (1963).
37. T. Mukaiyama, T. Fujisawa, and T. Hyungaji, *Bull. Chem. Soc. Jap.* **35**, 687 (1962); T. Mukaiyama, T. Fujisawa, H. Nohira, and T. Hyungaji, *J. Org. Chem.* **27**, 3337 (1962).
38. Y. Iwakura, M. Sakamoto, and Y. Awata, *J. Polym. Sci., Part A* **2**, 881 (1964); Y. Iwakura and M. Sakamoto, *J. Polym. Sci.* **47**, 277 (1960).
39. G. R. Griffin and L. J. Willworth, *Ind. Eng. Chem., Prod. Res. Develop.* **1**, 265 (1962).
40. H. C. Beachell and J. C. P. Buck, *J. Polym. Sci. Part A-1* **7**, 1873 (1969).
41. A. D. Delman, *J. Macromol. Sci., Rev. Macromol. Chem.* **3**, 281 (1969).
42. J. Hollander, F. D. Trischler, and E. S. Harrison, *Polym. Prepr., Amer. Chem. Soc., Div. Polym. Chem.* **8**, 1149 (1967).
43. F. J. T. Fildes and A. V. Tobolsky, *J. Polym. Sci., Part A-1* **10**, 151 (1972).
44. A. V. Tobolsky and A. Rembaum, *J. Appl. Polym. Sci.* **8**, 307 (1964).
45. G. Borkent and J. J. van Aartsen, *Polym. Prepr., Amer. Chem. Soc., Div. Polym. Chem.* **13**, 567 (1972).
46. R. A. Jerussi and C. M. Orlando, *J. Appl. Polym. Sci.* **16**, 2853 (1972).
47. A. W. Levine and J. Fech, Jr., *J. Org. Chem.* **37**, 1500 and 2455 (1972).

Chapter 9

THERMALLY STABLE POLYMERS

I. INTRODUCTION

The search for thermally stable resins usable at 500°C for long periods of time has been given great support as a result of the material needs of space and advanced aircraft industries.

Thermally stable polymers should have the following properties: (a) high melting or softening points; (b) low weight loss as determined by thermogravimetric analysis; (c) structures that are not susceptible to degradative chain scission or intra- or intermolecular bond formation; (d) chemically inert especially to oxygen, moisture, and dilute acids and bases.

Chemical structures which are thermally stable usually also have (a) a highly resonance-stabilized system; (b) an aromatic or other thermally unreactive ring structure as a major portion of the polymer composition; and (c) a high bond and cohesive energy density.

In view of these requirements, the most promising thermally stable polymers are the following: polyimides [1,1a], polybenzimidazoles [1,2,2a–2c], polyoxadiazoles [2a,3], polytriazoles [3,4], polybenzoxazoles [2a,4], polyazomethines [5], polyphenylenes [6,6a], polytriazines [7], polyquinoxalines [8,8a,8b], polythiazoles [4], polyphenylene ethers [9,9a–d], and polyaryl sulfones [1a,10] and related polymeric sulfur analogs. Many others are constantly being reported each day and space limitations will not permit them all to be described in this chapter. Several others are also mentioned in Section 10, Miscellaneous High-Temperature Polymers.

Most of these polymers are prepared by condensation polymerization techniques, with the exception of the polyphenylene ethers, which probably are prepared via a radical mechanism as described below.

Polymers which possess the desirable thermal properties are usually difficult to fabricate into products. Recently prepolymers which are workable before cyclization have been employed to extend the practical applications of these polymers.

Notable among polymers utilizing these prepolymers have been the polypyromellitimides [1a] and the polybenzimidazoles (2).

Several worthwhile reviews [11] have appeared and should be consulted for additional information. Recently the status of fibers useful at high temperatures has been reviewed [12].

CAUTION: Several high-temperature polymer starting materials such as benzidine, 3,3′-diaminobenzidene, and other aromatic amines should be handled with care since they are known to be carcinogenic [13].

Polycarbonates have already been described in Chapter 3 and will not be discussed further here.

2. POLYIMIDES

Polyimides were first reported in 1908 by Bogert and Renshaw [14], who described the condensation of 4-aminophthalic anhydride as shown in Eq. (1).

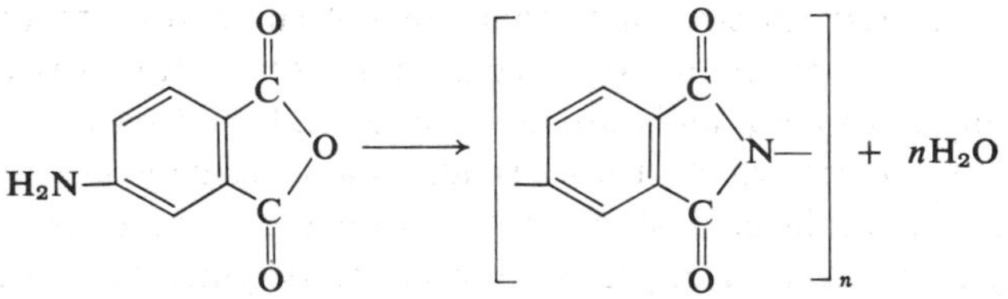

$$H_2N\text{–(4-aminophthalic anhydride)} \longrightarrow [\text{–(phthalimide)–N–}]_n + nH_2O \qquad (1)$$

Other early reports on polyimides were concerned with the preparation of aliphatic polyamide-imides [15,15a–15c]. However, these early reports [15b] also mentioned the use of aromatic systems. In these early systems pyromellitic dianhydride was reacted with ethanol to give a diester-diacid which was subsequently reacted with diamines to give the diester-amine acid salt. The

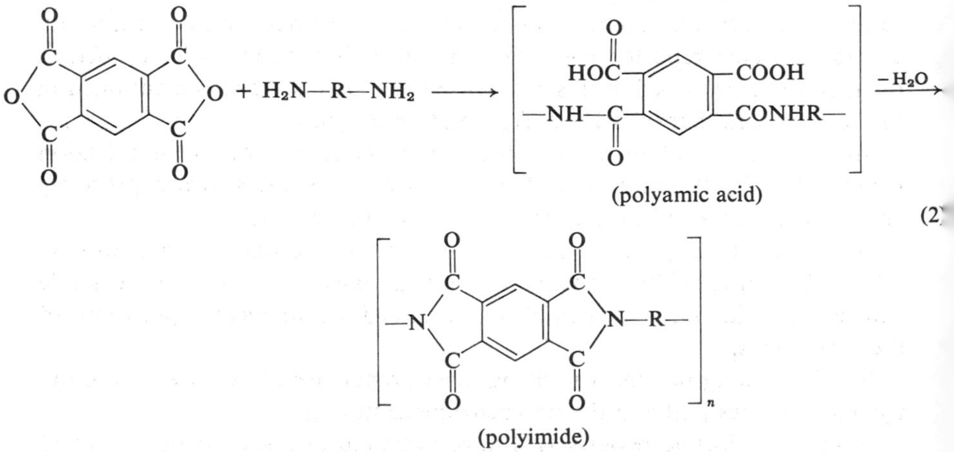

$$\text{(pyromellitic dianhydride)} + H_2N\text{—R—}NH_2 \longrightarrow \left[\begin{array}{c} HOC(=O)\text{–}C_6H_2\text{–}COOH \\ \text{—NH—}C(=O)\text{–}C_6H_2\text{–}CONHR\text{—} \end{array}\right] \xrightarrow{-H_2O}$$

(polyamic acid)

$$\left[\text{—N(pyromellitimide)N—R—}\right]_n \qquad (2)$$

(polyimide)

(R = examples shown in Table I)

latter salt is heated in the melt to give the polyimide. Recently this procedure has been modified to allow direct polymerization of pyromellitic dianhydride with aromatic diamines in polar solvents (DMF, DMAC, etc.) to give polyimides [1a,16].

Edwards and co-workers described the conditions of reaction of bis(4-aminophenyl) ether and pyromellitic anhydride, and the data are shown in Table II. The properties of the polyamide are also described in Table II.

2-1. *Preparation of Poly[N,N′-(p,p′-oxydiphenylene)pyromellitimide]* [1a]

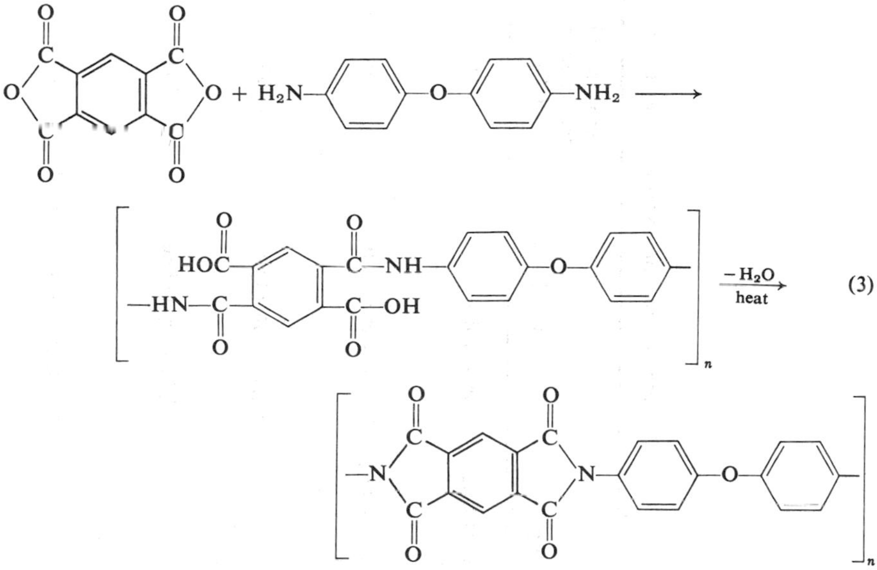

$$\text{pyromellitic dianhydride} + H_2N-C_6H_4-O-C_6H_4-NH_2 \longrightarrow$$

$$\left[\begin{array}{c}\text{HOC(=O)}, \text{C(=O)}-NH-C_6H_4-O-C_6H_4- \\ -HN-\text{C(=O)}, \text{C(=O)}-OH\end{array}\right]_n \xrightarrow[\text{heat}]{-H_2O} \qquad (3)$$

$$\left[-N(\text{C=O})_2C_6H_2(\text{C=O})_2N-C_6H_4-O-C_6H_4-\right]_n$$

(*a*) *Preparation of polyamic acid.* To an anhydrous 500 ml three-necked flask equipped with a mechanical stirrer (ground glass or mercury seal), nitrogen inlet, and drying tube is added 10.0 gm (0.05 mole) of bis(4-aminophenyl) ether dissolved in 160 gm of dry dimethylacetamide (distilled from P_2O_5 at 30 mm Hg). Then 10.90 gm (0.05 mole) of pyromellitic dianhydride is added over a 2–3 min period by means of a ground glass solid addition adapter [17] while the solution is vigorously agitated. The reaction is slightly exothermic (temperature increases to 40°C) and then drops to room temperature. These polyamic acid solutions (10% solids) can be stored at −15°C in dry sealed bottles.

(*b*) *Preparation of polyamic acid films.* Some of the polyamic acid solution is poured onto dry glass plates and doctored into a thin layer (10–25 mils). The layers are dried at 80°C for 20 min in a forced draft oven and additionally dried under reduced pressure at room temperature (see Fig. 1).

TABLE I

PROPERTIES[a] OF POLYPYROMELLITIMIDES (R = Diamine Component)[a]

R	Solubility	Crystallinity	Zero strength temperature (°C)	Thermal stability in air[b]	
				275°C	300°C
$-C_6H_4-$ (meta)	Amorphous conc. H_2SO_4; crystalline insol.	Crystallizable	900	>1 yr	>1 month
$-C_6H_4-$ (para)	Amorphous conc. H_2SO_4; crystalline insol.	Crystallizes readily	900	>1 yr	—
$-C_6H_4-C_6H_4-$	Fuming HNO_8	Highly crystalline	>900	—	1 month
$-C_6H_4-CH_2-C_6H_4-$	Conc. H_2SO_4	Slightly crystalline	800	—	7–10 days
$-C_6H_4-C(CH_3)_2-C_6H_4-$	Conc. H_2SO_4	Crystallizable with difficulty	580	—	15–20 days

$-C_6H_4-S-C_6H_4-$	Fuming HNO_3	Crystallizable	800	10–12 months (estimated)	6 weeks
$-C_6H_4-O-C_6H_4-$	Fuming HNO_3	Crystallizable	850	> 1 yr	> 1 month
$-C_6H_4-SO_2-C_6H_4-$	Conc. H_2SO_4	—	—	—	> 1 month
$-C_6H_4-SO_2-C_6H_4-$ (meta)	Conc. H_2SO_4	—	—	—	> 1 month

[a] Reprinted from C. E. Sroog, S. V. Abramo, C. E. Berr, W. M. Edwards, A. L. Endrey, and K. L. Olivier, *Polym. Prepr., Amer. Chem. Soc., Div. Polym. Chem.* **5**, 132 (1964). Paper presented at the American Chemical Society Meeting, Philadelphia, Pa., April 1964. Copyright 1964 by the American Chemical Society. Reprinted by permission of the copyright owner.

[b] As measured by retention of film creasability.

TABLE II

EFFECT OF TEMPERATURE OF POLYMERIZATION ON INHERENT VISCOSITY OF POLYAMIC ACID [REACTION OF BIS(4-AMINOPHENYL) ETHER AND PYROMELLITIC DIANHYDRIDE][a]

Solvent	Bis(4-aminophenyl) ether		Pyromellitic dianhydride		Solids	Temp.	Time at temp.	
	(gm)	(moles)	(gm)	(moles)	(%)	(°C)	(min)	η_{inh}[b]
DMAc	10.00	0.05	10.90	0.05	10.0	25	120	4.05
DMAc	10.00	0.05	10.90	0.05	10.0	65	30	3.47
DMAc	20.00	0.10	21.80	0.10	10.6	85–88	30	2.44
DMAc	20.00	0.10	21.80	0.10	10.7	115–119	15	1.16
DMAc	20.00	0.10	21.80[c]	0.10	10.3	125–128	15	1.00
DMAc	20.00	0.10	21.80[d]	0.10	15.7	135–137	15	0.59
N-Me caprolactam	10.00	0.05	10.90	0.05	14.2	150–160	2	0.51
N-Me caprolactam	10.00	0.05	10.90	0.05	12.9	175–182	1–2	Only partly soluble
N-Me caprolactam	20.00	0.10	21.80	0.10	15	200	1	Insoluble

[a] Reprinted from C. E. Sroog, S. V. Abramo, C. E. Berr, W. M. Edwards, A. L. Endrey, and K. L. Olivier, *Polym. Prepr., Amer. Chem. Soc., Div. Polym. Chem.* **5**, 132 (1964). Paper presented at the American Chemical Society Meeting, Philadelphia, Pa., April 1964. Copyright 1964 by the American Chemical Society. Reprinted by permission of the copyright owner.

[b] Determined at 0.5% concentration in the particular solvent at 30°C.

[c] Increment of 0.35 gm pyromellitic dianhydride added before determination of η_{inh}.

[d] Increment of 0.25 gm pyromellitic dianhydride and then 0.21 gm of bis(4-aminophenyl) ether added before determination of η_{inh}.

(*c*) *Preparation of polyimide film.* The polyamic acid films are peeled off the plates and are now at 65–75% solids. The films are clamped onto frames, heated in a forced draft oven to 300°C over a 45 min period, then held at 300°C for 1 hr to give the polyimide whose IR is shown in Fig. 2.

The reaction can be followed by observing the IR spectrum of the polyamic ester, Fig. 1, and waiting for it to change to the spectrum the fully formed polyimide of Fig. 2.

The thermal stability of the polyimides is described by the data in Figs. 3 and 4, which indicate the weight loss in helium. The data in Table III describe the properties of the polyimide poly[*N*,*N'*-(*p*,*p'*-oxydiphenylene)pyromellitimide] (H-film) in air [18].

The polyimides generally have particularly good stability in air. They have good dielectric and mechanical properties at elevated temperature and are relatively resistant to ionizing radiation. Furthermore the polyimides are

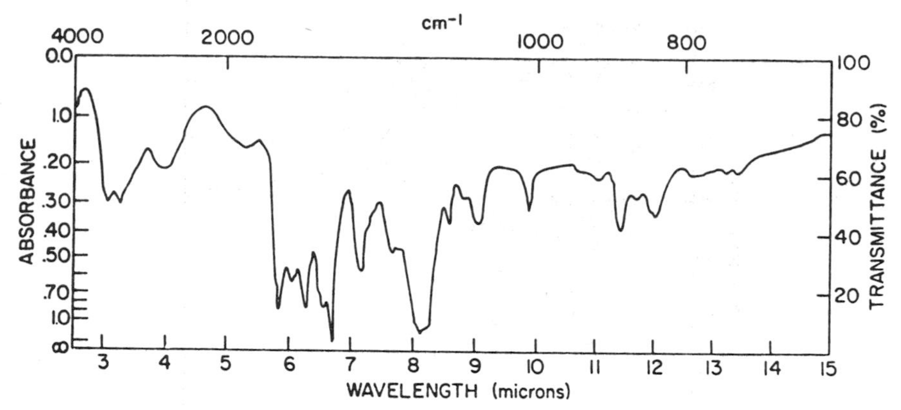

FIG. 1. Polyamic acid from bis(4-aminophenyl) ether and pyromellitic dianhydride. Reprinted from C. E. Sroog, S. V. Abramo, C. E. Berr, W. M. Edwards, A. L. Endrey, and K. L. Olivier, *Polym. Prepr., Amer. Chem. Soc., Polym. Chem.* **5**, 132 (1964). Paper presented at the American Chemical Society Meeting, Philadelphia, Pennsylvania, April, 1964. Copyright 1964 by the American Chemical Society; reprinted by permission of the copyright owner.

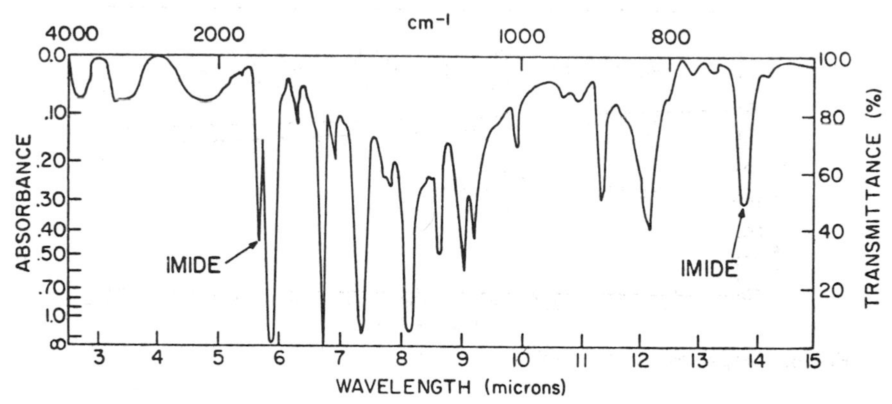

FIG. 2. Infrared spectrum of 0.1 mil poly[*N,N'*(*p,p'*-oxydiphenylene)pyromellitimide] (POP-PI) film. The film was heated to 300°C over a period of 45 min, then held at 300°C for 1 hr. Reprinted from C. E. Sroog, S. V. Abramo, C. E. Berr, W. M. Edwards, A. L. Endrey, and K. L. Olivier, *Polym. Prepr., Amer. Chem. Soc., Polym. Chem.* **5**, 132 (1964). Paper presented at the American Chemical Society Meeting, Philadelphia, Pennsylvania, April, 1964. Copyright 1964 by the American Chemical Society; reprinted by permission of the copyright owner.

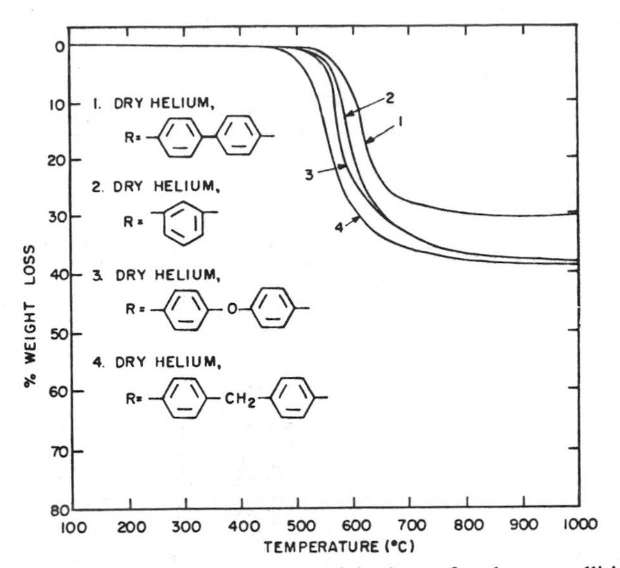

FIG. 3. Effect of diamine structure on weight loss of polypyromellitimides during constant rise in temperature (3°C/min). Reprinted from C. E. Sroog, S. V. Abramo, C. E. Berr, W. M. Edwards, A. L. Endrey, and K. L. Olivier, *Polym. Prepr., Amer. Chem. Soc., Polym. Chem.* **5**, 132 (1964). Paper presented at the American Chemical Society Meeting, Philadelphia, Pennsylvania, April, 1964. Copyright 1964 by the American Chemical Society; reprinted by permission of the copyright owner.

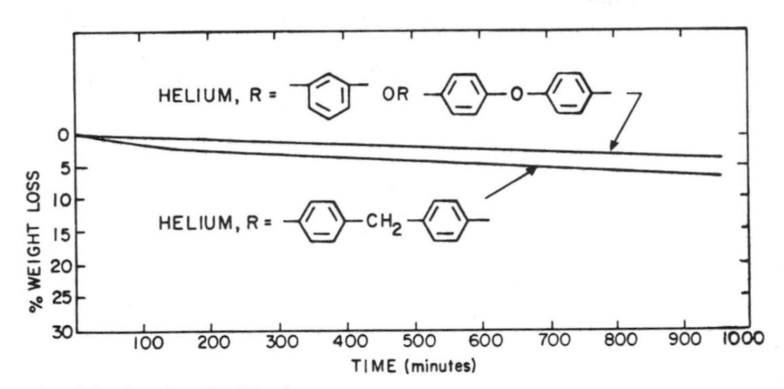

FIG. 4. Weight loss at 450°C of

$$\left[-R-N\begin{matrix} C(=O) \\ C(=O) \end{matrix} C_6H_2 \begin{matrix} C(=O) \\ C(=O) \end{matrix} N- \right]_n$$

Reprinted from C. E. Sroog, S. V. Abramo, C. E. Berr, W. M. Edwards, A. L. Endrey, and K. L. Olivier, *Polym. Prepr., Amer. Chem. Soc., Polym. Chem.* **5**, 132 (1964). Paper presented at the American Chemical Society Meeting, Philadelphia, Pennsylvania, April, 1964. Copyright 1964 by the American Chemical Society; reprinted by permission of the copyright owner.

TABLE III

PROPERTIES OF H FILM[a]

Thermal stability in air[b]	
250°C	10 year (extrapolated)
275°C	1 year
300°C	1 month
400°C	1 day
Melting point	None to 900°C
Zero strength temp.	815°C
Solvents	None
Density	1.42
Glass transition temp.	> 500°C

Mechanical properties	4°K	25°C	200°C	300°C	500°C
Tensile strength (lb/in^2)	Flexible	25,000	17,000	10,000	4,000
Elongation (%)		70	90	120	60
Tensile modulus		400,000	260,000	200,000	40,000

[a] Reprinted from C. E. Sroog, S. V. Abramo, C. E. Berr, W. M. Edwards, A. L. Endrey, and K. L. Olivier, *Polym. Prepr., Amer. Chem. Soc., Div. Polym. Chem.* **5**, 132 (1964). Paper presented at the American Chemical Society Meeting, Philadelphia, Pa., April, 1964. Copyright 1964 by The American Chemical Society. Reprinted by permission of the copyright owner.

[b] Retention of film flexibility.

greatly resistant to solvent attack and are soluble only in sulfuric or nitric acid (with decomposition).

Polyimides have found great use in films, coatings, varnishes, and laminating resins.

The high glass-transition temperature of the polyimides makes them suitable as structural adhesives and fibers [19].

The other dianhydride that is important for thermally stable polyimides is 3,4,3′,4′-benzophenonetetracarboxylic acid dianhydride (BTDA). Polyimides

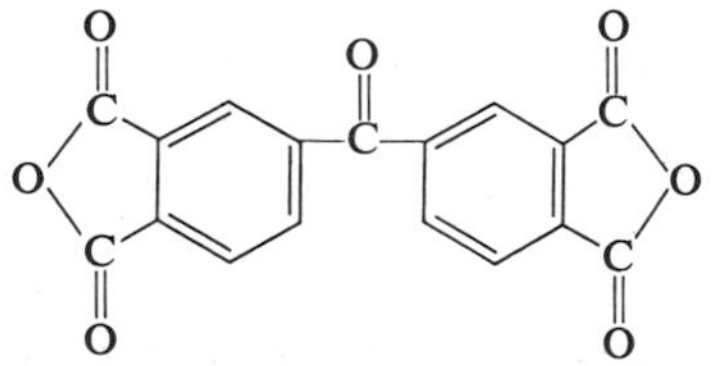

from BTDA can be cross-linked [20] by reaction of the carbonyl group [21] with amine [22], hydrazine, hydrazide, and other groups [23]. A thermosetting polyimide based on BTDA and *m*-phenylenediamine has recently been described [24].

Polyimides have been prepared which contain other functional groups such as amides, heterocyclic groups, and esters, as described in the following section, Miscellaneous Polyimide Preparations. The chemistry of polyimides has recently been reviewed [25].

Miscellaneous Polyimide Preparations

1. Low-temperature conversion of polyamic acids to polyimides using acetic anhydride in pyridine [26].
2. Polyimides by the reaction of dianhydrides with diisocyanates [15a].
3. Polyimides containing heterocyclic groups [27].
4. Preparation of polyamide-imides from the reaction of trimellitic anhydride acid chloride with diamines [28].
5. Polyester-imides [29].
6. Preparation of a polyimide devoid of hydrogen groups [30].
7. Preparation of aliphatic-aromatic polyimides by the reaction of aromatic tetraacids and derivatives with aliphatic diamines [15b,c,31].
8. Free-radical polymerization of bismaleimides [32].
9. Preparation of epoxy-imide resins [33].
10. Polyimides by the reaction of diamines with bismaleimide [34].
11. Preparation of polyimides via the Diels-Alder reaction [35].
12. Polyimide-amides from maleopimaric acid [36].
13. Preparation of polyimides by the reaction of benzalazine with bismaleimides [37].
14. Polyimides by the reaction of hydrazine hydrate with aromatic dianhydrides [38].
15. Polyimides from the reaction of pyromellitimide with bisepoxides [39].
16. Polyimides with hydroxy and acetoxy groups [40].
17. Heat-resistant polybenzoxazole imides [41].
18. Polyimides prepared by a hydrogen-transfer polymerization between divinyl sulfone and diimides of tetracarboxylic acids [42].
19. Polyimides by the reaction of pyromellitic diimide with dihalo compounds [43,44].
20. Preparation of polyimides in glycol [45].
21. Polyimide-polyethers [46].
22. One-step synthesis of polyimides [47].

3. POLYBENZIMIDAZOLES

Polybenzimidazoles (PBI) were first reported by Brinker and Robinson, who prepared them by the polycondensation reaction of bis(*o*-diaminophenyl) compounds with aliphatic dicarboxylic acids [2].

Vogel and Marvel [2b] later reported the preparation of fully aromatic polybenzimidazoles by the condensation of aromatic tetramines with aromatic dicarboxylic acid or derivatives as shown in Eq. (4). Furthermore phenyl

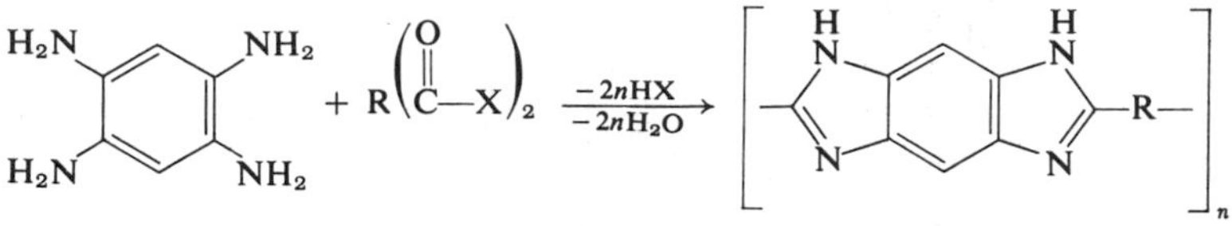

$$(H_2N)_2C_6H_2(NH_2)_2 + R\left(\overset{O}{\overset{\|}{C}}-X\right)_2 \xrightarrow[-2nH_2O]{-2nHX} \left[\text{benzobisimidazole}-R-\right]_n \qquad (4)$$

($X = OH, Cl, OCH_3$, or OC_6H_5)

3,4-diaminobenzoate could also be polymerized to a high molecular weight polymer as shown in Eq. 5.

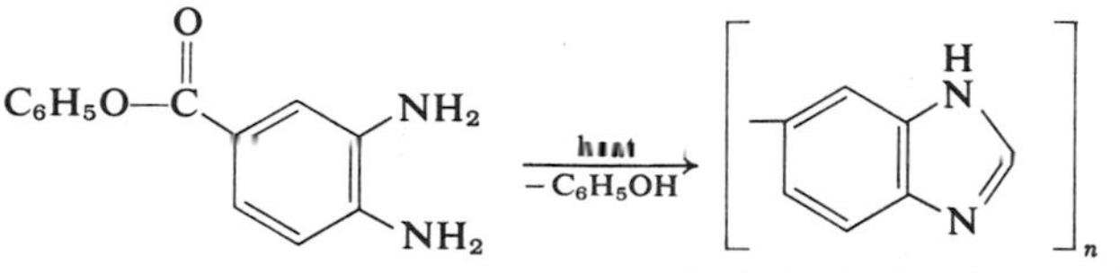

$$C_6H_5O-\overset{O}{\overset{\|}{C}}-C_6H_3(NH_2)_2 \xrightarrow[-C_6H_5OH]{\text{heat}} \left[\text{benzimidazole}\right]_n \qquad (5)$$

Poly-2,5(6)-benzimidazole

Vogel and Marvel [2b] also reported that 3,3′-diaminobenzidine can also be reacted with aromatic dibasic acids and derivatives to give polybenzimidazoles as shown in Eq. (6).

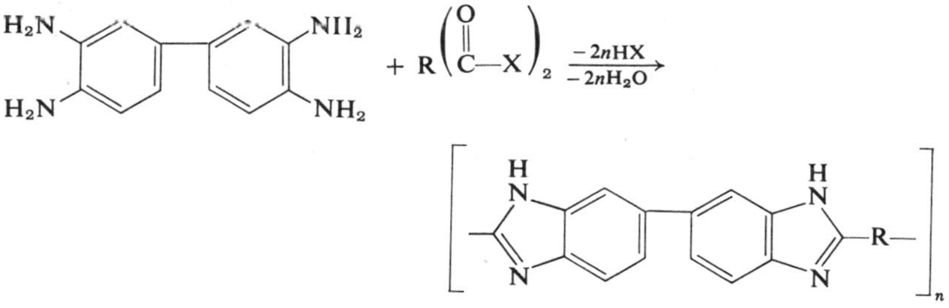

$$(H_2N)_2C_6H_3-C_6H_3(NH_2)_2 + R\left(\overset{O}{\overset{\|}{C}}-X\right)_2 \xrightarrow[-2nH_2O]{-2nHX} \left[\text{bibenzimidazole}-R-\right]_n \qquad (6)$$

The polybenzimidazoles are usually prepared by a melt polymerization utilizing the phenyl esters of carboxylic acids. The use of methyl esters or the free acids gave inferior results. The tetraamino compounds are heated in an inert atmosphere with the diphenyl esters to give high molecular weight polymers. Some representative examples are shown in Table IV.

3-1. Preparation of Poly-2,5(6)-benzimidazole [2b]

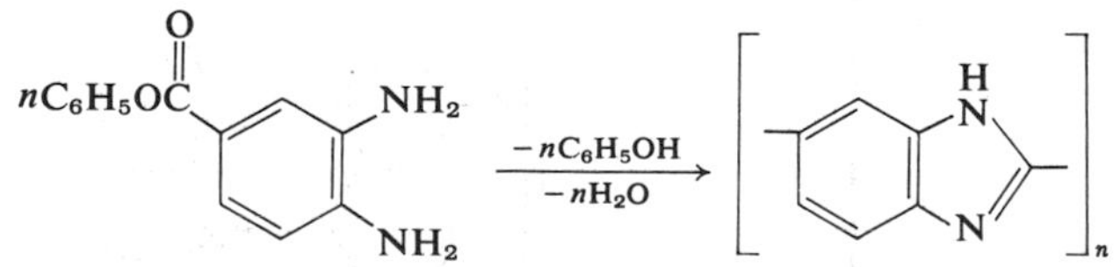

$$nC_6H_5O\overset{O}{\overset{\|}{C}}-C_6H_3(NH_2)_2 \xrightarrow[-nH_2O]{-nC_6H_5OH} \left[\text{benzimidazole}\right]_n \qquad (7)$$

A 50 ml single-necked, round-bottomed flask containing 3.0 gm (0.0131 mole) of phenyl 3,4-diaminobenzoate is purged with nitrogen by repeated

TABLE IV

PREPARATION AND PROPERTIES OF POLYBENZIMIDAZOLES[a]

Polymer repeat unit	Preparative reaction conditions		Polymer properties			Total wt. loss over 5 hr heating for 400°–600°C
	Temp. (°C)	Reaction time (hr)	η_{inh}[b]	Wt. loss in % after 1 hr at 400°C	Wt. loss in % after 1 hr at 600°C	
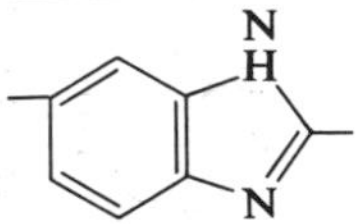	280–500	4	1.27	1.1	5.0	7.3
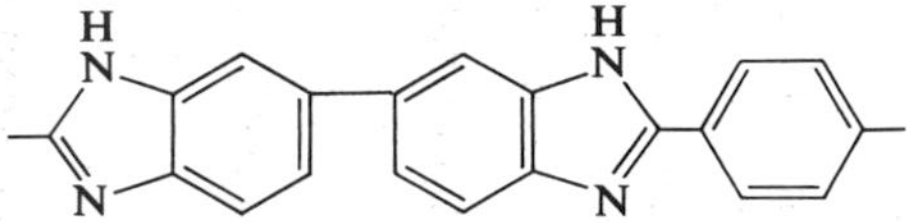	240–400	5	1.00	1.0	1.0	4.7
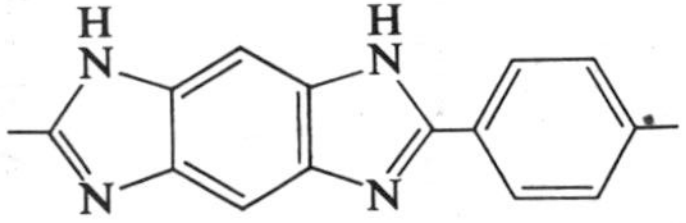	280–360	4.5	0.80	2.8	3.0	10.2

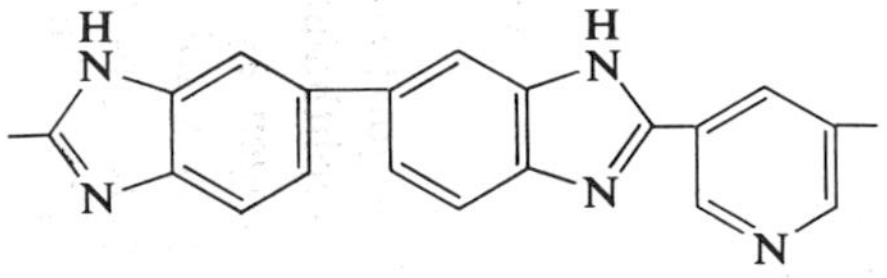	260–400	5	1.43	0.2	2.7	5.6
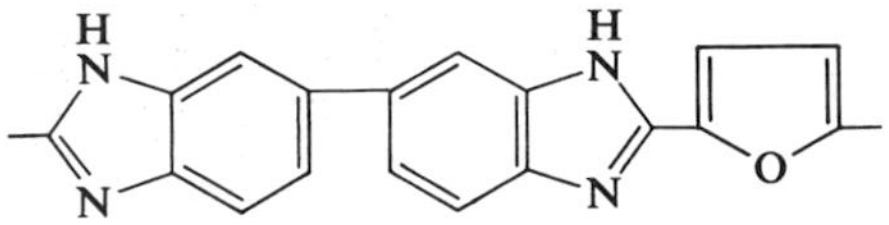	260–400	5	0.74	1.4	2.0	10.0
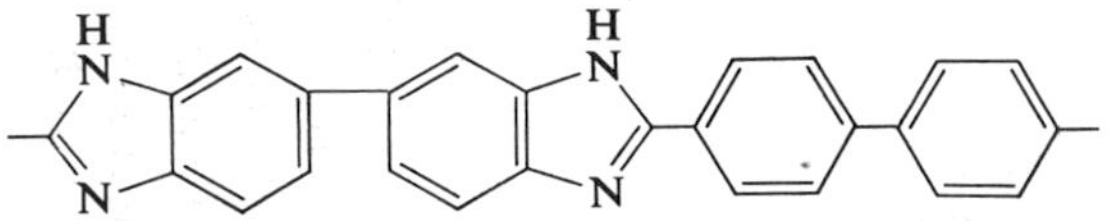	250–400	6	0.86[c]	0.3	2.1	3.5

[a] Data taken from paper of H. Vogel and C. S. Marvel, *J. Polym. Sci.* **50**, 511 (1961).
[b] 0.2% in formic acid or in DMSO (0.4 to 1.1%).
[c] 0.2% in H_2SO_4.

evacuation and refilling. The flask is placed in a silicone oil bath and heated under nitrogen for 1 hr at 290°C, producing a pale yellow solid mass, η_{inh} = 0.3 (0.2% conc. in formic acid). The polymer is pulverized and reheated under reduced pressure for $1\frac{1}{2}$ hr at 350°C and $1\frac{1}{2}$ hr at 400°C to give a final polymer 1.52 gm (100%), inherent viscosity, 1.27.

3-2. Preparation of Poly[2,2′-(m-phenylene)-5,5′-bibenzimidazole] [48]

H_2N, NH_2, H_2N, NH_2 + $COOC_6H_5$, $COOC_6H_5$ $\xrightarrow[-2nH_2O]{-2nC_6H_5OH}$

H, N, H, N, N, N]$_n$ (8)

To a polymer tube is added 2.39 gm (0.0075 mole) of diphenyl isophthalate and 2.18 gm (0.0075 mole) of 3,3′-diaminobenzidine. The tube is heated for 1 hr in a Wood's metal bath at 300°C while a stream of nitrogen slowly passes over the sample. At the end of this time the pressure is reduced to 0.5 mm Hg and heating is continued for $\frac{1}{2}$–1 hr at 300°C. The polymer tube is cooled and the black flaky polymer is isolated, 2.1 gm (99%), softening point 250°–255°C. The infrared spectrum of the polymer is identical to that reported by Vogel and Marvel [2c].

3-3. Preparation of Poly[2,2′-(p-phenylene)-5,5′-bibenzimidazole] [2b]

H_2N, NH_2, H_2N, NH_2 + $C_6H_5OC(=O)$–C_6H_4–$C(=O)$–C_6H_5 $\xrightarrow[-2nH_2O]{-2nC_6H_5OH}$

H, N, H, N, N, N]$_n$ (9)

To a 50 ml single-necked, round-bottomed flask is added 2.14 gm (0.01 mole) of diaminobenzidine and 3.18 gm (0.01 mole) of diphenyl terephthalate and the contents melted together. The flask is purged with nitrogen as in Preparation 3-1 and then heated for 10 min at 250°C to give a yellow solid. The contents are next heated under reduced pressure (0.1 mm Hg) for $\frac{1}{2}$ hr at 260°C, cooled, powdered, and reheated to 400°C for $4\frac{1}{2}$ hr to give a final polymer weighing 2.8 gm (100%), η_{inh} = 1.00.

The polymerization mechanism proposed for polybenzimidazole formation from the reaction of 3,3′,4,4′-tetraaminobiphenyl and diphenyl isophthalate has been described [49]. Several other workers have shown that the elimination of water is the final step in the aromatization of the benzimidazole ring and this process is not complete below 400°C.

Marvel and co-workers in a series of papers (a) have reported several other types of polybenzimidazoles, as shown in Table V [2c, 51–53].

3-4. Preparation of Poly[2,2′-(1,3-phenylene)sulfonyl-5,5′-bibenzimidazole] [53]

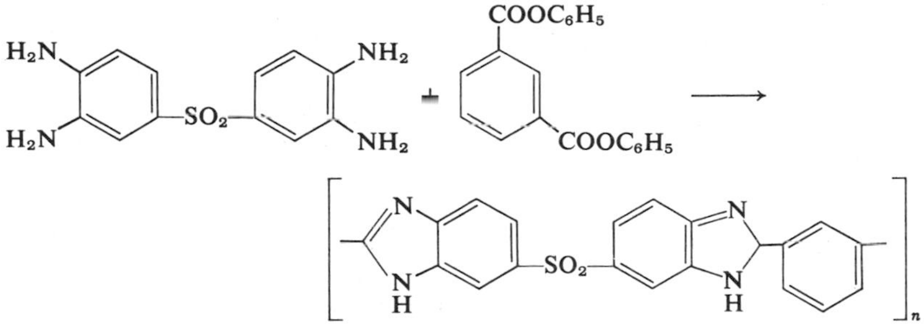

(10)

To a 50 ml round bottomed, single necked flask is added 0.5184 gm (0.0019 mole) of 3,3′,4,4′-tetraaminodiphenyl sulfone and 0.5929 gm (0.0018 mole) of diphenyl isophthalate. The flask is heated in a nitrogen atmosphere for $\frac{1}{2}$ hr at 280°C. The heating is continued for 1 hr under reduced pressure, and the reaction mixture is cooled, powdered, and reheated to 400°C under reduced pressure for 5 hr. The polymer is isolated by dissolving in dimethylsulfoxide (DMSO) and reprecipitating in benzene to afford 0.4314 gm (60%) with $\eta_{inh} = 0.14$ at 30°C (0.24% conc. in DMSO) and $\eta_{inh} = 0.38$ at 30°C (0.13% conc. in formic acid). At room temperature the polymer is completely soluble in dimethylacetamide (DMAC) and DMSO.

3-5. Preparation of Poly[2,2′-(1,3-phenylene)oxy-5,5′-bibenzimidazole] [52]

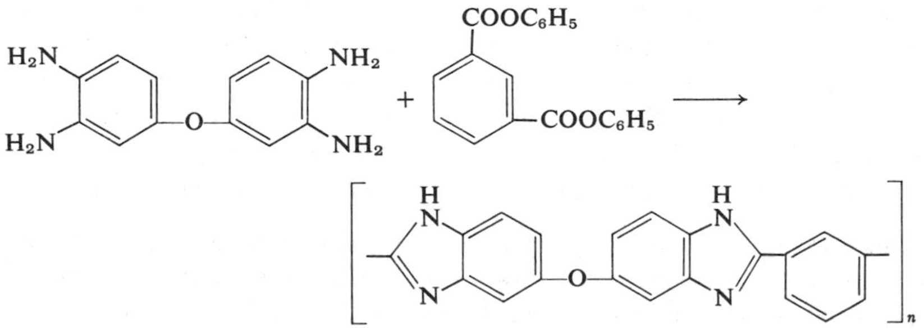

(11)

TABLE V

Reactants (equimolar)	
Dicarboxylic acid or deriv.	Diamine
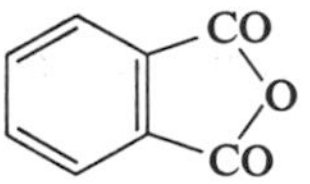 (phthalic anhydride: CO, O, CO)	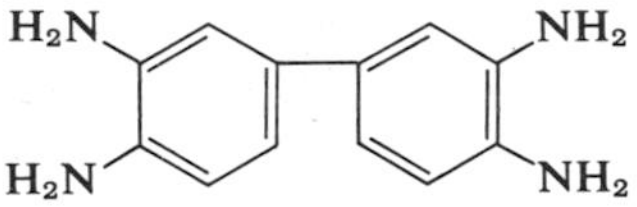 H_2N, H_2N, NH_2, NH_2 (3,3′,4,4′-tetraaminobiphenyl)
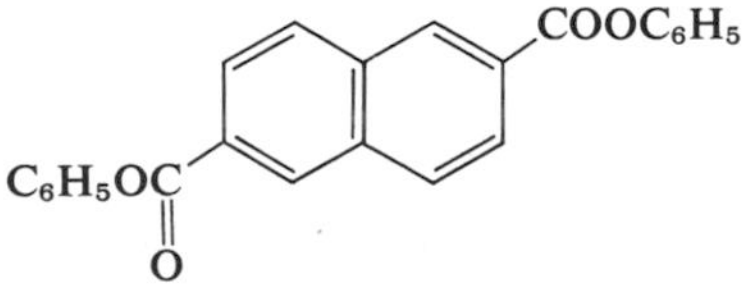 $COOC_6H_5$, $C_6H_5OC(=O)$ (naphthalene diester)	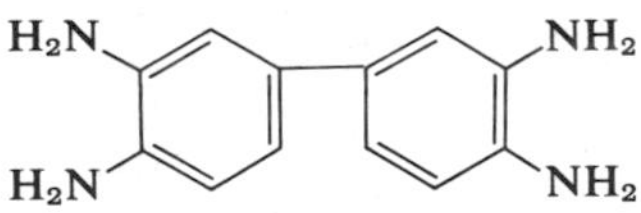 H_2N, H_2N, NH_2, NH_2 (3,3′,4,4′-tetraaminobiphenyl)
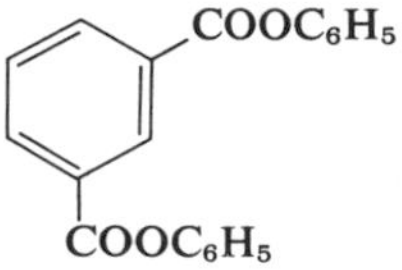 $COOC_6H_5$, $COOC_6H_5$ (diphenyl isophthalate)	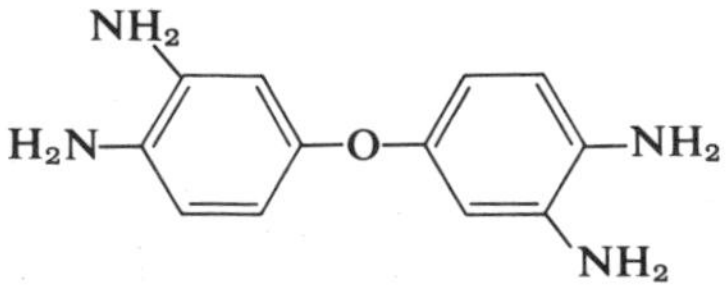 NH_2, H_2N–, –O–, –NH_2, NH_2 (tetraaminodiphenyl ether)
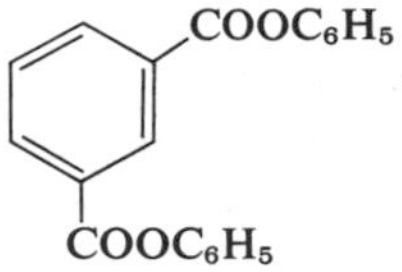 $COOC_6H_5$, $COOC_6H_5$ (diphenyl isophthalate)	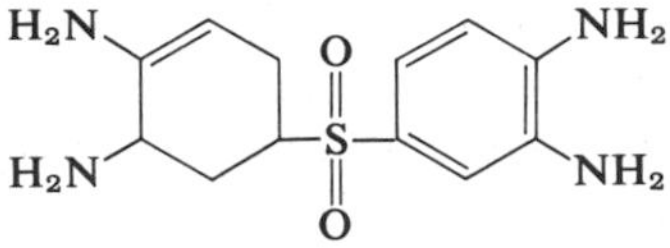 H_2N, H_2N, O, S, O, NH_2, NH_2 (tetraaminodiphenyl sulfone)

To a 70 mm × 200 mm polymerization tube with a 45/50 ground glass joint is added 7.1 gm (0.0307 mole) of 3,3′,4,4′-tetraaminodiphenyl ether and 10.1 gm (0.0319 mole) of diphenyl isophthalate. The tube is evacuated and filled with nitrogen repeatedly and then heated for $\frac{1}{2}$ hr at 260°C. During this heating period water and phenol are eliminated. The heating is continued under reduced pressure (0.025 mm) and the temperature slowly raised to 340°C over a 4 hr period. The heating is discontinued but kept under reduced pressure for 12 hr to afford 9.2 gm (100%) of polymer, $\eta_{inh} = 0.5\text{–}0.6$ at 30°C (0.5 gm/100 ml DMSO), approx. 90–100% soluble in DMSO or DMAC.

Iwakura, Uno, and Imai [54] reported that polybenzimidazoles can be prepared by the solution polycondensation of 3,3′-diaminobenzidine hydrochloride and aromatic dicarboxylic acid derivatives in polyphosphoric acid.

PREPARATION OF POLYBENZIMIDAZOLES

Polym. temp. (°C)	Time (hr)	Structure of polybenzimidazole	Ref.
200	½		2c
250	½		
270 (0.1 mm Hg)	½		
270–400 (0.1 mm Hg)	5		
300	1		51
400	5		
300	1		52
400 (0.01 mm Hg)	5		
280	½		53
280 (0.02 mm Hg)	1		
400 (0.01 mm Hg)	5		

The dicarboxylic acid derivatives dimethyl ester, diamide, and dinitric or free acid, give the best results. The diphenyl ester or the diacid chloride decomposed in polyphosphoric acid. Earlier it was shown by Hein and co-workers [55] that polyphosphoric acid is useful to aid in the condensation of *o*-phenylenediamine with carboxylic acid derivatives, RCO—X, where X = OH, OR, NH_2, or RCN.

3-6. *Preparation of Poly[2,2′-(m-phenylene)-5,5′-bibenzimidazole] by the Solution Polycondensation Process* [54]

To a 100 ml three-necked flask equipped with a stirrer, nitrogen inlet, and calcium chloride drying tube is added 60 gm of 116% polyphosphoric acid and 2.0 gm (0.0056 mole) of 3,3′-diaminobenzidine tetrahydrochloride. The

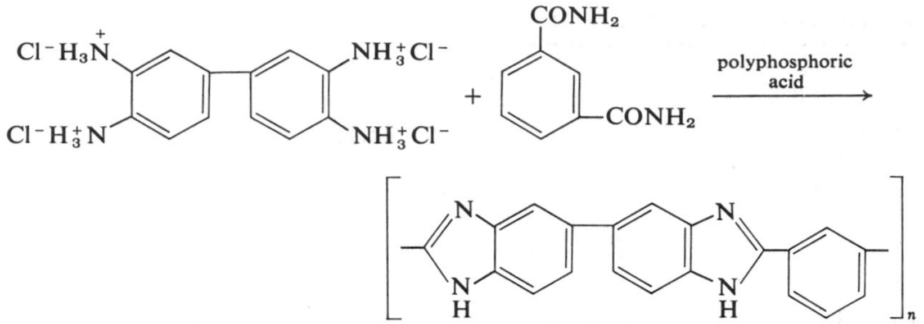

(12)

mixture is gradually heated while bubbling in nitrogen to 140°C to eliminate the hydrogen chloride gas. Then 0.80 gm (0.0049 mole) of isophthalamide is added and the mixture heated for 12 hr at 200°C. During the reaction the mixture develops a violet fluorescence and gradually becomes viscous as the polymerization progresses. The polymer is obtained by adding the hot solution to water, washing by decantation with water, allowing to sit overnight in dilute sodium bicarbonate, and washing again with water and finally with methanol. The dried polymer weighs 1.4 gm (100%), $\eta_{sp}/c = 1.68$ at 30°C (0.2 gm/100 ml in conc. H_2SO_4).

It should be noted that the benzimidazole nucleus contains an active hydrogen which should undergo reaction with bases such as sodium hydride. Reactions of the sodium salts with alkyl halides and other reagents should afford 1-substituted polybenzimidazoles.

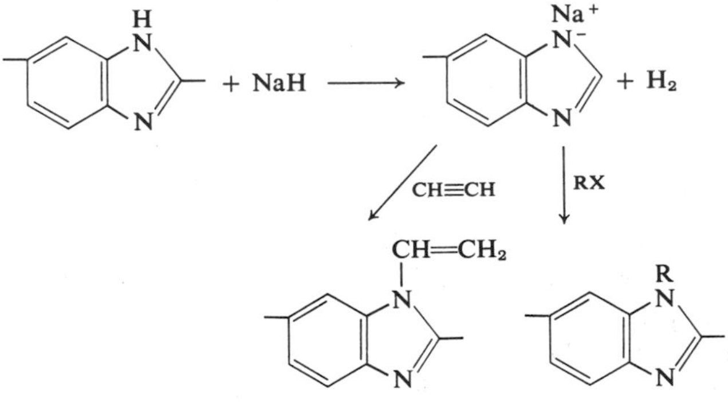

(13)

For related studies see one of the author's recent works on the vinylation of carbazole [56].

Miscellaneous Polybenzimidazole Preparations

1. Polybenzimidazoles containing adamantane rings [57].
2. Preparation of polybenzimidazoquinazolines [58].

3. Preparation of poly[2,6-(*m*-phenylene)-3,5-dimethyldiimidazolebenzene] [59].
4. Preparation of polybenzimidazoles by the condensation of 3,3′,4,4′-tetraminobiphenyl and 1,4-diacetylbenzene [60].
5. Polyamide-polybenzimidazoles [61].
6. Preparation of polyalkylenebenzimidazoles [62].

4. POLYQUINOXALINES

Sorenson [63] was the first to suggest extending Körner's [64] and Hinsberg's [65] work (Eq. 14) on the cyclocondensation of aromatic *o*-diamines and 1,2-dicarbonyl compounds to give polyquinoxalines as shown in Eq. (15).

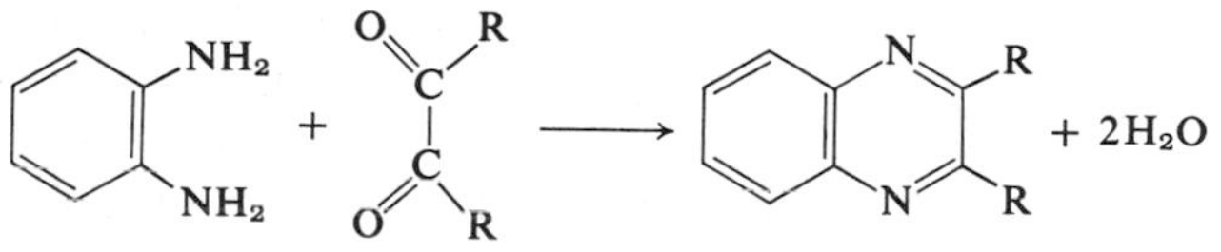

$$+ 2H_2O \qquad (14)$$

2,3-Disubstituted quinoxalene

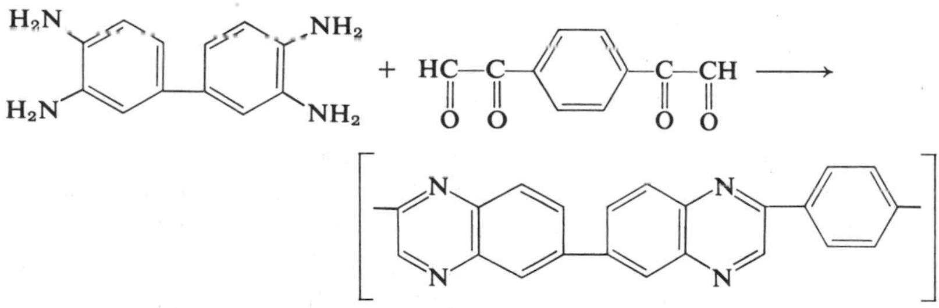

$$(15)$$

Poly[2,2′-(1,4-phenylene)6,6′-diquinoxaline]
TGA = 460°C

Stille, who already had work in progress on preparing polyquinoxalenes, published several papers shortly later [66,66a,67]. de Gaudemaris and Sillion [68] also published related work on the preparation of polyquinoxaline using 3,3′-diaminobenzidine with *p,p*′-oxybis(phenyleneglyoxal hydrate) [68].

The methods which have been used to carry out the polymerization of 3,3′-diaminobenzidine with 1,4-diglyoxalylbenzene dihydrate are described in Preparation 4-1.

4-1. Preparation of Poly[2,2′-(1,4-phenylene)-6,6′-diquinoxaline] [66]

CAUTION: Tetraaminobenzidines and other aromatic tetramines are carcinogenic and should be handled only with gloves in a well-ventilated hood.

$$n\mathrm{HC(=O)-C(=O)-C_6H_4-C(=O)-CH(=O)\cdot 2H_2O} + \mathrm{(H_2N)_2C_6H_3-C_6H_3(NH_2)_2} \longrightarrow \left[\text{-quinoxaline-quinoxaline-}C_6H_4\text{-}\right]_n + 6n\mathrm{H_2O} \quad (16)$$

(*a*) *Melt polymerization.* To a polymer melt tube is added a well-mixed mixture of 2.0 gm (0.0094 mole) of 3,3′-diaminobenzidine and 2.1 gm (0.0094 mole) of 1,4-diglyoxalylbenzene dihydrate. The mixture is heated to 180°C to cause it to melt and held at this temperature for 1 hr. After this time the melt solidifies at 180°C and the temperature is raised to 250°C and kept there for 3 hr. After this time the pressure is reduced to 0.1 mm Hg and the sample is heated at 375°C for 3 hr while rotating the tube. The use of 8 mm steel ball bearings helps to facilitate good mixing while rotating at all temperatures. The polymer tube is cooled, broken open, and the dark red polymer isolated in nearly quantitative yield. The polymer is completely soluble in H_2SO_4 but only 55% soluble in hexamethylphosphoramide (HMP). Viscosity in HMP, $[\eta]_{inh} = 1.28$ dl/gm at 25°C and 0.04–0.031 gm/100 ml solvent. Thermal gravimetric analysis (TGA) in N_2 at 800°C gave a 20% loss in weight and in air at 470°C gave a break in the TGA curve.

(*b*) *Solution polymerization in dioxane.* To an Erlenmeyer flask containing a warm (95°C) solution of 2.8 gm (0.013 mole) of 3,3′diaminobenzidine in 50 ml of purified dioxane is added with rapid stirring a warm (95°C) solution of 2.97 gm (0.013 mole) of 1,4-diglyoxalylbenzene dihydrate in 100 ml of purified (deoxygenated) dioxane. After 6 hr refluxing the precipitated polymer is filtered off in nearly quantitative yield. The polymer is heated in a rotating flask at 180°C for 1 hr and then at 250°C for 5 hr. The polymer did not melt during the latter heating periods. The isolated polymer was orange in color and had a similar solubility as the melt-polymerized sample. The inherent viscosity is $[\eta]_{inh} = 0.40$ dl/gm at the same concentration used in part (a). The TGA data are identical to those of the melt-polymerized sample.

4-2. *Preparation of Poly[2,2′-p,p′-oxydi(p″-phenoxy)phenylene-6,6′-diquinoxaline]* [8b]

CAUTION: Tetraaminobenzidines and other aromatic tetramines are carcinogenic and should be handled only with gloves in a well-ventilated hood.

To a three-necked, round-bottomed flask equipped with a heating mantle, condenser, and mechanical stirrer, and containing a slurry of 2.14 gm (0.010

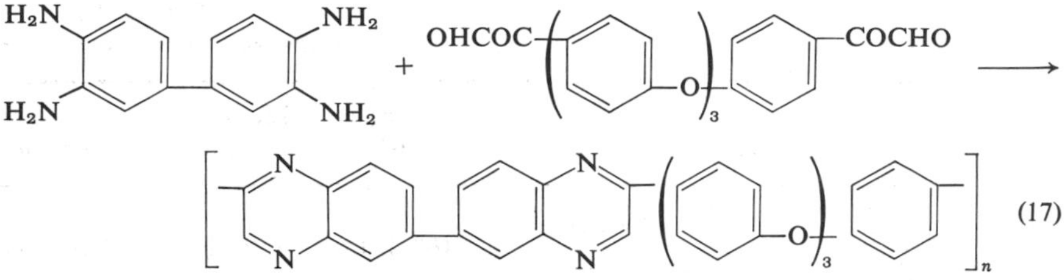

(17)

mole) of 3,3′-diaminobenzidine in 30 ml of *m*-cresol at 25°C under an argon atmosphere is added 5.03 gm (0.010 mole) of bis[*p*-(*p*′-glyoxalylphenoxy)-phenyl] ether dihydrate. An additional 34.5 ml of *m*-cresol is added to rinse down the remaining glyoxal and to adjust the solids content to 10%. The yellowish orange mixture is stirred for 1 hr at 25°C followed by heating slowly to 200°C over a 4 hr period and stirring at 200°C for 1 hr. The viscous red-brown solution is cooled and poured into an *explosion-proof* blender containing methanol to precipitate a fibrous yellow solid. Note that the precipitation can also be carried out in a hood with good stirring in a jar containing methanol. The polymer is added to fresh methanol, boiled, filtered, and dried under reduced pressure at 150°C for 4 hr to afford 6.0 gm (98.5%) of a yellow intermediate polymer, $\eta_{inh} = 0.62$ (0.5% H_2SO_4 at 25°C), polymer softening temp. (clear melt), approx. 240°C. The polymer readily forms a solution in *m*-cresol (20%) and a film that has good flexibility and toughness when cast. The molecular weight of the polymer can be advanced by adding the polymer to a tube and heating under an argon atmosphere at 350°C for 1 hr. The polymer softens and foams slightly to give a final polymer $\eta_{inh} = 1.27$ (same conditions as the earlier sample); $\lambda_{max} = 284$ mμ, $\varepsilon_{max} = 30.0 \times 10^3$; $\lambda_{max} = 333$ mμ, $\varepsilon_{max} = 46.0 \times 10^3$. TGA at 400°C in air gives a 50% wt. loss after 15 hr and complete loss after 24 hr. The most stable polyquinoxaline studied by these authors [8b] is shown below.

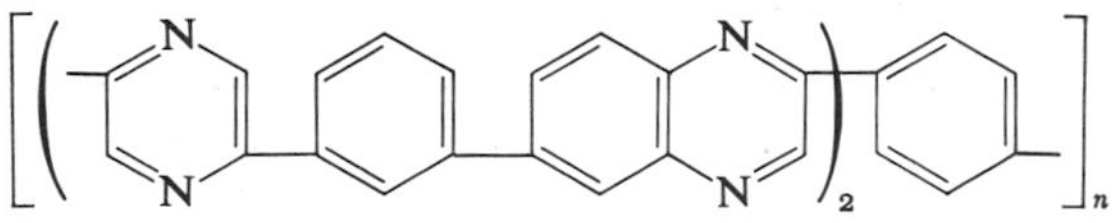

To obtain polymers stable in air at temperatures above 400°–500°C further work will have to be done to modify existing structures of polyquinoxalines.

The glyoxal derivatives were prepared from the bisacetyl compounds by selenium dioxide oxidation in refluxing (18 hr) acetic acid. Some typical diacetyl compounds used are shown in Table VI and the oxidation to bis-glyoxals is shown in Table VII. Consult ref. 8b for typical experimental procedures in the preparation of each of these materials.

Several other reports on the synthesis of polyquinoxalines are worthwhile consulting [8,69]. Recently polyphenylquinoxalines [8a,70,70a–70c] have been

TABLE VI

DIACETYL COMPOUNDS[a]

Compound	M.p. (°C)	Lit. m.p. (°C)	Formula
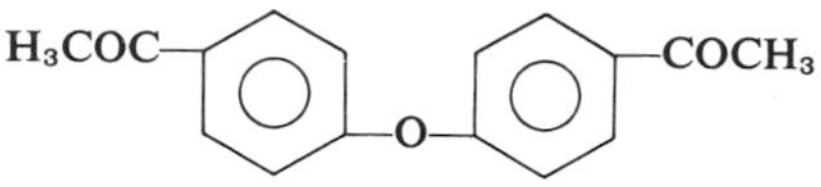 $H_3COC{-}C_6H_4{-}O{-}C_6H_4{-}COCH_3$	100–101.5	100–101[b]	$C_{16}H_{14}O_3$
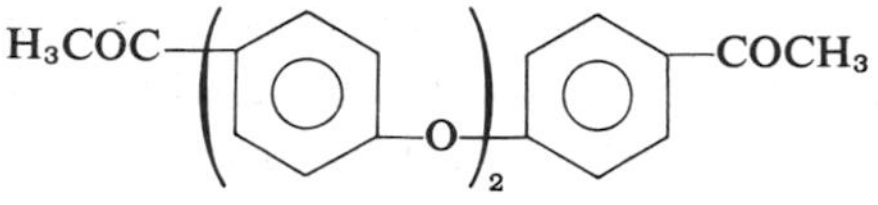 $H_3COC{-}(C_6H_4{-}O)_2{-}C_6H_4{-}COCH_3$	179–181	183.5[c]	$C_{22}H_{18}O_4$
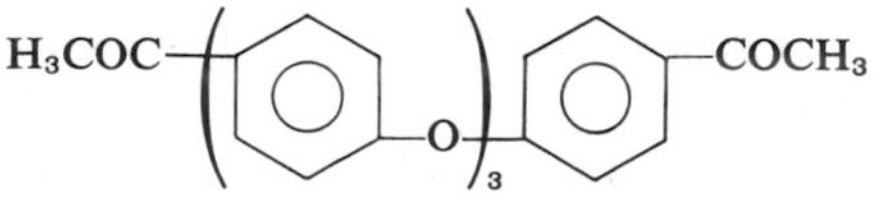 $H_3COC{-}(C_6H_4{-}O)_3{-}C_6H_4{-}COCH_3$	180–181.5	185[c]	$C_{28}H_{22}O_5$
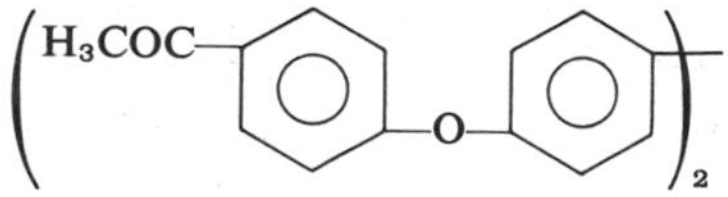 $(H_3COC{-}C_6H_4{-}O{-}C_6H_4{-})_2$	222–224	225[d]	$C_{28}H_{22}O_4$
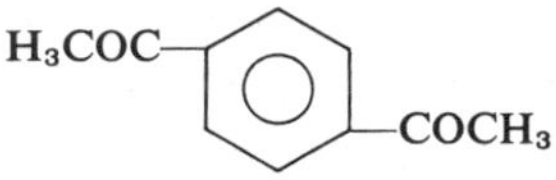 $H_3COC{-}C_6H_4{-}COCH_3$	113–114	114[e]	$C_{10}H_{10}O_2$
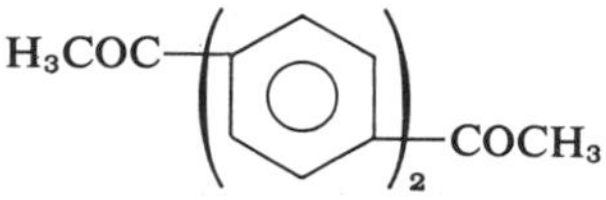 $H_3COC{-}(C_6H_4)_2{-}COCH_3$	189–190	190–191[f]	$C_{16}H_{14}O_2$
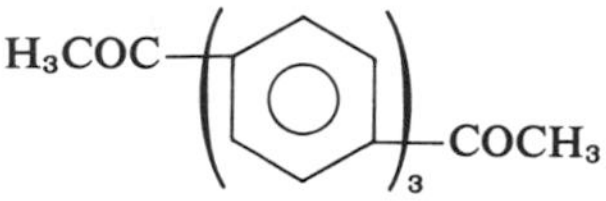 $H_3COC{-}(C_6H_4)_3{-}COCH_3$	282–285	268–269[g]	$C_{22}H_{18}O_2$

[a] Reprinted from P. M. Hergenrother and D. E. Kiyohara, *Macromolecules* **3**, 387 (1970). Copyright 1970 by the American Chemical Society. Reprinted by permission of the copyright owners.

[b] W. Dilthey, E. Bach, H. Grutering and E. Hausdorfer, *J. Prakt. Chem.* [2] **117**, 350 (1927).

[c] R. G. Neville and R. W. Rosser, *Makromol. Chem.* **123**, 19 (1969).

[d] R. W. Rosser and R. G. Neville, *J. Appl. Polym. Sci.* **13**, 215 (1969).

[e] J. Ingle, *Ber. Deut. Chem. Ges.* **27**, 2527 (1894).

[f] C. V. Ferris and E. E. Turner, *J. Chem. Soc., London* **117**, 1147 (1920).

[g] S. Patai, S. Schoenberg, and L. Rajbenbach, *Bull. Res. Coun. Isr., Sect. A* **5**, 257 (1956).

TABLE VII

BISGLYOXAL MONOMERS[a]

Compound	M.p. (°C)	Recrystal. solvent	Formula
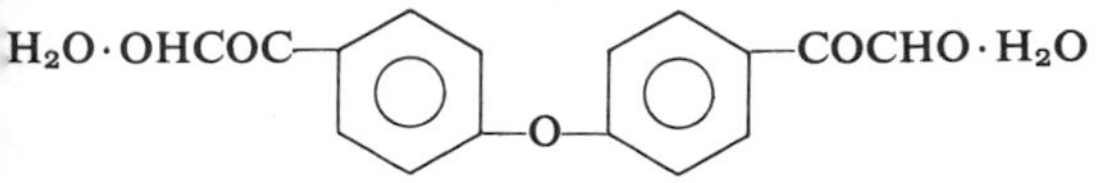 $H_2O\cdot OHCOC-C_6H_4-O-C_6H_4-COCHO\cdot H_2O$	141–143[b] (dec)	Dioxane + H_2O	$C_{16}H_{14}O_7$
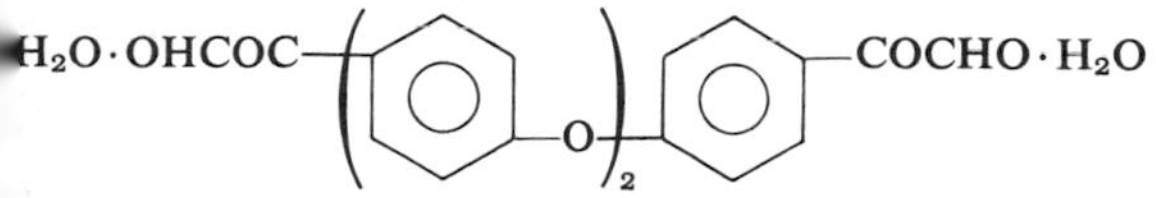 $H_2O\cdot OHCOC-(C_6H_4-O)_2-C_6H_4-COCHO\cdot H_2O$	157–160 (dec)	Dioxane + H_2O	$C_{22}H_{18}O_8$
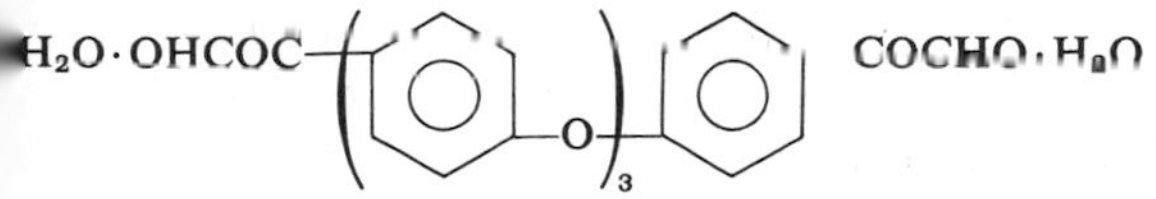 $H_2O\cdot OHCOC-(C_6H_4-O)_3-C_6H_4-COCHO\cdot H_2O$	154–156.5 (dec)	AcOH + H_2O	$C_{28}H_{22}O_9$
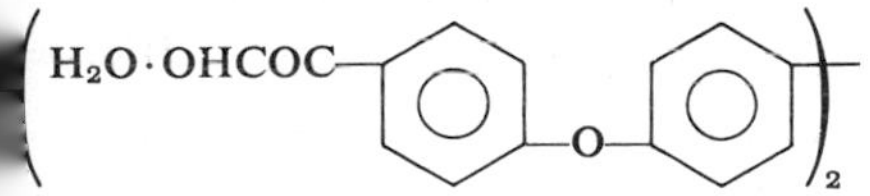 $(H_2O\cdot OHCOC-C_6H_4-O-C_6H_4)_2$	178–180 (dec)	AcOH + H_2O	$C_{28}H_{22}O_8$
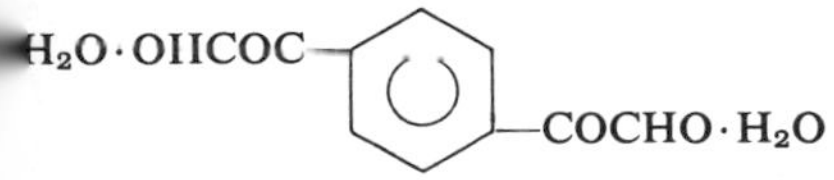 $H_2O\cdot OHCOC-C_6H_4-COCHO\cdot H_2O$	162–164[c] (dec)	H_2O	$C_{10}H_{10}O_6$
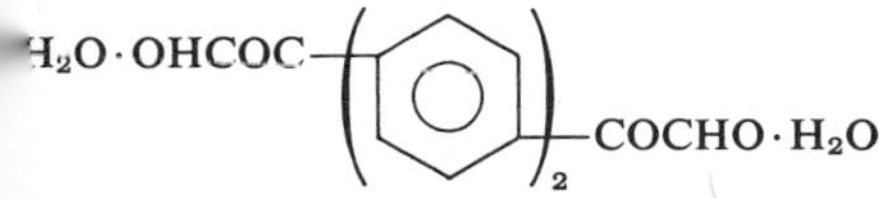 $H_2O\cdot OHCOC-(C_6H_4)_2-COCHO\cdot H_2O$	163–166 (dec)	AcOH + H_2O	$C_{16}H_{14}O_6$
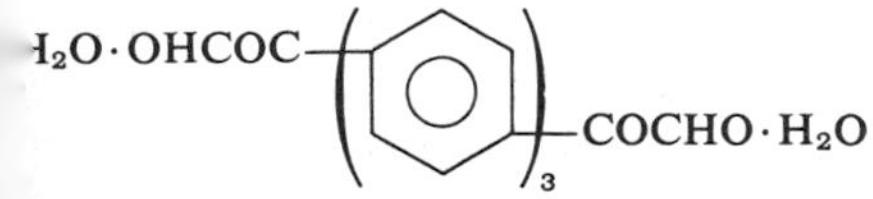 $H_2O\cdot OHCOC-(C_6H_4)_3-COCHO\cdot H_2O$	211–214 (dec)	AcOH + H_2O	$C_{22}H_{18}O_6$

[a] Reprinted from P. M. Hergenrother and D. E. Kiyohara, *Macromolecules* **3**, 387 (1970). Copyright 1970 by the American Chemical Society. Reprinted by permission of the copyright owners.

[b] P. Ruggli and E. Gassenmeier [*Helv. Chim. Acta* **22**, 496 (1939)] report m.p. 122°C.

[c] Lit. m.p. 110°–111°C; see footnote *b*.

described to exhibit superior solubility, thermal, and processing properties as compared to the nonphenylated analogs. The introduction of flexibilizing ether groups into the main polymer chain affords polymers with improved flow characteristics [71] but lowers the oxidative-thermal stability [70c]. A similar effect was also found to occur when the ether group was replaced by sulfone or carbonyl groups [Eq. 18] [8a].

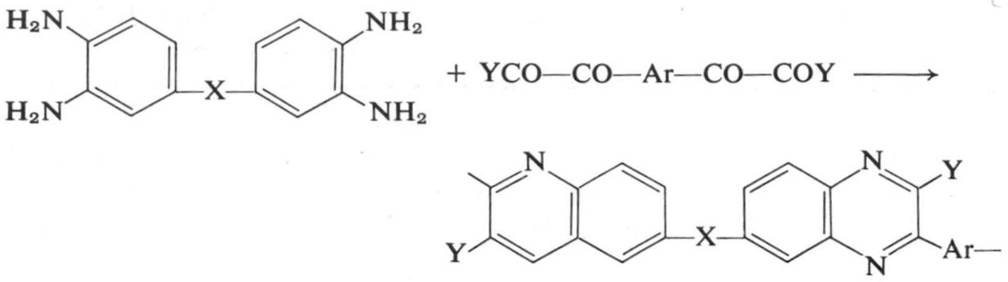

$$H_2N, H_2N\text{-}C_6H_3\text{-}X\text{-}C_6H_3\text{-}NH_2, NH_2 + YCO\text{—}CO\text{—}Ar\text{—}CO\text{—}COY \longrightarrow \quad (18)$$

X = CO, SO_2
Y = H, C_6H_5
Ar = *p*-phenylene
4,4′-diphenyl ether
4,4″-diphenyl sulfone

Bisbenzils (Y = C_6H_5) were reacted with 3,3′,4,4′-tetraaminodiphenyl sulfone or 3,3′,4,4′-tetraaminobenzophenone at room temperature to yield high molecular weight soluble final polyphenylquinoxalines [8a]. Bisglyoxals (Y = H) required more severe reaction conditions to react with the tetramines.

The characterization of some polyquinoxalines containing flexibilizing groups is described in Table VIII. The effect of structure on T_g for some selected polyphenylquinoxalines is shown in Table IX.

TABLE VIII
CHARACTERIZATION OF POLYMERS[a]

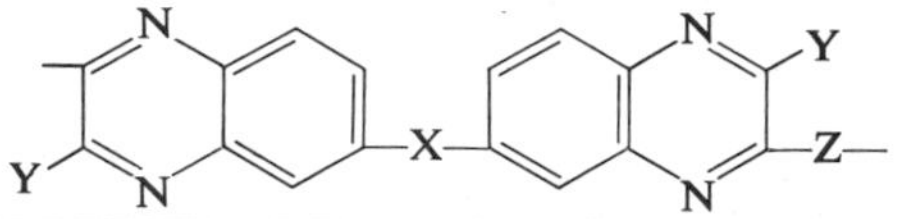

X	Y	Z	Inherent[b] viscosity	PDT[c]
SO_2	C_6H_5	*p*-Phenylene	1.68	490
SO_2	C_6H_5	*p,p′*-Diphenyl ether	1.47	490
SO_2	C_6H_5	*p,p′*-Diphenyl sulfone	0.76	490
CO	C_6H_5	*p*-Phenylene	2.06	520
CO	C_6H_5	*p,p′*-Diphenyl ether	1.45	510
CO	C_6H_5	*p,p′*-Diphenyl sulfone	1.03	510
SO_2	H	*p*-Phenylene	0.73	450
SO_2	H	*p,p′*-Diphenyl ether	0.95	430
CO	H	*p*-Phenylene	0.87	465
CO	H	*p,p′*-Diphenyl ether	0.55	440

[a] Reprinted from W. Wrasidlo and J. M. Angl, *Macromolecules* **3**, 544 (1970). Copyright 1970 by the American Chemical Society. Reprinted by permission of the copyright owner.
[b] Determined in 98% sulfuric acid, 0.5% concentrated at 30°C.
[c] Determined by TGA at a heating rate of 5°/min under vacuum.

An example of the improved thermal stability of the polyphenylquinoxalines is shown in Table X.

Dilute solution characterization data of polyphenylquinoxalines in *m*-cresol have recently been reported [72].

The use of polyquinoxalines as films, fibers, composites, and adhesives has been described [8], and, recently, soluble imide-quinoxaline copolymers have been prepared [73].

TABLE IX

EFFECT OF STRUCTURE ON T_g OF RELATED POLYPHENYLQUINOXALINES[a]

	T_g (°C)
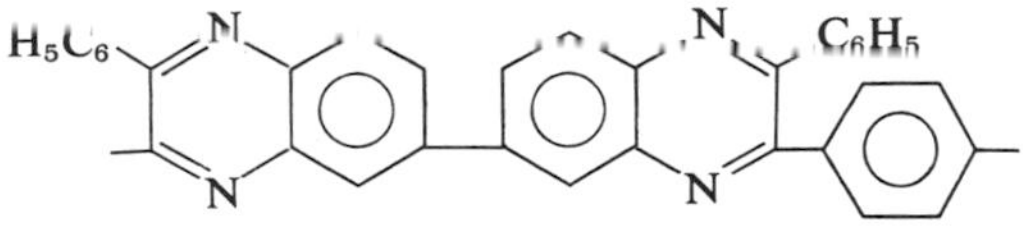 H_5C_6 N N N N C_6H_5	317
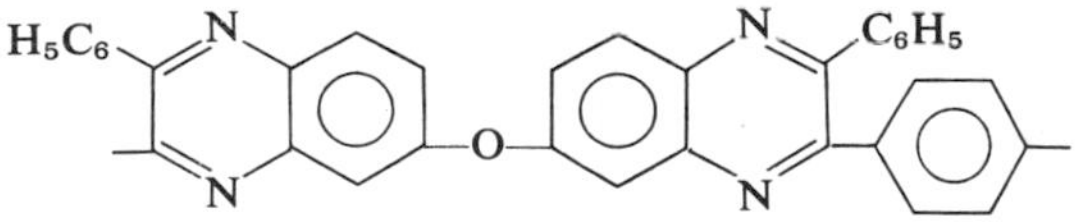 H_5C_6 N N O N N C_6H_5	298
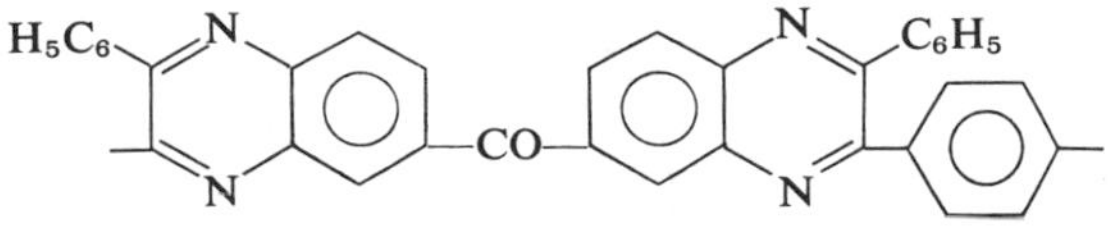 H_5C_6 N N CO N N C_6H_5	258
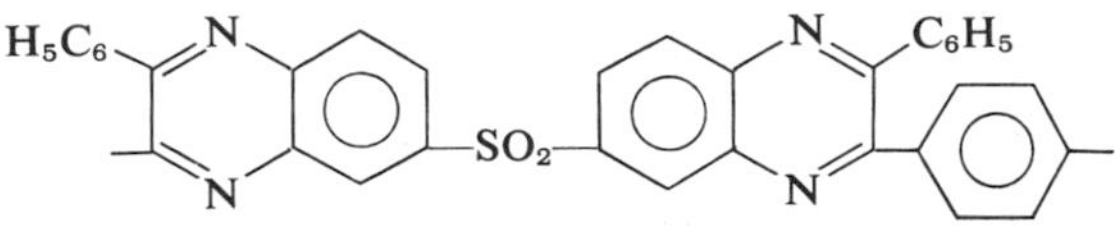 H_5C_6 N N SO_2 N N C_6H_5	212

[a] Reprinted from W. Wrasidlo and J. M. Angl, *Macromolecules* **3**, 544 (1970). Copyright 1970 by the American Chemical Society. Reprinted by permission of the copyright owner.

5. POLYPHENYLENE ETHERS

Hunter [74–78] was the first to prepare a poly(phenylene oxide) by the ethyl iodide-induced decomposition of anhydrous silver 2,4,6-tribromophenoxide [74a,75]. The polymer was isolated as a white, amorphous powder with a softening point above 300°C. The molecular weight was estimated to be about 6,000–12,000. Hunter [76] reported in a series of papers the polymerization of other trihalophenols and established that iodine was displaced

TABLE X

THERMAL STABILITY OF POLYPHENYLQUINOXALENES

Polyquinoxaline	η_{inh}[a] (dl/gm)	T_g[c] (°C)	Polymer decomp. temp. in air (°C)	% Wt. retention in air at 371°C after		Ref.
				100 hr	200 hr	
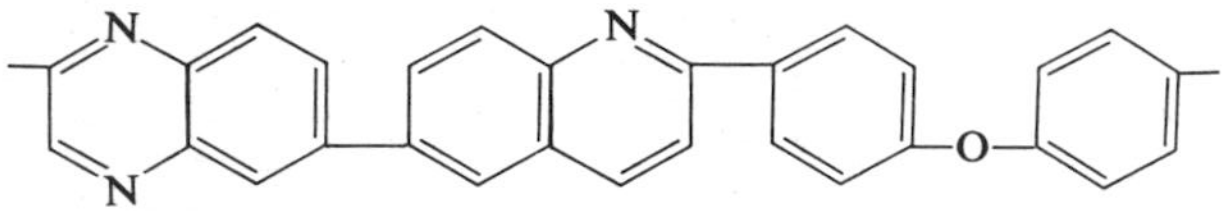	2.06	388	510	60	0	70
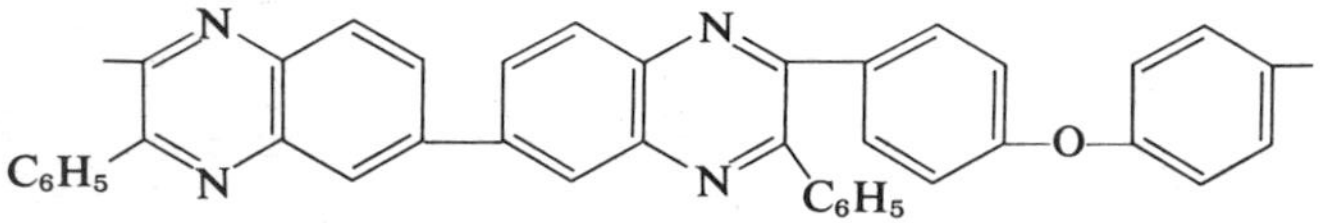	2.05	420	520	80	40	70
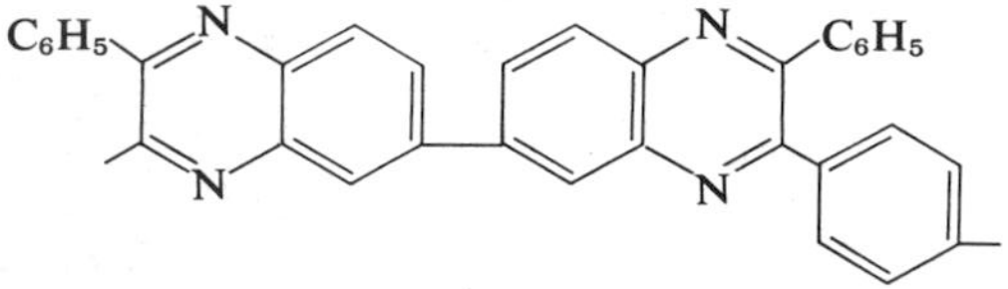	2.1[b]	317	540	—	94	70b

[a] Conc., 0.5% H_2SO_4 at 25°C.
[b] Conc., 0.5% in *m*-cresol at 30°C.
[b] T_g, determined from dielectric properties of polymer.

more readily than bromine and the latter more rapidly than chlorine. The *p*-halogen reacted more rapidly than the *o*-chlorine group [77] and only halogen with a free ionizable phenolic group could be displaced [78].

5-1. Preparation of Poly(2,6-dimethyl-1,4-phenylene oxide) by Polymerization of Silver 4-Bromo-2,6-dimethylphenolate [79]

$$\text{Br}-\text{C}_6\text{H}_2(\text{CH}_3)_2-\text{OH} \longrightarrow \text{Br}-\text{C}_6\text{H}_2(\text{CH}_3)_2-\text{OAg} \longrightarrow$$

$$\text{Br}-\text{C}_6\text{H}_2(\text{CH}_3)_2-\text{O}-\left[\text{C}_6\text{H}_2(\text{CH}_3)_2-\text{O}\right]_n-\text{C}_6\text{H}_2(\text{CH}_3)_2-\text{OH} + \text{AgBr} \quad (19)$$

(*a*) *Preparation of silver salt* [74,79]. To an Erlenmeyer flask containing 100 ml of distilled water is added 20.1 gm (0.1 mole) of 4-bromo-2-6-dimethylphenol and 5.6 gm (0.1 mole) of potassium hydroxide dissolved in 100 ml of distilled water. The resulting solution is filtered and to the filtrate 1% acetic acid is added dropwise to give a faint turbidity. A few drops of dilute silver nitrate are then added and any resulting silver bromide filtered. Then to the filtrate is added a solution of 17.0 gm (0.1 mole) of silver nitrate in 100 ml of water to give a voluminous precipitate of the silver salt. After storage in the dark for 1–2 hr the precipitate is filtered, washed with water, and dried under reduced pressure to give an almost quantitative yield of the silver salt.

(*b*) *Polymerization of silver salt* [79]. To a flask is added 4.0 gm (0.013 mole) of silver 4-bromo-2,6-dimethylphenolate and 100 ml of benzene. The suspension is refluxed for 1 hr to give a precipitate of silver bromide. The benzene solution is added to 300 ml of methanol to precipitate the polymer. The polymer is dried under reduced pressure to afford 0.79 gm (50%) with a softening point of 210°–220°C and $[\eta] = 0.06$ dl/gm in $CHCl_3$ at 25°C. The mol. wt. of 1600 is determined by an ebuliometric method in benzene or chloroform.

Price [9c] reinvestigated this problem and studied the polymerization of 4-bromo-2,6-dimethylphenol with several oxidizing agents such as ferricyanide in PbO_2, I_2, O_2, and light. The soluble polymer had molecular weights in the range 2,000–10,000. Sterically hindered halogenated phenols (4-bromo-2,6-di-*tert*-butylphenol, bromodurenol, and pentabromophenol) failed to polymerize under the same conditions. A radical mechanism [see Eq. (20)] was suggested for this reaction, and was later confirmed.

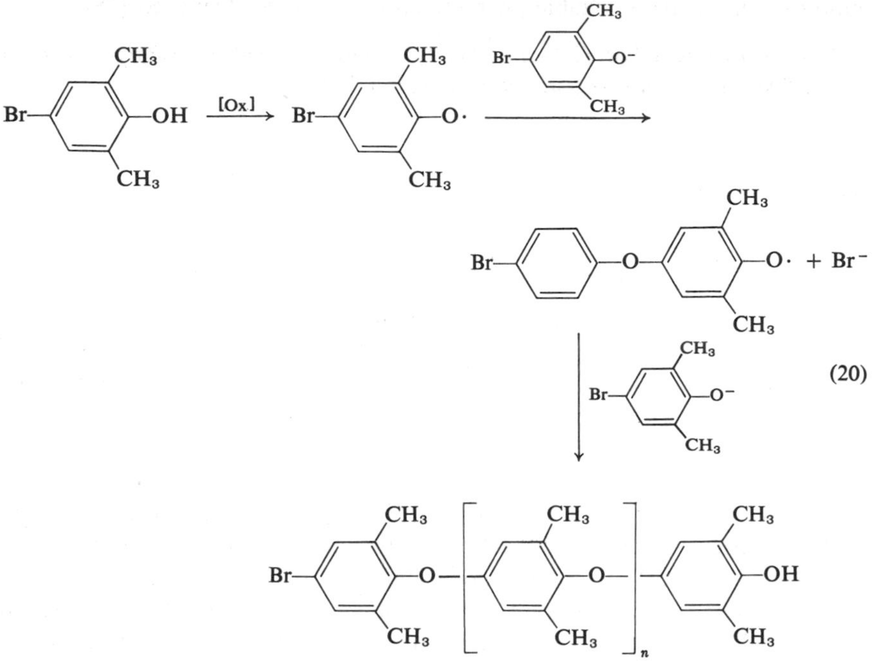

(20)

5-2. Preparation of Poly(2,6-dimethyl-1,4-phenylene oxide) Using Lead Dioxide–KOH [9c]

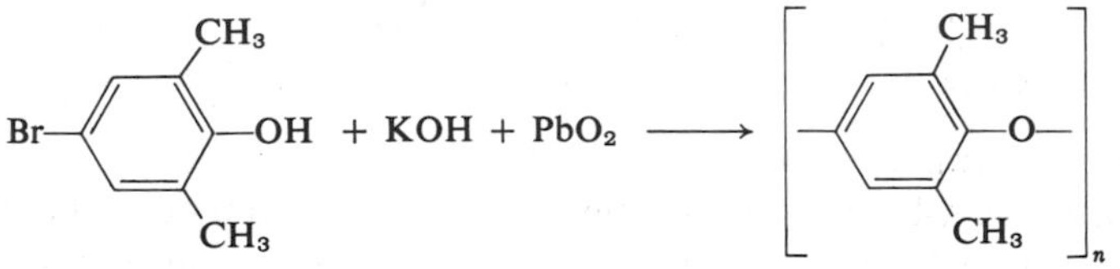

$+ KOH + PbO_2 \longrightarrow$ (21)

To a flask containing a solution of 96.4 gm (0.47 mole) of 2,6-dimethyl-4-bromophenol in 2.3 liters of benzene is added 26.4 gm (0.47 mole) of potassium hydroxide in 2.3 liters of water, and 1.13 gm (0.0047 mole) of lead dioxide. The mixture is stirred vigorously for 45 hr at room temperature. The benzene layer is separated, concentrated to 0.5 liter, and filtered into 15 liters of methanol to afford 46.5 gm (81%) of polymer, $[\eta] = 0.11$ (benzene, 25°C).

5-3. Preparation of Poly(2,6-dimethyl-1,4-phenylene oxide) by the Action of KOH–O_2 [80]

To a flask containing 2.01 gm (0.01 mole) of 4-bromo-2,6-dimethylphenol in 100 ml of water is added 0.56 gm (0.01 mole) of potassium hydroxide and

$$\text{Br}-C_6H_2(CH_3)_2-OH + KOH \xrightarrow{O_2} \left[-C_6H_2(CH_3)_2-O-\right]_n \tag{22}$$

then 100 ml of benzene. The reaction mixture is stirred for 3 days and then the benzene layer is separated. Petroleum ether is added to the benzene layer and the solution cooled to 0°C to precipitate 0.63 gm (52%) of polymer, $[\eta] = 0.205$, m.p. 232°–244°C. No polymer is obtained when the same reaction is carried out under a nitrogen atmosphere. The reaction can be carried out in 24 hr to give a 78% yield of polymer ($\eta = 0.302$, m.p. 215°–233°C) if 1 mole % of potassium ferricyanide is added dropwise over a $\frac{1}{2}$–$1\frac{1}{2}$ hr period.

Earlier [81] polyphenylene oxides were also reported to be prepared by the pyrolysis of diazoxides and by the pyrolysis of phenol mercuriacetates. Price suggested that a similar radical process can also be used to explain the reactions shown in Eqs. (23) and (24).

$$\overset{+}{N_2}-C_6H_2X_2-O^- \xrightarrow{\text{heat}} \left[-C_6H_2X_2-O-\right] + N_2 \tag{23}$$

$$CH_3COOHg-C_6H_2R_2-OH \xrightarrow{200^\circ C} \left[-C_6H_2R_2-O-\right]_n + Hg + CH_3COOH \tag{24}$$

Hay [9,82] and co-workers were the first to report that oxidative coupling of 2,6-disubstituted phenols gave high molecular weight 2,6-disubstituted 1,4-phenylene ethers (for R = CH_3). When the substituents are bulky, such as R = *tert*-butyl, then only carbon–carbon coupling occurs to give diphenoquinones as the sole product. When R = isopropyl then both reactions shown in Eqs. (25) and (26) occur.

In order to overcome the steric reduction in rate for C—O coupling with 2,6-diphenylphenol, the reaction was carried out at 60°C using a Cu–*N,N,N′,N′*-tetramethyl-1,3-butanediamine catalyst to give polymers in high yield having a molecular weight ($\overline{M}_n$) of approx. 300,000 [83,83a]. Only 3% of the diphenoquinone was formed [83,83a]. The polymerization of various other 2,6-diarylphenols has also been reported [84].

$$n\,\text{(2,6-R}_2\text{C}_6\text{H}_3\text{)—OH} + \tfrac{1}{2}n\text{O}_2 \xrightarrow{\text{Cu-amine}} \left[\text{—C}_6\text{H}_2\text{R}_2\text{—O—}\right]_n + n\text{H}_2\text{O} \qquad (25)$$

$$\downarrow \tfrac{1}{4}n\text{O}_2$$

$$\tfrac{1}{2}n\text{HO—C}_6\text{H}_2\text{R}_2\text{—C}_6\text{H}_2\text{R}_2\text{—OH} \xrightarrow{\tfrac{1}{4}n\text{O}_2} \tfrac{1}{2}n\text{O=C}_6\text{H}_2\text{R}_2\text{=C}_6\text{H}_2\text{R}_2\text{=O} + n\text{H}_2\text{O} \qquad (26)$$

Hay [84] reported that the oxidative polymerization of 2,6-diphenylphenol could be effected with either oxygen–amine copper salts [83a], lead dioxide [85], or silver oxide [86].

Block copolymers of polyphenylene ethers starting with 2,6-dimethylphenol and 2,6-diphenylphenol were prepared using a catalyst system consisting of cuprous bromide, *N,N,N′,N′*-tetramethyl-1,3-butanediamine, and magnesium sulfate [9b]. Recently NMR data have been presented to show that a random copolymer is produced using a cuprous chloride–pyridine catalyst at 25°C [9d,87].

$$\left[\text{—C}_6\text{H}_2(\text{CH}_3)_2\text{—O—}\right]_x \left[\text{—C}_6\text{H}_2(\text{C}_6\text{H}_5)_2\text{—O—}\right]_y$$

5-4. Preparation of Poly(2,6-dimethyl-1,4-phenylene oxide) Using CuCl–Pyridine–O_2 [9,88]

$$n\,\text{(2,6-(CH}_3)_2\text{C}_6\text{H}_3\text{)—OH} + \tfrac{1}{2}n\text{O}_2 \xrightarrow[\text{pyridine, C}_6\text{H}_5\text{—NO}_2]{\text{CuCl}} \left[\text{—C}_6\text{H}_2(\text{CH}_3)_2\text{—O—}\right]_n + n\text{H}_2\text{O} \qquad (27)$$

To a 50 ml flask connected to an oxygen buret (an oxygen addition tube can also be used in place of a buret) is added 15 ml of nitrobenzene, 4.5 ml of pyridine, and 0.020 gm (0.0002 mole) of finely divided cuprous chloride. The mixture is stirred rapidly for 15 min to give a dark green solution. At this point 0.49 gm (0.004 mole) of 2,6-xylenol is added and stirring is continued. After 20 min the theoretical amount of oxygen uptake is complete and the orange reaction mixture is stirred for an additional 20 min (total oxygen uptake, 101.5%). At this time the orange color is almost gone and the solution is added dropwise to 65 ml of methanol acidified with 0.65 ml of conc. HCl. The precipitated polymer is filtered and washed with additional acidified methanol. The polymer is reprecipitated twice from solution in 10 ml of benzene by adding to 40 ml of cold methanol. Finally the polymer is dissolved in 20 ml of benzene, washed with water, and freeze-dried at 1 mm Hg pressure to afford 0.40 gm (84%) of polymer, $\eta_{\text{intrinsic}} = 0.95$ to 1.0 dl/gm; osmotic molecular weight, 28,000; and softening point, 240°C. The infrared spectrum of the polymer shows no absorption due to OH groups.

The same polymer is obtained when one starts with 2,4-dimethyl-4-halophenol (halogen = Br or Cl) but equimolar amounts of cuprous chloride are required.

The linear polyphenylene ethers from the amine–copper-catalyzed polymerization have a DP greater than 200, with softening points above 240°C [89].

The polymerization reaction has been shown to involve a free-radical step-growth process. The final polymer may be prepared from either the monomer, dimer, trimer, or other isolated low molecular weight intermediates [80]. The oxygen has the function of keeping the copper ions in the active divalent state [79,90].

The use of excess amine in the amine–Cu complex gives a more active catalyst with higher rate yields of C—O products rather than those formed by C—C bonding, as shown in Tables XI and XII. This evidence supports the suggestion that the catalyst using pyridine probably can be represented by Eq. (28).

$$\text{Cu}^{\text{I}}\text{Pyr} + \text{Pyr} \xrightarrow{O_2} \text{Cu}^{\text{II}}\text{Pyr} + \text{Pyr} \rightleftharpoons \text{Cu}^{\text{II}}\text{Pyr} - \text{Pyr} \qquad (28)$$

Price and Nakoaka [91] recently studied the kinetics of the polymerization and found that the oxidative coupling reaction is first order in oxygen pressure, first order in copper catalyst concentration, and independent of the phenol concentration. However, the rate of phenol oxidation is influenced by its type of substituents. The sequence of reactions shown in Scheme 1 was suggested to fit the experimental data for catalyst initiation by oxygen consumption.

Scheme 1 also helps to explain the favorable effect of electron-donor groups in the phenol. With two pyridines/copper the aryloxy radicals will probably be more readily dissociated to a "free" aryloxy radical to undergo C—O coupling via the quinol ether mechanism. The role of strongly basic ligands at copper in promoting C—O coupling rather than C—C coupling may be the result of a decreased bonding of the aryloxy radical to the copper ion.

TABLE XI

EFFECTS OF VARYING PYRIDINE CONCENTRATION[a,b] ON THE POLYMERIZATION OF 2,6-DIMETHYLPHENOL

Pyridine (M)	Ligand ratio, N/Cu	$R_{max} \times 10^3$ (mole liter^{-1} min^{-1})	Oxygen absorbed (%)	Fractional yields f_{C-O}	f_{C-C}	Intrinsic viscosity (dl/gm)
0.0033	0.67	0.091	107	0.072	0.56	*d*
0.0050	1.0	0.206	105	0.16	0.49	*d*
0.0100	2.0	0.662	96.5	0.40	0.34	0.086
0.0150	3.0	1.26	100	0.51	0.26	0.097
0.050	10	5.20	99	0.75	0.10	0.17
0.50	100	11.4	98.5	0.86	0	0.49
2.79	558	7.70	99	0.82	0	0.725
9.00	1800	1.30	108	0.785	0	0.71
9.00	1800	1.33	111	0.79	0	0.76
12.1[c]	2420	0.666	109	0.80	0	0.94

[a] Conditions: 2,6-Dimethylphenol 0.2 M; copper(I) chloride 0.005 M; *o*-dichlorobenzene solvent; 30°C. At ligand ratios 0.2 to 100, anhydrous magnesium sulfate was also present at 0.2 mole/liter.

[b] Reprinted from G. F. Endres, A. S. Hay, and J. W. Eustance, *J. Org. Chem.* **28**, 1300 (1963). Copyright 1963 by the American Chemical Society. Reprinted by permission of the copyright owner.

[c] Pyridine solvent.

[d] Insufficient sample for determination.

TABLE XII

OXIDATION OF 2,6-DIMETHYLPHENOL WITH PYRIDINE AND DERIVATIVES[a,b]

Ligand	Ligand ratio, N/Cu	$R_{max} \times 10^3$ (mole liter^{-1} min^{-1})	Oxygen absorbed (%)	Fractional yields f_{C-O}	f_{C-C}	Intrinsic viscosity (dl/gm)
Pyridine	1.0	3.27	101	0.465	0.18	0.16
	55.8	21.3	100	0.85	0	1.35
Quinoline	1.0	0.935	105	0.48	0.30	0.13
	55.8	0.53	104	0.82	0	1.15
2,6-Lutidine	1.0	1.80	118	0.29	0.375	0.08
	10.0	1.54	107.5	0.73	0	0.205
	55.8	1.98	103	0.93	0	0.55

[a] Conditions: 2,6-Dimethylphenol, 0.2 M; copper(I) chloride 0.05 M; anhydrous magnesium sulfate, 0.2 mole/liter; *o*-dichlorobenzene solvent; 30°C.

[b] Reprinted from G. F. Endres, A. S. Hay, and J. W. Eustance, *J. Org. Chem.* **28**, 1300 (1963). Copyright 1963 by the American Chemical Society. Reprinted by permission of the copyright owner.

SCHEME 1

A: $(Pyr)_2(ArOH)Cu^{I}(\mu\text{-}Cl)_2Cu^{I}(HOAr)(Pyr)_2$ $\xrightarrow[\text{(slow)}]{O_2}$ B: $(Pyr)_2(ArOH)Cu^{II}(\mu\text{-}Cl)_2(\mu\text{-}O_2)Cu^{II}(HOAr)(Pyr)_2$

B $\xrightarrow{-2ArO\cdot \text{ fast}}$ C: $(Pyr)_2(HO)Cu^{II}(\mu\text{-}Cl)_2Cu^{II}(OH)(Pyr)_2$

C $\xrightarrow[-2H_2O\cdot \text{ (fast)}]{+2ArOH}$ D: $(Pyr)_2(ArO)Cu^{II}(\mu\text{-}Cl)_2Cu^{II}(OAr)(Pyr)_2$

D $\xrightarrow[-2ArO\cdot]{+2ArOH, \text{ (fast)}}$ A

Reprinted from C. C. Price and K. Nakoaka, *Macromolecules* **4**, 364 (1971); copyright 1971 by the American Chemical Society. Reprinted by permission of the copyright owner.]

Miscellaneous Polyphenylene Ether Preparations

1. Oxidative polyarylation of di-1-naphthoxy alkanes [92].
2. Decomposition of phenoxycopper(II) complexes [93].
3. Preparation of poly-*m*-phenoxylenes [94].
4. Polymerization of 2,6-xylenol using a catalyst prepared from a vanadium pentoxide adduct with H_3PO_4 and BuOH and a tertiary amine (pyridine) [95].
5. Redistribution of polyhydroxyarylene ethers [96].
6. Preparation of metalated poly-2,6-dimethyl-1,4-phenylene ethers [97].

6. POLY-1,3,4-OXADIAZOLES AND POLY-1,2,4-TRIAZOLES

Polyhydrazides obtained by the condensation of dicarboxylic acid chlorides with hydrazine or arylene dihydrazides can be converted either to poly-1,3,4-oxadiazoles or poly-1,2,4-triazoles as shown in Eq. (29) [3].

*6-1. Preparation of Poly[N,N′-(2,6-pyridinediimodyl)-N″,N-‴ isophthaloyldihydrazine]**

The following preparation is performed under conditions to exclude moisture using solvent and reactants of high purity. A cold solution of isophthaloyl chloride (4.060 gm, 0.020 mole) in hexamethylphosphoramide (24 ml) is added during 0.5 hr under nitrogen to a vigorously stirred slurry of 2,6-pyridinediyldihydrazidine (3.862 gm, 0.020 mole), anhydrous sodium acetate (3.280 gm, 0.040 mole), and anhydrous lithium chloride (2.50 gm) in hexamethylphosphoramide (47 ml) at 5°–7°C. After complete addition, the

* From P. M. Hergenrother, *Macromolecules* **3**, 10 (1970). Copyright 1970 by the American Chemical Society. Reprinted by permission of the copyright owner.

(29)

viscous yellow reaction mixture is stirred at 5°C for 3 hr followed by stirring at ambient temperature for 18 hr. The resulting, very viscous, yellow reaction mixture is poured into water in a Waring blender to precipitate a fibrous yellow solid which is thoroughly washed with water and methanol to yield poly-*N*-acylhydrazidine as a yellow solid (6.5 gm, 95.5% yield based upon the monohydrate), $\eta_{inh} = 0.71$. The precursor polymer is also prepared in the following manner by interfacial polycondensation. A solution of isophthaloyl chloride (2.030 gm, 0.010 mole) in methylene chloride (150 ml) is added during 10 min to a vigorously stirred solution of 2,6-pyridinediyldihydrazidine (1.932 gm, 0.010 mole) in water (500 ml) at 20°C. A yellow precipitate appears upon the initial addition of the diacid chloride. After complete addition, the reaction mixture is stirred at ambient temperature for 1 hr followed by filtration. The isolated yellow solid is thoroughly washed with water and methanol to afford a quantitative yield of poly-*N*-acylhydrazidine ($\eta_{inh} = 0.21$).

*6-2. Preparation of Poly[3,3′-(2,6-pyridinediyl)-5,5′-(m-phenylene)di(1,2,4-triazole)]**

Poly-*N*-acylhydrazidine (1.0 gm) in *m*-cresol (20 ml) under nitrogen is heated to the reflux temperature during 1.5 hr and maintained at the reflux

* From P. M. Hergenrother, *Macromolecules* **3**, 10 (1970). Copyright 1970 by the American Chemical Society. Reprinted by permission of the copyright owner.

temperature for 18 hr. The resulting cooled yellow suspension is diluted with methanol and filtered to afford a quantitative yield of poly-1,2,4-triazole, $\eta_{inh} = 0.65$. The polymer readily adsorbs water to form a hydrate where one molecule of water is attached per polymer unit.

An alternate means of converting the poly-*N*-acylhydrazidine to poly-1,2,4-triazole is accomplished by solid-state advancement. A test tube with a side arm containing poly-*N*-acylhydrazidine (1.0 gm) under nitrogen is introduced into a preheated oil bath at 280°C. The temperature is increased to 350°C during 1 hr and maintained at 350°C for 1 hr. During this thermal treatment, the original pale yellow solid sinters slightly and turns a slight beige color but exhibits no evidence of melting. A weight loss of 15.6% (15.9% theoretical weight loss for hydrated poly-*N*-acylhydrazidine) is recorded. The polymer exhibits an $\eta_{inh} = 0.61$.

6-3. *Preparation of Poly[2,2′-(2,6-pyridinediyl)-5,5′-(m-phenylene)di(1,3,4-oxadiazole)]**

A yellow solution of poly-*N*-acylhydrazidine (1.0 gm) in trifluoroacetic acid (20 ml) under nitrogen is refluxed for 18 hr. The resulting orange solution is poured into water to precipitate a white solid which is isolated by filtration and washed successively with aqueous sodium carbonate, water, and methanol. Poly-1,3,4-oxadiazole is obtained in quantitative yield as a white solid which exhibits an η_{inh} of 0.63. The poly-1,3,4-oxadiazole does not exhibit as strong a tendency to adsorb water as the poly-1,2,4-triazole.

Frazer and co-workers [98,99] earlier described the preparation of poly(1,3,4-oxidazoles) by the thermal cyclodehydration of polyhydrazides (Eq. 30). The polyoxadiazoles are reasonably stable at 300°–400°C in air but degrade severely at 450°C.

$$R-\left[-\underset{\underset{O}{\|}}{C}-NH-NH-\underset{\underset{O}{\|}}{C}-R-\underset{\underset{O}{\|}}{C}-NH-NH-\underset{\underset{O}{\|}}{C}-\right] \longrightarrow \left[-R-C\overset{N-N}{\underset{O}{\langle\rangle}}C-R-C\overset{N-N}{\underset{O}{\langle\rangle}}C-\right]_n \qquad (30)$$

6-4. *Preparation of Poly[1,3-phenylene-2,5-(1,3,4-oxadiazole)]* [100]

To a flask containing 48.5 gm (0.245 mole) of isophthaloyl dihydrazide in 325 ml of *N*-methyl-2-pyrrolidinone is added 15 gm of lithium chloride. The

* From P. M. Hergenrother, *Macromolecules* 3, 10 (1970). Copyright 1970 by the American Chemical Society. Reprinted by permission of the copyright owner.

$$H_2N{-}NH{-}\overset{O}{\overset{\|}{C}}{-}C_6H_4{-}\overset{O}{\overset{\|}{C}}{-}NH{-}NH_2 + \underset{Cl}{\overset{|}{C}}{=}O{-}C_6H_4{-}\overset{O}{\overset{\|}{C}}{-}Cl \longrightarrow$$

$$\left[-\overset{}{\underset{O}{\underset{\|}{C}}}{-}NH{-}NH{-}\underset{O}{\underset{\|}{C}}{-}C_6H_4{-} \right]_n \xrightarrow[-H_2O]{heat} \left[-\overset{N{-}N}{\underset{O}{\langle\ \rangle}}{-}C_6H_4{-} \right]_n \qquad (31)$$

mixture is cooled while 50.75 gm (0.249 mole) of isophthaloyl chloride is added with stirring. The poly(isophthaloyl hydrazide) is isolated by precipitation in water, washing with methanol, and drying to afford 81 gm (100%), m.p. 370°C $[\eta] = 0.50$ (in dimethyl sulfoxide at 25°C). The polyhydrazide is converted into the corresponding poly-1,3,4-oxadiazole by heating for 8 hr at 283°C at 1 mm Hg to afford 71 gm (100%), m.p. >400°C, $[\eta] = 0.45$ (in H_2SO_4 at 25°C).

Recently Rode and co-workers [101] reported on the thermal stability of poly 1,2,4-oxadiazoles and reported that they were of lower thermal stability than the isomeric poly 1,3,4-oxadiazoles.

Lilyquist and Holsten [102,103] were able to convert several aromatic polyhydrazides to aromatic 1,2,4-triazoles by reaction with aromatic amines in polyphosphoric acid at 200°C. Thermally stable fibers and films were prepared from formic acid solutions of poly(*m,p*-phenylene)phenyl-1,2,4-triazole and the latter polymer and had maximum zero-strength temperatures of 465° and 490°, respectively. The differential and thermogravimetric analyses showed that the polymer was stable in nitrogen up to 512°C.

6-5. Preparation of Poly[(m,p-phenylene)-4-phenyl-1,2,4-triazole] [103]

To a 1 liter flask equipped with a mechanical stirrer, condenser, nitrogen inlet, and dropping funnel is added approximately 300 gm of polyphosphoric acid and it is heated to 150°C under a nitrogen atmosphere. While stirring, 75.1 gm (0.81 mole) of aniline is added dropwise at such a rate to keep the temperature at 190°C. The temperature is lowered to 175°C and then 9.72 gm (0.03 mole) of poly[(*m,p*-phenylene) hydrazide] [from the reaction of isophthalic hydrazide, 97.0 gm (0.5 mole), at 50°C in hexamethylphosphoramide with 101.5 gm (0.5 mole) of terephthaloyl chloride at 3°–5°C for 2 hr] ($\eta_{inh} = 2.39$–2.94) is added with stirring and the mixture heated for 96 hr at 180°–182°C. The polymer is isolated by pouring into hot deionized water in a blender. Caustic pellets are cautiously added to precipitate the polymer in

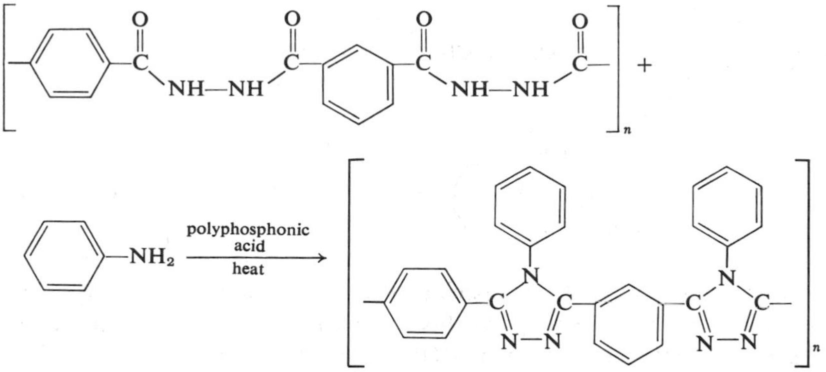

polyphosphonic acid, heat (32)

almost quantitative yield. The polymer is filtered, washed with water, dried, and extracted with ethanol using a Soxhlet extractor. The dried polymer residue is dissolved in formic acid and the number average molecular weight is approximately 21,000–28,000.

7. POLYAROMATIC SULFONES

The poly(ether-sulfone) derived from 4,4′ dichlorodiphenyl sulfone and the disodium or dipotassium salt of 2,2-bis(4-hydroxyphenyl)propane (bisphenol A) was recently announced [104,104a]. The polymeric sulfone shown above is thermoplastic and has glass-transition temperatures ranging from 180° to 250°C depending on the value of n (50–80). The polymer may be used continuously in the range of $-100°$ to 175°C. The polymer rapidly degrades in air at temperatures above 460°C to give sulfur dioxide and other products [104b].

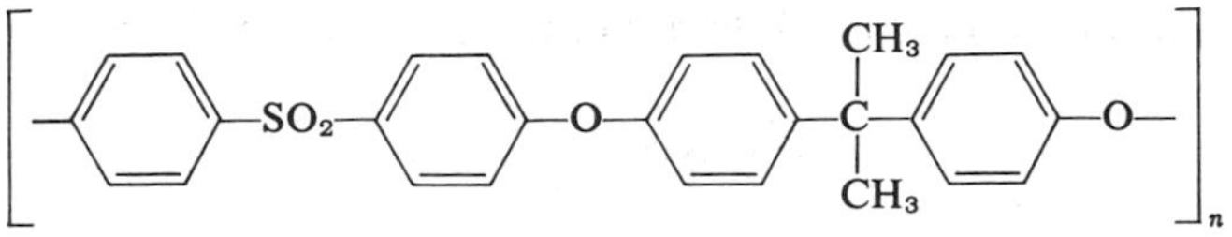

7-1. Preparation of an Aryl Sulfone Ether Polymer [104c]

To a 1 liter resin kettle equipped with a mechanical stirrer, dropping funnel, Dean and Stark trap, condenser, and inert gas (N_2) sparge tube is added 51.35 gm (0.225 mole) of 2,2-bis(4-hydroxyphenyl)propane (bisphenol A), 115 gm of dimethyl sulfoxide (DMSO), and 330 gm of chlorobenzene. The reaction mixture is heated to 60°–80°C, flushed with nitrogen, and exactly 35.86 gm (0.450 mole) of 50.2% solution of aqueous sodium hydroxide is

$$NaO-C_6H_4-C(CH_3)_2-C_6H_4-ONa + Cl-C_6H_4-SO_2-C_6H_4-Cl \xrightarrow{-2NaCl}$$

$$\left[-C_6H_4-C(CH_3)_2-C_6H_4-O-C_6H_4-SO_2-C_6H_4-O-\right]_n \qquad (33)$$

added with stirring over a 10 min period. The reaction mixture is refluxed and the water of reaction collected, and this is continued until the temperature of the reaction mixture reaches 155°–160°C. To the clear solution is added 64.61 gm (0.225 mole) of a 50% solution of 4,4′-dichlorophenyl sulfone in dry chlorobenzene at 110°C over a 10 min period. The excess solvent is allowed to distill at such a rate as to hold the reaction temperature at 160°C. The temperature should not be allowed to be lowered below 150°C, otherwise sodium-terminated polymer will precipitate on the reactor walls. Furthermore, the temperature should not be allowed to go above 160°C, otherwise discoloration or gelation may occur. Polymerization is continued for about 1 hr at 160°C until the reduced viscosity in chlorobenzene (0.2 gm/100 ml) at 25°C is about 1.0 or higher.

The polymerization is terminated by adding methyl chloride gas and a color change is usually complete in about 5 min. The reaction mixture is diluted with about 700 gm of chlorobenzene, filtered to remove sodium chloride, and the filtrate added to cold ethanol to precipitate the polymer. The polymer is dried in a vacuum oven at 135°C to give 74.2 gm (90%), softening point approx. 200.

An alternate procedure has been described in ref. [104a].

Other aryl sulfone ether polymers are available by polymerization of the monosodium salt of 4-chloro-4′-hydroxydiphenyl sulfone [105].

$$Cl-C_6H_4-SO_2-C_6H_4-ONa$$

Several other bisphenol A-type polysulfones can be prepared by reaction of 4,4′-dichlorodiphenyl sulfone with the sodium or potassium salts, as shown in Table XIII [104c,105a].

Miscellaneous Polyaryl Sulfone Preparations

1. Preparation of sodium diphenyl ether 4-sulfonate and heating with polyphosphoric acid to give a linear polysulfone [105b].

TABLE XIII
EFFECT OF T_g OF POLYSULFONES ON THE STRUCTURE OF THE STARTING DIHYDROXY AROMATIC COMPOUND

Dihydroxy aromatic starting material	Polymer T_g (°C)	Ref.
Hydroquinone	210	105a
4,4′-Dihydroxydiphenyl	230	105a
4,4′-Dihydroxydiphenyl sulfone	245	105a
Bis(4-hydroxyphenyl)-1,1-cyclohexane	205	105a
4,4′-Dihydroxydiphenyl ether	210	104c
4,4-Dihydroxydiphenylmethane	180	105a
2,2′-Dichlorobisphenol A	205	104c
2,6,2′,6′-Tetramethylbisphenol A	235	104c

2. Polysulfone by heating 4,4′-oxydibenzenesulfonic acid or 4,4′-thiodibenzenesulfonic acid with $(C_6H_5)_2O$, $(C_6H_5)_2S$, or $(C_6H_5)_2Si(Me)_2$ under a variety of conditions [106].

3. Polysulfone by the Lewis acid-catalyzed reaction of diphenyl ether with phenyl ether *m,m′*-bis(sulfonyl chloride) by $FeCl_3$ in nitrobenzene [107].

4. Autocatalytic free-radical reaction of SO_2 with bicyclo[2.2.1]hept-2-ene to give a polysulfone [108].

5. Heat-resistant polysulfone by polymerizing bis(4-mercaptophenyl) ether disodium salt and bisphenol A with bis(4-chlorophenyl) sulfone [109].

6. Preparation of new heat-resistant polyaryl sulfone using the sodium salt α,α'-bis(4-hydroxyphenyl)*p*-diisopropylbenzene with bis(4-chlorophenyl) sulfone [110].

8. POLYBENZOTHIAZOLES

Hergenrother, Wrasidlo, and Levine [111,111a] reported that polybenzothiazoles can be prepared by the reaction of 3,3′-dimercaptobenzidine with aromatic dicarboxylic acids and derivatives as shown in Eq. (34).

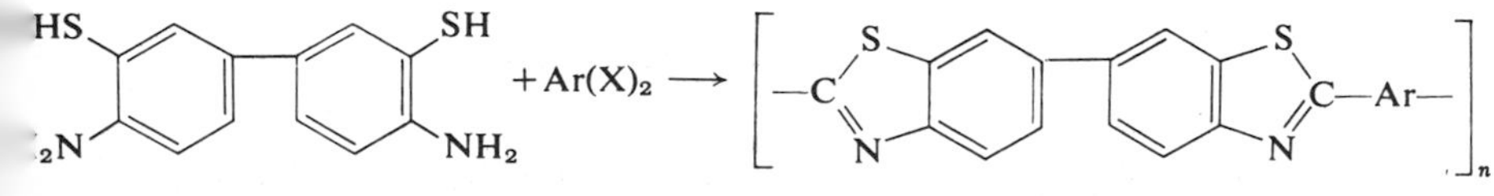

$$\text{HS, SH, } H_2N\text{, } NH_2 \text{ (3,3'-dimercaptobenzidine)} + Ar(X)_2 \longrightarrow \left[-C{\overset{S}{\underset{N}{\smash{\lessgtr}}}}\!\!\text{—(biphenylene)—}{\overset{S}{\underset{N}{\smash{\gtrless}}}}C\text{—Ar—} \right]_n \tag{34}$$

X = COOH, COOR, $CONH_2$, COCL, $\overset{NH}{\overset{\|}{C}}—NH_2$ CN, etc.

AR = aromatic rings such as benzene, pyridine, diphenyl ether, benzophenone, etc.

Using polyphosphoric acid as a condensing medium gave high molecular weight polymers. Isophthalic acid reacted with 3,3′-dimercaptobenzidine in phosphoric acid at 225°C and after $\frac{1}{2}$ hr gave a polymer with an inherent viscosity of 0.29; raising the temperature to 250°C gave a polymer of η_{inh} = 0.50 in polyphosphoric acid.

The preparation of several polybenzothiazoles is shown in Table XIV along with the experimental details and properties of the polymers. Table XIV also shows that the polymers have good thermal stability since the weight losses are very low at temperatures up to 1100°F.

8-1. *Preparation of Polybenzothiazole by the Condensation of Isophthalic Acid and 3,3′-Dimercaptobenzidine in Polyphosphoric Acid* [111]

HS, SH, HCl·H₂N, NH₂·HCl + COOH, COOH →

(35)

To a flask containing 100 ml of polyphosphoric acid is added 32.1 gm (0.1 mole) of 3,3′-dimercaptobenzidine dihydrochloride. The reaction mixture is heated and stirred under an argon atmosphere until a clear yellow solution results. Then 16.6 gm (0.1 mole) of isophthalic acid is added and the mixture heated for 20 hr at 220°C in order to give a blue-green fluorescent solution. The polymer is isolated by pouring the reaction mixture into a blender containing hot water. The polymer is washed several times with sodium carbonate solution and distilled and then dried to afford 30.8 gm (90%) of a yellow polymer, η_{inh} = 0.33 (0.5% in H_2SO_4 at 25°C).

9. POLYBENZOXAZOLES

The preparation of polybenzoxazoles from aliphatic dicarboxylic acids and bis-*o*-aminophenols was reported in a patent in 1959 [112]. In 1964 the preparation of fully aromatic polybenzoxazoles of high molecular weight was reported [113]. Moyer [114] described the polymerization of 3,3′-dihydroxybenzidine with phenyl esters of phthalic, isophthalic, terephthalic, and

5-chloroisophthalic acids to afford aromatic polybenzoxazoles with good thermal stability (Eq. 36). Melt polymerization of the starting materials gave quantitative yields of polymers with inherent viscosities in the range of 0.20 to 1.04. Homopolymerization of 3-amino-4-hydroxybenzoic acid and 4-amino-3-hydroxybenzoic acid also afforded polybenzoxazoles (Eq. 37).

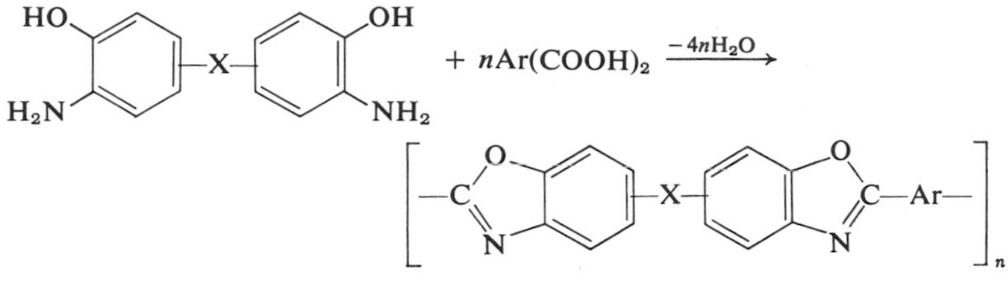

HO, H_2N —X— OH, NH_2 + nAr(COOH)$_2$ $\xrightarrow{-4nH_2O}$ [—C(O,N)—X—(O,N)C—Ar—]$_n$ (36)

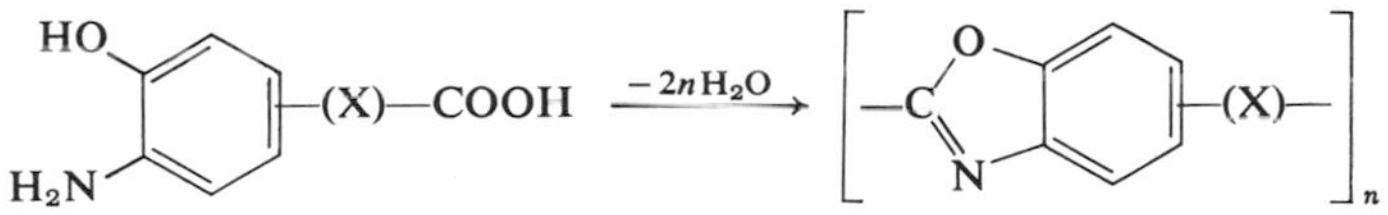

HO, H_2N —(X)—COOH $\xrightarrow{-2nH_2O}$ [—C(O,N)—(X)—]$_n$ (37)

(X = aromatic group or a single bond)

Polymerizing at reduced pressure had little effect on rate or final molecular weight. The polymerization, for the most part, proceeded in the solid state.

9-1. Preparation of Poly-2,5-benzoxazole [114,115]

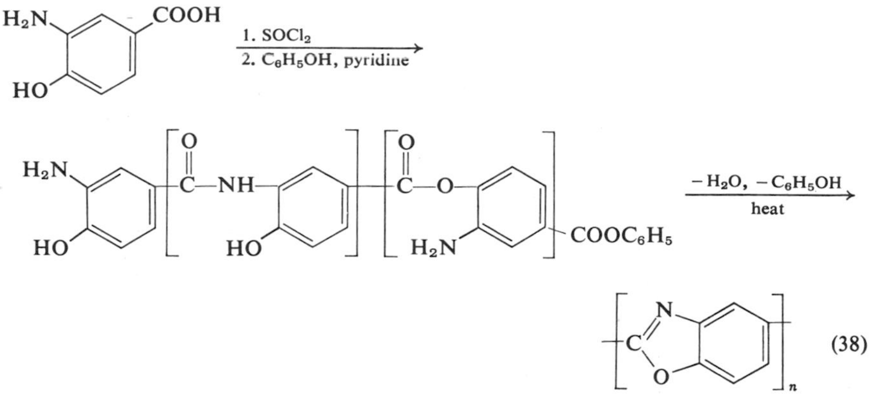

H_2N, HO, COOH $\xrightarrow[2.\ C_6H_5OH,\ pyridine]{1.\ SOCl_2}$

H_2N, HO —[C(=O)—NH— HO]—[C(=O)—O— H_2N]—$COOC_6H_5$ $\xrightarrow[heat]{-H_2O,\ -C_6H_5OH}$

[—C(N,O)—]$_n$ (38)

To a flask is added 5.0 gm (0.033 mole) of 3-amino-4-hydroxybenzoic acid and 25 ml (0.033 mole) of thionyl chloride. The mixture is heated together for 3 hr while HCl and SO_2 are evolved. The excess thionyl chloride is removed under reduced pressure and the yellow residue dispersed in 25 ml of benzene. The dispersion is slowly added to a solution of 3.5 gm (0.037 mole) phenol and 3.0 gm (0.038 mole) of pyridine in 25 ml of benzene. The

TABLE XIV

BENZOTHIAZOLE POLYMERS FROM 3,3′-DIMERCAPTOBENZIDINE[a]

Carboxylic reactant	Reaction conditions			Postreaction polymerization conditions	Color	Maximum η_{inh}[b]	TGA[c]; % wt. loss up to	
	Medium	Temp. (°C)	Time (hr)				1000°F	1100°F
PhO_2C–(1,3-phenylene)–CO_2Ph	Et_2N-Ph	215	5	2 hr 400°C	Yellowish brown	0.48	2.0	4.0
NC–(1,3-phenylene)–CN	Et_2N-Ph	215	17	1 hr 400°C	Yellow	0.30	2.4	7.3
PhO_2C–(1,4-phenylene)–O–(1,4-phenylene)–CO_2Ph	Et_2N-Ph	215	25	1 hr 400°C	Green	0.50	2.1	4.8
PhO_2C–(1,4-phenylene)–O–(1,4-phenylene)–CO_2Ph PhO_2C–(3,5-pyridinediyl)–CO_2Ph	Et_2N-Ph	215	5	1 hr 400°C	Brown	Insol.	3.1	6.9

$HO_2C-C_6H_4-CO_2H$ (meta)	PPA	220	20	—	Yellow	0.33	0	6.1
$H_2NOC-C_6H_4-CONH_2$ (meta)	PPA	250	18	—	Yellow	0.49	0	1.0
$HO_2C-C_5H_3N-CO_2H$ (3,5-pyridine)	PPA	270	21	1 hr 400°C	Brown	0.41	3.2	6.4
$HO_2C-C_6H_4-O-C_6H_4-CO_2H$	PPA	270	21	—	Light green	0.62	1.8	3.6
$MeO_2C-C_6H_4-C(=O)-C_6H_4-CO_2Me$	PPA	160	2	1 hr 400°C	Orange	0.36	3.9	8.8

[a] Reprinted from P. M. Hergenrother, W. Wrasidlo, and H. H Levine, *Polym. Prepr., Amer. Chem. Soc., Div. Polym. Chem.* **5**, 153 (1964). Copyright 1964 by the American Chemical Society. Reprinted by permission of the copyright owner.

[b] 0.5% H_2SO_4 solution.

[c] $\Delta T = 3.2°C\ min^{-1}$ in air.

resulting mixture is refluxed for 1 hr, cooled, the solid filtered, washed with benzene and water, and dried to afford 3.3 gm, $\eta_{inh} = 0.075$ (0.2 gm/100 ml conc. H_2SO_4 at 25°C), m.p. >300°C. This intermediate polymer shows IR absorption bands at 5.75, 6.05, 6.22, 6.46, 6.60, and 6.90 indicative of ester, amide, and benzoxazole groups. It is converted to the final polymer by heating in an inert atmosphere at 270°C for 16 hr to give a polymer, 2.6 gm (68%), with the following properties: $\eta_{inh} = 1.04$ (0.2 gm/100 ml H_2SO_4 at 25°C).

9-2. Preparation of Poly-2,2′-(m-phenylene)-6,6′-bibenzoxazole [114,115]

HO, OH, H_2N, NH_2 + $COOC_6H_5$, $COOC_6H_5$ ⟶ [—C(O)(N)— ... —(O)(N)C— ...]$_n$ (39)

To a polymer tube with a nitrogen capillary is added a finely ground mixture of 2.16 gm (0.010 mole) of 3,3′-dihydroxybenzidine and 3.18 gm (0.010 mole) of diphenyl isophthalate. The mixture is heated at 370°C (Wood's metal) in a stream of nitrogen for 75 min and then for 70 min at 300°C. The reaction mixture melts and then resolidifies during the heating period to give a yellow resin. The phenol and water are completely removed during the reaction period. The resin is additionally heated for 4 hr at 300°–320°C and then cooled to afford 3.10 gm (100%) of the yellow polymer, $\eta_{inh} = 0.43$ (0.4 gm/100 ml H_2SO_4).

Recently polybenzoxazoles have also been prepared by the homocondensation of 3-amino-4-hydroxybenzonitrile and the condensation of 3,3′-dihydroxybenzidine with isophthalamide [116]. These results tend to indicate that the other functional groups listed in Eq. (40) should be effective in under-

HO, OH, H_2N, NH_2 + $Ar(X)_2$ ⟶ [—C(O)(N)— ... —(O)(N)C—Ar—]$_n$ (40)

X = COOH, COOR, $CONH_2$, $C(=NH)—NH_2$, COCl, CN, etc.

going condensation with 3,3-dihydroxybenzidine (or related structures) to give polybenzoxazoles.

10. MISCELLANEOUS HIGH-TEMPERATURE POLYMERS

1. Preparation of aromatic polyamide by the solution polycondensation of *m*-phenylenediamine and isophthaloyl chloride [117].
2. Preparation of polyphenyls and polyphenylenes [118].
3. Preparation of polyazoles [4].
4. Preparation of poly-*p*-xylylenes [119–123].
5. Preparation of polytetraazopyrenes [124].
6. Preparation of aromatic polysiloxanes [125].
7. Preparation of polyacrylonitrile [126].
8. Preparation of polyaromatic heterocyclic polymers [124].
9. Preparation of polybenzyls and polyphenethyls [127].
10. Novel polymers via the photocondensation method [128].
11. Diels-Alder step-growth polymerization of bis cyclopentadienones with bis acetylenes to give polyphenylenes [6a].
12. Preparation of poly(phenyl-*cis*-triazines) [7,129].
13. Preparation of polybithiazoles [130].
14. Preparation of aromatic polyamide (Nomex) [117].
15. Preparation of "step-ladder" polymers [131].
16. Preparation of polymeric azomethines [132].
17. Preparation of poly-α,α-2,3,5,6-hexafluoro-*p*-xylene [133].
18. Preparation of polyimidazopyrrolones [134].
19. Preparation of polyimidazopyrrolones and related polymers [135].
20. Ablation resistance of some thermally stable polymers [136].
21. Synthesis of polyimides by the photoaddition of bismaleimidestobenzene [137].
22. Polymerization of diphenylamine with Friedel-Crafts oxidation catalysts to afford thermally stable polymers [138].
23. Preparation of amide-quinoxaline copolymers [139].
24. Fibers from aromatic copolyamides of limited order [140].
25. Aromatic copolyamides containing pendant carboxyl groups [141].
26. Preparation of polyamide-imide fibers [142].
27. Preparation of polybenzoxazole fibers [143].
28. Preparation of azo polymers [144].
29. Preparation of poly(phenylene-1,3,4-oxadiazoylbenzoxazole) fibers [145].
30. Preparation of benzheterocycle-imide and amide-imide fibers [146].
31. Fluorine containing polyimides [147].
32. Thermally stable pyrrolidone polyamides [148].

REFERENCES

1. R. Phillips and W. W. Wright, *J. Polym. Sci., Part B* **2**, 47 (1964).
1a. C. E. Sroog, A. L. Endrey, S. V. Abramo, C. E. Berr, W. M. Edwards, and K. L. Olivier, *J. Polym. Sci., Part A* **3**, 1373 (1965).
2. K. C. Brinker and I. M. Robinson, U.S. Patent 2,895,984 (1959).
2a. N. M. Koton, *Russ. Chem. Rev.* **31**, 81 (1962).
2b. H. Vogel and C. S. Marvel, *J. Polym. Sci.* **50**, 511 (1961).
2c. H. Vogel and C. S. Marvel, *J. Polym. Sci., Part A* **1**, 1531 (1963).
3. P. M. Hergenrother, *Macromolecules* **3**, 10 (1970).
4. V. V. Korshak and M. M. Teplyakov, *J. Macromol. Sci., Rev. Macromol. Chem.* **5**, 409 (1971).
5. G. F. D'Aleilo and R. E. Schoenig, *J. Macromol. Sci., Rev. Macromol. Chem.* **3**, 108 (1969).
6. C. S. Marvel and G. E. Hartzell, *J. Amer. Chem. Soc.* **81**, 488 (1959); C. S. Marvel and N. Tarkoy, *ibid.* **79**, 6000 (1957); J. G. Speight, P. Kovacic, and F. W. Koch, *J. Macromol. Sci., Rev. Macromol. Chem.* **5**, 295 (1971).
6a. J. K. Stille and G. K. Noren, *Macromolecules* **5**, 49 (1972).
7. W. Wrasidlo and P. M. Hergenrother, *Macromolecules* **3**, 548 (1970).
8. P. M. Hergenrother, *J. Macromol. Sci., Rev. Macromol. Chem.* **6**, 1 (1971).
8a. W. Wrasidlo and J. M. Augl, *Macromolecules* **3**, 544 (1970).
8b. P. M. Hergenrother and D. E. Kiyohara, *Macromolecules* **3**, 387 (1970).
9. A. S. Hay, H. S. Blanchard, G. F. Endres, and J. W. Eustance, *J. Amer. Chem. Soc.* **81**, 6335 (1959).
9a. G. D. Cooper, H. S. Blanchard, G. F. Endres, and H. L. Finkbeiner, *J. Amer. Chem. Soc.* **87**, 3996 (1965).
9b. J. G. Bennett, Jr. and G. D. Cooper, *Macromolecules* **2**, 101 (1969).
9c. G. D. Staffin and C. C. Price, *J. Amer. Chem. Soc.* **82**, 3632 (1960).
9d. G. D. Cooper and J. G. Bennett, Jr., *J. Org. Chem.* **37**, 441 (1972).
10. *Chem. & Eng. News* **43**, 48 (1965); Union Carbide, Netherlands Patent Appl. 6,604,731 (1966).
11. C. L. Segal, ed., "High Temperature Polymers." Dekker, New York, 1967; J. E. Mulvaney, *Encycl. Polym. Sci. Technol.* **7**, 478 (1967).
12. J. Preston, *Chem. Technol.* p. 664 (1971).
13. C. E. Searle, *Chem. & Ind.* (*London*) p. 111 (1972); "The Carcinogenic Substances Regulations," No. 879. HM Stationery Office, London, 1967.
14. T. M. Bogert and R. R. Renshaw, *J. Amer. Chem. Soc.* **30**, 1140 (1908).
15. E. I. du Pont de Nemours & Co., British Patent 570,858 (1945).
15a. W. M. Edwards and I. M. Robinson, U.S. Patent 2,710,853 (1955).
15b. W. F. Gresham and M. A. Naylor, U.S. Patent 2,731,447 (1956).
15c. W. M. Edwards and I. M. Robinson, U.S. Patent 2,900,369 (1959).
16. E. I. du Pont de Nemours & Co., French Patent 1,239,491 (1960); Australian Patent Appl. 58,424 (1960); J. I. Jones, F. W. Ochynski, and F. A. Rackley, *Chem. & Ind.* (*London*) p. 1686 (1962); C. E. Sroog, S. V. Abramo, C. E. Berr, W. M. Edwards, A. L. Endrey, and K. L. Olivier, *Polym. Prepr., Amer. Chem. Soc., Div. Polym. Chem.* **5**, 132 (1964) (Paper presented at the American Chemical Society Meeting, Philadelphia, Pa., April 1964).
17. I. Scheer, *J. Chem. Educ.* **35**, 281 (1958) [available at Kontes Glass Co., Vineland, New Jersey (Stock No. K15550)].

18. Trademark, E. I. du Pont de Nemours & Co.; L. E. Androski, *Ind. Eng. Chem., Prod. Res. Develop.* **2**, 189 (1963).
19. J. Preston and W. B. Black, *J. Appl. Polym. Sci.* **9**, 107 (1969); J. Preston, W. F. DeWinter, W. B. Black, and W. L. Hofferbert, Jr., *J. Polym. Sci., Part A-1* **7**, 3027 (1969).
20. A. H. Markhart and R. E. Kass, U.S. Patent 3,190,856 (1965).
21. D. J. Parish, U.S. Patent 3,492,270 (1970).
22. J. R. Chalmers and C. Victorius, U.S. Patent 3,416,994 (1968).
23. E. F. Hoegger, U.S. Patent 3,436,372 (1969).
24. G. M. Bower, J. H. Freeman, E. J. Traynor, L. W. Frost, H. A. Burgman, and C. R. Ruffing, *J. Polym. Sci., Part A-1* **6**, 877 (1968).
25. J. Preston, "Kirk-Othmer Encyclopedia of Chemical Technology," 2nd ed., The Interscience Encyclopedia, Inc. New York, Suppl., pp. 746–773, 1971; C. E. Sroog, *Encycl. Polym. Sci. Technol.* **11**, 247 (1969).
26. E. I. du Pont de Nemours & Co., Inc., British Patent 903,271 (1962).
27. J. Preston and W. B. Black, *J. Polym. Sci., Part B* **3**, 845 (1965); *J. Polym. Sci., Part A-1* **5**, 2429 (1967); J. Preston, W. F. DeWinter, and W. B. Black, *ibid.* **7**, 283 (1969).
28. S. R. Sandler, F. R. Berg, and G. Kitazawa, U.S. Patent 3,531,436 (1970); C. W. Stevens, U.S. Patent 3,049,518 (1962); L. W. Frost and G. M. Bower, U.S. Patent 3,179,635 (1965); E. Lavin, A. H. Markhart, and J. O. Santer, U.S. Patent 3,260,691 (1966); M. Suzuki, M. Waki, E. Hosokawa, and M. Fukushima, German Patent 2,029,078 (1971); W. M. Alvino and L. W. Frost, *J. Polym. Sci., Part A-1* **9**, 2208 (1971).
29. G. M. Bower and L. W. Frost, *J. Polym. Sci., Part A* **1**, 3135 (1963).
30. S. S. Hirsch, *J. Polym. Sci., Part A-1* **7**, 15 (1969).
31. W. M. Edwards and I. M. Robinson, U.S. Patent 2,880,230 (1959).
32. F. Grundschafer and J. Sambeth, U.S. Patent 3,380,964 (1968); N. E. Searle, U.S. Patent 2,444,536 (1948).
33. J. O. Salyer and O. Glasgow, U.S. Patent 3,481,813 (1969).
34. P. Kovacic and R. W. Hein, *J. Amer. Chem. Soc.* **81**, 1187 (1959); Rhome-Poulenc S.A., French Patent 1,555,564 (1969); R. C. P. Cubbon, *Polymer* **6**, 419 (1965).
35. J. K. Stille and L. Plummer, *J. Org. Chem.* **26**, 4026 (1961); J. K. Stille and T. Anyos, *J. Polym. Sci., Part A* **3**, 2397 (1965); E. A. Kraiman, U.S. Patents 3,074,915 (1963); 3,890,206 (1959).
36. W. H. Schulle, R. V. Lawrence, and B. M. Culbertson, *J. Polym. Sci., Part A-1* **5**, 2204 (1967).
37. J. K. Stille and R. A. Morgan, *J. Polym. Sci., Part A* **3**, 2397 (1965).
38. R. A. Dine-Hart, *J. Polym. Sci., Part A-1* **6**, 2755 (1968).
39. Y. Iwakura and F. Hayano, *J. Polym. Sci. Part A-1* **7**, 597 (1969).
40. Y. Iwakura, K. Kurita, and F. Hayano, *J. Polym. Sci. Part A-1* **7**, 609 (1969).
41. M. Kwihara, A. Kobayashi, and N. Yoda, Japanese Patent 71/24,250 (1971).
42. M. Russo and L. Mortillaro, *J. Polym. Sci. Part A-1* **7**, 3337 (1969).
43. S. Nishizaki, *Kogyo Kagaku Zasshi* **68**, 574 (1965).
44. S. Nishizaki and F. Fukami, *Kogyo Kagaku Zasshi* **68**, 383 (1965).
45. J. Strickrodt and U. Konig, U.S. Patent 3,607,838 (1971).
46. L. A. Laius, M. I. Bessonov, and F. S. Florinskii, *Vysokomol. Soedin., Ser. A* **13**, 2006 (1971).
47. V. V. Korshak, G. M. Tseitlin, and V. I. Azarov, *Vysokomol. Soedin., Ser. B* **13**, 603 (1971).

48. Author (S.R.S.) Laboratory.
49. W. Wrasidlo and H. H. Levine, *J. Polym. Sci., Part A-2* **5**, 4795 (1967).
50. D. N. Gray and G. P. Shulman, *Polym. Prepr., Amer. Chem. Sci., Div. Polym. Chem.* **6**, 773 (1965); V. V. Korshak, T. M. Frunze, V. V. Kurashev, and A. A. Izyneev, *Dokl. Akad. Nauk SSSR* **149**, 104 (1963).
51. L. Plummer and C. S. Marvel, *J. Polym. Sci., Part A-2* **5**, 2559 (1964).
52. R. T. Foster and C. W. Marvel, *J. Polym. Sci., Part A-3* **3**, 417 (1965).
53. T. V. L. Narayan and C. S. Marvel, *J. Polym. Sci., Part A-3* **5**, 1113 (1967).
54. Y. Iwakura, K. Uno, and Y. Imai, *J. Polym. Sci., Part A-3* **2**, 2605 (1964).
55. D. W. Hein, R. J. Alheim, and J. J. Leavitt, *J. Amer. Chem. Soc.* **79**, 427 (1957).
56. S. R. Sandler, U.S. Patent 3,679,700 (1972); *Pap., 164th Nat. Meet., Amer. Chem. Soc., New York*, Organic Section Paper No. 168 (1972).
57. S. Moon, A. L. Schwartz, and J. K. Hecht, *J. Polym. Sci., Part A-1* **8**, 3665 (1970).
58. M. Soga, M. Hochchama, and T. Shono, *J. Polym. Sci., Part A-1* **8**, 2265 (1970).
59. K. Mitsuhashi and C. S. Marvel, *J. Polym. Sci., Part A-3* **3**, 1661 (1965).
60. D. N. Gray, L. L. Ronch, and E. L. Strauss, *Polym. Prepr., Amer. Chem. Soc., Div. Polym. Chem.* **8**, 1138 (1967).
61. Y. Iwakura, K. Uno, Y. Imai, and M. Fukui, *Makromol. Chem.* **77**, 41 (1964).
62. Y. Iwakura, K. Uno, and Y. Imai, *Makromol. Chem.* **77**, 33 (1964).
63. W. R. Sorenson, *Polym. Prepr., Amer. Chem. Soc., Div. Polym. Chem.* **2**, 226 (1961).
64. V. Körner, *Ber. Deut. Chem. Ges.* **17**, 573 (1884).
65. O. Hinsberg, *Ber. Deut. Chem. Ges.* **17**, 318 (1884).
66. J. K. Stille and J. R. Williamson, *J. Polym. Sci., Part A* **2**, 3867 (1964).
66a. J. K. Stille and J. R. Williamson, *J. Polym. Sci., Part B* **2**, 209 (1964).
67. J. K. Stille, J. R. Williamson, and F. E. Arnold, *J. Polym. Sci., Part A* **3**, 1013 (1965); J. K. Stille and F. E. Arnold, *J. Polym. Sci., Part A-1* **4**, 551 (1966).
68. G. P. de Gaudemaris and B. J. Sillion, *J. Polym. Sci., Part B* **2**, 203 (1964); G. P. de Gaudemaris, B. J. Sillion, and J. Preve, *Bull. Soc. Chim. Fr.* p. 1763 (1964).
69. J. K. Stille and E. Mainea, *J. Polym. Sci., Part B* **4**, 39 (1966); *J. Polym. Sci.* **5**, 665 (1967); H. Jadamus, F. DeSchryver, W. F. DeWinter, and C. S. Marvel, *J. Polym. Sci., Part A-1* **4**, 2831 (1966); F. DeSchryver and C. S. Marvel, *ibi:.* **5**, 545 (1967); J. K. Stille and M. E. Freeburger, *J. Polym. Sci., Part B* **5**, 989 (1967).
70. P. M. Hergenrother and H. H. Levine, *J. Polym. Sci., Part A-1* **5**, 1453 (1967).
70a. P. M. Hergenrother, *J. Polym. Sci., Part A* **6**, 3170 (1968); W. Wrasidlo and J. M. Augl, *J. Polym. Sci., Part B* **7**, 281 (1969).
70b. W. Wrasidlo and J. M. Augl, *J. Polym. Sci., Part A* **7**, 3393 (1969).
70c. W. Wrasidlo, *J. Polym. Sci., Part A-1* **8**, 1107 (1970).
71. W. Wrasidlo and M. J. Augl, *Polym. Prepr., Amer. Chem. Soc., Polym. Div.* **10**, 1353 (1969).
72. G. L. Hagananer and G. Mulligan, *Polym. Prepr., Amer. Chem. Soc., Polym. Div.* **13**, 128 (1972).
73. J. M. Augl and J. V. Duffy, *J. Polym. Sci., Part A-1* **9**, 1343 (1971).
74. W. H. Hunter, A. Olson, and E. A. Daniels, *J. Amer. Chem. Soc.* **38**, 1761 (1916).
74a. A. Hantzsch, *Ber. Deut. Chem. Ges.* **40**, 4875 (1907).
75. H. A. Torrey and W. H. Hunter, *J. Amer. Chem. Soc.* **33**, 194 (1911).
75a. W. H. Hunter and G. H. Woollett, *J. Amer. Chem. Soc.* **43**, 131, 135, and 1761 (1921).
76. W. H. Hunter and M. A. Dahlen, *J. Amer. Chem. Soc.* **54**, 2459 (1932).
77. W. H. Hunter and F. E. Joyce, *J. Amer. Chem. Soc.* **39**, 2640 (1917).
78. W. H. Hunter and M. A. Dahlen, *J. Amer. Chem. Soc.* **54**, 2456 (1932).

79. H. S. Blanchard, H. L. Finkbeiner, and G. A. Russell, *J. Polym. Sci.* **58**, 469 (1962).
80. C. C. Price and N. S. Chu, *J. Polym. Sci.* **61**, 135 (1962).
81. M. J. S. Dewar and A. N. James, *J. Chem. Soc., London* p. 917 (1958).
82. A. S. Hay, French Patents 1,322,152 (1963); 1,384,255 (1965); Belgian Patents 635,349 and 635,350 (1964); U.S. Patents 3,262,892 (1966); 3,432,466 (1969).
83. A. S. Hay and D. M. White, *Polym. Prepr., Amer. Chem. Soc., Div. Polym. Chem.* **10**, 92 (1969).
83a. A. S. Hay, *Macromolecules* **2**, 107 (1969).
84. A. S. Hay and R. F. Clark, *Macromolecules* **3**, 533 (1970).
85. H. M. V. Dort and C. R. H. I. de Jonge, U.S. Patent 3,400,100 (1968).
86. B. O. Lindgren, *Acta Chem. Scand.* **14**, 1203 (1960).
87. G. D. Cooper and J. G. Bennett, Jr., *Polym. Prepr., Amer. Chem. Soc., Div. Polym. Chem.* **13**, 551 (1972).
88. General Electric Co., British Patent 930,993 (1963); W. A. Butte, Jr. and C. C. Price, *J. Amer. Chem. Soc.* **84**, 3567 (1962).
89. A. S. Hay, *Polym. Prepr., Amer. Chem. Soc., Div. Polym. Chem.* **2**, 319 (1961); G. F. Endres and J. Kwiatek, *J. Polym. Sci.* **58**, 593 (1962).
90. C. R. G. R. Bacon and H. A. O. Hill, *Quart. Rev., Chem. Soc.* **19**, 114 (1965); E. Ochiai, *Tetrahedron* **20**, 1831 (1964); A. I. Scott, *Quart. Rev., Chem. Soc.* **19**, 2 (1965).
91. C. C. Price and K. Nakoaka, *Macromolecules* **4**, 364 (1971).
92. R. G. Feasey, J. H. Freeman, P. C. Daffurn, and A. Turner-Jones, *Polym. Prepr., Amer. Chem. Soc., Div. Polym. Chem.* **13**, 562 (1972).
93. J. F. Harrod and B. Carr, *Polym. Prepr., Amer. Chem. Soc., Div. Polym. Chem.* **13**, 557 (1972).
94. G. P. Brown and A. Goldman, *Polym. Prepr., Amer. Chem. Soc., Div. Polym. Chem* **5**, 195 (1964).
95. S. Nakashio and T. Sanoh, Japanese Patent 71/21,741 (1971).
96. G. D. Cooper, A. R. Gilbert, and H. L. Finkbeiner, *Polym. Prepr., Amer. Chem. Soc., Div. Polym. Chem.* 7, 166 (1966); D. M. White, *ibid.*, p. 178; D. A. Bolon, *ibid.*, p. 173.
97. A. J. Chalk and A. S. Hay, *J. Polym. Sci., Part A-1* **7**, 691 (1969).
98. A. H. Frazer and F. T. Wallenberger, *J. Polym. Sci., Part A* **2**, 1147 and 1171 (1964).
99. A. H. Frazer, W. Sweeney, and F. T. Wallenberger, *J. Polym. Sci., Part A* **2**, 1157 (1964).
100. A. H. Frazer, U.S. Patent 3,238,183 (1966).
101. V. V. Rode, E. M. Bondarenko, V. V. Korshak, A. L. Rusanov, E. S. Kronganz, D. A. Bochvar, and I. V. Stankevich, *J. Polym. Sci., Part A-1* **6**, 1351 (1968).
102. M. R. Lilyquist and J. R. Holsten, *Polym. Prepr., Amer. Chem. Soc., Div. Polym. Chem.* **4**, 6 (1963).
103. J. R. Holsten and M. R. Lilyquist, *J. Polym. Sci., Part A* **3**, 3905 (1965).
104. *Chem. & Eng. News* **43**, 48–49 (1965); Union Carbide Corp., Netherlands Patent Appl. 3,408,130 (1965).
104a. Union Carbide Corp., Netherlands Patent Appl. 6,604,731 (1966).
104b. W. F. Hale, A. G. Farnham, R. N. Johnson, and R. A. Clendinning, *J. Polym. Sci., Part A-1* **5**, 2399 (1967).
104c. R. N. Johnson, A. G. Farnham, R. A. Clendinning, W. F. Hale, and C. N. Merriam, *J. Polym. Sci., Part A-1* **5**, 2375 (1967).
105. S. R. Shulze, *Abstr. Pap., 155th Meet., Amer. Chem. Soc., San Francisco* p. L-090 (1968).

105a. R. N. Johnson, *Encycl. Polym. Sci. Technol.* **11**, 447 (1969).
105b. Y. Iwakura, K. Uno, S. Hara, and T. Takiguchi, *Yuki Gosei Kagaku Kyokai Shi* **25**, 797 (1967).
106. S. M. Cohen, M. Young, and H. Raymond, Jr., U.S. Patent 3,418,277 (1968).
107. H. Vogel, U.S. Patent 3,406,149 (1968).
108. N. L. Zutty, U.S. Patent 3,313,785 (1967).
109. R. Gabler and J. Studinka, German Patent 1,909,441 (1969).
110. R. J. Cornell, *Polym. Prepr., Amer. Chem. Soc., Div. Polym. Chem.* **13**, 607 (1972).
111. P. M. Hergenrother, W. Wrasidlo, and H. H. Levine, *Polym. Prepr., Amer. Chem. Soc., Div. Polym. Chem.* **5**, 153 (1964).
111a. P. M. Hergenrother, W. Wrasidlo, and H. H. Levine, *J. Polym. Sci., Part A* **3**, 1665 (1965).
112. E. I. du Pont de Nemours & Co., British Patent 811,758 (1959).
113. T. Kubota and R. Nakanishi, *Polym. Lett.* **2**, 655 (1964).
114. W. W. Moyer, Jr., C. Cole, and T. Anyos, *J. Polym. Sci., Part A* **3**, 2107 (1965).
115. Borg-Warner Corp., French Patent 1,365,114 (1963); W. W. Moyer, Jr., U.S. Patent Appl. 214,838 (1962).
116. R. D. Stacy, N. P. Loire, and H. H. Levine, *Polym. Prepr., Amer. Chem. Soc., Div. Polym. Chem.* **7**, 161 (1966).
117. S. L. Kwolek, P. W. Morgan, and W. R. Sorenson, U.S. Patent 3,063,966 (1962).
118. J. G. Speight, P. Kovacic, and F. W. Koch, *J. Macromol. Sci., Revs. Macromol. Chem.* **5**, 295 (1971).
119. L. A. Errede and M. Szwarc, *Quart. Rev., Chem. Soc.* **12**, 301 (1959).
120. W. F. Gorham, *J. Polym. Sci.* **4**, 3027 (1966).
121. *Chem. & Eng. News* **43**, 48–49 (1965).
122. F. W. Ortunng, Jr., R. W. Kluiber, H. G. Gilch, and W. L. Wheelwright, U.S. Patent 3,240,722 (1966).
123. W. F. Gorham, U.S. Patent 3,288,728 (1966).
124. C. S. Marvel, *Polym. Prepr., Amer. Chem. Soc., Div. Polym. Chem.* **5**, 167 (1964).
125. W. J. Bailey, J. Economy, and M. E. Hermes, *J. Org. Chem.* **27**, 3295 (1962).
126. W. J. Brulant and J. L. Parsons, *J. Polym. Sci.* **22**, 249 (1956); W. G. Vosburgh, *J. Text. Res.* **30**, 882 (1960).
127. J. E. Moore, *Polym. Prepr., Amer. Chem. Soc., Div. Polym. Chem.* **5**, 203 (1964).
128. J. G. Higgins and E. A. McCombs, *Chem. Technol.* p. 176 (1972).
129. P. M. Hergenrother, *J. Polym. Sci., Part A-1* **7**, 945 (1969).
130. D. T. Longone and H. H. Un, *J. Polym. Sci., Part A* **3**, 3117 (1965).
131. R. L. Van Deusen, O. R. Goins, and A. J. Sicree, *J. Polym. Sci., Part A-1* **6**, 1777 (1968).
132. G. F. D'Alelio and R. E. Schoenig, *J. Macromol. Sci., Rev. Macromol. Chem.* **3**, 105 (1969).
133. W. P. Norris, *J. Org. Chem.* **37**, 147 (1972).
134. V. L. Bell, *Encycl. Polym. Sci. Technol.* **11**, 240 (1969).
135. G. M. Bower and L. W. Frost, *J. Appl. Polym. Sci.* **16**, 345 (1972).
136. P. P. Misevichyus, A. N. Machyulis, B. I. Liogon'kii, G. M. Shamrayev, A. A. Berlin, and V. M. Grigorovskaya, *Vysokomol. Soedino, Ser. A* **12**, 2091 (1970).
137. Y. Musa and M. P. Stevens, *J. Polym. Sci.* **10**, 319 (1972).
138. B. Ellis and J. V. Stevens, *J. Polym. Sci., Part A-1* **10**, 554 (1972).
139. J. V. Duffy and J. M. Augl, *J. Polym. Sci., Part A-1* **10**, 1123 (1972).
140. J. Preston, R. W. Smith, and S. M. Sun, *Amer. Chem. Soc., Div. Org. Coatings Plast. Chem., Pap.* **32**, 1 (1972).

141. H. E. Hinderer, R. W. Smith, and J. Preston, *Amer. Chem. Soc., Div. Org. Coatings Plast. Chem., Pap.* **32**, 7 (1972).
142. R. P. Pigeon, *Amer. Chem. Soc., Div. Org. Coatings Plast. Chem., Pap.* **32**, 1 (1972).
143. D. E. Montgomery and A. Judge, *Amer. Chem. Soc., Div. Org. Coatings Plast. Chem., Pap.* **32**, 14 (1972).
144. H. C. Bach and H. E. Hinderer, *Amer. Chem. Soc., Div. Org. Coatings Plast. Chem., Pap.* **32**, 15 (1972).
145. S. C. Temin and D. A. Allen, *Amer. Chem. Soc., Div. Org. Coatings Plast. Chem., Pap.* **32**, 15 (1972).
146. J. Preston, W. F. DeWinter, and W. B. Black, *Amer. Chem. Soc., Div. Org. Coatings Plast. Chem., Pap.* **32**, 24 (1972).
147. *Chem. & Eng. News*, p. 16 (July 30, 1973).
148. Y. Avny, N. Saghian, and A. Zilkha, *Is. J. Chem* **10**, 949 (1972).

Chapter 10

POLYMERIZATION OF ACRYLATE AND METHACRYLATE ESTERS

1. INTRODUCTION

Acrylic polymers have taken on increasing importance in a wide range of applications. Methacrylate glazing, protective coatings, paint bases, binders, adhesives, plastic bottles, elastomers, floor polishes, plastic films, and leather finishes are among the areas of continued growth of the use of these materials.

Therefore the synthetic methods of polymerization—in bulk, solution, suspension, and emulsion—need to be discussed in some detail. Furthermore, the information in the present chapter is directly applicable to the addition polymerization of a wide variety of other vinyl monomers.

Methods such as the anaerobic polymerization, which appear to be confined to the methacrylate esters, are also presented.

From the point of view of the synthetic organic chemist, the suspension and emulsion techniques are perhaps the best methods of preparing a large variety of polymers and copolymers in reasonable quantity since the procedures resemble familiar laboratory syntheses in apparatus and manipulation.

2. GENERAL CONSIDERATIONS

The esters of acrylic and methacrylic acids as well as those of other α-substituted acrylic acids (e.g., α-cyanoacrylic, α-chloroacrylic, and α-ethylacrylic acids) are frequently referred to by the term "acrylic esters." This generalized nomenclature is also used in this section since the polymerization procedures described frequently apply to all acrylic esters. To be sure, simple esters of α-cyanoacrylic acid readily and characteristically undergo polymerization upon initiation by moisture and the dimethacrylate of the higher polyethylene glycols may be subjected to "anaerobic polymerization," yet the methods of free-radical addition polymerization emphasized here are usually applicable to all members of this class of compounds.

With minor modifications, the preparations outlined here may be applied to many of the other common "vinyl" monomers, such as styrene, vinyl acetate, vinyl chloride, vinylidene chloride, acrylonitrile, and acrylamide. Therefore, the general principles of the preparation of poly(acrylic esters) are discussed in some detail here.

Since a large number of acrylic esters are available, polymers with a range of properties are readily prepared by a variety of procedures. Naturally, by resorting to co- and terpolymerization techniques, the properties of the resulting polymers may be modified further so that the time-worn phrase "tailor-making" resins for specific applications does indeed have considerable validity. Unfortunately, the literature does not develop the principles necessary to obtain desired properties in a polymer in a systematic manner. Much information may be found in the patent literature. However, patents usually attempt to protect as broad a range of compositions as possible, and thereby make these guides difficult to follow.

A similar criticism may be leveled against sales information offered by suppliers of raw materials; however, where definite experience substantiates such information, we draw freely on such data.

Furthermore, the variety of properties which a research group may be called upon to build into a product is so great that it is difficult to generalize on the methods of obtaining the desired results. A very important additional guide to achieving a particular balance of properties is the instinct which the experimenter develops with experience.

Despite the foregoing statement, we shall attempt to suggest a few considerations useful in developing specific properties in poly(acrylic esters):

1. Acrylate esters, with the exception of methyl acrylate, generally produce elastomeric polymers, whereas methacrylates of corresponding alcohols are usually more brittle or more resinous in nature. As a guide, the glass-transition temperature, T_g, is the physical constant which must be considered when flexibility and other elastomeric properties are of importance. Roughly speaking, materials with a T_g below 0°C are elastomeric.

The glass-transition temperature is at a minimum in the case of poly(octyl acrylate) [for the closely related poly(2-ethylhexyl acrylate), $T_g = -85$°C] [1]. In the methacrylate series, the minimum glass-transition temperature is found to be that of poly(lauryl methacrylate), $T_g = -65$°C [1]. The "brittle points" follow the glass-transition temperatures. The minimum film formation temperature of emulsion systems also seems roughly related to the glass-transition temperature.

2. The average molecular weight, the molecular weight distribution, and the degree of chain branching affect properties. Thus, for example, one may anticipate differences in injection molding properties between a resin with a normal Gaussian distribution of the molecular weight of the polymer chains and one with the same average molecular weight, but consisting of a blend with two or more resins of different individual average molecular weights which have been compounded to give the same composite average molecular weight. Until the advent of gel permeation chromatography (GPC) the determination of the molecular weight distribution of a resin sample was difficult and tedious. It is fortunate that GPC involves equipment which has recently experienced a considerable reduction in cost and bulk.

3. The chemical nature of the monomers affects properties of their polymers in a great variety of ways. Mention has already been made of the influence of the chain length of the alcohol portion of an acrylic ester on the glass-transition temperature and the brittle point. Similarly, the percent elongation

generally increases with increasing this chain length while the tensile strength decreases. As expected, the percent elongation of polyacrylates is substantially greater than that of polymethacrylates as a class [2]. This may be related to "legginess" of polyacrylate-based adhesive solutions. On the other hand, the tensile strengths of polymethacrylates are generally higher than those of the corresponding polyacrylates [2].

Oxidation resistance and cross-linking characteristics are, naturally, related to the chemical nature of the monomers. While acrylic polymers generally exhibit excellent oxidation resistance, differences may be expected between acrylates and methacrylates.

Variation in cross-linking properties may be related to the length of the chain of a glycolic moiety separating two methacrylate units, or to whether di-, tri-, or tetramethacrylates are involved. A two-step cross-linking reaction is conceivable with monomers such as allyl methacrylate; cross-linking with external reagents such as formaldehyde is possible with monomers such as acrylamide and methacrylamide. Polymers containing acrylic acid may be cross-linked under certain circumstances with magnesium oxide, zinc oxide, aluminum oxide, or a variety of other polyvalent ions.

Monomers containing other functional groups, such as 2-hydroxyethyl methacrylate, 2-dimethylaminoethyl acrylate, acrylic acid, methacrylic acid, and a host of others contribute a variety of properties to polymers containing them. For example, solubility in or sensitivity to alkaline solutions are conferred on a polymer by the presence of the polymerizable acids. On the other hand, monomers containing amino groups lead to sensitivity to acids. Both amine- and carboxylic acid-containing monomers may enhance the adhesive character of a polymer. Antistatic properties and dye-acceptability may also be conferred by these functional monomers [3].

The presence of halogen in a monomer naturally affects the refractive index of the monomer as well as the refractive index of the corresponding polymers. Halogen may also improve the flame retardance of a polymer. Both water and hydrocarbon solvent resistance may be observed with polymers of monomers such as heptafluorobutyl acrylate.

The freeze–thaw resistance and mechanical stability of latex polymers is frequently enhanced by incorporating small quantities of acrylic or methacrylic acids in a copolymer.

The foregoing is merely a brief indication of the range of properties desired from time to time in a resin and the effect of the chemical nature of the monomer on these properties.

4. Chemical modifications of polymers influence properties. Here again, such a host of possibilities exists that only a few notions can be mentioned. As a field of investigation, the chemical modification of polymers is still in its infancy.

Some cross-linking of polymer systems by external reagents has already been mentioned above. Polymers derived from α-hydroxyethyl acrylic esters may be cross-linked with a variety of polyfunctional reagents which react with alcohols such as diisocyanates. Glycidyl acrylic ester polymers may react under conditions similar to those used with epoxy resins.

Among the problems encountered in subjecting a polymer to chemical reactions is the fact that not all functional groups on the polymer chain are accessible to attack by the coreagent. Thus, in the hydrolysis of poly(vinyl) acetate, to cite a well known case in point, a variety of conditions have been developed to perfect controlled partial conversion to "polyvinyl alcohols" of various degrees of hydrolysis. Since chain scission may also take place during the hydrolysis, a large variety of polyvinyl alcohols have been produced. Many of these have found specific applications in industry. One may anticipate similar phenomena where other polymers are subjected to classical chemical reactions.

5. While an extensive discussion of co- and terpolymerization is beyond the scope of this chapter, it should be mentioned here that properties of polymers may be modified by the judicious copolymerization of several monomers. The sequence of monomers influences the properties of the polymer. Control of this arrangement of monomer units depends, in part, on the copolymerization ratio of pairs of monomer. To some extent the constraints placed on the monomer sequence may be overcome by a variety of techniques. By special techniques, "block" and "graft" copolymers may be produced.

6. Stereospecific polymers may be produced under certain conditions. Such products may exhibit significantly different properties from their amorphous atactic counterparts.

7. Blends of different polymers may be produced.

8. A host of reinforcing fillers, plasticizers, surfactants, solvents, modifiers, foaming agents, etc., may be compounded with a parent polymer to modify and control desired properties.

9. The method of polymerization also is of importance in tailor-making a product. For example, if the end product is to be used as a coating material, a latex may be desirable. Therefore an emulsion polymerization technique will be selected. On the other hand, poly(methyl methacrylate) glazing may have to be prepared by bulk polymerization in a suitable mold.

The polymerization techniques will affect the molecular weight, the molecular weight distribution, and the degree of branching of the polymer. In emulsion polymerizations, particle-size distribution, latex stability, etc., are controlled by the method of polymerization.

From the standpoint of industrial production of a polymer, the techniques employed become extremely important. For example, on a laboratory scale, bulk polymerizations may be relatively easily carried out. However, in an industrial reactor, heat transfer and the method of removing the product from the reactor can present insurmountable problems. These considerations are superimposed on those normally encountered in the scale-up of chemical production. The average molecular weight of a polymer produced on a large scale may be substantially different from that formed in the laboratory.

3. REACTANTS AND REACTION CONDITIONS

A. Inhibitors

As normally supplied, acrylic esters are inhibited to enhance their storage stability. The most common inhibitors used with these monomers are hydroquinone (HQ) and *p*-methoxyphenol (MEHQ). The concentration level of these inhibitors is usually in the range of 50–100 parts per million by weight.

Other phenolic compounds are also used as inhibitors from time to time, and other types of inhibitors, e.g., copper, copper oxide, methylene blue, phenothiazine, or *p*-phenylenediamine, may occasionally be used as inhibitors of specialty acrylic monomers.

Mass-produced acrylic esters are also usually aerated since dissolved oxygen is said to help reduce the possibility of "popcorn" polymer formation [2].

Unlike compounds such as benzoquinone, the phenolic compounds are not considered true inhibitors of polymerization. They normally act to retard the onset of polymerization. For this reason, the removal of hydroquinone from an acrylic ester is not always necessary. A modest increase in the amount of the initiator used in the polymerization frequently overcomes the retardation by the inhibitor. However, this practice may lead to the formation of a slightly discolored polymer.

3-1. General Procedure for Removal of Phenolic Inhibitors

The removal of phenolic inhibitors such as hydroquinone is usually readily accomplished by alkaline extraction. Since most monomers are normally inhibited with very small amounts of inhibitors, three extractions with convenient volumes of 0.5% aqueous sodium hydroxide is satisfactory. In the case of hydroquinone, the progress of the procedure is readily followed since the solution of the sodium salt of hydroquinone is intensely colored. Therefore, caustic extractions are simply continued until the aqueous layer is colorless. The monomer is washed with water until the water layer has the pH normal for the water. Finally, the monomer is dried over magnesium sulfate and filtered. The inhibitor-free acrylic esters should be used reasonably promptly.

In the case of MEHQ removal, color changes cannot be used as a guide to the course of the procedure. Evidently the sodium salt of MEHQ gives colorless solutions. It is therefore the practice to remove this inhibitor simply by three or four extractions with 0.5% aqueous sodium hydroxide followed by water washes and drying. Higher concentrations of sodium hydroxide are not recommended for the extraction of phenolic compounds, particularly in the case of MEHQ removal.

From time to time, the caustic treatment of acrylic esters leads to emulsion formation. These emulsions may frequently be broken by the addition of a substantial quantity of a saturated aqueous solution of sodium sulfate.

Obviously when the monomer under consideration is water-soluble (e.g., α-hydroxyethyl acrylate), the inhibitor cannot be removed by treatment with an aqueous alkaline solution. Passage of the monomer through a column of a basic ion-exchange resin or through moderately coarse activated alumina may be helpful in removing the inhibitor.

The separation of hydroquinone or *p*-methoxyphenol from the acrylic monomer by distillation is difficult since the inhibitors seem to codistill with the monomers. A further problem, of course, is the one associated with handling warm monomer vapors freed of inhibitor in a distillation column and in the distillation head. This does not mean that distillation is beyond the realm of possibility; however, careful techniques are required.

The removal of dissolved oxygen from monomers is of particular importance in careful theoretical studies. As a matter of fact, it is very possible that many of the contradictory and irreproducible observations in polymer kinetics may be attributed to variations in the quantity of residual oxygen in the monomers under study. The method of removing dissolved oxygen is not particularly difficult, although it is troublesome.

3-2. Deaeration of Acrylic Monomers

The uninhibited acrylic monomer is placed in a heavy-walled vessel fitted with a suitable stopcock which in turn is attached to a high-vacuum train. The stopcock is closed and the monomer is frozen in a Dry Ice–acetone bath. Then the stopcock is opened to the high-vacuum system and the vessel evacuated with an oil-diffusion pump while maintaining the monomer in a frozen condition. When the high-vacuum gauge indicates a constant low pressure in the system, the stopcock is closed and the monomer is allowed cautiously to thaw. After reaching room temperature and when ebullution has subsided, the monomer is refrozen. Then the stopcock is opened to the vacuum system again and the space above the solid monomer is again evacuated. This procedure is repeated at least three times. After finally returning the monomer to room temperature, it is transferred to its reaction apparatus under conditions which will no longer expose it to oxygen. It is best to protect the monomer from excessive exposure to daylight throughout this procedure.

B. Initiators

In the polymerization of acrylic monomers, diacyl peroxides and azo compounds are most commonly used in bulk, suspension, and organic solution processes. Combinations of oxidizing and reducing agents such as hydroperoxides and the formaldehyde-bisulfite addition product, or persulfates and bisulfites find application in emulsion and aqueous solution polymerization. Hydroperoxides are also of importance in "anaerobic" polymerizations, while acyl peroxides in conjunction with tertiary aromatic amines are used in "self-curing" monomer–polymer doughs. Photopolymerizations frequently utilize ultraviolet absorbers, at times in the presence of peroxides.

The purification of acyl peroxides is discussed in Chapter 14 of this volume. One point which should be emphasized is that when the methanol–chloroform recrystallization method is used, care must be taken to free the acyl peroxide of residual traces of chloroform. Chloroform is a very active chain-transfer agent in acrylic polymerizations and it may accelerate the process by a kinetic chain-branching reaction.

The selection of an initiator is to a large extent an empirical matter. With the exception of emulsion polymerization procedures, acrylic ester polymerizations normally require initiators which are soluble in the monomer. Particularly in suspension polymerizations, the initiator must have a very low solubility in water to reduce the possibility of latex formation. On the other hand, in emulsion polymerizations classically the initiating system had to be low in monomer solubility but high in water solubility (e.g., the sodium metabisulfite–ammonium persulfate redox system). More recently, quite satisfactory redox initiation systems have been devised for emulsion processes in which monomer-soluble hydroperoxides are incorporated in the monomer while a water-soluble reducing agent is used. Hydrogen peroxide, which has a surprisingly high solubility in certain monomers, has also been used in conjunction with water-soluble reducing agents to produce latexes.

The degree of polymerization (i.e., the average molecular weight) is dependent on both the effective initiator concentration and the temperature. Plant or processing considerations may impose restrictions on the operating temperatures. Consequently the half-life of the initiator becomes an important factor. These data must be used with some caution since there frequently are significant solvent effects on the thermal decomposition of peroxides. The peroxides which have significantly short half-lives below approximately 50°C may be classified as thermally unstable and, with the possible exception of diisopropyl percarbonate, are of limited practical consequence. If polymerizations are to be carried out at temperatures below 60°C at a reasonable rate, the usual practice is to use "accelerators" (e.g., tertiary amines, possibly along with cobalt salts or one of a variety of metal "dryers") in conjunction with organic peroxides [4]. Such accelerated systems are capable of initiating polymerizations below 0°C. The components of such systems must never be premixed because of explosive hazards.

2,2′-Azobis(isobutyronitrile) (AIBN) and, potentially, other azo compounds may be used at relatively low temperatures (60°–80°C). AIBN has the added advantage that it is not an oxidizing agent and therefore is useful in the production of colorless polymers. Since its thermal decomposition is relatively simple, polymers formed from its free radicals are thought to be free of branched chains.

The problems involved in the selection of initiators has been reviewed by Walling [5]. The factors of importance include the following:

1. The initiator must produce free radicals at a reasonably constant rate during the polymerization process. If necessary, initiators of low thermal stability may have to be added cautiously during the process.
2. The active radicals have to be "available" to initiate polymerization. The homolytic decomposition of initiators to pairs of free radicals is such that a portion of these products may recombine before they can react with monomer molecules ("cage effect"). Thus not all radicals are available for initiation. If the decomposition of the initiator involves the breaking of further bonds, as in the case of azo compounds, the possibility of the cage effect is reduced.
3. Initiators must be stable toward induced decomposition in the media involved since, in effect, induced decompositions consume initiator molecules without generating new free radicals. Since, however, many monomers react more rapidly with free radicals than the free radicals act on the initiator molecule, this factor is usually minor. The attack of growing polymer chains on initiators is a chain-transfer process which reduces the concentration of available initiator and tends toward the formation of polymers of low average molecular weight. This effect is said to be particularly noticeable with hydroperoxides.
4. Free radicals should initiate chain polymerization efficiently. Some peroxides, e.g., dialkyl peroxides and peresters, tend to abstract hydrogen more readily than they react with monomers. Consequently their efficiency as initiators is reduced.

The rate of polymerization of monomers is generally proportional to the square root of the initiator concentration over a wide range of concentrations [6]. Since the molecular weight distribution of a polymer is related to the reciprocal of the rate of polymerization [7], lower levels of initiators tend to produce polymers of higher molecular weight. In view of the square-root relationship between the rate of polymerization and the initiator concentration, the effect on the molecular weight is observable only when substantial changes in the amount of initiator concentration are made. Since high concentrations of such compounds as dibenzoyl peroxide in a monomer system may represent a serious safety hazard, due precautions must be taken.

The *initial* polymerization rate is also proportional to the square root of the initiator concentration. Furthermore, these rates at a given temperature and concentration are also proportional to the square root of the thermal decomposition rates of the initiator [6]. Thus, in effect, at a predetermined temperature and concentration the rate of polymerization may be controlled by a judicious selection of an initiator on the basis of its half-life at that temperature. These factors should, however, be considered only as a preliminary guide to the selection of an initiating system since many other parameters may mask these effects. In particular, it should be kept in mind that much of the theoretical research on polymerization is based on studies which were carried out to a very low degree of conversion, while many practical polymerizations run to high conversions. For example, the viscosity of the system may have a profound influence—particularly in regard to the Trommsdorff effect (also known as the "gel effect" or "autoacceleration"). Cross-linking and/or chain-branching may take place. Solvents, inhibitors, retarders, impurities, or deliberately added chain-transfer agents may cause chain-transfer reactions.

The foregoing discussion is of direct application to the selection of initiators for bulk and suspension polymerizations by free-radical mechanisms.

In other processes, the general principles indicated are still of value. In the case of photoinitiated polymerization, for example, diacyl peroxides may be added to a light-sensitized system. In this case, the actinic decomposition of the initiator may be the primary source of free radicals for initiation.

In a procedure in which doughlike mixtures of poly(methyl methacrylate) and monomeric methyl methacrylate are caused to react, the reaction seems to be initiated by high molecular weight polymeric free radicals which have persisted in the mass of the polymer since their termination reactions were prevented by reduced mobility of chain ends in the viscous resin. Such systems may be accelerated by the addition of peroxides and/or accelerating amines or metallic dryers.

The initiating systems used in emulsion polymerizations may be water-soluble compounds which generate free radicals by thermal decomposition or combinations of oxidizing and reducing agents. In the latter procedure, one or the other of the reagents may be insoluble in water but soluble in the monomer.

Ionic polymerizations of acrylic esters are currently of minor industrial interest, although stereospecific and crystallizable polymers have been produced. Among the anionic initiators are sodium in liquid ammonia, other alkali metals, alkaline earth metals, alkaline methylates, alkali amides, butyllithium, Grignard reagents, 9-fluorenyllithium, and other organometallic compounds [8].

C. Chain-Transfer Agents

A variety of compounds may act to reduce the average molecular weight of the polymer produced by a chain-transfer mechanism during polymerization. As indicated above, solvents may act as chain-transfer agents, although their activity is usually low. The most commonly used agents are mercaptans, particularly the higher molecular weights ones such as dodecyl mercaptan. Naturally, such reagents may give rise to serious odor problems.

Halogenated compounds such as carbon tetrachloride and chloroform have particularly high chain-transfer constants. However, these compounds must be used with extreme caution since explosive polymerizations have been observed.

The activity of chain-transfer reagents is a function of the reaction temperature, concentration, and monomer type. Table I lists a number of typical chain-transfer agents.

TABLE I
TYPICAL CHAIN-TRANSFER AGENTS

Acetic acid	Dodecyl mercaptan
Acetone	Ethyl acetate
Benzene	Ethylbenzene
n-Butyl alcohol	Ethylene dibromide
Isobutyl alcohol	Ethylene dichloride
sec-Butyl alcohol	Ethyl thioglycolate
tert-Butyl alcohol	Mercaptoethanol
tert-Butylbenzene	Methyl isobutyl ketone
n-Butyl chloride	Methylcyclohexane
n-Butyl iodide	Methyl isobutyrate
n-Butyl mercaptan	Methylene chloride
tert-Butyl mercaptan	Methyl ethyl ketone
Carbon tetrabromide	Pentaphenylethane
Carbon tetrachloride	Propylene chloride
Chlorobenzene	Isopropylbenzene
Chloroform	Isopropyl mercaptan
Diethyl ketone	Tetrachloroethane
Diethyl dithioglycolate	Thio-β-naphthol
Diethyl disulfide	Thiophenol
Dioxane	Toluene
Diphenyl disulfide	Triphenylmethane

D. Reaction Temperature

The reaction temperature of a polymerization has a profound influence on the course of the reaction. Other factors being equal, the higher the reaction temperature, the lower the average molecular weight of the polymer produced.

However, the temperature also has an affect on the rate of decomposition of initiators, the number of effective free radicals produced, and the reactivity of these free radicals. Temperature also effects chain-transfer agent activity. These considerations may offset the simple reciprocal relation between temperature and molecular weight.

The viscosity of the reacting system, of course, is also temperature-dependent. Thus, the diffusion of monomers and of growing polymer chains as well as the heat-transfer properties of the system are modified.

4. POLYMERIZATION PROCEDURES

Most of the common acrylic esters may be homopolymerized by relatively few procedures. Variations in the methods may be made because of requirements related to the final application of the polymer, limitations set by laboratory or plant equipment, the reactivity of the monomers, and the physical state of the monomer or of the polymer.

While an attempt is made in the present discussion of polymerization procedures to describe homopolymerizations, the nature of the polymer literature is such that, from time to time, information had to be drawn from copolymerization procedures to illustrate satisfactory preparations. This approach has the advantage that it shows the reader that, from the preparative standpoint, many similarities exist between homo- and copolymerizations and that the same principles usually apply.

A. Bulk Polymerization

The conversion of a monomer to a polymer in the absence of diluents or dispersing agents is termed a "bulk" polymerization.

From the laboratory standpoint, this procedure may be the simplest method since it basically consists of heating a solution of an initiator in the monomer. To prepare a small sample of a polymer in a test tube, this method is commonly used (although our personal preference for preparation of a small preliminary polymer sample happens to be a variation of the photoinitiated method).

Simple polymethacrylates are usually rigid and therefore either slide out of a test tube or can be isolated by breaking the test tube. The polyacrylates, however, tend to be elastomeric and frequently adhere to glass surfaces. Therefore it is good practice to coat surfaces with "parting agents" prior to introduction of the monomer into a reaction vessel. A variety of parting agents are available for such purposes—coatings of soap solution, films deposited by evaporation of poly(vinyl alcohol) solutions, silicone coatings, or fluorocarbon coatings. If the reaction is carried out at sufficiently low temperatures, polyethylene or Teflon equipment may be used. Lining suitable apparatus with polyethylene films or sheets may also be satisfactory. We have found a spray coating of fluorocarbon particularly effective, provided that the properties of the surface of the polymer are of no consequence. Naturally, if, for example, the adhesion to an acrylic polymer surface is to be studied, traces of the parting agent, particularly a fluorinated one, will interfere.

Several other factors must be kept in mind, particularly in bulk and suspension polymerizations:

1. Polymerizations of acrylic and methacrylic esters are highly exothermic (e.g., $\Delta H_{\text{polymerization}}$ of ethyl acrylate is 13.8 kcal/mole [8]). Generally, the heats of polymerization of acrylates are greater than those of methacrylates.

2. Frequently, even if as little as 20% of the monomer has polymerized, an autoaccelerating polymerization effect will take place. This may manifest itself in an increase in the heat evolved as the process nears completion. Particularly in large-scale, industrial polymerizations, this effect, known as the "Trommsdorff effect" or "gel effect," may be quite dangerous. In fact serious explosions have been attributed to it [9]. The effect is associated with a simultaneous increase in the average molecular weight of the polymer. It is assumed that as polymerization progresses, the termination step of the chain process is prevented because of the increasing viscosity of the system. Further, we presume that the increased viscosity also reduces the heat-transfer rate of the system.

3. Since the density of a polymer is substantially higher than that of the corresponding monomer, there is a considerable shrinkage of the volume of the material. In the case of methyl methacrylate this shrinkage, at 25°C, amounts to 20.6–21.2% [2]. Obviously, in the preparation of sheets of acrylic polymers, shrinkage is a significant problem. It may be overcome to some extent by use of molds which have flexible spacers, by preparing solutions of polymers in monomer ("casting syrups"), or by using cross-linking agents which form rigid networks of polymer even though very low levels of the monomer have been converted. The percent shrinkage is readily estimated by use of the equation:

$$\%\ \text{Shrinkage} = 100(D_p - D_m)/D_p$$

where D_m is the density of the monomer at 25°C and D_p is the density of the polymer at 25°C.

TABLE II

PERCENT SHRINKAGE OF ACRYLIC MONOMERS[a]

	% Shrinkage calculated[b]	Measured
Acrylates		
Methyl	24.8	—
Ethyl	20.6	—
Butyl	15.7	—
2-Ethylhexyl	10.8	—
Methacrylates		
Methyl	21.1	21.2[c], 20.6[d]
Ethyl	18.0	17.8[c], 18.4[d]
Butyl	14.1	14.3[c], 15.5[d]
Isobutyl	14.0	12.9[c], 15.2[d]
Dodecyl	7.8	—
Octadecyl	5.8	—
Dimethylaminoethyl	13.4	—
tert-Butylaminoethyl	11.1	—
Hydroxyethyl	18.8	—
Hydroxypropyl	16.7	—

[a] Based on Rohm and Haas Company, "Volume Shrinkage During Polymerization of Acrylic Monomers," Tech. Rep. TMM-11 12/64. Special Products Department, Philadelphia, Pennsylvania, 1964.

[b] Calculated on the basis of the equation:

$$\% \text{ Shrinkage} = \frac{k \times D_m \times 100}{\text{mol. wt. of monomer}}$$

where $k \cong 22.5$.

[c] F. S. Nichols and R. G. Flowers, *Ind. Eng. Chem.* **42,** 292 (1950).

[d] J. W. Crawford, *J. Soc. Chem. Ind.* **68,** 201 (1949).

Table II lists the percent shrinkage of a representative group of acrylic monomers and presents an alternative method of estimating the percent shrinkage.

4. In most cases, a small amount of unreacted monomer remains in the polymer. Frequently this residual monomer may be converted by a post-treatment of the polymer at elevated temperatures [9,10] or by exhaustive warming under reduced pressure [11].

4-1. Bulk Copolymerization of Trifluoroethyl Acrylate and Acrylonitrile [12]

In an ampoule, a solution of 9.5 gm of 2,2,2-trifluoroethyl acrylate, 0.5 gm of acrylonitrile, and 0.01 gm of benzoyl peroxide is flushed with nitrogen.

The ampoule is sealed, placed in a protective sleeve, and heated at 65°C in an oil bath for 60 hr. After this period, the ampoule is heated an additional 8 hr at 100°C. After cooling, the ampoule is broken open. Yield of elastomeric product approximately 9 gm.

In a recent patent [13], equipment and procedures are described for the bulk polymerization of methyl methacrylate on a commercial scale in which the molecular weight distribution of the polymer is closely controlled. The procedure involves continuously passing a solution of methyl methacrylate, containing co-monomers if desired, a high-temperature initiator such as cumene hydroperoxide, and, if necessary, a chain-transfer agent, between two heated surfaces in a layer from 0.2 to 20 cm thick. The initiator and temperatures are so adjusted that within 20–80 minutes, 40–60% of the monomer is polymerized. The temperature is controlled in the range of 145° to 165°C. The syrup is then passed into a flash evaporator where the monomeric methyl methacrylate is removed at 220°C, under reduced pressure, for recycling. The residual polymer is removed and is said to be particularly suitable for extrusion processes. Since the molecular weight distribution of the polymer is fairly narrow, good optical clarity of the product is obtained. The polymer is also said to have improved resistance to discoloration on heating.

In the casting of polymer sheets, bulk polymerization of initiator-containing monomers may be used, although "prepolymers" (i.e., solutions of polymers in their monomers, usually prepared by bulk polymerizing a monomer until the viscosity of the mixture has reached a desired level) [14] are more commonly used commercially to reduce problems arising from shrinkage, not to mention leakage of monomer through the flexible gasketing usually used. The method of making polymer sheets has been reviewed in some detail by Beattie [15]. While his review shows the method of producing large castings, the technique does lend itself to use in the laboratory.

4-2. Generalized Procedure for the Preparation of Polymer Sheets

A casting mold is constructed by clamping a length of polyethylene tubing between three sides of two sheets of carefully cleaned plate glass with spring paper clips of suitable size. It is most convenient to make the mold of plate glasses cut to the same width but unequal lengths (e.g., one piece 10 × 10 cm and one 10 × 15 cm) and forming the gasket along three sides. The excess glass section will facilitate the filling of the mold.

The mold is supported in an inclined position with the larger side forming the lower part of the mold. A solution of methyl methacrylate containing 0.5% of benzoyl peroxide is carefully poured into the mold to fill approximately two-thirds of the available space.

The mold is then supported in an upright position, preferably in a shallow dish. The top of the mold may be closed by forcing a length of tubing along the open edge. The assembly is then placed in an *explosion-proof*, high-velocity air oven and heated at 70°C for 72 hr. The curing time varies with the overall size of the casting. In the case of thick-cross-section castings, the curing time is considerably longer.

After the polymerization has been completed, the casting is cooled gradually. The sheet is removed by removing the glass plates. Because of the inhibiting effect of air, the top portion of the sheet may be somewhat soft. The plastic sheet may be finished by conventional plastic shaping techniques.

Similar molding techniques may be used with casting syrups, although great care must be exercized to allow trapped bubbles to rise prior to the final curing stages. With sizable glass molds, special arrangements have to be made to prevent bulging of the glass plates under the hydrostatic pressure of the monomer.

B. Photopolymerization

The initiation of polymerization by ultraviolet radiation is of particular interest in the study of free-radical processes [16,17]. However, technical applications of photopolymerization processes are currently being uncovered.

Much of the research in the field involves initiation of monomers suitably sensitized to form free radicals upon irradiation. Common sensitizers are biacetyl and benzoin [17,18]. We have also found benzil a suitable, albeit a less active, sensitizer. With these sensitizers, initiation of polymerization stops when ultraviolet radiation is turned off. This is followed by rapid decay of the polymerization process. In the laboratory, the source of radiation is a high-intensity mercury lamp. For simple test purposes, we have found out-of-door light quite suitable. Evidently even a lightly overcast sky furnishes sufficient ultraviolet radiation to permit photoinduced polymerization, although, naturally, direct bright sunlight is more effective.

Polymerization is somewhat more rapid when a conventional chemical initiator is present in the monomer along with the UV-sensitizer. We presume that an ultraviolet-induced decomposition of the peroxide assists the polymerization process at temperatures well below those normally used with peroxides.

4-3. Photopolymerization of Methyl Methacrylate, A Test Tube Demonstration [19]

In a Pyrex test tube is placed a solution of 0.5 gm of benzoin and 0.5 gm of benzoyl peroxide in 10 gm of a commercial grade of methyl methacrylate. The test tube is exposed to direct sunlight for approximately 3 hr at 35°C. The resultant polymer is not entirely free of trapped bubbles.

Polymers of α-methylbenzyl methacrylate prepared by this method at about room temperature are said to be atactic, while photopolymerization in the presence of benzoin at −65°C results in syndiotactic polymers [20]. A preparation of syndiotactic poly(isopropyl acrylate) at −105°C in toluene solution using a 15-watt Sylvania black-light bulb (wavelength, 320–420 mμ, 100% transmission at 360 mμ) has been described in detail [21].

Among other sensitizers are riboflavin, fluorescein [22], benzophenone and its derivatives [23], and tertiary amines such as *N,N*-dimethylbenzylamine, triethylamine, and *N,N*-dimethylaniline [24]. Trimethylaluminum is also said to have application in photopolymerization [24a].

Recent interest in γ-radiation-induced polymerization has found industrial application. This source of radiation is of interest also in graft polymerization [25]. This form of radiation is particularly effective in initiating the polymerization of acrylate esters rather than methacrylate esters.

C. Suspension Polymerization

A sharp distinction must be drawn between suspension (or slurry) and emulsion polymerization processes. Unfortunately, particularly the patent literature hedges on this point by vague references to "dispersion polymerizations."

We use the term "suspension polymerization" to refer to the polymerization of macroscopic droplets in a medium (usually aqueous) to produce macroscopic polymer particles which are readily separated from the reaction medium. The polymerization kinetics are essentially those of a bulk polymerization, with the expected adjustments associated with carrying out a number of bulk polymerizations in small particles more or less simultaneously and in reasonably good contact with a heat exchanger (i.e., the reaction medium) to control the exothermic nature of polymerization. Usually the process is characterized by the use of monomer-soluble initiators and of suspending agents.

On the other hand, emulsion polymerization processes involve the formation of colloidal polymer particles which are essentially permanently suspended in the reaction medium. The reaction mechanism involves the migration of monomer molecules from liquid monomer droplets to sites of polymerization which originate in micelles consisting of surface-active agent molecules surrounding monomer molecules. For present purposes it is enough to state here that emulsion polymerizations are usually characterized by the requirement of surfactants during the initiation of the process and the use of water-soluble initiators. This process also permits good control of the exothermic nature of the polymerization.

Suspension polymerizations are among the most convenient laboratory as well as plant procedures for the preparation of polymers. The advantages of the method include wide applicability (it may be used with most water-insoluble or partially water-soluble monomers), rapid reaction, ease of temperature control, ease of preparing copolymers, ease of handling the final product, and control of particle size.

In this procedure, the polymer is normally isolated as fine spheres. The particle size is determined by the reaction temperature, the ratio of monomer to water, the rate and efficiency of agitation, the nature of the suspending agent, the suspending agent concentration, and, of course, the nature of the monomer.

With increasing levels of suspending agent, the particle size decreases. It is therefore a good policy, when first experimenting with a given system, to have a measured quantity of additional suspending agent ready at hand. Then, if incipient agglomeration of particles is observed, additional suspending agent is added rapidly. In subsequent preparations, this additional quantity of suspending agent may be added from the start. If excess suspending agent is used, emulsification of the monomer may take place and a polymer latex may be produced by the emulsion process along with polymer beads.

Common suspending agents are polyvinyl alcohols of various molecular weights and degrees of hydrolysis, starches, gelatin, calcium phosphate (especially freshly precipitated calcium phosphate dispersed in the water to be used in the preparation), salts of poly(acrylic acid), gum arabic, gum tragacanth, etc. In some cases surfactants have been used to suspend the monomer (especially in the case of fluorinated acrylates which usually are difficult to polymerize in emulsion processes).

The initiators usually used are benzoyl peroxide, lauroyl peroxide, 2,2′-azobis(isobutyronitrile) and others which are suitable for use in the temperature range of approximately 60°–90°C. Water-soluble persulfates have been suggested occasionally. However, there is an ever-present problem that the incorporation of sulfate-type fragments on a growing polymer chain may produce a surfactant molecule which, in turn, will lead to latex formation. This, of course, is undesirable when the desired product is to consist of filterable polymer particles.

The hazard of agglomeration is greatest when acrylates are polymerized. The products tend to be elastomers and, in the course of the polymerization of these monomers, they tend to go through a "sticky stage." However, the proper selection of the suspending agent frequently prevents agglomeration.

The suspension process may be carried out not only with compositions consisting of a solution of the initiator in the monomer, but also with complex mixtures which incorporate plasticizers, pigment particles, chain-transfer agents, modifiers, etc.

Normally either suspension or emulsion polymerizations are carried out in 3-liter glass resin kettles equipped with a condenser, nitrogen inlet, thermometer, addition funnel, and a stirrer. In many cases we have found that a stainless steel stirrer fitted through a polyethylene stopper which has been drilled to fit the rod snugly forms a satisfactory greaseless stirrer bearing and closure. The temperature of the reaction is usually controlled with a hot water bath which can be raised or lowered as required.

Small quantities of suspension polymers may be made by shaking or tumbling an appropriate composition in a partially filled capped bottle which is surrounded with a suitable metal sleeve [26]. While this method has been widely used, particularly where a large variety of samples of copolymers are desired, it is somewhat hazardous and not entirely satisfactory as far as the quality of the product is concerned.

4-4. Suspension Copolymerization (Bottle Polymerization) [27]

To a beverage bottle of approximately 170 ml capacity is added a solution of 22.5 gm of 2,2,2-trifluoroethyl acrylate, 1.25 gm of methoxymethyl acrylate, 1.25 gm of acrylonitrile, and 0.025 gm of lauroyl peroxide, 30 ml of a 1.25% solution of polyvinyl alcohol in water, 30 ml of an 0.21% solution of sodium hydrosulfite in water, and 30 ml of an 0.21% solution of potassium persulfate in water. The bottle is sealed with a crown cap, inserted in a protective sleeve and shaken in a water bath at 50°C for 48 hr. The bottle is then cooled and opened. The flask contents are filtered off, washed with water, and dried under reduced pressure at 50°C. Yield: 83%.

4-5. Suspension Polymerization (Stirred Flask Method) [12]

To a vessel arranged for continuous stirring, as described in Section C, and containing 1 gm of soluble starch dissolved in 400 ml of distilled water is added a solution of 95 gm of 2,2,2-trifluoroethyl acrylate, 5 gm of *N-tert*-butylacrylamide, and 0.1 gm of benzoyl peroxide. The stirred mixture is heated on a water bath to 80°C. As the reaction temperature becomes self-sustaining the water bath is lowered. After the exothermic phase of the reaction has gone to completion, the reaction mixture is heated to reflux. The excess monomer is removed by steam distillation. Then the mixture is cooled, the stirrer is stopped, and the polymeric beads are filtered through a nylon-chiffon cloth. After washing the product in boiling water, the product is dried at a modest temperature under reduced pressure. Yield: 89 gm (89%).

The suspension polymerization of methyl methacrylate using a buffered sodium polyacrylate solution has been described [28]. A method of continuous suspension polymerization in a nonaqueous system has also been described [29]. This procedure (described for maleic anhydride copolymers) depends on the insolubility of the monomers in the organic liquid used.

D. Solution Polymerization

The polymerization of acrylic monomers in solution is usually carried out quite simply [2]. It must be kept in mind that since the viscosity of a polymer solution increases with increasing molecular weight, limitations are imposed on this method of polymerization by the handling problems involving high-viscosity systems. Furthermore, reference to Table I will show that many solvents for the monomer also may act as chain-transfer agents. Consequently the development of very high molecular weights in a solution process is difficult. Polymer solutions find application in protective coatings, adhesives, viscosity modifiers, etc.

In organic solvents, "oil-soluble" initiators, such as benzoyl peroxide, are used. In aqueous solutions, ammonium persulfate has been suggested.

4-6. Solution Copolymerization of Glycidyl Methacrylate and Styrene [30]

To 228 gm of xylene maintained at 136°C is added with stirring over a 3 hr period a solution of 453 gm of styrene, 80 gm of glycidyl methacrylate, and 11 gm of di-*tert*-butyl peroxide. After the addition has been completed, the solution is heated and stirred for an additional 3 hr at 136°C (to 100% conversion). After cooling, the polymer solution may be diluted to 54% solids by the addition of 228 gm of methyl isobutyl ketone. The relative viscosity of the copolymer (1% solution in 1,2-dichloroethylene at 25°C) is 1.175.

While many solution polymers are utilized directly in the solvent in which they are prepared, the polymer itself may be isolated by such techniques as evaporation of the solvent, usually in thin layers, or by precipitation of the polymer with nonsolvent. For example, an alcoholic solution of an acrylic polymer may be treated with petroleum ether to isolate the resin [31]. This type of technique is used in fractional precipitations to study the molecular weight distribution of a polymer sample. Generally, it is best to flow a thin stream of the polymer solution slowly into a vigorously stirred container of the nonsolvent.

The photopolymerization of isopropyl acrylate in solution at −105°C has been mentioned [21]. The procedure leads to a syndiotactic polymer. In a hexane solution at 40°C syndiotactic poly(methyl methacrylate) is said to form by polymerization with a sodium dispersion, while *n*-butyllithium or phenylmagnesium bromide at −40°C in toluene leads to isotactic poly(methyl methacrylate) [20,32–34].

Polymerization of methyl methacrylate by radical initiation at 25°C in the presence of preformed isotactic poly(methyl methacrylate) in dimethylformamide solution has resulted in the formation of syndiotactic polymers [35].

The above discussion dealt with nonaqueous solution systems. The ordinary esters of acrylic and methacrylic acids have only limited solubility in water. Even so, very dilute aqueous solutions of methyl methacrylate have been polymerized with a potassium persulfate–sodium bisulfite system in the absence of surfactants. The resultant product was a polymeric hydrosol [36,37]. The mechanism of hydrosol formation is said to be distinctly different from that of emulsion polymerization. Particle formation appears to be thermodynamically rather than kinetically controlled, as is the case in the formation of emulsions.

Monomers such as 2-hydroxyethyl methacrylate are distinctly soluble in water. Their polymerization in aqueous systems is of considerable importance. From the industrial standpoint, perhaps the most valuable application arises from the fact that the polymer, when slightly cross-linked, may be prepared in such a manner that it traps the water in which the reaction is carried out to produce a soft elastic gel. Such systems find application in the preparation of soft contact lenses and as carriers for the localized application of medications. The example given here to illustrate the method of preparation has been patented and is for reference only.

4-7. Preparation of a Cross-linked Polymer Gel [38]

To a dispersion of 150 gm of acrylamide, 100 gm of 2-hydroxyethyl methacrylate, 0.1 gm of ethylene dimethacrylate, and 750 gm of water is added 10 ml of a 2% aqueous solution of sodium thiosulfate and 15 ml of a 2% aqueous solution of ammonium persulfate.

The mixture is poured into an appropriate mold and allowed to polymerize at room temperature. The product is described as a soft, elastic material which may be washed in running water for several hours to remove water-soluble materials. To preserve the elastic nature of the material, it should be stored in contact with water.

Another important application of the use of aqueous solution polymerizations is the technique of embedding tissue samples in polymers to facilitate microtoming. The procedure involves successive exposure of a tissue sample to a series of aqueous solutions of 2-hydroxyethyl methacrylate with increasing concentrations of the monomer until the tissue has been soaked in a solution which is nearly pure monomer with additions of benzoin and 2,2′-azobis(isobutyronitrile). In some techniques, the initiator-bearing solution may consist of 2-hydroxyethyl methacrylate, butyl methacrylate, methyl methacrylate, aqueous ammonium persulfate, and 2,2′-azobis(isobutyronitrile). The tissue and the final embedding solution are then placed in a gelatin capsule and the assembly is polymerized at about 3°C with ultraviolet light [39–43].

E. The Monomer-Polymer System

The polymerization of methyl methacrylate monomer in the presence of its polymer may be considered a special case of solution polymerization. However, the behavior of the system is such that it is best to consider these two cases separately.

Mention has already been made of the preparation of "casting syrups" by partial polymerization of methyl methacrylate. Such fluids are usually treated with inhibitors to confer a certain degree of storage stability. The situation is different when poly(methyl methacrylate) is dispersed in monomeric methyl methacrylate. For one thing, compositions may be prepared which are so high in polymer that they have the consistency of a dough. Suitably pigmented, such compositions find application as dental filling materials [44]. More fluid compositions may be used to embed specimens for decorative or other display purposes [44a]. Quite recently use of the monomer–polymer system to prepare protective coatings has been patented [45].

This system may represent a novel extension of the Trommsdorff effect [9]. It will be recalled that one explanation of the effect is based on the theory that in highly viscous systems the possibilities of termination reactions are reduced. Consequently, a finished polymer may still contain reactive, polymeric free radicals. When fresh monomer is supplied, further polymerization can take place. Since there is a high level of polymer present in these compositions, shrinkage is substantially reduced.

In actual fact, so-called "dental methacrylates" are usually specially prepared so that some benzoyl peroxide may also be trapped in the suspension polymer spheres. To control the rate of polymerization in dental materials, accelerators are usually incorporated in the monomer. For example, tertiary aromatic amines are used for this purpose. Since the color of the final polymer is quite important in dental materials, considerable empirical work on the effects of various amines has been reported [46–48].

Common accelerators are *N,N*-dimethyl-*p*-toluidine, and, if color is particularly critical, *N,N*-dimethyl-3,5-dimethylaniline. The use of high molecular weight tertiary aromatic amine accelerators [48] is still in the early stages of exploration.

The monomer–polymer system of polymerization seems to be effective only if both the polymer and the monomer contain some methacrylates. Polymerization of an acrylate monomer in the presence of a methacrylate polymer does not usually take place in a reasonable time at room temperature. The compatibility of the monomer with the polymer may be a factor. However, cross-linking agents such as ethylene dimethacrylate, plasticizers, fillers, or pigments may be incorporated successfully.

4-8. Polymerization of a Monomer–Polymer Slurry [46]

At 22°C, 1 gm of a dental grade of poly(methyl methacrylate) containing 1% of benzoyl peroxide is rapidly mixed with 1 ml of a solution of methyl methacrylate monomer inhibited with 0.006% of hydroquinone and containing 2% of *N,N*-dimethyl-*p*-toluidine. At an initial temperature of 22°C, the composition sets within 9 minutes.

By varying the ratio of monomer to polymer, the level and type of accelerator, and the concentration and type of inhibitor, the setting time may be varied over a wide range.

The embedding of specimens in a block of poly(methyl methacrylate) requires considerable art. Biological specimens usually have to be dehydrated and impregnated with inhibitor-free monomers by techniques similar to those described above for the preparation of 2-hydroxyethyl methacrylate-impregnated specimens. Metallic objects, which tend to act as inhibitors of acrylic polymerization, have to be given a protective coating (e.g., with cellulose acetate, clear enamel, sodium silicate, or polyvinyl alcohol). The sample is then suspended in an appropriate monomer–polymer slurry. Polymerization may have to be carried out under pressure to reduce the possibility of bubble formation [44,44a]. The selection of the proper grade of the polymer and the use of inhibitor-free monomers may be critical in obtaining perfectly colorless castings. The skill with which some of these embedded articles have been made is extraordinary.

Since the monomer–polymer system involves the initiation of polymerization by preformed macromolecular species at least in part, if the polymer is, for example, poly(ethyl methacrylate) and the monomer is butyl methacrylate, it is reasonable to presume that the final polymer is a block copolymer of the type:

—E—E—E—E—B—B—B—

where E represents ethyl methacrylate units and B, butyl methacrylate units.

The structure of the polymer is, in all probability, not quite this simple. It is conceivable that the poly(ethyl methacrylate) free radical may be terminated by an entering monomer molecule which in the process becomes a free radical (chain-transfer mechanism). This butyl methacrylate radical may then propagate in a normal fashion to give poly(butyl methacrylate). Or, the free peroxide which is frequently said to be present in the polymer may simply initiate a butyl methacrylate polymerization. In either of these two situations, a tangle of different polymer chains may result; these have been called rather picturesquely, "snake-pit polymers." Other reaction mechanisms and, consequently, polymer structures may also be proposed. There would seem to be room for considerably more investigational work in this field. Reference

35 should probably be considered in relation to such research since it involves the radical polymerization of methyl methacrylate in a DMF solution containing isotactic poly(methyl methacrylate). The product contained syndiotactic poly(methyl methacrylate) associated with isotactic polymer in stereo-complexes, as well as isotactic–syndiotactic stereoblock polymers.

F. Anaerobic Polymerization

In 1953 a patent was issued to Burnett and Nordlander [49] which stated, in part, that when an organic peroxide is added to a dimethacrylate of one of the higher glycols like tetraethylene glycol and air or oxygen is bubbled through such a composition, the mixture becomes "activated" after several hours. This activation manifested itself by the fact that once the air flow was stopped, the monomer composition polymerized rapidly. However, as long as the air flow continued through the composition, it inhibited the polymerization. In effect, air had a twofold function—to "activate" the monomer and to inhibit the polymerization of the activated monomer. In addition, there was some impression that certain metals catalyzed the polymerization process after the air flow had stopped. Since the polymerization takes place when air no longer bubbles through the monomer composition, this original process was called "anaerobic."

An extension of this concept involved the elimination of the cumbersome aeration step of the original work of Burnett and Nordlander. It was found that surprisingly large levels of the organic hydroperoxides could be dissolved in the dimethacrylates of the polyethylene glycols and, while exposed to air and protected from excessive exposure to light, remained fluid for prolonged periods of time. Such compositions polymerized when subjected to a high vacuum or confined in such a manner that only a limited amount of oxygen could diffuse through the liquid, particularly in the presence of metals.

This phenomenon, which has extremely valuable industrial applications, unfortunately has not been subjected to extensive scientific study and publication. References [50–65] are a selection of pertinent material. It will be noted that most of the information available covers compositional factors. These are summarized here for reference only since much of this information is covered by patents.

4-9. Factors in the Composition of Anaerobic Monomers

(*a*). *Monomers.* The most suitable monomers are dimethacrylates of polyethylene glycols containing at least one ether linkage. The dimethacrylate of neopentyl glycol and monomethacrylates of glycols are also operative.

(*b*). *Inhibitors.* The monomer is usually inhibited with hydroquinone in the concentration range of 50–100 ppm by weight.

(*c*). *Initiator*. High concentrations such as 10% by weight hydroperoxides which may be considered substituted *tert*-butyl hydroperoxide (e.g., cumene hydroperoxide) are dissolved in the monomer. Hydrogen peroxide has also been suggested as an initiator.

(*d*). *Miscellaneous additives*. Accelerators such as tertiary amines to adjust the polymerization rate and additives which adjust the viscosity may be incorporated. Other additives to modify the adhesive properties of the polymer may also be compounded into the finished composition.

(*e*). *Packaging*. Since the composition is inhibited by air, storage in a shallow open dish is possible. More practical for long-term storage is an opaque polyethylene bottle which is partially filled so that the top surface area is substantial in relation to the depth of the fluid. Since oxygen can diffuse through polyethylene, this plastic is the ideal material for a storage vessel.

The primary application of anaerobic compositions is as metal-to-metal adhesives for threaded parts such as nuts and bolts. The confined space in the thread reduces contact with air when filled with the composition. The metal then seems to catalyze polymerization. Suitable metals seem to include most of the common metals such as copper, brass, iron, various types of steel, chrome-plated metal, several types of aluminum, silver-plated metal, and nickel. Even platinum and mercury may cause the polymerization of anaerobic monomers under reduced pressure. Of the common metals, zinc, cadmium, and tin seem to inhibit this polymerization. These metals may be specially treated to activate the polymerization.

It is an interesting theoretical question whether the catalytic effect of metals is caused by the free metal, metal oxides, or metallic ions on the surface. It is conceivable that the "anaerobic" phenomenon may not be a variation of free-radical polymerizations but, rather, a unique process.

G. Emulsion Polymerization

In the section on suspension polymerization we have indicated the differentiation between suspension and emulsion (or latex) polymerization. Emulsion polymers usually are formed with the initiator in the aqueous phase, in the presence of surfactants, and with polymer particles of colloidal dimensions, i.e., on the order of 0.1 μm in diameter [66]. Generally the molecular weights of the polymers produced by an emulsion process are substantially greater than those produced by bulk or suspension polymerizations. The rate of polymer production is also higher. Since a large quantity of water is usually present, temperature control is most often simple.

It is beyond the scope of this work to discuss theoretical aspects of emulsion polymerization. References [66–78a] contain material pertinent to the theoretical and practical aspects of this topic.

While most of these references deal primarily with the Harkins–Smith–Ewart approach to the interpretation of the emulsion polymerization process, alternative mechanisms have been proposed [66]. Of particular interest in this connection is a recent examination of the role that the surfactant plays in the process. It has been known to most workers in the field that the surfactant influences the final properties and applications of a latex. However, rarely has the paradox been discussed of how a negatively charged initiator like the sulfate radical ion ($SO_4^{\dot{-}}$) can possibly enter a micelle consisting of an assemblage of negatively charged surfactant anions and solubilized monomer molecules. One would expect coulombic repulsive forces to prevent this from taking place and therefore prevent initiation of polymerization. This matter is currently under investigation [78a].

Typical emulsion polymerization recipes involve a large variety of ingredients. Therefore the possibilities of variations are many. Among the variables to be considered are the nature of the monomer or monomers, the nature and concentration of surfactants, the nature of the initiating system, protective colloids and other stabilizing systems, cosolvents, chain-transfer agents, buffer systems, "short stops," and other additives for modification of latex properties to achieve the desired end properties of the product.

The ratio of total nonvolatiles to water (usually referred to as "% solids") is also important. Usually it is best when starting experimental work in emulsion polymerization to develop the techniques required to prepare 35–40% solid latexes without the formation of coagula. Latexes with higher solids content are more difficult to prepare. The geometry of close-packing of uniform spheres imposes a limit on the percent nonvolatiles in the range of approximately 60–65%. Dissolved nonvolatile components and the judicious packing of spheres of several diameters may permit the formation of more concentrated latexes in principle.

The ratio of the various ingredients to each other (which Gerrens calls "Flottenverhältnis") [70] as well as the reaction temperature are significant in the development of an emulsion recipe.

In the preparation of a polymer latex, the initial relationship of water, surfactant, and monomer concentration determines the number of particles present in the reaction vessel. Further addition of monomer does not change the number of latex particles. If such additional monomer polymerizes, the additional polymer is formed on the existing particles. As expected, the smaller initial particles imbibe more of the additional monomer than the larger ones. Consequently a procedure in which monomer is added to preformed latex polymer tends to produce a latex with a uniform particle size, i.e., a "monodispersed latex." Since the stability of the latex is dependent to a major extent on the effective amount of surfactant on a particle surface, considerable increase of the volume of the latex particles is possible with minor increases of the surface area purely on geometric grounds (an increase of the volume of a sphere by a factor of 8, increases the surface area by a factor of 4, while the particle diameter only doubles). These considerations have many practical applications, not the least of which is the possibility of preparing latex particles started with one co-monomer composition to which a different co-monomer solution is added.

From the preparative standpoint there are two classes of initiating systems:

1. The thermal initiator system, in which use is made of water-soluble materials which produce free radicals at a certain temperature to initiate polymerization. The most commonly used materials for such thermal emulsion polymerizations are potassium persulfate or ammonium persulfate.
2. The activated or redox initiation systems. Since these systems depend on the generation of free radicals by the oxidation–reduction reactions of water-soluble compounds, initiation near room temperature is possible. In fact, redox systems operating below room temperature are available (some consisting of organic hydroperoxides dispersed in the monomer and a water-soluble reducing agent). A typical redox system consists of ammonium persulfate and sodium metabisulfite. There is some evidence, particularly in the case of redox polymerizations, that traces of iron salts catalyze the generation of free radicals. Frequently these iron salts are supplied by impurities in the surfactant (quite common in the case of surfactants specifically manufactured for emulsion polymerization) or by stainless steel stirrers used in the apparatus. In other recipes, the iron salts may be supplied in the form of ferrous ammonium sulfate, or, if the pH is low enough, in the form of ferric salts.

While emulsion polymers have been prepared by shaking all components in a sealed ampoule [79] or by stirring all components together [80], the slightly more complex procedure given in Procedure 4-10 is usually more satisfactory.

4-10. Emulsion Polymerization of Ethyl Acrylate (Thermal Initiation) [2]

In a 2 liter Erlenmeyer flask, in a hood, to 800 ml of deionized water is added in sequence 96 gm of Triton X-200, 1.6 gm of ammonium persulfate, and 800 gm of ethyl acrylate. The contents is thoroughly mixed to form a monomer emulsion.

In a 3 liter resin kettle fitted with a stainless steel stirrer, a thermometer which extends well into the lowest portion of the reactor, a reflux condenser, and a dropping funnel a mixture of 200 ml of deionized water and 200 ml of the monomer emulsion is heated in a water bath while stirring at a constant rate in the range of 160–300 rpm to an internal temperature of 80°–85°C. At this temperature refluxing begins and vigorous polymerization starts (frequently signalized by the appearance of a sky-blue color at the outer edges of the liquid). The temperature may rise to 90°C. Once refluxing subsides, the remainder of the monomer emulsion is added from the dropping funnel over a 1 to 2 hr period at such a rate as to maintain an internal temperature between 88° and 95°C by means of the hot water bath, if necessary. After the addition has been completed and the reaction temperature has subsided, the temperature of the latex is raised briefly to the boiling point. Then the reaction mixture is cooled to room temperature with stirring. After removing stopcock grease from all joints, the latex is strained through a fine-mesh nylon chiffon. A negligible amount of coagulum should be present.

In particular, if a latex is to be used for coatings, adhesives, or film applications, no silicone-base stopcock greases should be used on emulsion polymerization equipment. While hydrocarbon greases are not completely satisfactory either, there are very few alternatives.

In the above examples, it will be noted that the monomer is added to the reaction system as an oil-in-water emulsion. Many emulsion polymerizations are more simply carried out by adding pure monomer to an aqueous dispersion of surfactant and initiators. This procedure permits a more rigid control of the number of particles in the aqueous phase.

4-11. Generalized Emulsion Polymerization Procedure for Acrylic Esters (Thermal Initiation, Fixed Number of Particles)

In a 3 liter resin kettle, fitted as in Procedure 4-10, is placed 1 liter of deionized water, 96 gm of Triton X-200, and 1.6 gm of ammonium persulfate. With constant stirring at 160–300 rpm, the solution is warmed in a water bath until the internal temperature reaches 80°–85°C. From the addition funnel, 10% of the total monomer is added all at once. Once the polymerization of this portion of the monomer has subsided, the remaining monomer is added at a constant rate such that the temperature remains about 85°C. After the addition has been completed and the reaction temperature has subsided, the temperature is raised briefly to the boiling point to complete the polymerization of residual monomer. The latex is then cooled, with stirring, to room temperature and strained as described in Procedure 4-10.

A recent patent claims the preparation of latexes with percent nonvolatile content in the range of 60% or greater by adding a water-in-oil emulsion of water in monomer, heavily loaded with surfactants, to a redox initiating system [81].

Redox emulsion polymerizations have been carried out by shaking an appropriate charge in a sealed bottle [82], but such an approach is unsafe and the resultant latex frequently is not reproducible as to properties. A procedure similar to that outlined in 4-11 is preferable.

The redox initiation system may consist of a large variety of reagents. Usually it consists of a water-soluble oxidizing agent such as ammonium persulfate and a reducing agent such as sodium metabisulfite, sodium hydrosulfite, or sodium formaldehyde sulfoxylate and an activator, usually a salt of a polyvalent metal, particularly ferrous sulfate or ferrous ammonium sulfate. As we have indicated before, traces of metallic ions such as one would obtain from metal stirrers may be sufficient to act as activators. While water-soluble oxidizing agents are usually used, monomer-soluble hydroperoxides have also been used in redox systems along with a variety of reducing agents including reducing sugars and metal activators. The paper of Bovey *et al.* [Ref. 67, Chapter 11], while it reviews the World War II experience with the polymerization of butadiene–styrene elastomers, is still a useful introduction to this area of redox polymerization.

4-12. Redox Emulsion Polymerization of Ethyl Acrylates [74]

In a 1 liter resin kettle fitted as in Procedure 4-10 except that a nitrogen inlet tube is also attached to the apparatus is prepared a solution of 376 ml of deionized water and 24 gm of Triton X-200. While the nitrogen flow is on, with stirring, 200 gm of uninhibited ethyl acrylate, 4 ml of a solution freshly prepared by dissolving 0.3 gm of ferrous sulfate heptahydrate in 200 ml of deionized water, and 1 gm of ammonium persulfate are added. After stirring for 30 min, the mixture is cooled to 20°C and 1 gm of sodium metabisulfite and 5 drops of 70% *tert*-butyl hydroperoxide are added. The polymerization starts rapidly and the temperature in the flask rises to nearly 90°C. After approximately 15 min, the polymerization is complete. The latex is cooled and passed through a nylon chiffon strainer.

If inhibited monomers are used, an induction period is observed in this process although the rate of polymerization remains about the same. Variations of procedure such as the gradual addition of monomer to a seed polymer as in Procedure 4-11 are also possible. The gradual addition of reducing agent solution along with gradually added monomer may also be used as a technique of redox latex polymerization.

With the higher methacrylates, such as stearyl methacrylate, special techniques are required to produce satisfactory latexes. One technique involves the use of a cosolvent such as acetone in the aqueous medium [83].

In Table III the synthesis of a representative group of monomers by a variety of techniques is outlined. The methods indicated in this table should not be considered as the only means by which polymers of a given monomer may be produced. Attention is directed to the interesting monomer α-methylbenzyl methacrylate; its optical isomers as well as the racemic mixture have been polymerized under a variety of conditions to produce tactic as well as atactic polymers. To be noted is that tacticity refers to the polymer chain and not to the optical activity of the ester side groups. The optical activity of the asymmetric carbon in the side chain is a matter which must be considered separate and distinct from the structure of the polymeric backbone.

Table IV gives the polymerization conditions for a range of approximately 40 monomers under a variety of free-radical conditions with the glass-transition temperatures (T_g) and the number average molecular weight (M_n) of the polymers produced. Unfortunately the exact details of the polymerizations were not given beyond the statement that all were carried out with 2,2′-azobis(isobutyronitrile) except for those cases in which the polymerization took place while the purified monomer was stored in the freezer and for the sample which had been prepared in emulsion. The monomers were generally very highly purified. In the table, the polymerization temperature, T, the solvent, the monomer concentration in the solvent, and the conversion are given. This conversion is not to be considered the maximum obtainable, but rather the extent to which polymerization was carried out. These data are said to be sufficient to define the tacticity of the polymer, the incorporation of impurities, the molecular weight distribution, and the degree of branching of the product.

H. Anionic Polymerization

Particularly in the sections dealing with solution polymerization, passing reference to anionic polymerizations of acrylic monomers has been made since these reactions are normally carried out in a suitable solvent [7,20,21, 33,34,84–90]. In recent years, anionic polymerizations of acrylic monomers have received considerable attention, although commercial application of stereospecific polymers of acrylic monomers produced by such procedures seem to be lacking.

The anionic polymerization of methyl methacrylate was probably observed long before the nature of the process was recognized since polymer formation is rampant when an attempt is made to prepare the higher methacrylate esters by transesterification of a high molecular weight alcohol with methyl methacrylate in the presence of sodium methylate.

Anionically initiated polymerizations frequently lead to polymers which contain a sufficiently long sequence of stereoregular structure that, on purification, samples with high degrees of crystallinity may be isolated. While even free-radically initiated acrylic polymers may exhibit a considerable tacticity, these are normally not crystallizable.

TABLE III

SYNTHESIS OF REPRESENTATIVE POLYMERS

Monomer	Polymerization Time (hr)	Polymerization Temp. (°C)	Initiator (% on monomer)[n]	Regulator (% on monomer)	Method of polym.	Solvent (quantity)	Conversion (%)	Mol. wt.	Remarks	Ref.
Ethyl α-acetoacrylate	—	60	Bz_2O_2 (0.1)	—	Bulk (Sheet)	—	—	—	Polymer insoluble in most solvents	*a*
	24	50	Bz_2O_2 (0.25)	—	Solution	Dioxane (100 gm/100 gm of monomer)	—	—	More soluble polymer	*a*
α-Methylbenzyl methacrylate[b]	—	−60	Butyllithium	—	Solution	Toluene	—	—	Isotactic polymer	*c*
	—	−65	Benzoin, UV	—	Photo	—	—	—	Syndiotactic polymer	*c*
	—	35	Benzoin, UV	—	Photo	—	—	—	Atactic polymer	*c*
Isopropyl acrylate	2	−105	Benzoin, UV (1)	—	Photo	Toluene (15 gm/15 gm of monomer)	10	$[\eta]$ = 1.4 dl/gm	Syndiotactic polymer	*d*
2-Hydroxyethyl or hydroxpropyl methacrylate	—	40–60	Thermal UV, ammonium persulfate (1)	—	Solution	Water	—	—	—	*f*
					Bulk					
			Bz_2O_2 (1)		Solution	Organic solvents				

Methyl methacrylate	2½–3	76–78	Bz_2O_2	—	Suspension	Sodium poly(acrylate) and phosphonate buffer in water	90–95	$[\eta]$ = 1.87–1.90 dl/gm	—	*e*
	—	−40	Butyllithium or phenylmagnesium bromide	—	Solution	Toluene	—	—	Isotactic polymer	*g*
	—	40	Sodium dispersion	—	Solution	Hexane	—	—	—	*g*
	—	60	AIBN (1)	*n*-Butyl mercaptan (0.1[illegible])	Bulk	—	—	100,000 (number average)	—	*h*
	—	60	AIBN (1)	*n*-Butyl mercaptan (2.5)	Solution	Toluene (70 gm/30 gm of monomer)	—	5,000 (number average)	—	*h*
Ethyl α-chloroacrylate (33 mmoles)	—	—	Grignard reagents	—	—	—	—	—	Low conversions	*i*
	22	−78	Butyllithium (0.45 mmole)	—	Solution	THF (30 ml)	43	16,700 (number average)	—	*i*
	29	0	Fluorenyllithium (0.45 mmole)	—	Solution	Toluene (30 ml)	39	8,500 (number average)	—	*i*
	48	−78	Fluorenyllithium (0.45 mmole)	—	Solution	Toluene	14	14,200 (number average)	—	*i*
	17	−78	Fluorenyllithium (0.45 mmole)	—	Solution	THF (30 ml)	26	19,000 (number average)	—	*i*

(*cont.*)

TABLE III (*cont.*)

Monomer	Polymerization		Initiator[n] (% on monomer)	Regulator (% on monomer)	Method of polym.	Solvent (quantity)	Conversion (%)	Mol. wt.	Remarks	Ref.
	Time (hr)	Temp. (°C)								
	47	−78	Fluorenyllithium (0.45 mmole)	—	Solution	THF (10 ml) and DEDE[j] (20 ml)	25	8,900 (number average)	—	*i*
	3.5	0	Fluorenyllithium (0.45 mmole)	—	Solution	DEDE[j] (30 ml)	80	9,030 (number average)	—	*i*
	49	25–30	None	—	Bulk	—	23.5	24,500 (number average)	—	*i*
Ethyl acrylate	1.5	82–94	Ammonium persulfate (0.2)	—	Emulsion	Water (1,000 ml/800 gm of monomer) Triton X-200 (12% on monomer)	98	—	—	*k*
Methyl acrylate (10 gm)	4	50	Potassium thiosulfate (0.03 gm)	—	Emulsion (in sealed tube)	Water (10 gm) Sodium lauryl sulfate (0.5 gm)	100	4.6×10^6	—	*l*

Isopropyl acrylate (38.3 gm)	4	−70 to −80	Phenylmagnesium bromide (0.024 equiv. in ether)	—	Solution	Toluene (525 ml)	76–86	$[\eta]$ = 5 to 10 dl/gm	Isotactic polymer	*m*
sec-Butyl acrylate, *t*-Butyl acrylate, Cyclohexyl acrylate	—	—	Prepared by similar techniques	—	—	—	—	—	—	*m*

[a] T. M. Laakso and C. C. Unruh, *Ind. Eng. Chem.* **50**, 119 (1958).
[b] *L*, *D*, and racemic isomers were polymerized.
[c] K. J. Liu, J. S. Szuty, and R. Ullman, *Polym. Prepr., Amer. Chem. Soc., Div. Polym. Chem.* **5** (2), 761 (1964).
[d] C. F. Ryan and J. J. Gormley, *Macromol. Syn.* **1**, 30 (1963).
[e] D. P. Hart, *Macromol. Syn.* **1**, 22 (1963).
[f] Reference [32], based on recommendations of M. Rosenberg, P. Bart, and J. Lesko, *J. Ultrastruct. Res.* **4**, 298 (1960).
[g] R. G. Bauer and N. C. Bletso, *Polym. Prepr., Amer. Chem. Soc., Div. Polym. Chem.* **10**, 632 (1969); see also C. F. Ryan and J. J. Gormley, *Macromol. Syn.* **1**, 30 (1963).
[h] "Use of Acrylic Monomers in the Preparation of Low Number Average Molecular Weight Polymers," Tech. Rep. TMM-23, 3/65. Special Products Department, Rohm and Haas Company, Philadelphia, Pennsylvania, 1965.
[i] B. Wesslen and R. W. Lenz, *Polym. Prepr., Amer. Chem. Soc., Div. Polym. Chem.* **11**, 105 (1970).
[j] Diethyleneglycol diethyl ether.
[k] "Emulsion Polymerization of Acrylic Monomers," Bull. CM-104 J/cg. Rohm and Haas Company, Philadelphia, Pennsylvania.
[l] H. W. Burgess, H. B. Hopfenberg, and V. T. Stannet, *Polym. Prepr., Amer. Chem. Soc., Div. Polym. Chem.* **10**, 1067 (1969).
[m] W. E. Goode, R. P. Fellmann, and F. H. Owens, *Macromol. Syn.* **1**, 25 (1963); See also C. Schwerch, W. Fowells, A. Yamada, F. A. Bovey, F. P. Hood, and E. W. Anderson, *Polym. Prepr., Amer. Chem. Soc., Div. Polym. Chem.* **5**, 1145 (1964).
[n] Bz_2O_2 = Dibenzoyl peroxide; AIBN = 2,2′-azobis(isobutyronitrile).

TABLE IV

SYNTHESIS OF POLYMERS FOR GLASS TEMPERATURE MEASUREMENTS[a]

Polymer of	Polymerization Conditions: Reaction temp. (°C)	Monomer (wt. % in solvent)	Solvent	Conversion (%)	T_g (°C)	$M_n \times 10^{-3}$
Acrylates						
Isopropyl	44.1	43	Methyl propionate	30	−5	121
sec-Butyl	60	48	Benzene	27	−17	250
3-Pentyl	60	50	Toluene	33	−6	114
Neopentyl	60	48	Toluene	46	−22	97.7
2-Phenylethyl	60	50	Benzene	43	−3	570
2-Cyanoethyl	60	20	DMF	78	4	59.8
3-Chloro-2,2-bis(chloromethyl)-propyl	60	55	Benzene	33	46	1020
Benzyl	60	50	Toluene	32	6	256
Phenyl	60	50	Toluene	19	55	300
p-Cyanobenzyl	60	40	DMF	71	44	92.8
p-Cyanophenyl	60	20	Acetone	64	94	71
p-Cyanophenyl	60	50	Acetone	75	90	277
o-Chlorophenyl	65	14	Benzene	50	53	—
2,4-Dichlorophenyl	64.5–65	32	Benzene	60	60	115
Pentachlorophenyl[b]	64–66	20	Benzene	70	145	—
Rerun				—	147	—
p-Methoxyphenyl	60	33	Acetone	68	48	310
m-Dimethylamino-phenyl	60	45	Benzene	—	47	46.6
o-Carbomethoxyphenyl	60	—	Bulk	60	46	—
m-Carbomethoxyphenyl	60	50	Benzene	>50	38	184
p-Carbomethoxyphenyl	60	16	Benzene	76	67[c]	—
o-Carboethoxyphenyl	60	—	Bulk	22	30	246
m-Carboethoxyphenyl	60	50	Benzene	30	24	700
p-Carboethoxyphenyl	60	40	Acetone	74	37[d]	197
p-Carbobutoxyphenyl	60	44	Benzene	5	13	688
Methacrylates						
Isopropyl	44	22	Methyl propionate	20	78	153
Isobutyl	60	—	Bulk	8	48	912
sec-Butyl	60	—	Bulk	6	60 ± 2	160
tert-Butyl	44.1	—	Bulk (pre-polymerized)	9	107	250
2-Phenylethyl	60	33	Chloroform	—	26	—
2-Ethylsulfinylethyl	60	36	Acetone	33	25	54

TABLE IV (*cont.*)

Polymer of	Polymerization Conditions: Reaction temp. (°C)	Monomer (wt. % in solvent)	Solvent	Conversion (%)	T_g (C°)	$M_n \times 10^{-3}$
tert-Butylaminoethyl	60	—	Bulk (prepolymerized	9	33	—
2-Cyanoethyl	60	62	Acetonitrile	30	91	985
Dimethylaminoethyl	−20	—	Bulk[d]	—	19	—
	44.1	61	Benzene	—	17	242
2-Bromoethyl	60	—	Bulk	—	52	431
3-Hydroxyethyl[e]	60	—	Bulk	4	55	—
Benzyl	60	—	Bulk	5	54	400
Phenyl	—	—	Bulk	High	110	816
p-Cyanophenyl	60	25	Benzene	34	155	848
p-Carbomethoxyphenyl	60	28	Acetone	27	106±3	129
p-Cyanomethylphenyl	60	42	Acetone	53	128±3	700
Ethyl α-ethylacrylate	50	—	Emulsion	50	27	40

[a] Based on S. Krause, J. J. Gormley, N. Roman, J. A. Shetter, and W. H. Watanabe, *J. Polym. Sci., Part A* **3**, 3573 (1965).

[b] First run: a slight second transition was observed at 35° ± 10°C. No second transition was observed in the rerun.

[c] X-Ray diffraction pattern indicated crystallinity. Curvature of the volume–temperature plot from 140° to 190°C may indicate m.p. >190°C.

[d] Monomer had polymerized in freezer.

[e] Monomer contained some methacrylic acid and ethylene dimethacrylate.

Depending on the solvent used, the initiator, and the temperature, three types of stereoregular poly(methyl methacrylates) have been produced [91,92]. For convenience these types of polymers are designated as Type I (syndiotactic), Type II (isotactic), and Type III (stereoblocks of long alternating segments of syndiotactic and isotactic structure). Table V summarizes the polymerization conditions for the formation of these polymers.

In general these polymerizations are carried out in systems from which oxygen and water has been rigorously excluded. The monomer is freed of inhibitor prior to use.

The syndiotactic polymers are produced in highly solvating media with 9-fluorenyllithium while the isotactic polymers are formed by organolithium compounds in hydrocarbon solvents or by Grignard reagents in toluene. The

TABLE V

POLYMERIZATION CONDITIONS FOR THE ANIONIC POLYMERIZATION OF METHYL METHACRYLATE[a]

Solvent	Conc. of monomer (moles/liter)	Initiator (moles/liter)	Reaction temp. (°C)	Conversion (%)	$M_v \times 10^{-3}$[b]	Polymer type (see text)
		9-Fluorenyllithium				
THF	1.66	(0.02)	−70	100	16.1	I
1,2-Dimethoxyethanol	1.66	(0.02)	−60	100	105	I
Toluene	1.66	(0.02)	−60	67.0	152	II
Diethyl ether	1.66	(0.02)	−65	31.4	29.5	III
		n-Butyllithium				
	0.5	(0.02)	−70	87.3	160	II
		9-Fluorenylsodium				
Toluene	0.5	(0.02)	−60	28.6	150	?
	0.5	(0.02)	+2	48.6	33.5	III
		Dipotassium stilbene				
	0.5	(0.02)	0	98.8	100	III
		Di-*n*-butylmagnesium				
	0.5	(0.02)	−60	53.5	99	III
	0.5	(0.02)	+2	14	—	III
		Diphenylmagnesium				
	0.5	(0.02)	−60	30	220	III
	0.5	(0.02)	+2	72	834	III

[a] W. E. Goode, F. H. Owens, R. P. Fellmann, W. H. Snyder, and J. E. Moore, *J. Polym. Sci.* **46**, 317 (1960).

[b] Molecular weight computed from the intrinsic viscosity of the polymer.

syndiotactic–isotactic stereoblocks form in moderately solvating media (ether or toluene containing ether or dioxane) with 9-fluorenyllithium, dibutylmagnesium, or diphenylmagnesium in toluene at 0° or at −70°C [8]. It would seem that these preparative systems are such that the products formed are really mixtures of several of the stereoregular types in which one or the other predominates and lends itself to more or less ready separation.

4-13. Preparation of Isotactic Poly(methyl methacrylate) [91]

In a 500 ml four-necked flask with suitable adapters to carry a mechanical stirrer, addition funnel, thermometer, inlets and outlets for dry, oxygen-free nitrogen, and an addition port covered with a rubber serum-bottle cap, is placed 300 ml of anhydrous toluene. Inhibitor-free, dry methyl methacrylate (15 gm, 0.15 mole) is placed in the addition funnel and nitrogen is bubbled through both the monomer and the solvent for 3 hr. After cooling the flask contents to 3°C with an ice bath, 3.6 ml of a 3.3 *M* solution of phenylmagnesium bromide in diethyl ether is injected by means of a hypodermic syringe and, with moderate stirring, the monomer is added. Stirring is continued for 4 hr. The viscous solution is then poured slowly into 3 liters of vigorously stirred petroleum ether. The solid is collected on a filter and dried under reduced pressure. Inorganic impurities are removed by digestion with aqueous methanol containing hydrochloric acid. Yield, 11.2 gm (74.6%) of isotactic poly(methyl methacrylate), viscosity average molecular weight 480×10^3. After annealing at 115°C, crystalline characteristics of the polymer can be demonstrated by X-ray diffraction.

By similar procedures, isopropyl acrylate, ethyl acrylate, and methyl acrylate have been polymerized [92]. Detailed directions for the preparation of isotactic poly(isopropyl acrylate) and of syndiotactic poly(isopropyl acrylate) are given in *Macromolecular Syntheses* [21,84] and by Sorenson and Campbell [93] and Kiran and Gillham [94]. The effects of various catalysts and reaction conditions are discussed in References [85]–[90].

5. MISCELLANEOUS METHODS

1. Initiation of methyl methacrylate by Nylon 6 [95].
2. Polymerization of methyl methacrylate activated by sulfur dioxide [96].
3. Radiation-induced solid-state polymerization of barium methacrylate [97].
4. Preparation of sulfonium polyelectrolytes derived from methylthioethyl acrylate [98].
5. Graft polymerization of acrylic esters to polysaccharides [99].

6. Polymerization of methyl methacrylate initiated by arylcarbene [100].

7. Preparation of 3,5-di-*tert*-butyl-4-hydroxybenzyl methacrylate and its utilization as a stabilizing monomer which inhibits premature polymerization but serves as a co-monomer after the initiator system has been activated [101].

8. Structure and properties of polymeric materials possessing spatial gradients [102].

9. Charge-transfer complex polymerization of methacrylates (see Chapter 11 on acrylonitrile) [103].

10. Copolymerization involving bisphenol-A dimethacrylate [104].

11. Polymerization of butyl acrylate initiated by an aluminum–mercuric chloride catalyst [105].

12. Polymerization of methyl methacrylate initiated by calcium dispersion in THP [106].

13. Copolymerization of methacrylate esters initiated by sodium metal [107].

14. Formation of long-living methyl methacrylate polymer radicals in phosphoric acid medium (UV-initiated polymerization) [108].

15. Precipitation polymerization of methyl methacrylate [109].

16. Kinetics of emulsion copolymerization of methyl acrylate–methyl methacrylate and acrylic acid [110].

17. Preparation of syndiotactic poly(methyl methacrylate) with triethylaluminium and titanium tetrachloride [111].

18. Free-radical polymerization of methyl methacrylate in presence of chromium acetylacetonate-triisobutylaluminum [112].

19. Stereospecific polymerization of methyl methacrylate with π-cyclopentadienyl-π-cyclopentenylnickel [113].

20. Anionic oligomerization by metal alkoxides [114].

21. Graft copolymerization of methyl methacrylate onto starch and triethylaminostarch [115].

22. Biomedical applications of hydrogels, [116].

23. Separation of stereospecific poly(methyl methacrylate) by TLC [117].

24. Preparation of fluorinated alkyl methacrylates and their polymers [118].

REFERENCES

1. "Glass Temperature Analyzer," Tech. Bull. SP-222, 3/65. Rohm and Haas Company, Philadelphia, Pennsylvania, 1965.
2. E. H. Riddle, "Monomeric Acrylic Esters." Van Nostrand-Reinhold, Princeton, New Jersey, 1954.
3. "Or-Chem Topics," No. 18. Rohm and Haas Company, Philadelphia, Pennsylvania, 1964.
4. W. H. Brinkman, L. W. J. Damen, and S. Maira, *Mod. Plast.*, p. 167 (1968).

5. C. Walling, *Polym. Prepr., Amer. Chem. Soc., Div. Polym. Chem.* **11** (2), 721 (1970).
6. P. J. Flory, "Principles of Polymer Chemistry," p. 114ff. Cornell Univ. Press, Ithaca, New York, 1953.
7. P. J. Flory, "Principles of Polymer Chemistry," p. 132ff. Cornell Univ. Press, Ithaca, New York, 1953.
8. L. S. Luskin and R. J. Myers, *Encycl. Polym. Sci. Technol.* **1**, 246–328 (1964).
9. E. Trommsdorff, H. Köhle, and P. Lagally, *Makromol. Chem.* **1**, 169 (1948); M. S. Matheson, E. E. Auer, E. B. Bevilacqua, and E. J. Hart, *J. Amer. Chem. Soc.* **71**, 497 (1949); G. Odian, M. Sobel, A. Rossi, and R. Klein, *J. Polym. Sci.* **55**, 663 (1961); V. E. Shashoua and K. E. Van Holde, *ibid.* **28**, 395 (1958); A. T. Guertin, *J. Polym. Sci., Part B* **1**, 477 (1963); K. Horie, I. Mita, and H. Kambe, *J. Polym. Sci., Part A-1* **6**, 2663 (1968); G. Henrici-Olivé and S. Olivé, *Makromol. Chem.* **27**, 166 (1958); *Kunstst.-Plast.* (*Solothurn*) **5**, 315 (1958); G. V. Schulz, *Z. Phys. Chem* **8**, 290 (1956); M. Gordon and B. M. Grieveson, *J. Polym. Sci.* **17**, 107 (1955); G. V. Korolev *et al.*, *Vysokomol. Soedin.* **4** (10), 1520; (11), 1654 (1962).
10. T. M. Laakso and C. C. Unruh, *Ind. Eng. Chem.* **50**, 1119 (1958).
11. R. H. Wiley and G. M. Brauer, *J. Polym. Sci.* **3**, 647 (1948).
12. B. D. Halpern, W. Karo, and P. Levine, U.S. Patent 2,834,763 (1958).
13. E. Fivel, U.S. Patent 3,637,545 (1972).
14. "The Manufacture of Acrylic Polymers," Tech. Bull. SP-233. Rohm and Haas Company, Philadelphia, Pennsylvania, 1962.
15. J. O. Beattie, *Mod. Plast.* **33**, 109 (1956).
16. P. J. Flory, "Principles of Polymer Chemistry," p. 149ff. Cornell Univ. Press, Ithaca, New York, 1953.
17. C. E. Schildknecht, "Vinyl and Related Polymers," p. 207ff. Wiley, New York, 1952.
18. W. E. F. Gates, British Patent 566,795 (1945).
19. Authors Laboratory (WK).
20. K. J. Liu, J. S. Szuty, and R. Ullman, *Polym. Prepr., Amer. Chem. Soc., Div. Polym. Chem.* **5** (2), 761 (1964).
21. C. F. Ryan and J. J. Gormley, *Macromol. Syn.* **1**, 30 (1963).
22. H. L. Needs, *Polym. Prepr., Amer. Chem. Soc., Div. Polym. Chem.* **10** (1), 302 (1969).
23. A. Ledwith and A. R. Taylor, *Polym. Prepr., Amer. Chem. Soc., Div. Polym. Chem.* **11** (2), 1013 (1970).
24. K. Yokota, H. Tomioka, T. Ono, and F. Kuno, *J. Polym. Sci., Part A-1* **10, 1335** (1972).

24a. P. E. M. Allen, B. O. Batenp, and B. A. Casey, *Eur. Polym. J.* **8** (3), 329 (1972).

25. I. S. Unger, J. F. Kirchner, W. B. Gager, F. A. Sliemes, and R. I. Leininger, *J. Polym- Sci., Part A* **1**, 277 (1963).
26. E. F. Jordan, Jr., W. E. Palm, L. P. Witnauer, and W. S. Port, *Ind. Eng. Chem.* **49**, 1695 (1957).
27. B. D. Halpern and W. Karo, U.S. Patent 2,947,732 (1960).
28. D. P. Hart, *Macromol. Syn.* **1**, 22 (1963).
29. G. R. Barrett, U.S. Patent 2,838,475 (1958).
30. J. A. Simms, *J. Appl. Polym. Sci.* **5**, 58 (1961).
31. F. C. Merriam, U.S. Patent 2,870,129 (1953).
32. "2-Hydroxyethyl Methacrylate (HEMA) and Hydroxypropyl Methacrylate (HPMA)," Bull. SP-216 8/64. Rohm and Haas Company, Philadelphia, Pennsylvania, 1964.

33. R. G. Bauer and N. C. Bletso, *Polym. Prepr., Amer. Chem. Soc., Div. Polym. Chem.* **10** (2), 632 (1969).
34. B. Wesslén and R. W. Lenz, *Polym. Prepr., Amer. Chem. Soc., Div. Polym. Chem.* **11** (1), 105 (1970).
35. R. Buter, Y. Y. Tan, and G. Challa, *J. Polym. Sci., Part A-1* **10**, 1031 (1972).
36. R. M. Fitch, M. B. Prenosil, and K. V. Sprick, *Polym. Prepr., Amer. Chem. Soc., Div. Polym. Chem.* **7** (2), 707 (1966).
37. R. M. Fitch and T-J. Chen, *Polym. Prepr., Amer. Chem. Soc., Div. Polym. Chem.* **10** (1), 424 (1969); H. P. Beardsley and R. N. Selby, *J. Paint Technol.* **40**, 263 (1968).
38. O. Wichterle and D. Lim, U.S. Pat. 3, 220, 960 (1965).
39. M. E. Gettner and L. Ornstein, *Phys. Tech. Biol. Res.* **3**, 627 (1956).
40. B. D. Halpern, "Glycol Methacrylate (GMA) Water Soluble Embedding Media," Data Sheet. Polysciences, Inc., Rydal, Pennsylvania, 1963.
41. J. D. McLean and S. J. Singer, *J. Cell Biol.* **20**, 518 (1964).
42. P. Bartl, *J. Histochem. Cytochem.* **14**, 741 (1966).
43. D. N. Rampley, *Lab. Pract.* **16**, 591 (1967).
44. B. D. Halpern, *Ann. N.Y. Acad. Sci.* **146**, 106 (1968).
44a. "Plexiglass Design, Fabrication, and Molding Data," Bull. No. 169. Plastics Department, Rohm and Haas Company, Philadelphia, Pennsylvania, 1953.
45. P. S. Pinkney, U.S. Patent 3,637,559 (1972).
46. G. M. Brauer, R. M. Davenport, and W. C. Hansen, *Mod. Plast.* **33**, 158 (1956).
47. R. L. Bowen and H. Argentar, *J. Amer. Dent. Ass.* **75**, 918 (1967).
48. R. L. Bowen and H. Argentar, *J. Dent. Res.* **51**, Part 2 of 2 parts, 473 (1972).
49. R. E. Burnett and B. W. Nordlander, U.S. Patent 2,628,178 (1953).
50. J. Dickstein, E. Blommers, and W. Karo, *in* "Symposia of Adhesives for Structural Applications" (M. J. Bodnar, ed.), p. 77. Wiley (Interscience), New York, 1962.
51. J. Dickstein, E. Blommers, and W. Karo, *J. Appl. Polym. Sci.* **6**, 193 (1962).
52. W. Karo and B. D. Halpern, U.S. Patent 3,125,480 (1964).
53. Borden Chemical Co., British Patent 1,067,433 (1967).
54. W. Karo, U.S. Patent 3,180,777 (1965).
55. W. Karo, U.S. Patent 3,239,477 (1966).
56. Leicester, Lovell and Co., Ltd., British Patent 1,015,868 (1966).
57. L. W. Kalinowski, U.S. Patent 3,249,656 (1966).
58. Henkel and Cie, G.m.b.H., French Patent 1,581,314 (1969).
59. Rohm and Haas G.m.b.H., British Patent 1,013,708 (1965).
60. J. W. Gorman and B. W. Nordlander, U.S. Patent 3,300,547 (1967).
61. A. S. Toback and W. E. Cass, German offen. 1,817,916 (1969).
62. E. Frauenglass, U.S. Patent 3,547,851 (1970).
63. The American Sealant Co., Australian Patent 233,683 (1958).
64. V. K. Krieble, U.S. Patent 2,895,950 (1959).
65. V. K. Krieble, U.S. Patent 3,218,305 (1965).
66. F. W. Billmeyer, Jr., "Textbook of Polymer Science," 2nd ed. Wiley (Interscience), New York, 1971.
67. F. A. Bovey, I. M. Kolthoff, A. J. Medalia, and E. J. Meehan, "Emulsion Polymerization." Wiley (Interscience), New York, 1955.
68. E. W. Duck, *Encycl. Polym. Sci. Technol.* **5**, 801 (1966); J. G. Brodnyan, J. A. Cala, T. Konen, and E. L. Kelley, *J. Colloid Sci.* **18**, 73 (1963).
69. H. Fikentscher, H. Gerrens, and H. Schuller, *Angew. Chem.* **72**, 856 (1960).
70. H. Gerrens, *Fortschr. Hochpolym.-Forsch.* **1**, 234 (1959).
71. H. Gerrens, *Ber. Bunsenges. Phys. Chem.* **67**, 741 (1963).

72. W. D. Harkins, *J. Amer. Chem. Soc.* **69**, 1428 (1947).
73. G. Odian, "Principles of Polymerization." McGraw-Hill, New York, 1970.
74. "Emulsion Polymerization of Acrylic Monomers," Bull. CM-104 J/cg. Rohm and Haas Company, Philadelphia, Pennsylvania.
75. W. V. Smith and R. H. Ewart, *J. Chem. Phys.* **16**, 592 (1948).
76. J. W. Vanderhoff, *in* "Vinyl Polymerization" (G. E. Ham *et al.*, eds.), Vol. 1, Part II, p. 1. Dekker, New York, 1969.
77. P. J. Flory, "Principles of Polymer Chemistry." Cornell Univ. Press, Ithaca, New York, 1953; M. S. Guillod and R. G. Bauer, *J. Appl. Polym. Sci.* **16**, 1457 (1972).
78. C. P. Roe, *Ind. Eng. Chem.* **60**, 20 (1968).
78a. A. S. Dunn, *Chem. Ind.* (*London*) p. 1406 (1971).
79. H. W. Burgess, H. B. Hopfenberg, and V. T. Stannet, *Polym. Prepr., Amer. Chem. Soc., Div. Polym. Chem.* **10**, 1067 (1969).
80. W. C. Mast and C. H. Fisher, *Ind. Eng. Chem.* **41**, 790 (1949).
81. R. C. Christena, U.S. Patent 3,637,563 (1972).
82. P. Fram, A. J. Szlachtun, and F. Leonard, *Ind. Eng. Chem.* **47**, 1209 (1955).
83. "The Emulsion Polymerization of Lauryl and Stearyl Methacrylate," Tech. Rep. TMM-29 3/66. Special Products Department, Rohm and Haas Company, Philadelphia, Pennsylvania, 1966.
84. W. E. Goode, R. P. Fellmann, and F. H. Owens, *Macromol. Syn.* **1**, 25 (1963).
85. C. Schuerch, W. Fowells, A. Yamada, F. A. Bovey, F. P. Hood, and E. W. Anderson, *Polym. Prepr., Amer. Chem. Soc., Div. Polym. Chem.* **5** (2), 1145 (1964).
86. N. S. Wooding and W. C. E. Higginson, *J. Chem. Soc., London* p. 774 (1952).
87. A. Zilkha, P. Neta, and M. Frankel, *J. Chem. Soc., London* p. 3357 (1960).
88. D. M. Wiles and S. Bywater, *Polym. Prepr., Amer. Chem. Soc., Div. Polym. Chem.* **4** (2), 317 (1963).
89. S. P. S. Yen and T. G. Fox, *Polym. Prepr., Amer. Chem. Soc., Div. Polym. Chem.* **10** (1), 1 (1969).
90. J. Trekoval and P. Kratochvil, *J. Polym. Sci., Part A-1* **10**, 1391 (1972).
91. W. E. Goode, F. H. Owens, R. P. Fellmann, W. H. Snyder, and J. E. Moore, *J. Polym. Sci.* **46**, 317 (1960).
92. W. E. Goode, F. H. Owens, and W. L. Myers, *J. Polym. Sci.* **47**, 75 (1960).
93. W. R. Sorenson and T. W. Campbell, "Preparative Methods of Polymer Chemistry," 2nd ed., pp. 284–286. Wiley (Interscience), New York, 1968.
94. E. Kiran and J. K. Gillham, *Polym. Prepr., Amer. Chem. Soc., Div. Polym. Chem.* **13** (2), 1212 (1972).
95. M. Imoto, *Polym. Prepr., Amer. Chem. Soc., Div. Polym. Chem.* **11** (2), 1003 (1970).
96. P. Ghost and F. W. Billmeyer, Jr., *Advan. Chem. Ser.* **91**, 75 (1969).
97. M. J. Bowden, J. H. O'Donnell, and R. D. Sothman, *Macromolecules* **5**, 269 (1972).
98. F. E. Bailey, Jr. and E. M. LaCombe, *Polym. Prepr., Amer. Chem. Soc., Div. Polym. Chem.* **10** (2), 711 (1969).
99. Imperial Chemical Industries, Ltd., British Patent 809,745 (1959).
100. M. Imoto, T. Nakaya, T. Tomomato, and K. Ohasi, *Polym. Lett.* **4**, 955 (1966).
101. R. L. Bowen and H. Argentar, *J. Dent. Res.* **51** (4), 1071; (6), 1614 (1972).
102. M. Shen and M. B. Bever, *J. Mater. Sci.* **7**, 741 (1972).
103. N. G. Gaylord and A. Takahashi, *Advan. Chem. Ser.* **91**, 94 (1969).
104. M. Atsuta, N. Nakabayashi, and E. Masuhara, *J. Biomed. Mater. Res.* **6**, 479 (1972).
105. F. X. Werber and R. N. Chadha, U.S. Patent 3,183,218 (1965).

106. C. Mathis and B. François, *Makromol. Chem.* **156**, 7–16 (1972).
107. J. C. Bevington and B. W. Malpass, *Eur. Polym. J.* **1**, 85 (1965).
108. E. E. Garina, E. G. Lagutkina, V. P. Zubov, and V. A. Kabanov, *Vysokomol. Soedin., Ser. B* **14**, 563 (1972); *Chem. Abstr.* **77**, 165142a (1972).
109. N. Shinmon, Y. Saitou, and T. Imoto, *Nippon Kagaku Kaishi*, p. 1544 (1972); *Chem. Abstr.* **77**, 165130v (1972).
110. K. P. Zabotin, V. V. Dudorov, and S. A. Ryabov, *Tr. Khim. Khim. Tekhnol.* No. 1, p. 131 (1971); *Chem. Abstr.* **77**, 165141z (1972).
111. M. Mihailov, S. Dirlikov, N. Peeva, and Z. Georgieva, *Macrinikecules* **6**, 511 (1973); H. Abe, K. Imai, and M. Matsumoto, *J. Polym. Sci., Part C*, **23**, 469 (1968).
112. A. B. Deshpande, S. M. Kale, and S. L. Kapur, *J. Polym. Sci., Part A-1*, p. 1307 (1973).
113. E. V. Bykova, and G. A. Berezhok, *Vysokomol. Soedin., Ser. B.* **15** (4), 226 (1973); *Chem. Abstr.* **79**, 66875b (1973).
114. S. Freireich, and A. Zilkha, *J. Macromol. Sci.*; *Chem. Abstr.*, p. 1383 (1972).
115. T. Nishuichi, and K. Ozaki, *Nippon Kagaku kaishi*, p. 728 (1972); *Chem. Abstr.* **78**, 30435y (1973).
116. S. D. Bruck, *J. Biomed. Mater. Res.* **7**, 387 (1973).
117. H. Inagaki, and F. Kamiyami, *Macromolecules* **6**, 107 (1973).
118. M. Hauptschein and R. B. Hager, U.S. Patent 3,544,663 (1970); R. B. Hager and A. H. Fainberg, U.S. Patent 3,532,659 (1970); C. S. Rondestvedt, Jr., U.S. Patent 3,655,732 (1972).

Chapter 11

POLYMERIZATION OF NITRILE MONOMERS

I. INTRODUCTION

The polymerization of acrylonitrile is of considerable economic importance particularly since polyacrylonitrile and certain related copolymers may be drawn into fibers. Polyacrylonitrile fibers (e.g., Orlon) have successfully competed with wool in a variety of textile applications. We recall that several years ago a small change in the price of New Zealand wool resulted in a significant boom in the construction of acrylonitrile manufacturing facilities in the United States along with significant shortages of the monomer as the demand for acrylic and modacrylic fibers rose.

The use of butadiene–acrylonitrile copolymers (Buna rubber, Perbunan, etc.) as rubber substitutes goes back to pre-World War II days. Because of their resistance to various solvents, these nitrile rubbers have found wide application on their own merit. Copolymers of acrylonitrile and styrene (SAN resins) as well as terpolymers of acrylonitrile, butadiene, and styrene and blends containing ABS resins are useful molding compounds. Elastomers of acrylonitrile, styrene, and acrylic monomers are known as ASA elastomers. Latexes of copolymers containing acrylonitrile have a wide variety of applications.

Fairly recently acrylonitrile has been used as the starting material for the preparation of one variety of graphite fibers. These are important in the production of high-performance reinforced plastics.

Methacrylonitrile has become of some importance relatively recently when a commercially useful method of manufacture was developed. Its behavior in polymerization is somewhat analogous to that of acrylonitrile. The major characteristic which differentiates the polymerization processes of methacrylonitrile from those of acrylonitrile is that polyacrylonitrile is virtually insoluble in monomeric acrylonitrile, while polymethacrylonitrile, like acrylic polymers in relation to their monomers, is soluble in methacrylonitrile. The softening point of polymethacrylonitrile is low, as compared to that of polyacrylonitrile. Consequently, polymethacrylonitrile is not suitable for the production of fibers.

Since solvent and oil resistance of the nitrile polymers is associated with the nitrile function, a monomer with a higher concentration of this grouping per molecule is naturally desirable. For this reason, vinylidene cyanide is studied from time to time and, in fact, may be used in certain commercial copolymers. However, this molecule has such violent physiological effects that work with it should be carried out with extreme precautions (see Sandler and Karo [1]). The monomer polymerizes readily in the presence of moisture. The residual monomer in a sample of poly(vinylidene cyanide), even after approximately 2 weeks in a normal laboratory atmosphere, was sufficiently potent to cause a relapse of asthmalike symptoms which one of our associates had contracted upon initial exposure to a trace of the monomer.

The esters of α-cyanoacrylic acid also polymerize under basic conditions. In particular, methyl α-cyanoacrylate is commercially produced as a metal-to-metal adhesive.

A number of other nitrile-containing monomers, such as fumaronitrile, have been studied. However, their utility appears to be limited.

The literature on nitrile monomers is voluminous. The producers of the monomers do supply useful reviews on acrylonitrile [1a] and on methacrylonitrile [2,2a]. An extensive review will also be found in the *Encyclopedia of Polymer Science and Technology* [2b].

A symposium on the science and technology of polymers of acrylonitrile was held at the Chicago meeting of the American Chemical Society, August 26–31, 1973. The reader is referred to the abstracts of this symposium for additional data on the chemistry of polyacrylonitrile.

At one time, it was difficult to obtain reliable information on the inhibitor used in acrylonitrile. There were indications that an aromatic amine may have been used; consequently separation of the inhibitor from the monomer involved either distillation from orthophosphoric acid or percolation through a silica gel column. Current sales literature indicates that both acrylonitrile [3] and methacrylonitrile [2] are inhibited with 40 ± 5 ppm by weight of *p*-methoxyphenol (MEHQ). Procedure 1-1 outlines a method for the removal of inhibitor from acrylonitrile. A similar procedure should be effective for the treatment of methacrylonitrile, the major difference in procedure arising from the fact that the melting point of methacrylonitrile is −35.8°C [2]. Consequently, the technique of freezing out the residual water at Dry Ice temperature is tricky. On the other hand, since it is only a matter of removing between 1% and 2% of water, standard drying agents should suffice.

1-1. Removal of Inhibitor from Acrylonitrile [4]

In a well-ventilated hood, with the operator wearing protective gloves, goggles, and protective clothing, 2 liters of acrylonitrile is shaken with 500 ml of 5% aqueous sodium hydroxide solution. After the aqueous layer has been separated, 500 ml of a 5% aqueous solution of orthophosphoric acid is cautiously added. After vigorous shaking, the aqueous layer is removed and the acrylonitrile is transferred to a steam distillation apparatus fitted with a spray trap to prevent the acidic solution from being mechanically transferred with the distillate.

The acrylonitrile is steam-distilled out of the mixture. The distillate is cooled to 25°C and the upper, acrylonitrile, layer is separated. This layer contains approximately 3.4% water.

To dry the monomer further, the moist monomer is placed in a Dry Ice–acetone slurry and filtered while cold with suction. The resultant product contains approximately 0.40% of water.

The remainder of the water may be removed by azeotropic distillation (b.p. 71°C, composition of azeotrope 88% of acrylonitrile) or by percolation through silica gel.

TABLE I

PHYSICAL PROPERTIES OF ACRYLONITRILE AND METHACRYLONITRILE

Property	Acrylonitrile[a]	Methacrylonitrile[b]
Molecular weight	53.06	67.09
Freezing point (°C)	-83.55 ± 0.05	-35.8
Refractive index, n_D^{25}	1.3888	1.3977
Surface tension (dynes/cm) (24°C)	27.3	24.4
Boiling point (°C/760 mm)	77.3°	90.3°
(°C/50 mm)	8.7°	—
Density (gm/ml at 25°C)	0.8004	d_4^{25} 0.805
Azeotropes		
Benzene	B.p. 73.3°C, 47% by wt. acrylonitrile	—
Isopropanol	B.p. 71.7°C, 56% by wt. acrylonitrile	—
Water	B.p. 71°C, 88% by wt. acrylonitrile	B.p. 77°C, 84% by wt. methacrylonitrile
Entropy of polymerization kcal/mole		-34 ± 2
Activation energy of polymerization (kcal/mole)	—	5.4
Heat of polymerization (bulk) (kcal/mole)	-17.3 ± 0.5	-13.5 ± 0.2 (-15.3 ± 1)
Softening temperature of polymer	Decomposes above 200°C	100°–130°C Depolymerizes at 250°C (amorphous polymer)

	% H_2O in acrylonitrile	% Acrylonitrile in water	% H_2O in methacrylonitrile	% Methacrylonitrile in water
Solubility in water				
at 0°C	2.1	7.2	1.06	2.89
20°C	3.1	7.35	1.62	2.57
40°C	4.8	7.9	2.38	2.59
50°C	6.3	8.3	2.83	2.69
80°C	10.8	11.0		

[a] From Reference [1a].

[b] From Reference [2].

Table I gives physical properties of acrylonitrile and methacrylonitrile. It will be noted that while the acrylonitrile–water azeotrope at 71°C contains 88% acrylonitrile, on cooling to 20°C, the distillate from an azeotropic distillation consists of two layers, one of acrylonitrile containing 3.1% of water, the other of water containing 7.35% of acrylonitrile. These factors may be used to account for the two-phase distillates which are frequently observed on distilling moist acrylonitrile.

The free-radical polymerization of acrylonitrile is effectively inhibited by oxygen [5], a fact which had not been recognized until the late 1940's. Consequently, many of the earlier kinetic studies on the polymerization of this monomer are of questionable value. In general, polymerizations should therefore be carried out in an inert atmosphere. However, our own experience has been that in preparative emulsion copolymerization systems which contain low levels of acrylonitrile, this precaution is not absolutely necessary.

To summarize, the factors which differentiate the free-radical polymerization characteristics of acrylonitrile from those of many other monomer systems are the inhibition by oxygen, the rather high heat of polymerization, the fact that the polymer is insoluble in its monomer, the high solubility of the monomer in water, and the extensive solubility of water in the monomer. In regard to most of these particular factors, methacrylonitrile behaves more like typical acrylic monomers.

As a matter of safety, all of the nitrile monomers should be handled as highly toxic materials. There are indications that acrylonitrile, methacrylonitrile, and other aliphatic nitriles are toxic by ingestion, inhalation, and by absorption of the liquids and vapors through the intact skin and mucous membranes [2]. Furthermore, some monomers of this group have other physiological actions. While few, if any, serious accidents have been reported, all of these monomers must be handled with all due precautions (as a minimum, well-ventilated hoods, protective gloves, protective clothing, and safety goggles must be used). Disposal of the monomers into the environment may present serious problems. The manufacturer's recommendations should be consulted as guides to the safety handling and the appropriate disposal of these materials.

2. POLYMERIZATION OF ACRYLONITRILE AND METHACRYLONITRILE

A. Bulk Polymerization

The basic bulk polymerization procedures for these monomers are similar to those described in Chapter 10 for acrylate esters. However, in the case of acrylonitrile, only relatively small quantities of the monomer can be polymerized in bulk. It will be recalled from the discussion above that polyacrylonitrile is insoluble in monomeric acrylonitrile. Therefore, at even low conversions, heavy slurries form, from which heat transfer is particularly difficult. Soon after the polymerization process is initiated, auto-acceleration takes place.

Typical initiators which have been used on a small scale to initiate acrylonitrile polymerizations have included benzoyl peroxide [6], 2,2′-azobis-(isobutyronitrile) and 2,2′azobis(2,4-dimethylvaleronitrile) [7], tetraalkyldiarylethanes [8], silver salts dissolved in monomer [9], and hydrogen peroxide [10]. Polymerizations have also been carried out in the vapor phase [11].

Electron microscopic observations during the course of bulk polymerization showed that the precipitating polymer forms initially in roughly spherical particles of nearly uniform particle size. The number of particles per unit volume remains reasonably constant for conversions up to about 25% (at which point, the suspension became too difficult to handle for this type of study). The number of particles is approximately proportional to the square root of the initiator concentration. The rate of polymerization increases with the initiator concentration to the 0.89 power when the initiator concentration is 10^{-4} *M* but becomes the initiator concentration to the 0.33 power when the initiator concentration is 10^{-2} *M*. These results are consistent with the assumption that in a situation like the polymerization of acrylonitrile, where polymers containing between five and ten monomer units will already precipitate, the probability of chain termination between the two chains which have

TABLE II
SPECIFIC SURFACE OF PARTICLES FORMED IN BULK POLYMERIZATION OF ACRYLONITRILE AT 50°C (AT A 10–15% CONVERSION)[a]

Conc. of AIBN (moles × 10^5/liter)	Rate of polymerization (moles × 10^5/liter/sec)	Specific surface area (m^2/gm)
7.4	1.2	89
20	5.8	85
66	22	121
76	29	107
130	39	109
400	94	123
1000	180	98
2000	220	94

[a] Reprinted from O. G. Lewis and R. M. King, Jr., *Advan. Chem. Ser.* **91**, 28 (1969), by permission of the American Chemical Society.

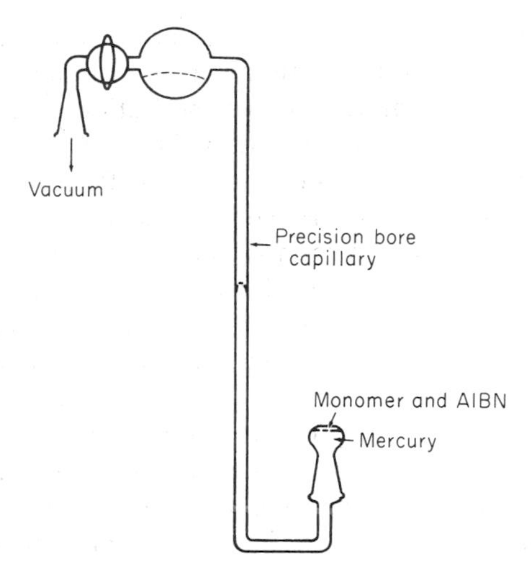

FIG. 1. Schematic diagram of a dilatometer. From O. G. Lewis and R. M. King, Jr., *Advan. Chem. Ser.* **91**, 30 (1969); by permission of the American Chemical Society.

been initiated by the two halves of the decomposing initiator (the so-called "geminate chains") is favored by the reduced chain mobility [12].

The precipitating polymer contains about 5% of unreacted monomer and has a glass-transition temperature of about 80°C. The polymerization appears to take place primarily on the surface of the polymer particles.

The specific surface of the particles decreases gradually with conversion. Since the particles are porous, this effect is probably the result of the interstices filling with polymerizing monomer as the process proceeds. The total surface does, however, increase as the polymerization progresses. At a constant degree of conversion, there appears to be no correlation between initiator concentration or rate of polymerization and the specific surface (Table II).

Preparation 2-1 is presented here to illustrate a procedure for studying the kinetics of polymerization not only of acrylonitrile but, with minor modifications, of many other monomers. The experiment is carried out in a special J-type dilatometer with a flattened reaction cell which confines the monomer in a uniform thin space between mercury and the glass wall. Figure 1 presents the dilatometer in schematic form.

2-1. Measurement of the Polymerization Kinetics of Acrylonitrile in Bulk [12]

The acrylonitrile is washed in turn with orthophosphoric acid and 10% aqueous sodium carbonate. After separation of the aqueous layer, the monomer is dried over calcium hydride and carefully distilled (b.p. 77.6–77.9°C at 760 mm Hg).

AIBN is recrystallized from cold methanol using safety precautions.

Solutions of AIBN in acrylonitrile (as in Table I) are prepared by volume at 25°C and delivered to the dilatometer cell with a micropipet (usually 1 ml except in the case of high rate of polymerization systems where only 0.1 ml is used). The cell is connected to the capillary, with mercury in the bulb, by means of a lightly greased ground glass joint. By careful manipulation, while maintaining the mercury in the bulb, the acrylonitrile is frozen with liquid nitrogen and the assembly is connected to a high-vacuum line for evacuation. After several cycles of thawing, freezing, and evacuation, the stopcock is closed and the dilatometer is disconnected from the high-vacuum line. The mercury in the bulb is warmed to approximately 50°C and, with appropriate turning of the instrument, the mercury is poured through the capillary tube into the cell in such a manner that the acrylonitrile is floated on the mercury as indicated in Fig. 1. This operation is probably most easily accomplished by turning the disconnected apparatus upside down and freezing the monomer against the appropriate surface with liquid nitrogen prior to pouring mercury into the cell.

The dilatometer is promptly mounted in a constant temperature bath at 50°C and the contraction of the mercury column height is taken at 5 min intervals with a cathetometer. The frequency of measurement will, of course, have to be varied with different rates of polymerization. The initial time is taken at the instant that the dilatometer is placed in the constant temperature bath.

To convert the contraction measured to percent conversion, a calibration constant has to be computed from direct measurement of the polymer formed in the dilatometer cell.

The free-radical bulk polymerization of methacrylonitrile resembles that of acrylic esters, since the polymer is soluble in the monomer, albeit the process is rather slow. Among the initiators reported for the bulk polymerization of methacrylonitrile are benzoyl peroxide [13] and 2,2′-azobis(isobutyronitrile) [14].

The slow conversion of free-radical polymerizations involving methacrylonitrile is carried over to copolymerizations involving this monomer. Table III is a selection of recent data on the copolymerization of methacrylonitrile with styrene and β,β-dideuterostyrene and of perdeuteromethacrylonitrile with β,β-dideuterostyrene [15].

B. Radiation-Initiated Polymerization

The polymerization of both acrylonitrile and methacrylonitrile has been initiated by ultraviolet radiation acting on pure monomer [16], in the presence of diacetyl [16a], in solution in the presence of ferric salts [17–19], in solutions

TABLE III
BULK PREPARATION AND COMPOSITION OF COPOLYMERS OF METHACRYLONITRILE AT 50°C USING AIBN INITIATOR[a]

Mole % of styrene or Monomer I in initial charge[b]	Polymerization time (hr)	% Conversion	Mole % of styrene or Monomer I in copolymer: By NMR analysis	Mole % of styrene or Monomer I in copolymer: Calculated
Styrene–Methacrylonitrile Copolymers[c]				
90	6	4.1	77.8 ± 0.8	78.2
50	10	8.3	50.7 ± 0.5	51.0
10	12	6.3	23.8 ± 0.5	23.3
Monomer I–Methacrylonitrile Copolymers[d]				
95	6	5.8	86.1 ± 0.3	88.8
50	10	9.8	52.1 ± 0.3	51.9
10	12	10.7	25.9 ± 0.2	22.8
Monomer I–Monomer II[b] Copolymers				
92	12	11.4	79.6[f]	84.5
56	19	9.9	55.6[f]	55.6
12	23	12.5	27.4[f]	26.2

[a] From Chang and Harwood [15].
[b] Monomer I = β,β-dideuterostyrene; Monomer II = perdeuteromethacrylonitrile.
[c] Reactivity ratios from this work, $r_s = 0.30 \pm 0.01$; $r_m = 0.24 \pm 0.01$.
[d] Reactivity ratios from this work, $r_s = 0.35 \pm 0.03$; $r_m = 0.25 \pm 0.01$.
[e] Reactivity ratios from this work, $r_s = 0.39 \pm 0.03$; $r_m = 0.25 \pm 0.02$.
[f] By IR analysis based on 4.48 μ and 6.24 μ bands.

containing hydrogen peroxide [20], and in the presence of triphenyl phosphite [21]. The latter system may very well involve the formation of an acrylonitrile–triphenyl phosphite charge-transfer complex which is sufficiently active to polymerize under conditions at which acrylonitrile alone does not polymerize.

The kinetics of the photopolymerization of acrylonitrile with light of wavelength 356, 546, and 405 mμ has been studied in the presence of ascorbic acid, sodium hydrogen orthophosphate, and one of the following dyes: acridine orange, rose bengal, dahlia violet, and pyronine G. It has been suggested that the radiation interacts with the dye and that the excited dye interacts with the complex formed between ascorbic acid and the buffer in the dark to produce ascorbic acid radicals which initiated the polymerization [22]. In the presence of certain dyes, reducing agents, and oxygen, acrylonitrile polymerizations may be initiated by visible light [23,24].

Other initiating systems include the use of γ-radiation [25,26], X-rays [27,28], high-energy electrons [29], and ultrasonic vibrations [30]. Cobalt-60-

initiated polymerizations of methacrylonitrile are free-radical between 10° and −40°C and ionic in nature between −40° and −196°C [30].

C. Suspension Polymerization

Since acrylonitrile has appreciable water solubility, suspension homopolymerizations of acrylonitrile are difficult. Because of this high solubility of the monomer in water, at best a mixture of solution and suspension polymerization may be expected.

The examples of homopolymerization of acrylonitrile in an aqueous medium which are found in the literature usually are not true suspension polymerizations as we define the process in Chapter 10, Polymerization of Acrylate and Methacrylate Esters. First of all, the level of acrylonitrile used is close to where one would expect nearly complete solubility of the monomer, there is an absence of a suspending agent, and the initiator system is totally water-soluble and resembles typical redox–emulsion polymerization systems. The resulting polymer–water mixture is described as a slurry, and the process has been termed "slurry polymerization." The process is best carried out at a pH below 5, preferably at about pH 3.2, and, indeed, the polymerization may be stopped short by raising the pH to 7–10 [31]. As in many redox systems, the presence of ferrous ions assists the initiation process [32], although, as Procedure 2-1 shows, this is not absolutely necessary.

2-2. Slurry Homopolymerization of Acrylonitrile [33]

In a four-necked 2 liter resin kettle fitted with a nitrogen inlet tube, addition funnel, stirrer, thermometer, and condenser is placed 900 gm of freshly boiled water and 0.029 gm of concentrated sulfuric acid. The stirrer is started and a rapid stream of nitrogen is passed into the solution for 20 min while the reactor is surrounded by a constant temperature bath maintained at 35°C.

Then 53.0 gm of acrylonitrile is added while maintaining a rapid nitrogen stream for an additional 10 min. The nitrogen flow is reduced and the gas inlet tube is raised to just above the surface of the stirred solutions.

A solution of 1.71 gm of ammonium persulfate in 50 gm of water is slowly added to the reactor, followed by the slow addition of a solution of 0.71 gm of sodium metabisulfite in 50 gm of water.

After about 3 min the solution becomes cloudy. Some heat is evolved during the first 30 min. Gradually a thick slurry develops. After 4 hr the stirring is stopped, and the polymer is filtered off and washed with 1 liter of water, followed by 175 ml of methanol. The polymer is dried overnight at 70°C, preferably under reduced pressure. Yield: 48 gm (92%).

On a slightly smaller scale a similar procedure, at 40°C, produced an 80–90% yield of polymer in approximately 3 hr [34].

Certainly when the ratio of acrylonitrile to another suitable monomer is low, no particular difficulty has been experienced in preparing true suspension polymers involving this monomer (see Preparation 3-1, Chapter 10). Another preparation is given here since it makes use of hydroxyapatite as a "suspending" agent. The use of such an inorganic powder is not well understood. It may be that a coating of this material during a "sticky" state of the polymerization prevents agglomeration. In any event, hydroxyapatite has been used in suspension procedures of a variety of monomers with considerable success. While a particle size of 0.03 to 0.06 μ for the suspending agent has been recommended [35], preparation *in situ* is simple enough and may even lead to more satisfactory results.

To prepare hydroxyapatite, $3Ca_3(PO_4)_2 \cdot Ca(OH)_2$, to a slurry of lime in water is added orthophosphoric acid until the mixture is only slightly basic. Obviously by varying concentrations, temperatures, addition rates, and the order of addition, the nature of the calcium phosphate complexes produced may be controlled.

2-3. Suspension Copolymerization of Styrene and Acrylonitrile [35]

To a stirred dispersion of 60 gm of deaerated water, 1 gm of powdered hydroxyapatite (see above), and 0.008 gm of sodium oleate are added under a nitrogen atmosphere, with stirring, a solution of 0.07 gm of benzoyl peroxide in 36 gm of styrene and 4 gm of acrylonitrile. The dispersion is heated at 90°C for 16 hr. The polymer is isolated by filtration, washed with warm water, and dried under reduced pressure at 45°–50°C.

In a recent patent [36], it is claimed that a homogeneous copolymer of acrylonitrile and styrene is obtained in a short reaction time, if the monomers are polymerized by the suspension technique at a controlled, predetermined temperature until the heating effect due to the Trommsdorff effect occurs. Then the temperature is allowed to rise. The whole point of the suspension technique is to control the Trommsdorff effect since the heat transfer from the small particles to a relatively large volume of water is more favorable than in the case of bulk polymerization. Consequently this patent requires further evaluation.

D. Solution Polymerization

Since acrylonitrile has appreciable water solubility, the boundaries between suspension, solution, and emulsion polymerization of this monomer are ill-defined. We have already indicated under the heading of suspension and slurry polymerization that it is quite common to start with an aqueous solution of acrylonitrile, and, because of the insolubility of the polymer, to produce a thick slurry.

Solvents for polyacrylonitrile are of considerable importance in the production of fibers and films from polyacrylonitrile. Table IV lists a variety of materials in which polyacrylonitrile is soluble [37].

It is to be noted that, as a rule, solvents for polyacrylonitrile are, with the exceptions of DMF and DMSO and their analogs, not the ordinary laboratory solvents customarily used. Particularly interesting is the fact that certain concentrated aqueous salt solutions are solvents for this polymer. Of these

TABLE IV
SOLVENTS FOR POLYACRYLONITRILE[a]

66.7% Lithium bromide in water (at 100°C)	γ-Butyrolactone
Concentrated aqueous solutions of sodium thiocyanate or zinc chloride	β-Hydroxypropionitrile
75% Sulfuric acid	1,3,3,5-Tetracyanopentane
Dimethylformamide	Dimethyl sulfoxide
Dimethyl sulfone	Tetramethylene sulfoxide
Sulfolane	2-Hydroxyethyl methyl sulfone
Dimethylacetamide	Methyl ethyl sulfone
Dimethylthioformamide	*m*-Nitrophenol
N-Methyl-β-cyanoethylformamide	*p*-Nitrophenol
α-Cyanoacetamide	Phenylene diamine (*o*-, *p*-, or *m*-)
Tetramethyloxamide	Methylene dithiocyanate
Malononitrile	Tetramethylene dithiocyanate
Fumaronitrile	Dimethyl cyanamide
Succinonitrile	Hot α-chloro-β-hydroxypropionitrile
Adiponitrile	Hot *N*,*N*-di(cyanomethyl)aminoacetonitrile
ε-Caprolactam	Hot ethylene carbonate
Bis(β-cyanoethyl) ether	Hot propiolactone
Hydroxyacetonitrile	Hot succinic anhydride
	Hot maleic anhydride

[a] From Schildknecht [37].

solvents, it must be pointed out, as will be discussed further below, that zinc chloride in anhydrous systems participates in the formation of charge-transfer complexes of the monomer. This fact may influence the course of polymerizations carried out in zinc chloride solutions.

Quite generally, since monomeric acrylonitrile is soluble in a wide variety of organic solvents, if polymerization of the monomer is carried out in one of the common solvents, the polymer precipitates soon after it forms. If the solution involves a good solvent for the polymer, naturally polymer solutions are formed. By way of contrast, the solubility behavior of methacrylonitrile and of polymethacrylonitrile is quite different. For example, a toluene solution of this polymer may be prepared by heating a solution of methacrylonitrile in toluene with AIBN at 70°C [38] and a solution polymer of methacrylonitrile in tertiary butanol has been prepared by straightforward solution polymerization with benzoyl peroxide at 79°C [39]. With solvents such as DMF or DMSO, inorganic initiators such as potassium persulfate or persulfate-bisulfite may be used as initiators for the polymerization of acrylonitrile.

In ethylene carbonate, under nitrogen, at 65°–70°C, the acrylonitrile polymerization has been initiated by AIBN, potassium persulfate, and by potassium persulfate–sodium bisulfite [40]. Polymerizations with AIBN have also been carried out in DMSO, DMF, dimethylacetamide [41,42], and in aqueous or aqueous alcohol solutions of thiocyanates [43].

2-4. Solution Polymerization of Acrylonitrile in Zinc Chloride Solution [44]

In a 250 ml flask, 72 gm of zinc chloride is dissolved in 18 gm of water. After displacing the air with nitrogen, and at 80°C, a solution of 0.002 gm of benzoyl peroxide in 10 gm of acrylonitrile is added with stirring. Heating and stirring at 80°C under nitrogen are continued until a highly viscous solution is produced. This solution may be extruded into water at 70°–80°C to produce a fiber whose tensile strength may be increased by stretching.

Dimethyl sulfoxide or DMSO-water mixtures have been used for the radiation-induced polymerization of acrylonitrile in the presence of potassium dioxalomanganate dihydrate. The resultant polymer is believed to contain carboxylate units along the polymer chain [45]. Photopolymerizations in the presence of hydrogen peroxide, AIBN, or 1,1′-azodicyclohexanecarbonitrile have also been reported [46].

Of the various techniques of preparing copolymers, the method involving initiation of a mixture of monomers of a composition calculated from the reactivity ratios of the monomers to produce a copolymer of predetermined composition, followed by gradual addition of a monomer solution of the ultimately desired composition is recommended not only because a polymer of homogeneous composition is produced but also because the molecular weight distribution remains virtually constant throughout the process. In many cases what seems to be a very complex process is actually quite easy. As indicated in the preparation cited here, a change in the boiling point of the reaction mixture indicates that a change in the unreacted monomer ratio has taken place. Addition of further monomer so as to maintain a constant boiling point in effect means that the initial co-monomer composition is maintained while a copolymer of the desired composition forms. In this particular case, the polymer precipitates from solution since the acrylonitrile content is quite high. Once the monomer addition has been completed, the process is stopped, and the polymer is isolated and worked up.

2-5. Solution Copolymerization of Acrylonitrile and Vinyl Acetate [47]

In a 5 liter, three-necked flask fitted with thermometer (in the vapor space), stirrer, reflux condenser, and addition funnel is placed 18.7 gm of vinyl acetate and 82.2 gm of acrylonitrile dissolved in 2850 ml of water. The solution is heated at 80°C. A solution of 12 gm of potassium persulfate in 150 ml of water, heated to 80°C, is added to the solution. Reaction starts almost immediately. As soon as the reflux temperature begins to rise, the gradual addition of a solution of 38.6 gm of vinyl acetate in 260.5 gm of acrylonitrile is started. The rate of addition is such as to maintain the reflux temperature at 80°C and the reflux rate remains constant. As reaction progresses, the rate of addition has to be increased progressively. When addition has been completed (about 63 min) the polymer is filtered off. The polymer is washed with water and dried under reduced pressure at 50°C.

In the Appendix to this volume, the mathematical foundations of this procedure are indicated.

Table V outlines other copolymerization systems based on this general principle.

E. Emulsion Polymerization

The homopolymerization of acrylonitrile by the emulsion technique is somewhat difficult, while copolymers, particularly those containing low levels of acrylonitrile, and methacrylonitrile polymers and copolymers form stable latexes more readily.

The difficulty of emulsion polymerizations of acrylonitrile may be attributed both to the insolubility of the polymer and to the water solubility of the monomer. In the emulsion polymerization procedure, part of the acrylonitrile may polymerize in aqueous solution while the remainder may tend to form a true latex. Obviously this situation will tend to form polymers which are not homogeneous as to molecular weight.

A recent study of the rate of emulsion polymerization of acrylonitrile has shown that the polymerization rate depends on the 0.16 power of the emulsifier concentration and on the square root of the initiator concentration, and generally increases with the monomer concentration above the saturation point. However, beyond a certain point, the rate of polymerization decreases with concentration. Reasonably stable latexes of polyacrylonitrile usually require rather high concentrations of an anionic surfactant [48,49].

2-6. Emulsion Polymerization of Acrylonitrile [50]

In a 5 liter, three-necked flask fitted with a nitrogen inlet tube, stirrer, thermometer, and condenser, supported in a constant temperature bath at

TABLE V

PREPARATION OF COPOLYMERS OF ACRYLONITRILE[a]

Initial vol. H_2O (ml)	Initial monomer charge		Reaction temp. (°C)	Initiator, gm potassium persulfate (in ml of water)	Composition of added co-monomers		Addition time (min)
	Acrylonitrile (gm)	Co-monomer (gm)			Acrylonitrile (gm)	Co-monomer (gm)	
2850	82.2	Vinyl acetate (18.7)	80	12 (150)	260.5	38.6	63
3400	85.7	Styrene (0.93)	85	10.5 (100)	345	75	48
950	50.4	Acrylamide (7.5)	80–85	3.0 (50)	69	23.1	—
550	20.5	*n*-Vinyl carbazole (0.97)	85	1.0 (50)	80	20	—

[a] From Chaney [47].

35°C, is placed 3100 ml of distilled water, 40 gm of sodium lauryl sulfate (weight on dry solids basis), and 20 ml of 0.1 *N* sulfuric acid. Nitrogen is bubbled through this solution rapidly for 15 min. Then the flow of nitrogen is reduced to approximately 3 bubbles/sec and 800 gm of acrylonitrile is added with stirring. After 10 min, a solution of 4.0 gm of ammonium persulfate in 50 ml of water and a solution of 1.82 gm of sodium metabisulfite in 50 ml of water is added.

During the first hour, the temperature in the flask is maintained at 35°C by adding ice to the water bath. Thereafter the bath needs to be heated. Within 3 hr the polymerization is substantially complete. The latex is filtered through a cheese cloth.

Characteristically, the emulsion polymerization of methacrylonitrile requires a longer period of time. The process can be carried out with a higher concentration of monomer in water than in the case of acrylonitrile.

2-7. Emulsion Polymerization of Methacrylonitrile [51]

In a sturdy 500 ml bottle is placed 225 gm of distilled water, 3.0 gm of sodium lauryl sulfate, 1.0 gm of *n*-dodecyl mercaptan, 100 gm of methacrylonitrile, and 0.5 gm of potassium persulfate. Nitrogen is bubbled through the mixture, the bottle is closed, placed in a protecting perforated metal sleeve, and shaken at 55°–70°C for 6–10 hr. The bottle is cooled to room temperature and cautiously opened. The latex is filtered through cheese cloth. Yield: greater than 90%.

Polymers containing 80% or more of a nitrile are said to have exceptional gas barrier properties. This has been ascribed to the packing characteristics or dipole interactions of the nitrile group. Polyacrylonitrile not only has a glass-transition temperature in the range of 140°C but on heating it does not melt. Inter- and intramolecular reactions take place which produce ladder and cross-linked polymers. Therefore it is not suitable for molding purposes. Polymethacrylonitrile, on the other hand, has a somewhat lower glass-transition temperature and melts without decomposition, thus making it more suitable as a molding resin. Preparation 2-8 gives the recipe for methacrylonitrile–methyl methacrylate copolymers which are of interest in the preparation of transparent barrier resins [52].

2-8. Preparation of Methacrylonitrile–Methyl Methacrylate Copolymer Latexes [52]

By techniques similar to those outlined in Preparation 2-7 the following materials were polymerized at 77°C for 6–8 hr at a pH initially adjusted to pH 7:

225 gm	demineralized water
80–100 gm	methacrylonitrile } to make 100 gm of
0–20 gm	methyl methacrylate } monomer mixture
3 gm	Garfac RE-610 Emulsifier
0.05 gm	Hampene K 4-100 (chelating agent)
0.5 gm	*n*-octyl, tetradecyl, or *n*-dodecyl mercaptan (type and amount of chain-transfer agent may be varied)
0.1 gm	potassium persulfate

If desired, up to 25% of a rubber latex may be added to this charge to produce a graft copolymer with improved toughness.

The emulsion copolymerization of butadiene with acrylonitrile is of considerable industrial importance. Since butadiene is a gas at room temperature, laboratory experiments with this system require pressure equipment. Bottle polymerizations have been described in considerable detail [53] but the hazards of such a procedure are self-evident. Coleman [53] should be consulted on the procedures for bottle polymerizations. A variation of the procedure which uses hydrogen peroxide (CAUTION: handle with care) as the source of free radical is described in Preparation 2-9.

2-9. *Preparation of Acrylonitrile–Butadiene Copolymers by Emulsion Polymerization* [54]

The preparation is carried out in a pressure vessel in an explosion-proof hood, with all due precautions, particularly in regard to explosive and fire hazards related to handling butadiene. To 250 gm of freshly boiled water, 5.0 gm of soap flakes (85% neutralized), 0.6 gm of diisopropyl xanthogen disulfide, 33.3 gm of acrylonitrile, and 66.7 gm of chilled liquid butadiene, is added 0.3 gm of hydrogen peroxide (added as 30% hydrogen peroxide solution) and 0.1 gm of a mixture prepared by grinding together 1.57 gm of sodium pyrophosphate, 0.28 gm of ferric sulfate, and 0.0014 gm of cobaltous chloride.

A small amount of the butadiene is allowed to vaporize to displace the air in the vessel. Then the vessel is closed, the protective covering is placed over the vessel, and the assembly is rotated in a bath at 30°C. When the conversion of monomers to polymer has reached 70%, a 1% aqueous solution of hydroquinone is injected into the reactor. The reactor is cooled in an ice bath and cautiously opened. The excess butadiene is allowed to evaporate in the hood.

In Preparation 2-5, the solution copolymerization of acrylonitrile and vinyl acetate is described in which gradual addition of mixed monomer to the reaction system is used to obtain homogeneity of the copolymer. In an emulsion polymerization system, a similar technique of adding a co-monomer mixture at such a rate as to maintain a constant reflux temperature may also be used. By this general technique, for example, acrylonitrile has been copolymerized in emulsion systems with vinyl acetate, with methacrylonitrile and vinyl acetate [55], with styrene, with α-methylstyrene [56], with vinyl chloride [57,58], and with vinylidene chloride [58]. Methacrylonitrile and vinylidene chloride have also been copolymerized by the gradual addition technique [58]. Table VI outlines typical procedures which use the gradual monomer addition process.

F. Anionic and Stereospecific Polymerization

Both acrylonitrile and methacrylonitrile may be polymerized in the presence of anionic initiators. Since the propagation step is susceptible to termination by protons, these reactions are normally carried out in aprotic solvents. Acrylonitrile does not give rise to useful products readily. Its polymers tend to be yellow, unless produced at very low temperatures, and, under the alkaline reaction conditions, side reactions involving cyanoethylation may take place.

A large variety of initiators for the anionic polymerization have been studied. Among these are alfin catalysts [59], alkoxides [60], butyllithium [61], metal ketyls [62], solutions of alkali metals [63,64], sodium malonic esters [65,66], sodium hydrogen sulfide and potassium cyanide [67], sodium in liquid ammonia or sodium amide [68–70], sodium cyanide [71], and calcium oxide [72].

2-10. Monosodium Benzophenone-Initiated Polymerization of Acrylonitrile [62]

(*a*) *Preparation of monosodium benzophenone.* In a three-necked flask equipped with a high-speed stirrer, provisions for introducing highly purified nitrogen, a condenser protected against atmospheric moisture and oxygen, and a thermometer, to 50 ml of highly purified tetrahydrofuran is added 1.48 gm of sodium wire and 1.15 gm of benzophenone with rapid stirring in a nitrogen atmosphere. A blue color appears almost instantaneously. Rapid stirring is continued for 15 min. There is then said to be no residual sodium.

(*b*) *Polymerization of acrylonitrile.* To the flask containing the monosodium benzophenone solution, enough purified THF is added to obtain 60 ml of solution. The stirred solution is cooled to 0°C and 10 ml (8 gm) of acrylonitrile is added at once. The temperature in the flask is maintained at 0°C for 1 hr. Then a mixture of 10 ml of concentrated hydrochloric acid and 10 ml of methanol is cautiously added. The precipitated polymer is filtered off and washed in turn with dilute hydrochloric acid, water, methanol, and ether. It is then dried under reduced pressure at 60°C.

A comparison of the anionic polymerization characteristics of acrylonitrile and methacrylonitrile may be found in Zilkha *et al.* [73]. Since methacrylonitrile does not have an α-hydrogen, the chain-termination and chain-transfer problems associated with acrylonitrile polymerizations are obviated. Since the methyl group permits the formation of an asymmetric carbon atom during polymerization, stereospecific polymers of methacrylonitrile can be prepared. In general, the anionic polymerization of methacrylonitrile is quite rapid. The amorphous polymethacrylonitriles prepared by anionic polymerization resemble polymers prepared by a free-radical process. They soften between 100° and 130°C and are soluble in organic solvents [20].

Among the catalyst systems used with methacrylonitrile are sodium in liquid ammonia [74], lithium in liquid ammonia [75], potassium in liquid ammonia [76], Grignard reagents and triphenylmethylsodium [74], quaternary ammonium hydroxides [73], diethylmagnesium and diphenylmagnesium [77], a variety of other organometallic compounds such as magnesium–aluminum alkyl, diethylmagnesium, and organomagnesium catalysts with magnesium–nitrogen bonds, etc. [78], and organoberyllium compounds [79].

From the work of Overberger *et al.* [75,76], it would appear that, unlike lithium amide, potassium amide initiates the polymerization as rapidly as potassium dissolved in liquid ammonia. At −75°C, the molecular weight increases with the methacrylonitrile concentration of the charge and decreases as the lithium concentration increases [75]. On the other hand, in the case of the potassium–liquid ammonia system, the molecular weight of the polymer is independent of the monomer and the potassium concentrations.

The preparation of isotactic polymethacrylonitrile is given in great detail by Joh [80]. In particular, details on the preparation of diethylmagnesium and of tetraethylaluminum magnesium, $Mg[Al(C_2H_5)_4]_2$, are carefully described there. Preparation 2-11 briefly outlines similar polymerization procedures.

2-11. Preparation of Stereospecific Polymethacrylonitrile [77]

In a 500 ml three-necked flask, flushed with oxygen-free nitrogen, 0.5 gm of diethylmagnesium is suspended in 370 ml of purified toluene. To this dispersion, with stirring, under nitrogen, while maintaining a constant temperature at 84°C, is added dropwise 30 ml of purified methacrylonitrile over a 4 hr period. Soon after the addition has begun, the mixture turns red.

After 4 hr, the reaction mixture is poured cautiously into methanol containing 5% hydrogen chloride. After stirring overnight at room temperature, the polymer is washed with methanol repeatedly and dried under reduced pressure. The conversion is 73.8%. Of the product 93.2% is insoluble in acetone. Table VII shows the effect of reaction temperature on conversion and tacticity of the product.

TABLE VI

EMULSION COPOLYMERIZATION OF ACRYLONITRILE AND OTHER MONOMERS

Water (gm)	Surfactant (gm)	Temp. (°C)	Gradually added mixed monomers[a]		Initiator, $K_2S_2O_8$ (gm)	Polym. time (hr)	% AN in polymer	Ref.
			Acrylonitrile (gm)	Co-monomer (gm)				
1,200	Sulfonated mahogany soap (4)	75	160	Vinyl acetate (240)	2 gm +2 gm after 8 hr	16	44.6	55
12,500	(12.5)	74–75	2100	Methacrylonitrile (275) +vinyl acetate (125)	2.5	2.5	—	55
2,000	(1.0)	80 ± 0.5	475	Vinyl acetate (25)	1	1.75	94.2	55
200	Alkyl·benzene-sulfonate (8)	80 ± 2	70	Styrene (30)	0.2	1.0	65	56
200	(8)	75 ± 2	60	Styrene (40)	0.2	4.5	55	56
200	(8)	80 ± 2	50	α-Methylstyrene (50)	0.2	—	41.5	56
200	(8)	80	35	(65)	0.2	4	28.7	56
200	(8)	80 ± 2	50	(50)	0.2	—	46.5	56

600	Acto 450 (0.6)	85 ± 2	180	Styrene (120) +0.6 gm Acto 450 +0.9 gm dodecyl mercaptan		1.0	3–5/6	—	56
600	(0.6)	80 ± 2	62.5	Styrene (187)		1.0	5.25	25.8	56
200	Di-2-ethylhexyl sodium sulfosuccinate (1.0)	90 ± 2	85	Styrene (15)		0.1 as 1% aq. soln. add 5 ml every hour	2.33	—	56
				Initial monomers[b]	Gradually added mixed monomers (gm)				
400	(1.5) Reaction carried out in autoclave	42	27	Vinyl chloride (73) +0.22 gm *tert*-dodecyl mercaptan	Acrylonitrile (33) +0.93 gm *tert*-dodecyl mercaptan	0.15	13.5	66.8	57
900	(2) Tertiary dodecyl mercaptan (0.2) Reaction carried out in autoclave	45	30	Vinyl chloride (70)	Acrylonitrile (43) + mercaptan (0.2)	0.5	2.7	62.8	57

(*cont.*)

TABLE VI (*cont.*)

Water (gm)	Surfactant (gm)	Temp. (°C)	Gradually added mixed monomers[a] Acrylonitrile (gm)	Co-monomer (gm)		Initiator, $K_2S_2O_8$ (gm)	Polymn. time, (hr)	% AN in polymer	Ref.
				Initial monomers[b]	Gradually added mixed monomers (gm)				
120	Sodium octyl sulfosuccinate (10)	25	Methacrylonitrile 10	Vinylidene chloride (10)	Acrylonitrile + vinylidene chloride in ratio of 1:1	0.1 gm + $K_2S_2O_8$, 1.0 gm	6	—	58
120	Sodium alkyl sulfosuccinate (10)	25	10	Vinylidene chloride (10)	Methacrylonitrile + vinylidene chloride in ratio 1.1	0.1 gm + $K_2S_2O_8$, 1.0 gm	—	—	58

[a] Polymerizations given in references 55 and 56 involve the gradual addition of the same mixed monomer composition throughout.
[b] The initial monomer composition is partially polymerized prior to the gradual addition of the second monomer composition.

TABLE VII

IONIC POLYMERIZATION OF METHACRYLONITRILE[a]

Temperature (°C)	Conversion (%)	Index of solubility (% of acetone-insoluble fraction)
Catalyst: 0.5 gm $Mg(C_2H_5)_2$		
Solvent: 370 ml toluene		
Monomer: Methacrylonitrile, 30 ml		
Polymerization time: 4 hr		
40	23.8	38.2
50	26.6	51.0
60	30.0	62.1
68	44.2	75.3
76	61.7	84.5
84	73.8	93.2
Catalyst: 0.5 gm $Mg(C_6H_5)_2$		
Solvent: 70 ml toluene		
Monomer: Methacrylonitrile, 30 ml		
Polymerization time: 4 hr		
85	62.1	86.0
64	33.3	72.5
40	19.2	67.5
20	23.3	25.0
0	17.5	14.3
−30	9.6	4.3
−78	trace	—

[a] From Joh *et al.* [77].

G. Polymerization of Charge-Transfer Complexes of Acrylonitrile

Highly polar monomers such as acrylic acid, methacrylic acid, acrylonitrile, methacrylonitrile, acrolein, esters of methacrylic acid, and vinyl acetate form charge-transfer complexes with Friedel-Crafts catalysts such as zinc chloride, aluminum chloride, halides of iron, boron, and titanium, ethylaluminum dichloride, and ethylaluminum sesquichloride $[(C_2H_5)_2AlCl\text{-}(C_2H_5)AlCl_2]$. Such complexes copolymerize with olefins such as butadiene, isoprene, and propylene, which homopolymerize at best with difficulty in the presence of the usual free-radical initiators. The resultant products usually appear to be alternating copolymers of the two monomers. Some systems form charge-transfer complexes without the intervention of Friedel-Crafts catalysts. An extensive review article is available which serves as a useful guide [80a].

Since the charge-transfer complexes vary in stability, some react to produce copolymers spontaneously (e.g., isobutylene and methyl α-cyanoacrylate at $-40°C$ [80b] and vinylidene cyanide and styrene at room temperature [80c]). Other systems require an input of energy in the form of heat, ultraviolet radiation, or free-radical initiators [80d].

The charge-transfer complexes may undergo three reaction paths: (1) intramolecular diradical coupling to produce the 1:1 adduct (of which the Diels-Alder reaction may be considered an example); (2) intermolecular diradical coupling to produce alternating 1:1 copolymers; and (3) no reaction until sufficient energy has been supplied to drive the reaction along path (1) or (2).

In the presence of zinc chloride or other catalysts, polar monomers may form charge-transfer complexes capable of further complexing either with the same monomer or with a nonpolar co-monomer. For this reason relatively small amounts of zinc chloride may bring about extensive copolymerization of two suitable monomers in the presence of a free-radical initiator, yet not every copolymer will correspond to a simple 1:1 copolymer of the two monomers.

A simple example of the process is illustrated for the case of the isoprene–acrylonitrile polymerization in Eqs. (1)–(4).

$$CH_2{=}CH{-}C{\equiv}N + ZnCl_2 \longrightarrow \left[\begin{array}{l} \;\;CH_2 \\ \;\;\| \\ HC{-}C{\equiv}N \rightarrow ZnCl_2 \end{array}\right] \qquad (1)$$

[Ref. 80b]

$$\begin{array}{l} \qquad\;\; CH_2 \\ H{-}C\!\!\diagup\!\!\diagup \\ \qquad | \\ CH_3{-}C\!\!\diagdown\!\!\diagdown \\ \qquad\;\; CH_2 \end{array} + \begin{array}{l} CH_2 \\ \| \\ C{-}C{\equiv}N \rightarrow ZnCl_2 \\ | \\ H \end{array} \longrightarrow \left[\begin{array}{l} \qquad\;\; CH_2^+ \\ H{-}C\diagup \\ \qquad \| \\ CH_3{-}C\diagdown \\ \qquad\;\; CH_2 \end{array} \qquad \begin{array}{l} {}^-CH_2 \\ \;\;| \\ \cdot C{-}C{\equiv}N \rightarrow ZnCl_2 \\ \;\;| \\ \;\;H \end{array}\right] \qquad (2)$$

$$\downarrow$$

$$\begin{array}{l} \qquad\;\; CH_3 \\ \qquad\;\; | \\ \cdot CH_2{-}C{=}CH{-}CH_2{-}CH_2{-}CH\cdot \\ \qquad\qquad\qquad\qquad\qquad\quad\;\; | \\ \qquad\qquad\qquad\qquad\qquad\quad\;\; C{\equiv}N \rightarrow ZnCl_2 \end{array} \qquad (3)$$

$$\swarrow$$

$$\begin{array}{l} \cdot CH_2{-}\overset{\displaystyle CH_3}{\overset{|}{C}}{=}CH{-}CH_2{-}CH_2{-}\underset{\displaystyle C{\equiv}N \rightarrow ZnCl_2}{\underset{|}{CH}}{-}\left[CH_2{-}\overset{\displaystyle CH_3}{\overset{|}{C}}{=}CH{-}CH_2{-}CH_2{-}\underset{\displaystyle C{\equiv}N \rightarrow ZnCl_2}{\underset{|}{CH}}\right]_n CH_2{-}\overset{\displaystyle CH_3}{\overset{|}{C}}{=}CH{-}CH_2{-}CH_2{-}\underset{\displaystyle C{\equiv}N \rightarrow ZnCl_2}{\underset{|}{CH}}\cdot \end{array} \qquad (4)$$

TABLE VIII
COPOLYMERIZATION OF ACRYLONITRILE WITH OLEFINS[a]

Olefin	Solvent (ml)	Catalysts: Free-radical initiator (moles)	Catalysts: $ZnCl_2$ (moles)	Temp. (°C)	Time (hr)	Yield (%)
2-Methyl-1-pentene	—	Bz_2O_2[b] (0.0017)	0	60	6.3	0
	—	Bz_2O_2[b] (0)	0.2	60	6.1	8.5
	—	Bz_2O_2 (0.0004)	0.2	60	1.1	60
Isoprene	—	—	0.2	70–37	3	3.3
	—	—	0.2	78–23	3	9.7
	Benzene (50)	AIBN[c] (0.15 gm on 0.2 mole of acrylonitrile)	0.2	55	0.16	30.3
	Heptane (50)	TBPP[d] (1.34 gm on 0.2 mole of acrylonitrile)	0.2	60–70	1.5	77.3
	Heptane (50)	Same as above	0	70	1.5	18.7
Butadiene	Heptane (50)	Same as above	0.2	60–70	1.5	84.5
	Heptane (50)	Same as above	0	60	1.5	4.4

[a] From Gaylord and Takahashi [80a].
[b] Bz_2O_2 = benzoyl peroxide.
[c] AIBN = 2,2′-azobis(isobutyronitrile).
[d] TBPP = *tert*-butyl peroxypivalate.

Equation (4) indicates the diradical coupling process which leads to the alternating polymer formation [80a]. Preparation 2-12 illustrates a process for the preparation of such a copolymer.

Table VIII illustrates the general requirements for the charge-transfer complex polymerization. It will be noted that virtually no polymer forms when the co-monomer mixture is treated with either benzoyl peroxide or zinc chloride alone, while a substantial yield of copolymer forms in the presence of both catalysts.

2-12. Preparation of an Isoprene–Acrylonitrile Copolymer [80a]

(*a*) *Preparation of the zinc chloride catalyst.* Zinc chloride is heated at 240°C for 1 hr under reduced pressure. It is then cooled in a dry box under oxygen-free nitrogen and, prior to use, is ground in the oxygen-free dry box.

(*b*) *Polymerization.* In the dry box, a 300 ml flask fitted with a thermometer, serum cap, and reflux condenser is charged with 27.3 gm (0.2 mole) of anhydrous zinc chloride and 50 ml of anhydrous benzene. The mixture is heated under nitrogen to 70°C and a solution of 10.6 gm (0.2 mole) of distilled acrylonitrile, 13.6 gm (0.2 mole) of isoprene, and 1.34 gm of a 75% solution of *tert*-butyl peroxypivalate in hexane is added slowly with a syringe through the serum cap. After 2 ml of the warm mixture has been added, the mixture solidifies. It is broken up with a spatula and the rest of the monomer is added. During the addition, the reaction temperature tends to drop. With external heating, the temperature is raised to 55°C. After the addition has been completed, the temperature is slowly raised to 65°C and refluxing is continued under nitrogen for 1.5 hr. Then the reaction mixture is poured into a Waring mixer containing dimethylformamide. The slurry is mixed with a large volume of methanol. The resultant white precipitate is filtered off and washed in turn with small quantities of dilute hydrochloric acid and of methanol. After drying at 50°C under reduced pressure, the methanol-insoluble copolymer, a 1 to 1 copolymer, weighs 13.6 gm (56.3% yield).

An 84.5% yield of copolymer is obtained from the copolymerization of butadiene with acrylonitrile using a similar procedure [80a].

The technique of adding the cool monomer mixture containing the peroxide to a warm zinc chloride suspension reduces the possibility of normal free-radical polymerization. However, if the reaction is carried out at too high a temperature or to too high a conversion, nonequimolar copolymers are obtained. Thus it would appear that simple free-radical polymerization may be taking place concurrently with charge-transfer complex polymerizations with one or the other reaction predominating, depending on the temperature [80e]. This point was perhaps again implied in a recent paper [80f], in which the photoinduced polymerization of co-monomer charge-transfer complexes was studied. It was found that in the copolymerization of styrene with acrylonitrile in the presence of triethylaluminum in tetrahydrofuran when the molar ratio of styrene to acrylonitrile was 1 and the ratio of THF to triethylaluminum was 1, a copolymer containing equimolar concentrations of the two monomers was produced. However, the copolymer composition was as anticipated from free-radical copolymerization theory when the ratio of THF to triethylaluminum was 4. In general, the solvent concentration, temperature, initiator, and the complexing catalyst affect the course of the reaction.

In the case of an acrylonitrile–α-methylstyrene copolymerization in the presence of zinc chloride, a mole ratio of zinc chloride to acrylonitrile of 0.1 to 1.0 as well as a 3 to 1 mole ratio of acrylonitrile to α-methylstyrene produced an alternating copolymer containing essentially 50 mole % of α-methylstyrene [80g]. The glass-transition temperature of this copolymer was found to be 122°C although predicted by the Fox equation to be 151°C. The lower T_g has been attributed to the sequence distribution of the monomeric units [80g].

Charge-transfer complex polymerizations with zinc chloride in tetramethylene sulfone were also carried out with the 1-bicyclobutanecarbonitrile–styrene and the methyl bis(cyclobutanecarboxylate)–styrene systems [80h].

3. POLYMERIZATION OF MISCELLANEOUS NITRILE MONOMERS

Besides acrylonitrile and methacrylonitrile, whose industrial importance is tremendous, many other nitrile monomers have been studied. Most of these are primarily of laboratory interest, particularly since many do not homopolymerize by free-radical procedures, although they may copolymerize. The major exceptions are the alkyl 2-cyanoacrylates, particularly methyl 2-cyanoacrylate.

A. Alkyl 2-Cyanoacrylates

Methyl 2-cyanoacrylate has found application as a metal-to-metal adhesive of considerable tensile strength [80i]. The monomer polymerizes rapidly when initiated by even weak bases such as water. Thus, the residual moisture on metal surfaces is sufficient to bring about polymerization. In operation as an adhesive, the low-viscosity material is rapidly applied to the smooth metal surface, the matching metal surface is brought into contact without applying undue pressure which would force the adhesive out from between the two parts and, within a short time, the two parts are bonded together by a poly(methyl-α-cyanoacrylate) film. The polar nature of the polymer is instrumental in achieving adhesive bonding.

Since the initiation of anionic polymerization is brought about by moisture, virtually any moist surface may be bonded. One area of application which has received considerable attention is the use of alkyl α-cyanoacrylates as tissue adhesives in surgical procedures [80j]. On biological substrates, in contrast with water substrates, the lower esters of α-cyanoacrylic acid spread only slightly. The higher esters, on the other hand, spread extensively and polymerize very rapidly. When the monomer is capable of spreading into thin layers, it comes in contact with relatively large amounts of water and consequently polymerizes rapidly. Monomers with higher interfacial tensions form lenses on the substrate and polymerize only on the outer surfaces to form protective coatings about the droplet which cannot polymerize further since no anionic initiator can penetrate to the monomer [80j].

While the alkyl α-cyanoacrylates have been studied in recent years mostly as anionically initiated materials, methyl α-cyanoacrylate can be polymerized by free-radical initiators. The anionic polymerization is normally suppressed by incorporating a boron trifluoride–acetic acid complex. Then, initiation with AIBN is possible. In bulk polymerizations an acceleration of the rate occurs rapidly at very low conversion [80k].

Solution free-radical polymerizations are difficult. In benzene, the polymer precipitates rapidly as a swollen gel; other solvents appear to be extremely difficult to free from traces of water. Carefully purified isobutyronitrile or nitromethane are solvents useful in solution polymerization of this monomer.

Homopolymers of methyl α-cyanoacrylate are soluble in propionitrile, pyridine, nitroethane, nitromethane, acetonitrile, dimethylformamide, ethylene carbonate, succinonitrile, and butyrolactone [80k].

B. Cinnamonitrile

Cinnamonitrile evidently does not homopolymerize under free-radical initiation conditions.

In copolymerizations, the nature of the unit attached to the end group undergoing copolymerization strongly influences the course of the copolymerization. This "penultimate effect" is also observed when styrene is copolymerized with benzylidene malononitrile, or ethyl benzylidene cyanoacetate, which, along with cinnamonitrile and atroponitrile, may be considered derivatives of styrene [80l].

C. Crotononitrile

trans-Crotononitrile fails to homopolymerize over a 10 hr period at 65°C with 0.02 mole % benzoyl peroxide. However, the carefully degassed monomer readily forms copolymers with styrene, methyl acrylate, and methyl methacrylate. Crotononitrile inhibits the polymerization of vinyl acetate but acts as an inert diluent toward the polymerization of acrylic acid [81,82].

In copolymerization there is a small penultimate effect observed [83].

D. 1- and 2-Cyano-1,3-butadienes

The nitrile analogs of the chloroprenes, 1-cyano-1,3-butadiene and 2-cyano-1,3-butadiene, may be designated 1-cyanoprene and 2-cyanoprene, respectively [84]. They may be polymerized with lithium dispersions or with aluminum alkyl catalysts. 1-Cyanoprene forms an amorphous polymer resulting from 1,4-additions. The polymers of 2-cyanoprene may also be the result of 1,4-additions but they are partially crystalline.

Table IX outlines the polymerization information for these two monomers.

TABLE IX
POLYMERIZATION OF THE CYANOPRENES[a]

Monomer (moles/liter)	Solvent	Initiator (moles/liter)	Polym. time, hr (Temp., °C)	Conversion (%)
1-Cyanoprene (2.4)	Benzene	Et_2AlCl (28×10^{-2}) +0.8 mole of [$Co(AcAc)_3$] per 5 ml of 1 *M* Et_2AlCl	72 (room temp.)	63.0
1-Cyanoprene (1.2)	Toluene	Li [dispersion containing equimolar mixture of Li, Cl and LiOBu (3.3 mmole of each reagent)] (18×10^{-2})	1 (room temp.)	93.0
2-Cyanoprene 1.0	Toluene	Et_2AlCl in hexane (9×10^{-2})	3 (−50)	20.8
2-Cyanoprene (2.8)	Ethyl chloride	Et_2AlCl in heptane (23×10^{-2})	16 (room temp.)	88.5

[a] From Wei and Milliman [84].

E. Fumaronitrile

Fumaronitrile is quite a hazardous material and must be handled in a hood while wearing protective clothing.

The monomer does not homopolymerize under free-radical initiation conditions, although copolymerizations with styrene and α-methylstyrene have been prepared [85,86]. On pyrolysis the monomer is thought to form poly(vinyltriazine) [87].

The photoinitiated copolymerization with divinyl ether seems to involve cyclopolymerizations (Eq. 5) [88].

$$CH_2{=}CH{-}O{-}CH{=}CH_2 + 2\,\underset{NC-CH}{\overset{\|}{CH}}{-}CN \longrightarrow \sim CH_2{-}\underset{NC-CH}{CH}\overset{O}{\frown}\underset{H_2C}{CH}\underset{CN}{CH}{-}\underset{CN}{CH}\sim \quad (5)$$

(ring: CH–O–CH, NC–CH, H_2C, CH bridged through CH bearing CN)

F. Vinylidene Cyanide

Vinylidene cyanide, in our experience, is an extremely hazardous material. It appears to be odorless, yet traces were detectable in the atmosphere because of the intensely irritating effect on the mucous membranes of the eyes, nose, and throat. One of our associates on exposure had to be hospitalized with symptoms resembling asthma. Upon his return to the laboratory he suffered a relapse even though it had been believed that the material had been cleaned up and the moisture of the room should have polymerized all residues in the time interval [89].

The polymerization of the monomer is initiated by anionic species such as water. This process is quite rapid while free-radical polymerization is sluggish. Water, alcohols, and amines, and ketones, dimethylformamide, and tetramethylene sulfone act as initiators [90].

4. MISCELLANEOUS METHODS

1. Polymerization of acrylonitrile by ceric ions in the presence of various amines [91].
2. Graft polymerization of acrylonitrile to cellulose [92].
3. Transport properties of nitrile polymers [93].
4. Review of copolymerization of styrene and acrylonitrile with cinnamonitrile, 2-phenyl-1,1-dicyanoethylene, and α-cyanocinnamate esters [94].
5. Acrylonitrile graft polymerization in poly(vinyl alcohol) solution [95].
6. Hydrogen-transfer polymerization of malononitrile [96].
7. Pressure-sensitive adhesives of cyanoalkyl acrylates [97].
8. Effect of ammonium salts on the polymerization of acrylonitrile in liquid sulfur dioxide [98,99].
9. Homopolymerization of ethylene cyanohydrin to a polyimidate resin [100].

10. Photo- and radiation-induced polymerization and copolymerization of acrylonitrile [101, 102].

11. Solution polymerization of acrylonitrile [103].

12. Photoinitiated cyclopolymerization of divinyl ether and fumaronitrile [104].

13. Ethyl α-cyanoacrylate and triethylene glycol dimethacrylate in quick-setting adhesive compositions [105].

REFERENCES

1. S. R. Sandler and W. Karo, "Organic Functional Group Preparations," Vol. 1, Chapter 17, p. 466. Academic Press, New York, 1968

1a. "The Chemistry of Acrylonitrile," 2nd ed. Petrochemicals Department, American Cyanamid Company, New York, 1959.

2. "Methacrylonitrile," Tech. Bull. Vistron Corporation, Cleveland, Ohio, 1968.

2a. "Methacrylonitrile," Patent Summary and Bibliography. Vistron Corporation, Cleveland, Ohio, 1967.

2b. C. H. Bamford, G. C. Eastward, A. Levovits, *Encycl. Polym. Sci. Technol.* **1**, 374 (1964).

3. "Handling, Storage, Analysis of Acrylonitrile." American Cyanamid Company, Wayne, New Jersey, 1970.

4. "Suggested Laboratory Procedure; Preparation of Inhibitor-Free Acrylonitrile," Tech. Bull. PD-1. American Cyanamid Company, New York, 1957.

5. K. C. Smeltz and E. Dyer, *J. Amer. Chem. Soc.* **74**, 623 (1952).

6. M. Imoto and K. Takemoto, *J. Polym. Sci.* **18**, 377 (1955).

7. L. Horner and W. Naumann, *Justus Liebigs Ann. Chem.* **587**, 93 (1954); S. Soenner-skog, *Acta Chem. Scand.* **8**, 579 (1954).

8. K. Ziegler, W. Deparade, and H. Külhorn, *Justus Liebigs Ann. Chem.* **567**, 151 (1950).

9. G. Salomon, *Rec. Trav. Chim. Pays-Bas* **68**, 903 (1949).

10. M. F. Shostakovskii and A. B. Bogdanova, *Zh. Prikl. Khim.* **24**, 495 (1951); *Chem. Abstr.* **46**, 1961 (1952).

11. Imperial Chemical Industries, Ltd., British Patent 567,778 (1945); *Chem. Abstr.* **41**, 2610 (1947); T. T. Jones and H. H. Melville, *Proc. Roy. Soc., Ser. A* **187**, 37 (1946).

12. O. G. Lewis and R. M. King, Jr., *Advan. Chem. Ser.* **91**, 25ff (1969).

13. W. Kern and H. Fernon, *J. Prakt. Chem.* **160**, 281 (1942).

14. W. K. Wilkinson, U.S. Patent 3,087,919 (1963).

15. R. C. Chang and H. J. Harwood, *Polym. Prepr., Amer. Chem. Soc., Div. Polym. Chem.* **12** (2), 338 (1971).

16. A. D. Jenkins, *Chem. Ind. (London)* No. 4, p. 89 (1955); E. Mertens and M. Fonteyn, *Bull. Soc. Chim. Belg.* **45**, 438 (1936); K. Nozaki, U.S. Patent 2,666,025 (1954).

16a. C. L. Agre, U.S. Patent 2,367,660 (1944).

17. M. G. Evans, M. Santappa, and M. Uri, *J. Polym. Sci.* **7**, 243 (1951).

18. M. G. Evans and M. Uri, *Nature (London)* **164**, 404 (1949).

19. M. Santappa, *J. Sci. Ind. Res.* **13**, 819 (1954).

20. L. T. Crews, U.S. Patent 2,495,214 (1950).

21. T. Ogawa and T. Taninaka, *J. Polym. Sci., Part A-1* **10**, 2005 (1972).

22. T. Nagabhushanam and M. Santappa, *J. Polym. Sci. Part A-1* **10**, 1511 (1972).
23. G. Oster, *Nature (London)* **173**, 300 (1954).
24. G. Oster and Y. Mizutani, *J. Polym. Sci.* **22**, 173 (1956).
25. I. A. Berstein, E. C. Farmer, W. G. Rothschild, and F. Spalding, *J. Chem. Phys.* **21**, 1303 (1953).
26. R. Bensson and A. Prevot-Bernas, *J. Chim. Phys. Physicochim. Biol.* **54**, 479 (1957).
27. E. Collinson and F. S. Dainton, *Discuss. Faraday Soc.* **12**, 212 (1952).
28. F. S. Dainton, *J. Chem. Soc., London* p. 1533 (1952).
29a. J. V. Schmitz and E. J. Lawton, *Science* **113**, 718 (1951).
29b. H. Sobue, T. Yoneho, and E. Oda, *Kagaku Zasshi* **64**, 378 (1961); *J. Polym. Sci.* **45**, 468 (1960).
30. O. Lindstrom and O. Lamm, *J. Phys. Colloid Chem.* **55**, 1139 (1951).
31. W. K. Wilkinson, *Macromol. Syn.* **2**, 78 (1966).
32. R. G. R. Bacon, *Trans. Faraday Soc.* **42**, 140 (1946).
33. J. A. Price, W. N. Thomas, and J. J. Padbury, U.S. Patent 2,626,946 (1953); *Chem. Abstr.* **47**, 670 (1953).
34. V. E. Shashoua, *in* "Preparative Methods of Polymer Chemistry" (W. R. Sorenson and T. W. Campbell, eds.), 2nd ed. Wiley (Interscience), New York, 1968. p. 235.
35. J. M. Grim, U.S. Patent 2,594,913 (1952); *Chem. Abstr.* **46**, 7822 (1952).
36. Y. Ashina, I. Oshima, and K. Sekine, U.S. Patent 3,701,761 (1972).
37. C. E. Schildknecht, "Vinyl and Related Polymers," pp. 270–271. Wiley, New York, 1952.
38. W. K. Wilkinson, U.S. Patent 3,087,919 (1963).
39. R. B. Parker and B. V. Moklar, U.S. Patent 3,161,511 (1964).
40. D. Feldman, *Mater. Plast.* **3**, 25 (1966).
41. H. Kiuchi, *Kobunshi Kagaku* **21**, 37 (1964); *Chem. Abstr.* **61**, 7107 (1964).
42. Union Rheinische Braunkohlen Kraftstoff, Belgian Patent 623,220 (1963); *Chem. Abstr.* **59**, 6582 (1963); German Patent 1,123,833 (1962); *Chem. Abstr.* **56**, 14483 (1962); J. Szafka and E. Turska, *Makromol. Chem.*, pp. 156, 297, and 311 (1972).
43. Courtaulds, Ltd., British Patents 796,294 (1958); *Chem. Abstr.* **52**, 19157 (1958); 831,049 (1960); *Chem. Abstr.* **54**, 15951 (1960).
44. E. L. Kropa, U.S. Patents 2,356,767 (1944); 2,425,192 (1947).
45. I. G. Murgulescu, T. Oncescu, and I. I. Vlagiu, French Patent 2,080,229 (1971); *Chem. Abstr.* **77**, 75804p (1972).
46. H. Miyama, N. Harumiya, and A. Takeda, *J. Polym. Sci., Part A-1* **10** (3), 943 (1972).
47. D. W. Chaney, U.S. Patent 2,537,031 (1951).
48. A. Tazawa, S. Omi, and H. Kubota, *J. Chem. Eng. Jap.* **5** (1), 44 (1972); *Chem. Abstr.* **77**, 88937j (1972).
49. R. G. Beaman, *in* "Preparative Methods of Polymer Chemistry" (W. R. Sorenson and T. W. Campbell, eds.), 2nd ed., p. 236. Wiley (Interscience), New York, 1968.
50. Research Division of American Cyanamid Co., *in* "Polymers and Copolymers of Acrylonitrile." American Cyanamid Co., New York, 1956.
51. Standard Oil of Ohio Research Laboratory, *in* "Methacrylonitric," Tech. Bull. Vistron Corporation, Cleveland, Ohio, 1968.
52. E. C. Hughes, J. D. Idol, J. T. Duke, and L. M. Wick, *Polym. Prepr., Amer. Chem. Soc., Div. Polym. Chem.* **10** (1), 403 (1969).
53. R. J. Coleman, *Macromol. Syn.* **2**, 63 (1966).
54. K. Tessmar, *Kunststoffe* **43**, 496 (1953); *Chem. Abstr.* **48**, 3061 (1954).
55. G. E. Ham, U.S. Patent 2,559,154 (1951).
56. E. C. Chapin and G. E. Ham, U.S. Patent 2,559,155 (1951).

57. A. T. Walter and G. H. Fremon, U.S. Patent 2,603,620 (1952).
58. H. W. Coover, Jr. and W. C. Wooten, Jr., U.S. Patent 2,831,826 (1958).
59. J. Furukawa, J. Tsuruta, and K. Morimoto, *Kogyo Kagaku Zasshi* **60**, 1402 (1957).
60. A. Zilkha, B. A. Feit, and M. Frankel, *J. Chem. Soc., London* p. 928 (1959).
61. M. Frankel, A. Ottolenghi, M. Albeck, and A. Zilkha, *J. Chem. Soc., London* p. 3858 (1959).
62. A. Zilkha, P. Neta, and M. Frankel, *J. Chem. Soc., London* p. 3357 (1960).
63. F. S. Dainton, D. M. Wiles, and A. N. Wright, *J. Polym. Sci.* **45**, 111 (1960).
64. J. L. Down, J. Lewis, B. Moore, and J. Wilkinson, *Proc. Chem. Soc., London* p. 209 (1957).
65. R. B. Cundall, J. Driver, and D. D. Eley, *Proc. Chem. Soc., London* p. 170 (1958).
66. R. B. Cundall, D. D. Eley, and R. Worrall, *J. Polym. Sci.* **58**, 869 (1962).
67. J. Ulbricht and R. Sourisseau, *Faserforsch. Textiltech.* **12**, 547 (1961).
68. L. Horner, W. Jurgeleit, and K. Klüpfel, *Justus Liebigs Ann. Chem.* **591**, 108 (1955).
69. D. C. Pepper, *Quart. Rev. Chem. Soc.* **8**, 88 (1954).
70. N. S. Wooding and W. C. E. Higginson, *J. Chem. Soc., London* p. 774 (1952).
71. E. F. Evans, A. Goodman, and L. D. Grandine, *in* "Preparative Methods of Polymer Chemistry" (W. R. Sorenson and T. W. Campbell, eds.), 2nd ed., p. 283. Wiley (Interscience), New York, 1968.
72. J. R. Schaefgen, *Macromol. Syn.* **2**, 81 (1966).
73. A. Zilkha, B. A. Feit, and M. Frankel, *J. Polym. Sci.* **49**, 231 (1961).
74. R. G. Beaman, *J. Amer. Chem. Soc.* **70**, 3115 (1948).
75. C. G. Overberger, E. M. Pearce, and N. Mayes, *J. Polym. Sci.* **34**, 109 (1959).
76. C. G. Overberger, H. Yuki, and N. Urakawa, *J. Polym. Sci.* **45**, 127 (1960).
77. Y. Joh, T. Yoshihara, Y. Kotake, F. Ide, and K. Nakatsuka, *J. Polym. Sci., Part B* **3**, 933 (1965).
78. Y. Joh, T. Yoshihara, Y. Kotake, F. Ide, and K. Nakatsuka, *J. Polym. Sci., Part B* **4**, 673 (1966).
79. G. Natta and G. Dall'Asta, *Chim. Ind.* (*Milan*) **46**, 1429 (1964); G. Natta, G. Mazzanti, and G. Dall'Asta, U.S. Patent 3,231,552 (1966).
80. Y. Joh, *in* "Preparative Methods of Polymer Chemistry" (W. R. Sorenson and T. W. Campbell, eds.), 2nd ed., p. 276. Wiley (Interscience), New York, 1968.
80a. N. G. Gaylord and A. Takahashi, *Advan. Chem. Ser.* **91**, 94 (1969).
80b. J. B. Kinsinger, private communication through M. Szwarc, *Makromol. Chem.* **35**, 132 (1959).
80c. H. Gilbert, F. F. Miller, S. J. Averill, E. J. Carlson, V. C. Folt, H. J. Heller, F. Stewart, R. F. Schmidt, and H. L. Trumbull, *J. Amer. Chem. Soc.* **78**, 1669 (1956).
80d. M. Imoto, T. Otsu, and M. Nakabayashi, *Makromol. Chem.* **65**, 194 (1963); M. Imoti, T. Otsu, and S. Shimizu, *ibid.*, p. 174.
80e. A. Takahashi and N. G. Gaylord, *Polym. Prepr., Amer. Chem. Soc., Div. Polym. Chem.* **10** (2), 546 (1969); B. Patnaik, A. Takahashi, and N. G. Gaylord, *Polym. Prepr., Amer. Chem. Soc., Div. Polym. Chem.* **10** (2), 554 (1969).
80f. N. G. Gaylord, S. S. Dixit, S. Maiti, and B. K. Patnaik, *7th Middle Atlantic Reg. Meet., Amer. Chem. Soc., Philadelphia*, Meeting Program and Abstracts, p. 101 (1972).
80g. N. W. Johnston, *Polym. Prepr., Amer. Chem. Soc., Div. Polym. Chem.* **13** (2), 1029 (1972).
80h. H. K. Hall, Jr. and J. W. Rhoades, *J. Polym. Sci., Part A-1* **10**, 1953 (1972).
80i. H. W. Coover, Jr. and T. H. Wicker, Jr., *Encycl. Polym. Sci. Technol.* **1**, 337 (1964).
80j. F. Leonard, J. A. Collins, and H. J. Porter, *J. Appl. Polym. Sci.* **10**, 1617 (1966).

80k. A. J. Canale, W. E. Goode, J. R. Kinsinger, J. R. Panchak, R. L. Kelso, and R. K. Graham, *J. Appl. Polym. Sci.* **4**, 231 (1960).
80l. M. Kreisel, U. Garbatski, and D. H. Kohn, *J. Polym. Sci., Part A* **2**, 105 (1964).
81. D. G. L. James and T. Ogawa, *J. Polym. Sci., Part B* **2**, 991 (1964).
82. Standard Oil (Ohio), Netherlands Patent Appl. 6,402,625 (1964); *Chem. Abstr.* **62**, 6649d (1965).
83. T. Tadokoro and H. Konishi, *Kogyo Kagaku Zasshi* **69**, 511 (1966); *Chem. Abstr.* **65**, 12295c (1966).
84. P. E. Wei and G. E. Milliman, *J. Polym. Sci., Part A-1* **7**, 2305 (1969).
85. R. G. Fordyce and G. E. Ham, *J. Amer. Chem. Soc.* **73**, 1186 (1951).
86. G. E. Ham, *Polym. Prepr., Amer. Chem. Soc., Dic. Polym. Chem.* **1** (2), 219 (1960).
87. H. S. Wildi and J. E. Katon, *J. Polym. Sci., Part A* **2**, 4709 (1964).
88. G. B. Butler and B. Zeeger, *Polym. Prepr., Amer. Chem. Soc., Div. Polym. Chem.* **12** (1), 420 (1971).
89. B. D. Halpern, W. Karo, L. Laskin, P. Levine, and J. Zomlefer, Tech. Rep. WADC TR 54-264. Wright Air Development Center, Air Research and Development Command, United States Air Force, Wright-Patterson Air Force Base, Ohio, 1954.
90. H. Gilbert, F. F. Miller, S. J. Averill, R. F. Schmidt, F. Stewart, and H. L. Trumbull, *J. Amer. Chem. Soc.* **76**, 1074 (1954).
91. S. K. Saha and A. K. Chandhuri, *J. Polym. Sci., Part A-1*, p. 797 (1972).
92. M. A. Siahkolah and W. K. Walsh, *Polym. Prepr., Amer. Chem. Soc., Div. Polym. Chem.* **13** (2), 716 (1972).
93. M. Salame, *Abstr., 164th Nat. Meet., Amer. Chem. Soc.*, New York ORPL No. 8 (1972).
94. D. H. Kohn, *Polim. Vehomarim. Plast.* No. 1, p. 10 (1971); *Chem. Abstr.* **77**, 102273y (1972).
95. A. M. Smirnova, P. I. Zubov, A. V. Uvarov, T. V. Raikova, and N. M. Kudryavtseva, *Vysokomol. Soedin., Ser. A*, p. 1169 (1972); *Chem. Abstr.* **77**, 62467f (1972).
96. N. Kawabata, C. K. Chen, and S. Yamasika, *Bull. Chem. Soc. Jap.* **45**, 1491 (1972); *Chem. Abstr.* **77**, 62357v (1972).
97. R. Maska, U.S. Patent 3,701,758 (1972).
98. M. Matsuda, *Makromol. Chem.* **65**, 224 (1963).
99. N. Tokura, M. Matsuda, and F. Yazaki, *Makromol. Chem.* **42**, 108 (1960).
100. S. R. Sandler, *J. Polymer Sci.* **11**, 2373 (1973).
101. C. Simionescu and C. Ungureaunu, Romanian Patent 52,689 (1972); *Chem. Abstr.* **78**, 30523a (1973); V. Nuta, D. Feldman, and C. Simionescu, *Rev. Roum. Chim.* 1755 (1972); *Chem. Abstr.* **78**, 30305f (1973).
102. le Doan Trung and A. Chapiro, *Eur. Polym. J.*, p. 455 (1973).
103. S. Matsumura and C. Kanemitsu, Japanese Patent, 72/37,713 (1972); *Chem. Abstr.* **78**, 16801h (1973).
104. D. E. Rice and C. L. Sandberg, *Polym. Prepr., Am. Chem. Soc., Div. Polym. Chem.* **12** (1), 396 (1971).
105. I. Sugiyama and K. Endo, *Japan Kokai* 72/16,539 (1972); *Chem. Abstr.* **79**, 54330g (1973).

Chapter 12

POLYACRYLAMIDE AND RELATED AMIDES

I. INTRODUCTION

As may be anticipated, a large variety of amides of acrylic and methacrylic acid have been studied. Of these, only acrylamide, methacrylamide, *N*-

methylolacrylamide (as 60% aqueous solution), *N,N'*-methylenebis(acrylamide) and diacetoneacrylamide have achieved a measure of industrial importance. Acrylamide itself is by far the most important monomer of this group. Acrylamide polymers and copolymers find application in a wide variety of fields such as adhesives, dispersants, flocculants, printing plates, viscosity modifiers and thickeners, fiber dying and modification, leather substitutes, paper sizing, protective colloids in photographic emulsions, surface coatings, textile treatments, gels for electrophoresis, improvement of cements, water purification, paper treatment, soil stabilization, well drilling, boiler water treatment, hair sprays, ion-exchange resins, pigment binders, and polyester-binding resins.

With the exception of *N*-methylolacrylamide, which is supplied as a 60% aqueous solution, the commercially available acrylic amides are all solids with substantial water solubility. The *N*-alkyl-*N*-1,1-dihydroperfluoroalkylacrylamides and methacrylamides are among the limited number of liquid amides [1].

Acrylamide, methacrylamide, and many of the simple nonfluorinated, related monomers are generally water-soluble. Usually they are also soluble in a large variety of organic solvents. Since they are amides, their solutions may be expected to have considerable compatibility or solvency for a variety of materials such as inorganic salts. Polyacrylamide, even when of high molecular weight, is readily dissolved in water. Under certain polymerization conditions, imide formation may take place [Eq. (1) and (2)]. Polymers which are partially imidized exhibit reduced water solubility. The intermolecular imide represents a thoroughly cross-linked polymer which is quite insoluble, while the intramolecular imides may exhibit only partial insolubility [2,3].

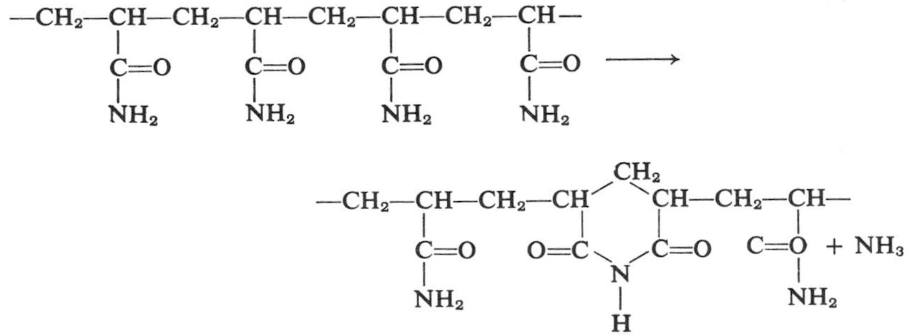

$$\text{—}CH_2\text{—}CH(CONH_2)\text{—}CH_2\text{—}CH(CONH_2)\text{—}CH_2\text{—}CH(CONH_2)\text{—}CH_2\text{—}CH(CONH_2)\text{—} \longrightarrow$$

$$\text{—}CH_2\text{—}CH(CONH_2)\text{—}CH_2\text{—}\underbrace{CH\text{—}CH_2\text{—}CH}_{\text{ring: } O{=}C\text{—}N(H)\text{—}C{=}O}\text{—}CH_2\text{—}CH(CONH_2)\text{—} + NH_3 \qquad (1)$$

Intramolecular imidization

The molecular weight and extent of imidization may be controlled by adjustment of the concentration of the monomer in a solvent, changes in the pH of the solution, control of the reaction temperature, and use of water-soluble alcohols as chain-transfer agents [4].

```
—CH2—CH—CH2—CH—CH2—CH—              —CH2—CH—CH2—CH—CH2—CH—
      |         |         |                |         |         |
     C=O       C=O       C=O              C=O       C=O       C=O
      |         |         |                |         |         |
     NH2       NH2       NH2              NH2       NH        NH2
                                                     |
              +                 ——→                 C=O              + NH3    (2)
                                                     |
—CH2—CH—CH2—CH—CH2—CH—              —CH2—CH—CH2—CH—CH2—CH—
      |         |         |                |                   |
     C=O       C=O       C=O              C=O                 C=O
      |         |         |                |                   |
     NH2       NH2       NH2              NH2                 NH2
```

Intermolecular imidization

Polymers of acrylamide frequently exhibit unusually high molecular weights. This phenomenon may be attributed to the ratio

$$k_p/k_t^{1/2}$$

(where k_p is the propagation rate constant and k_t is the termination rate constant)

being exceptionally high in an acrylamide free-radical polymerization. In fact, this ratio is said to be greater for this monomer than for most other monomers [5,6]. With the relatively low rate of terminations, the build-up of high molecular weight polymers can readily be visualized.

With a propagation rate at 25°C of 1.8×10^4 liter/mole/sec, a high rate of polymer formation may be expected. The heat of polymerization of the aqueous monomer is approximately 19.5 kcal/mole [7]; consequently adequate means of dissipating the evolving heat have to be found when more than a few grams of the monomer are to be polymerized. The heat of polymerization of methacrylamide is on the order of 13.5 kcal/mole.

As normally supplied, both acrylamide and methacrylamide are not inhibited. Consequently they should be stored in a cool location. Possible inhibitors which may be added, if required, are sodium nitrite, *tert*-butylhydroxyanisole, the ferrous-cupferron complex, chelating agents, etc.

The chemistry of acrylamide and related monomers has been reviewed [8–11]. Recent work on reactivity ratios based on three parameters—resonance stabilization energy, radical electronegativity, and monomer electronegativity—is reported by Hoyland [12].

The polymerization of acrylamide has been carried out by free-radical and anionic means using a variety of initiating systems. An unusual method of polymerization is the hydrogen-transfer process to form poly(β-alanine).

A. Safety Considerations

Since acrylamide is supplied as a white crystalline material (m.p. 84°–85°C) its appearance is similar to many well-crystallized, pure, and innocuous organic products.

As indicated above, the heat of polymerization of the monomer and its rate of polymerization are high. Consequently heat-transfer problems exist in handling the monomer.

Only relatively recently has it been discovered that acrylamide, and possibly also the substituted acrylamides, are toxic in a unique manner. Neuromuscular disorders of varying severity have been reported. The monomer may be absorbed through the intact skin, by inhalation, or by ingestion. There may be a degree of cumulative toxicity in animals [10]. It is suggested that reference [10] be consulted for recommendation of safety precautions that should be taken when handling acrylamide and related compounds.

Polyacrylamide, on the other hand, seems to be nontoxic, provided no residual monomer is present.

2. FREE-RADICAL POLYMERIZATION

A. Bulk and Solid-State Processes

For convenience, we are dividing the bulk polymerization processes into three subsections:

(1) catalytic polymerization
(2) radiation-induced solid-state polymerizations at ordinary pressure
(3) radiation-induced solid-state polymerization at high pressure.

a. Catalytic Polymerization

As indicated in the introductory section of this chapter, the heat of polymerization of acrylamide is quite high. Consequently, thermal polymerizations of this monomer should be attempted only on a very small scale with adequate precautions.

An early attempt to polymerize acrylamide as a melt at 150°C resulted in an insoluble polymer. Since ammonia was evolved during the process, presumably some imide formation had taken place [13].

Attempts to prepare polyacrylamide in bulk by heating the monomer with benzoyl peroxide also resulted in resins which were insoluble in water [14].

Three recent patents [15–17] claim that acrylamide may be polymerized in the presence of lithium perchlorate or lithium nitrate. It was postulated that only monomers with π-electron densities greater than ethylene (e.g., vinyl butyl ether, 1-methoxybutadiene, 2-vinyloxyethanol, styrene, α-methylstyrene, 4-vinylpyridine, acrylamide, methacrylamide, and *N*-vinyl monomers) polymerize in the presence of lithium perchlorate, while monomers with

π-electron densities less than ethylene (e.g., vinyl esters, methyl methacrylate, methyl vinyl ketone, maleic anhydride, and acrylonitrile) do not polymerize under these conditions [15]. A polymer produced by heating 1 mole of acrylamide with $\frac{1}{12}$ mole of lithium perchlorate in the presence of a small quantity of benzoyl peroxide between 105° and 130°C is described as rubberlike and of good optical clarity. During the polymerization the salt dissolved in the reaction mass [17].

In another example [16], 10 gm of acrylamide and 5 gm of anhydrous lithium nitrate were heated together at 82°C. After an induction period of 22 min, rapid polymerization took place. At 82°C, a mixture of 9 gm of acrylamide and 3 gm of anhydrous lithium nitrate polymerized after an induction period of 50 min [16].

b. Radiation-Induced Polymerization at Ordinary Pressures

The radiation-induced solid-state polymerization of acrylamide or of solid solutions of acrylamide in propionamide is characterized by considerable postirradiation reaction, the formation of amorphous polymers which nucleate as a second phase early in the process, and by the fact that crystal defects tend to be the sites of reaction. The polymerization takes place at the interface between the amorphous polymer and the monomer. The polymerization in the presence of propionamide is inhibited by oxygen [18,19]. It has been stated that these facts show that free radicals are certainly involved at the propagation state. Termination is, of course, restricted by the limited movement of radicals in the crystal lattice. The initiation is said to involve the scavenging of radiolytically produced hydrogen atoms by acrylamide to form free radicals [18].

Detailed directions for the polymerization of acrylamide in the crystalline state, initiated by the γ-rays from a cobalt-60 source, are given by Morawetz and Fadner [20].

In the presence of organic additives such as polynuclear hydrocarbons, aromatic amides, and urea derivatives, ionizing radiation is said to induce the formation of water-soluble polyacrylamides [21].

A recent study of the solid-state polymerization of acrylamide by ultraviolet radiation has shown that the polymerization is initiated primarily on the surfaces of the crystals [22]. It is interesting to speculate whether the capability of the penetration of γ-radiation into the crystal lattice as against the surface effects of ultraviolet radiation are significant factors in these processes.

Other references pertinent to this topic are [23–26]. See also Section 2,D, Photopolymerization and Related Polymerizations.

c. Radiation-Induced Polymerization at High Pressures

The solid-state polymerization induced by γ-radiation under conditions of very high pressure is of considerable theoretical interest. It has been demonstrated that the effects on the polymerization of solid acrylamide are the result of the pressure dependence of the activation enthalpy and of the activation entropy of the propagation step [27,28].

Since the equipment for this type of experimentation is beyond the scope of the average laboratory, detailed procedures are not given here. Table I cites references to a number of reports in this field [27–36].

TABLE I
HIGH-PRESSURE, SOLID-STATE POLYMERIZATION OF ACRYLAMIDE

Pressure range	Reaction temp. (°C)	Initiation by	Ref.
1–500 atm	0–70	γ-Radiation	27,28
1.5–3 × 10^4 atm	50	Shock waves induced by detonation	29
1.5–3.5 × 10^4 atm	−190 to +25	Shock waves induced by detonation	30–33
5 × 10^5 lb/in^2	25	Electron beam	34,35
10 kilobars	−196 to +25	With shear deformation	36

B. Suspension Polymerization

Since acrylamide and methacrylamide are quite water-soluble solids, the usual technique of suspension polymerization which consists of heating monomer droplets, suitably initiated, in a stirred aqueous medium cannot be used to prepare homopolymers of these compounds. This, of course, does not mean that copolymers cannot be prepared by standard methods provided the distribution coefficient of the acrylamide in the co-monomer and in water is favorable.

Alternative procedures involve adding a saturated aqueous solution of acrylamide converted to a water-in-oil emulsion by means of a suitable nonionic surfactant to an organic solvent, adding initiators and heating until beads of polymer form [37,38]. By this technique, the beads probably are highly hydrated polyacrylamide.

Preparation 2-1 is an example of this technique taken from the recent patent literature for reference purposes only. To be noted is the technique of removing the water in the course of the polymerization by azeotropic distillation.

2-1. Suspension Copolymerization of Acrylamide and Acrylic Acid [39]

$$n\,CH_2{=}CH{-}\overset{\overset{\displaystyle O}{\|}}{C}{-}NH_2 + m\,CH_2{=}CH{-}\overset{\overset{\displaystyle O}{\|}}{C}{-}OH \longrightarrow \left(\begin{array}{c}{-}CH_2CH{-}\\ |\\ C{=}O\\ |\\ NH_2\end{array}\right)_n \left(\begin{array}{c}{-}CH_2{-}CH{-}\\ |\\ C{=}O\\ |\\ OH\end{array}\right)_m \quad (3)$$

In an apparatus fitted for azeotropic removal of water and with a stirrer, to a refluxing mixture of 225 gm of benzene, 1 gm of carbon tetrachloride, and 14 gm of ethoxylated tallow fatty acid amides, is added over a period of approximately 2 hr a solution of 91.8 gm of acrylamide, 39.2 gm of acrylic acid, 16 gm of 25% aqueous solution of sodium vinylsulfonate, 10 gm of urea, 30 gm of water, 37 gm of 25% aqueous ammonia, and 0.02 gm of potassium persulfate. Heating is continued until all of the water has been removed. The bead polymer is filtered off and dried under reduced pressure at 50°C. Yield: 134 gm of a water-soluble copolymer.

Preparation 2-2, while describing a technique for the preparation of a series of cross-linked copolymers, is considered to be applicable to the preparation of non-cross-linked acrylamide polymers also. To be noted in this preparation is the use of *N,N'*-methylenebis(acrylamide) as a cross-linking agent. This compound, because of its obvious structural relationship to acrylamide, is frequently used to cross-link acrylamide polymers.

2-2. Suspension Copolymerization of Cross-linked Acrylamide and N-Acrylyl-4-aminomethylpyridine [40]

$$CH_2{=}CH{-}\overset{\overset{\displaystyle O}{\|}}{C}{-}NH_2 + CH_2{=}CH{-}\overset{\overset{\displaystyle O}{\|}}{C}{-}NHCH_2{-}C_5H_4N + CH_2{=}CH\overset{\overset{\displaystyle O}{\|}}{C}{-}NHCH_2NH\overset{\overset{\displaystyle O}{\|}}{C}{-}CH{=}CH_2 \longrightarrow$$

$$\cdots\left(\begin{array}{c}{-}CH_2{-}CH{-}\\ |\\ C{=}O\\ |\\ NH_2\end{array}\right)_n \left(\begin{array}{c}{-}CH_2{-}CH{-}\\ |\\ C{=}O\\ |\\ NH{-}CH_2{-}C_5H_4N\end{array}\right)_m \left(\begin{array}{c}{-}CH_2{-}CH{-}\\ |\\ C{=}O\\ |\\ NH{-}CH_2{-}NH{-}\overset{\overset{\displaystyle O}{\|}}{C}{-}(\underset{|}{C}H{-}CH_2{-})\end{array}\right)_p\cdots \quad (4)$$

To 3 ml of methanol is added 0.7 gm of Bentone 34 and 0.7 gm of Bentone 18C (National Lead Co.). After this suspending agent has been wetted thoroughly, it is stirred into 120 ml of xylene.

To this mixture is added a solution of 30 gm of acrylamide, the desired quantity of *N,N'*-methylenebis(acrylamide) (0.5 to 4%), and 3 gm of *N*-acrylyl-4-aminomethylpyridine in 27.6 ml of a pH 6.5 phosphate buffer solution (0.05 *M*). The mixture is deoxygenated for 30 min by passing a vigorous stream of nitrogen through it. The nitrogen flow is then reduced and, with very rapid stirring, 3 ml of a potassium persulfate solution (containing 25 mg of the salt per milliliter) is added. The stirrer speed is reduced to 60–80 rpm and the polymerization is initiated by introducing 3 ml of a solution of sodium metabisulfite (25 mg of the salt per milliliter).

The temperature of the reaction mixture rises rapidly to 70°C. Stirring is continued for 1 hr. Then 500 ml of acetone is added and stirring is stopped. The acetone assists in the settling of the beads. The beads are filtered off and dried at a moderate temperature under reduced pressure. By sieving, the dry beads may be separated into fractions of various diameters.

In the case of the *N*-alkyl-*N*-1,1-perfluoroalkylacrylamides, which are water-insoluble liquids, conventional suspension polymerization techniques may be used [1]. Suspension copolymers of *N*-1,1-dihydroheptafluorobutylacrylamide, a solid with m.p. 57.4°–57.6°C, may also be prepared by conventional procedures [41].

2-3. *Suspension Copolymerization of Trifluoroethyl Acrylate and N-1,1-Dihydroheptafluorobutylacrylamide* [41]

$$CH_2{=}CH{-}\overset{\overset{O}{\|}}{C}{-}OCH_2CF_3 + CH_2{=}CH{-}\overset{\overset{O}{\|}}{C}{-}NHCH_2(CF_2)_2CF_3 \longrightarrow$$

$$\left(\begin{array}{l}{-}CH_2{-}CH{-}\\ \qquad\quad |\\ \qquad\quad C{-}OCH_2CF_3\\ \qquad\quad \|\\ \qquad\quad O\end{array}\right)_n \left(\begin{array}{l}{-}CH_2{-}CH{-}\\ \qquad\quad |\\ \qquad\quad C{-}NH{-}CH_2{-}(CF_2)_2{-}CF_3\\ \qquad\quad \|\\ \qquad\quad O\end{array}\right)_m \qquad (5)$$

To 250 ml of a constantly stirred solution of 0.15% of poly(vinyl alcohol) in water, maintained at 80°C, is added a solution of 39 gm of *N*-1,1-dihydroheptafluorobutylacrylamide, and 0.025 gm of recrystallized benzoyl peroxide in 21.1 gm of trifluoroethyl acrylate. Constant stirring and heating at 80°C is continued for 3 hr. Steam is then passed into the dispersion to steam-distill the residual monomer off. The elastomeric product beads are filtered off and washed repeatedly with hot water. After drying at 60°C under reduced pressure the yield is 18 gm (72%). The product exhibits resistance to fungus growth. After conventional rubber processing with carbon black, paraffin, sulfur, and triethylamine, a cured flexible rubbery sheet with excellent solvent resistance is formed.

C. Solution Polymerization

From the standpoint of laboratory procedures, solution polymerizations of acrylamide and of related monomers are probably most satisfactory. Aqueous reaction media are widely used, generally with those free-radical initiators which are water-soluble. This implies the use of initiators which are commonly used in emulsion polymerizations of the water-insoluble monomers such as the persulfate salts or redox couples such as persulfate-metabisulfite. Since the polymers produced usually are of extremely high molecular weight, the processes are usually carried out in fairly dilute solutions (e.g., 10% acrylamide in water). If the viscosity of the system increases to ranges which make handling difficult, additional water may be added during the polymerization to keep the reaction conditions under control. The dilute reaction systems tend to prevent insolubilization of the polymer by imidization. They also permit control of the reaction temperature. The polymers may be isolated from the aqueous solution by precipitation with alcohols, by freeze-drying, or by evaporation of thin layers (if relatively small amounts of solid polymer are required).

Polymerizations at low pH tend to lead to imidization, while very high pH conditions may lead to hydrolysis of the amidic functional groups. A variety of additives have been used with the basic initiator systems. Among these are ammonia, ammonium chloride, and ammonia-producing compounds [42]. These reagents are said to reduce the induction period of the polymerization, increase the rate of polymerization, and produce polymers of higher molecular weight which are still readily dispersed in water. They are stable to degradation in dilute aqueous solution and yield polymers with superior flocculating properties when compared with polymers prepared in the absence of ammonia. Alcohols may be added to the initial reaction charge to act as chain-transfer agents [10]. Other chain-transfer agents, such as mercaptans, may also be used. Salts of the alkali metals or of the alkaline earth metals as well as 3,3′,3″-nitrilotris(propionamide) also enhance the rate of polymerization and control the chain length [43]. Preparation 2-4 is an example of the general procedure used for aqueous solution polymerizations. The method, taken from a patent, is given for reference only.

2-4. Polymerization of Acrylamide in Aqueous Solution [43]

In a 1 liter flask, to a solution of 27 gm of acrylamide and 54 gm of sodium sulfate in 210 ml of deoxygenated water under nitrogen is added with stirring 0.012 gm of 3,3′,3″-nitrilotris(propionamide) and 0.012 gm of ammonium persulfate. The reaction mixture is warmed at 60°C for 30 min. The solution is then cooled. In the product, 94% of the monomer has been converted to polymer.

In the field of soil stabilization, aqueous solutions of acrylamide containing *N,N'*-methylenebis(acrylamide) as a cross-linking agent are injected into poor types of soil along with an appropriate aqueous initiating system. Upon polymerization the soil quality is supposed to be improved. From time to time, soil stabilization is discussed both in relation to the use of acrylamide and of calcium or magnesium acrylate or methacrylate. Whether these approaches really have large-scale application is difficult to judge objectively. One thing is certain, the problems associated with long-term stability of the polymer–soil system under out-of-door conditions with various types of soils, bacteria, and weathering conditions must impose many severe constraints on this method.

A study of the kinetics of the polymerization initiated by potassium persulfate indicates that both the sulfate ion radical and the hydroxyl radical participate and that a cage effect may be operating to explain the reaction order [44,45].

The rate of polymerization (R_p) is given by

$$R_p = 0.108[K_2S_2O_8]^{0.5}[M]^{1.25}$$

where $[K_2S_2O_8]$ is the concentration of potassium persulfate in moles per liter and [M] is the concentration of acrylamide in moles per liter. R_p has the dimensions of moles per liter per minute [45].

The molecular weight of the polymer may be controlled by varying the potassium persulfate concentration. The empirical relation of molecular weight (M_n) to initiator concentration [I], for polyacrylamide in the concentration range of 10^{-4} to 2×10^{-5} mole/liter at 80°C in water is given by [46]

$$1.5 \times 10^{-4} M_n - 15 = [I]^{-1/2}$$

The well-known redox initiation couple, persulfate-metabisulfite, has been suggested for the polymerization of acrylamide [47,48]. The polymerization at pH 3, adjusted with phosphoric acid [48], produces a polymer solution which at 8.0% solids has a viscosity of approximately 2×10^4 cP. When this solution is injected slowly into 80% methanol, rigid particles of polyacrylamide are formed.

Initiating systems involving ceric ions as initiators are interesting since the polymer chains bind one cerium ion per polymer chain. It has been postulated that the incorporation of cerium involves the termination step of the polymerization. Ferric ions, which do not bring about initiation of acrylamide per se, also may add to the polymer chain at termination [49–53].

An interesting observation has been that a solution of acrylamide in water or other solvents may polymerize without the addition of further initiators. The effect has been attributed to air oxidation of the monomer to form peroxides or hydroperoxides which initiate further polymerization. Light accelerates the initiation process. While the rate of polymerization in water is high, in other anhydrous polar solvents, such as halides and alcohols, the rate of polymerization is lower [54].

Table II lists a selection of a variety of initiating systems which have been proposed for the polymerization of acrylamide in water in recent years [6,9,10,43–65].

In organic solvents, the polymerization of acrylamide may be carried out with the usual organic initiators. In many cases the polymer precipitates from the reacting system. Often the resultant products are water-insoluble. A recent study of the polymerization of acrylonitrile with styrene has shown that the solvent may have a marked effect on the reactivity ratio of the copolymerization and on the amount of imide formation which may take place [66].

A procedure for the preparation of polymethacrylamide in toluene is given here as an example of a technique which is said to produce a water soluble polymer.

2-5. Polymerization of Methacrylamide in Toluene [67]

In a 500 ml flask fitted with a nitrogen inlet tube, mechanical stirrer, reflux condenser, and an addition buret, thermostatted at 120° ± 1.5°C, to a solution of 15 gm of methacrylamide in 61 ml of toluene is added 9 ml of a 4% solution of benzoyl peroxide in toluene with stirring, while a stream of nitrogen is bubbled through the solution. At approximately 15 min intervals, further 9 ml portions of the 4% solution of benzoyl peroxide in toluene are added until 100 ml of the solution have been added (11 portions in a 3 hr period).

After the addition of initiator has been completed, heating and stirring is continued for 0.5 hr. The polymer is filtered off and purified by heating it under reflux three times with 30 ml portions of methylene chloride. After filtering from the methylene chloride, the polymer is dried to constant weight by repeatedly heating it to 100°C and cooling it under reduced pressure in the presence of calcium chloride and paraffin. Yield: 11.3 gm (75%), as a fine white powder.

This technique is particularly useful since the troublesome separation of unreacted monomer from its polymer is obviated.

A 20% solution of acrylamide in dioxane has been polymerized at 60°C in the presence of 0.3% of benzoyl peroxide. By nitrogen analysis of the polymer some imide formation was detected [68].

D. Photopolymerization and Related Polymerizations

The polymerization of acrylamide solutions, like that of many other monomers, may be initiated under suitable conditions by ultraviolet radiation

TABLE II

INITIATORS USED IN THE AQUEOUS SOLUTION POLYMERIZATION OF ACRYLAMIDE

Initiator	Additives	Reaction conditions: Temp. (°C)	Reaction conditions: Time (hr)	Ref.
Polymeric peroxide (from air oxidation)	—	—	—	54
Potassium persulfate	Isopropanol	75–80	2	10
Potassium persulfate	—	30–50	—	44–46
Ammonium persulfate	3,3′,3″-Nitrilotris-(propionamide)	20	8	10
Chlorate–sulfite redox couple	—	—	—	6
Persulfate–metabisulfite redox couple	—	—	—	47
Cesium(III) ion	3-Chloro-1-propanol	—	—	49
Ferric ion–bisulfite redox couple	—	—	—	55
Peroxy compounds, or redox couples	Neutral alkali metal or alkaline earth salts and nitrilobis(pro-pionamide)	0–100	0.5	43
$FeCl_2$	Oxazirane	−30	0.25	56
Persulfate–metabisulfite	Phosphoric acid	75	5–10	48
H^+, Cu^{2+}, Zn (metal)	—	50	0.5	57
Ceric ammonium nitrate	In dilute nitric acid	40	—	50,51
2,2′-Azobis(isobutyronitrile)	—	—	—	58,59
Perchloric acid, Co^{3+}	Dil. sulfuric acid	15–20	—	60
Persulfate–thiosulfate redox couple	—	30–50	—	61
Ce^{4+} and Fe^{3+}	In dil. nitric acid	40	—	51
Ce^{4+} and Cu^{2+}	In dil. nitric acid	40	—	51
1% Hydrogen peroxide	Isopropanol	90	—	9
Sodium thiosulfate–H_2O_2	—	—	—	62
Potassium permanganate–oxalic acid	$MnSO_4$	—	—	63
Pinacol–Ce^{4+}	Dilute nitric acid	15	1	52
Fe^{3+}	—	40	—	53
Thiourea–potassium bromate	Cationic or anionic surfactants	19–35	—	64
4,4′-Azobis-4-cyanopentanoic acid	—	—	—	65

or by visible light, not to mention by ultrasonic waves, X-rays, and γ-rays. (See also Sections 2, A, b and c on radiation-induced polymerizations.)

A relatively simple procedure for the preparation of small samples of polymer makes use of the procedure attributed to Jones [69]. In this procedure, as in many cases involving the polymerization of acrylamide, the process is carried out under carbon dioxide rather than nitrogen. While it has been postulated that carbon dioxide brings about a transitory decrease in the pH of the system which is thought to improve the polymerization process [70], the effect might simply be that it is easier to blanket a solution with carbon dioxide, which has a greater density than air, than with the lower density nitrogen.

2-6. *Ultraviolet Radiation-Induced Polymerization of Acrylamide* [68,69]

In a suitable ampoule is placed a solution of 2.5 gm of acrylamide and 0.010 gm of benzoyl peroxide in 10 ml of methanol. The solution is purged with carbon dioxide; the ampoule is then sealed and exposed for a period less than 24 hr to a General Electric AH-4 ultraviolet source. The polymer is then filtered off, extracted with acetone, washed with ether, and dried under reduced pressure over phosphorus pentoxide. An analysis of the polymer for nitrogen content indicated a level of 18.6% of nitrogen as against a theoretical value of 19.7% of nitrogen. The discrepancy has been attributed to imide formation (at a level which would not be detected by infrared spectroscopy). The polymer is, however, water-soluble.

Both homopolymers and copolymers of acrylamide have been prepared in liquid ammonia [71]. An outline for the patented procedure is given here for reference.

2-7. *Ultraviolet Radiation-Induced Polymerization of Acrylamide in Liquid Ammonia* [71]

At −30°C in a suitable apparatus to 550 gm of liquid ammonia is added 240 gm of precooled acrylamide and 2.5 gm of azobis(isobutyramidine) hydrochloride. This composition is irradiated for 4.5 hr with a 4 watt black-light fluorescent tube. A nontacky, water-soluble polymer may be isolated on evaporation of the ammonia. The polymer is said to exhibit an average molecular weight of approximately 4×10^6.

By similar procedures, copolymers of acrylamide with ammonium acrylate, vinyl chloride, and with diallyldimethylammonium chloride have been prepared.

With appropriate sensitizers, acrylamide solutions may be polymerized when irradiated by high-intensity visible light. The procedure has found application in the preparation of thin layers of polyacrylamide gels for electrophoretic studies, chromatography, and media for immunodiffusion in the biomedical field [70,72].

The procedure for the preparation of a thin layer of acrylamide–*N,N'*-methylenebis(acrylamide) copolymer involves equipment to prepare a gel of constant thickness. The monomer solution is placed on a glass plate floating on a layer of mercury inside a deep jar. Since the gel is somewhat difficult to cut, molds or templates made of poly(methyl methacrylate) coated with a parting agent (silicone oils, greases, or fluorocarbons) may also be used in this type of equipment).

2-8. Preparation of a Cross-linked Polyacrylamide Gel Thin Layer [70,72]

A solution of 13.98 ml of an aqueous solution containing 4.194 gm of acrylamide (i.e., a 30% solution of the monomer) and 0.03 gm of *N,N'*-methylenebis(acrylamide) (both monomers having been freshly reprecipitated), 0.74 ml of a 0.64% solution of sodium persulfate, 11.68 ml of an 0.005% solution of riboflavin, and 73.6 ml of water or a buffer are placed in the proper equipment (cf. preceding paragraph).

The outer jar is purged with carbon dioxide. The solution is then exposed for 6 min to the light of two well-centered 500 watt photoflood lamps mounted in appropriate reflectors with the bulbs placed at a distance of 40 cm from the surface of the liquid.

The concentration of the *N,N'*-methylenebis(acrylamide) may have to be varied to achieve the consistency of the gel required for specific applications.

A composition of 44.7 gm of acrylamide and 0.34 gm of *N,N'*-methylenebis(acrylamide) in 100 ml of water containing 2 ml of an 0.284% aqueous solution of sodium bisulfite, 2 ml of an 0.00350% aqueous solution of riboflavin-5-phosphate, 2 ml of an 0.060% aqueous solution of allyl thiourea, and 4 ml of a glycine-saline buffer (pH 7.3) is said to be photopolymerizable in the presence of oxygen [72].

The sensitization of the polymerization of acrylamide with various dyes has received considerable attention. Methylene blue is said to have the effect of increasing the rate of polymerization upon visible light irradiation of a melt of acrylamide beyond the rates achieved with ionizing radiation. The polymer produced is cross-linked [73]. In solution, the polymerization has been studied in the presence of methylene blue, new methylene blue N, methylene green, and eosin Y. An induction period due to oxygen was observed and the presence of a reducing agent was recommended. The rate of polymerization was at a maximum at pH 7 and approached zero below pH 4 [74]. The effects of dyes may be enhanced with triethanolamine, using a combination of cationic and anionic dyes, and using ethylene glycol as a solvent [75].

The role of oxygen in the dye-sensitized photopolymerization is still subject to further experimental investigation. For example, addition of cobaltous ions to a sensitized system changes its kinetic properties, e.g., the induction period is eliminated and the rate of polymerization is increased. This effect has been attributed to the cobaltous ion acting as a reversible oxygen carrier which acts as an "oxygen buffer" to furnish a required optimum level of oxygen to the system [76].

The use of diazidotetrammine cobalt(III) azide in photoinitiation has been investigated [77].

Application of sodium riboflavin-5′-phosphate and disodium fluorescein to the graft photopolymerization of acrylamide to rayon fibers has been reported [78].

Using electron irradiation, acrylamide, methacrylamide, or *N*-methylol-acrylamide has been grafted to polyethylene [79]. Similar grafts have been obtained with X-radiation [80].

The effect of pH on the γ-ray initiation (cobalt-60 source) has been studied [81].

The use of uranyl ions (UO_2^{2+}) as photosensitizers of acrylamide and methacrylamide polymerization has also been investigated [82].

3. ANIONIC POLYMERIZATION

The polymerization of acrylamide in the presence of simple anionic reagents such as sodium *tert*-butylate in the presence of free-radical inhibitors leads primarily to poly(β-alanine) by hydrogen-transfer polymerization, a topic to be discussed in the next section.

N-Substituted and *N*,*N*-disubstituted acrylamides in some cases may be polymerized by reagents such as metal alkyl–transition element halide catalysts or alkyl metals to produce crystalline polymers. Among the first crystalline materials produced by this means was poly(*N*-isopropylacrylamide) [83], followed by a series of crystalline *N*,*N*-disubstituted acrylamides [84]. The polymers have a certain degree of tacticity, although the polymer chains

$$\mathrm{Na} + \mathrm{CH_2{=}CH\overset{\overset{\displaystyle O}{\|}}{C}{-}NR_2} \longrightarrow \mathrm{Na^+} + \cdot\mathrm{CH_2{-}\underset{\underset{\displaystyle NR_2}{|}}{\underset{|}{\overset{-}{C}H}\!:\;C{=}O}} \xrightarrow{\text{monomer}}$$

$$\text{Radical}\underset{\substack{\text{atactic}\\ \text{block}}}{\sim\!\sim\!\sim}\mathrm{CH_2{-}\underset{\underset{\displaystyle CONR_2}{|}}{CH}}\overset{\text{isotactic block}}{\sim\!\sim\!\sim\!\sim}\text{anion} \quad (6)$$

are probably block copolymers of stereospecific polymer segments accompanied by free-radical chain segments. The polymerization process may be pictured as in Eq. (6) [84].

The catalysts suitable for anionic polymerization are based on the metals of the first and second groups of the Periodic Table in hydrocarbon media. In general, the lower the dielectric constant of the solvent, the greater the degree of tacticity of the polymer.

3-1. Anionic Polymerization of N,N-Dimethylacrylamide [84]

In a flask fitted with a stirrer and an argon inlet is placed a solution of *N,N*-dimethylacrylamide in toluene. The flask is thermostatted at room temperature, the air is displaced with argon, and a solution of ethyllithium is added. The temperature in the flask rises rapidly and within a few seconds the swollen polymer precipitates. The reaction is terminated by the addition of methanol. The product is filtered off and dried under reduced pressure. The yield is said to be quantitative. The product is said to be tactic and highly crystalline.

TABLE III
COMPARISON OF PROPERTIES OF TACTIC AND ATACTIC POLY(SUBSTITUTED ACRYLAMIDE)

Polymer of	M.p. (°C)	Approximate Vicat softening point (250 gm load) (%)	Solubility	Ref.
N-Isopropylacrylamide				
Atactic	100–125	—	Soluble in water, acetone, DMF	83
Tactic	170–200	—	Insoluble in water, acetone, DMF	83
N,N-Dimethylacrylamide				
Atactic	—	100	Soluble in benzene	84
Tactic	—	300	Insoluble in benzene	84
N,N-Di-*n*-butylacrylamide				
Atactic	—	60	Soluble in acetone	84
Tactic	—	350	Insoluble in acetone	84
N-Methyl-*N*-phenylacrylamide				
Atactic	—	180	Soluble in toluene	84
Tactic	—	330	Insoluble in toluene	84
N,N-Diisopropylacrylamide				
Atactic	—	120	Soluble in ethanol	84
Tactic	—	350	Insoluble in ethanol	84

A comparison of properties of stereospecific acrylamide-type polymers with the related free-radical-initiated atactic polymers is given in Table III.

4. HYDROGEN- (OR PROTON-) TRANSFER POLYMERIZATION

In a patent, Matlack [85] disclosed the formation of poly(β-alanine) when acrylamide was heated in the presence of bases. With subsequent investigations a novel method of polymerization was developed which has received much attention.

The incentive for extensive research in this field, aside from that generated by academic interest, may be attributed to the fact that poly(β-alanine), may

$$\left[-CH_2-CH_2-\overset{\overset{\displaystyle O}{\|}}{C}-NH-\right]_n$$

be considered nylon-3. Indeed it can be converted to fibers. The frequency of amide groups in the polymer chain is great, and the polymer is expected to resemble natural silk more closely than other nylons. However, to date it would appear that the molecular weight of polymers is still too low to permit commercial development.

The reaction of acrylamide to give poly(β-alanine) may be considered an example of the Michael reaction.

The initiation of the reaction may proceed by one of the two reactions represented by Eqs. (7) and (8) [86]. Equation (7) is the one favored by most investigators.

$$CH_2{=}CHCONH_2 + B^- \rightleftharpoons CH_2{=}CHCONH + BH \quad (7)$$

or

$$CH_2{=}CHCONH_2 + B^- \rightleftharpoons B{-}CH_2CHCONH_2 + BCH_2CH_2CONH^- \quad (8)$$

The propagation step also may be visualized as having two possible courses, the intramolecular propagation (Eq. 9) or the intermolecular propagation (Eq. 10):

$$CH_2{=}CHCONH^- + CH_2{=}CHCONH_2 \longrightarrow$$

$$CH_2{=}CH{-}CONHCH_2CHCONH_2 \longrightarrow CH_2CH_2CONHCH_2CH_2CONH \quad (9)$$

Intramolecular propagation

$$CH_2{=}CHCONH + CH_2{=}CHCONH_2 \longrightarrow$$

$$CH_2{=}CHCONHCH_2{-}CHCONH_2 + CH_2{=}CHCONH_2 \longrightarrow$$

$$CH_2{=}CHCONHCH_2CH_2CONH_2 + CH_2{=}CHCO\bar{N}H \quad (10)$$

Intermolecular propagation

The intermolecular propagation is currently considered the major factor in explaining the experimental details which have been accumulated [87,88].

In their extensive review of the hydrogen-transfer polymerization, Kennedy and Otsu [87] state that besides the step growth by the proton-transfer mechanism, anionic chain growth also takes place. The products therefore are either mixtures or copolymers formed by the two possible mechanisms. The difficulties in reaching clear-cut decisions as to the details of the process may arise from the fact that the polymerizations are carried out at temperatures above 100°C, so that it is difficult to differentiate between the various processes. Molecular weights and conversions are usually fairly low even upon prolonged heating.

In the preparation of polymers, it is usual to inhibit the possibilities of free-radical polymerization by incorporation of phenyl-β-naphthylamine. (CAUTION: may be carcinogenic.) (Obviously, hydroquinone would be unsuitable in a basic medium.)

4-1. Hydrogen-Transfer Polymerization of Acrylamide in Pyridine [86]

To a solution of 100 ml of specially dried pyridine containing 0.02 gm of phenyl-β-naphthylamine (CAUTION: may be carcinogenic), heated to 100°C, is added, with mechanical stirring, 10 gm of sublimed acrylamide. After the monomer has been dissolved, a solution of 0.1 gm of sodium in 10 ml of *tert*-butyl alcohol is added. Within a few minutes some polymer begins to separate. Heating and stirring is continued for 16 hr.

The polymer is then filtered off and extracted with boiling water for 1 hr. The insoluble portion is dried under reduced pressure at 80°C. Yield: 4.8 gm (48%).

The aqueous extract is neutralized with acetic acid and evaporated to dryness. The residue is dissolved in water and precipitated by stirring into methanol (yield 2.6 gm). From the mother liquor 0.4 gm of acrylamide and 1.2 gm of another solid may be isolated.

A similar procedure may be found in *Macromolecular Syntheses* [89].

A variation of Procedure 4-1 using toluene as a reaction medium is of interest since it involves not only the use of sodium but also magnesium, calcium, barium, and aluminum as catalysts along with pentanol or optically active 2-pentanol [88]. Table IV summarizes the polymerization procedures and results for the polymerization of acrylamide or methacrylamide which is outlined in Preparation 4-2.

4.2 Hydrogen-Transfer Copolymerization of Acrylamide with Methacrylamide in Toluene [88]

A solution of 10 gm of acrylamide (or 12 gm of methacrylamide) in 100 ml of toluene containing 0.4 gm of phenyl-β-naphthylamine (CAUTION: may be

TABLE IV [88]
POLYMERIZATION OF ACRYLAMIDE OR METHACRYLAMIDE

Catalyst system		Polym. time (hr)	Conversion (%) (% hydrogen-transfer units in polymer)	Polymer type[a]	Intrinsic viscosity[b] (optical rotation, α_D^{25} [c] in degrees)
Alcohol	Metal				
		Acrylamide			
1-Pentanol	Na	10	95 (68)	H	0.68 (0)
	Mg	10	8.4 (75)	H	—
	Ca	5	96 (73)	H	0.43 (0)
	Ba	5	95 (80)	H	0.86 (0)
	Al	20	14 (0)	A	—
2-Pentanol	Na	10	73 (81)	H	0.58 (0)
	Ca	5	94 (75)	H	0.30 (0)
	Ba	5	96 (77)	H	0.71 (0)
		Methacrylamide			
1-Pentanol	Na	10	75 (88)	H	0.079 (0)
	Mg	20	5.5 (0)	A	—
	Ca	10	68 (78)	H	0.087 (0)
	Ba	10	59 (70)	H	0.15 (0)
	Al	20	5.0 (0)	A	—
2-Pentanol	Na	20	53 (92)	H	0.071 (0)
	Ca	10	52 (73)	H	0.061 (1.1)
	Ba	10	49 (59)	H	0.15 (1.3)

[a] H, Hydrogen-transfer polymer; A, polymer of 1,2-addition polymerization ("vinyl" polymerization).

[b] Determined in formic acid at 30°C.

[c] Determined in formic acid at 25°C.

carcinogenic) and 4.67×10^{-3} mole of the catalyst (see Table IV for components) is stirred and heated at 100°C for 5–20 hr. The reaction mixture is cooled and treated with dilute hydrochloric acid. The product is then filtered off. The polymer is dissolved in formic acid and reprecipitated by addition to acetone. The resulting polymer is filtered, washed with acetone, and dried under reduced pressure. Note that in the case of methacrylamide polymers initiated with an optically active alcoholate, optically active polymers are formed.

As discussed at length by Kennedy and Otsu [87], hydrogen-transfer polymerizations may also be observed with a variety of other monomers. For example, maleimide and its derivatives give polymers that may be considered copolymers of the hydrogen-transfer polymer and vinyl polymer types [90]. The compound *p*-vinylbenzamide will undergo hydrogen-transfer polymerization under base-catalyzed conditions. Under certain conditions of solvent and in the presence of lithium chloride, this monomer will undergo vinyl type anionic polymerization, while under other conditions of solvent, salts have no effect on hydrogen-transfer polymerization [90–93]. Recently it was found that alkali salts of alkylxanthates, which initiate anionic processes in the case of many monomers, including *N*-phenylmaleimide (NPMI), bring about hydrogen-transfer polymerizations in the case of acrylamide [93a].

5. MISCELLANEOUS METHODS

1. Among the commercially available amide-type monomers is *N*-(1,1-dimethyl-3-oxybutyl)acrylamide, more commonly called diacetone acrylamide. This solid monomer is soluble in water, ethanol, hexanol, acetone, ethyl acetate, and benzene. It is somewhat soluble in heptane (about 1% solubility) and nearly insoluble in petroleum ether [94–96]. The monomer may be polymerized by most of the techniques discussed for acrylamide and methacrylamide. Its chemistry and applications are discussed in technical bulletins of the manufacturer [96a]. Related monomers also available are a hydroxymethylated diacetone acrylamide and 2-acrylamido-2-methylpropanesulfonic acid [96a].

2. Polymerization in an aqueous alcohol solution to produce polymer particles [97].

3. Cross-linked polyacrylamides by transamidation of polyacrylamides with polymeric diamines [98].

4. Polymerization of acrolein in the presence of acrylamide induced by pyridine in an aqueous system [99].

5. Graft copolymers of acrylamide on poly(vinyl alcohol) using ceric ammonium nitrate in dilute nitric acid as initiator [100].

6. Polymerization of acrylamide or methacrylamide with hydrogen atoms generated by reaction of zinc amalgam with hydrochloric acid [101].

7. Photo- and radiation-initiated polymerization of acrylamide [102].

REFERENCES

1. B. D. Halpern, W. Karo, and P. Levine, U.S. Patent 2,957,914 (1960).
2. W. O. Kenyon and L. M. Minsk, U.S. Patent 2,486,190 (1949); *Chem. Abstr.* **44**, 1750 (1950).
3. L. M. Minsk and W. O. Kenyon, U.S. Patent 2,486,192 (1949); *Chem. Abstr.* **44**, 1750 (1950).
4. L. M. Minsk, W. O. Kenyon, and J. H. Van Campen, U.S. Patent 2,486,191 (1949); *Chem. Abstr.* **44**, 1750 (1950).
5. F. S. Dainton and M. Tordoff, *Trans. Faraday Soc.* **53**, 499 and 666 (1957).
6. T. S. Suen, Y. Jen, and J. V. Lockwood, *J. Polym. Sci.* **31**, 481 (1958).
7. R. M. Joshi, *J. Polym. Sci.* **60**, 556 (1962).
8. C. E. Schildknecht, "Vinyl and Related Polymers," p. 314ff. Wiley, New York, 1952.
9. W. M. Thomas, *Encycl. Polym. Sci. Technol.* **1**, 177 (1964).
10. "Chemistry of Acrylamide." Process Chemicals Department, American Cyanamid Co., Wayne, New Jersey, 1969.
11. "Methacrylamide," SP-32. Rohm & Haas Company, Specialty Products Department, Philadelphia, Pennsylvania, 1954.
12. J. R. Hoyland, *J. Polym. Sci., Part A-1* **8**, 885 and 901 (1970).
13. H. Staudinger and E. Urech, *Helv. Chim. Acta* **12**, 1132 (1929).
14. G. D. Jones, J. Zomlefer, and K. Hawkins, *J. Org. Chem.* **9**, 500 (1944).
15. R. B. Hodgdon, Jr., U.S. Patent 3,239,494 (1966).
16. R. B. Hodgdon, Jr., U.S. Patent 3,214,419 (1965).
17. L. G. Gilman and R. I. Lait, U.S. Patent 3,236,705 (1966).
18. J. Adler and J. H. Petropoulos, *Polym. Prepr., Amer. Chem. Soc.; Div. Polym. Chem.* **5** (2), 1009 (1964).
19. G. Adler, *U.S. At. Energy Comm.* **BNL-9200** (1965).
20. H. Morawetz and T. A. Fadner, *Macromol. Syn.* **2**, 25 (1966).
21. A. D. Abkin, A. P. Sheinker, M. K. Yakovleva, L. P. Titova, V. Ya. Dudarev, S. N. Prosvirova, I. A. Yakabovich, and N. P. Pashkhin, U.S.S.R. Patent 183,389 (1966).
22. B. M. Baysal, H. N. Erten, and Ü. S. Ramelow, *J. Polym. Sci., Part A* **9**, 581 (1971).
23. H. Morawetz, *J. Polym. Sci., Part A-1* **4**, 2487 (1966).
24. I. Kaetsu, N. Saganes, K. Hayashi, and S. Okamura, *J. Polym. Sci., Part A-1* **4**, 2241 (1966).
25. M. K. Yakovlev, A. P. Sheinker, and A. D. Abkin, *Vysokomol. Soedin., Adgez. Polim.* **12**, 1103 (1970).
26. T. Matsuda, T. Higashim, and S. Okamura, *J. Macromol. Sci.* **4**, 1 (1970).
27. Y. Tabata and T. Suzuki, *Polym. Prepr., Amer. Chem. Soc., Div. Polym. Chem.* **5** (2), 997 (1964).
28. Y. Tabata and T. Suzuki, *Makromol. Chem.* **81**, 223 (1965).
29. G. A. Adadurov, I. M. Barkalov, V. I. Gol′danskii, A. N. Dremin, T. N. Ignatovich, A. M. Mikhailov, V. L. Tal′roze, and P. A. Yampol′skii, *Vysokomol. Soedin.* **7**, 180 (1965).

30. G. A. Adadurov, I. M. Barkalov, V. I. Gol'danskii, A. N. Dremin, T. N. Ignatovich, A. M. Mikhailov, V. L. Tal'roze, and P. A. Yampol'skii, *Dokl. Akad. Nauk SSSR* **165**, 851 (1965).
31. I. M. Barkalov, V. I. Gol'danskii, V. L. Tal'roze, and P. A. Yampol'skii, *Zh. Eksp. Teor. Fiz., Pis'ma Red.* 3, 309 (1966).
32. A. V. Barabe, A. N. Dremin, and A. M. Mikhailov, *Fiz. Goreniya Gzryva* **5**, 583 (1969).
33. T. N. Ignatovich, P. P. Tampol'skii, and L. M. Bragints, *Vysokomol. Soedin.* **12**, 506 (1970).
34. M. I. Prince and J. Hornyak, *J. Polym. Sci., Part B* **4**, 493 (1966).
35. M. I. Prince and J. Hornyak, *J. Polym. Sci., Part A-1* **5**, 531 (1967).
36. A. G. Kazakevich, A. A. Zharov, P. A. Yampol'skii, N. S. Enikolopyan, and N. S. Gol'danskii, *Dokl. Akad. Nauk. SSSR* **186**, 1348 (1969).
37. Morningstar-Paisley, Inc., British Patent 991,416 (1965).
38. Nalco Chemical Co., British Patent 1,000,307 (1965); J. W. Zimmermann and A. Kühlkamp, U.S. Patent 3,211,708 (1965).
39. H. Spoor, German Patent 2,064,101 (1972); *Chem. Abstr.* **77**, 127438u (1972).
40. J. I. Kau and H. Morawetz, *Polym. Prepr., Amer. Chem. Soc., Div. Polym. Chem.* **13**, (2), 819 (1972).
41. B. D. Halpern and W. Karo, U.S. Patent 2,951,830 (1960).
42. M. B. Goren, U.S. Patent 3,200,098 (1965).
43. American Cyanamid Co., Netherlands Patent Appl. 6,505,750 (1965).
44. J. P. Riggs and F. Rodriguez, *Polym. Prepr., Amer. Chem. Soc., Div. Polym. Chem.* **6** (1), 207 (1965).
45. J. P. Riggs and F. Rodriguez, *J. Polym. Sci., Part A-1* **5**, 3151 (1967).
46. E. Ureta and M. Salona, *Rev. Soc. Quim. Mex.* **10**, 153 (1966).
47. F. Rodriguez and R. D. Givey, *J. Polym. Sci.* **55**, 713 (1961).
48. J. F. Tellenzui, Japanese Patent 65/25,908 (1965).
49. G. Mino, S. Kaizerman, and E. Rasmussen, *J. Polym. Sci.* **38**, 393 (1959).
50. H. Narita and S. Machida, *Makromol. Chem.* **97**, 209 (1966).
51. H. Narita, S. Okamoto, and S. Machida, *Makromol. Chem.* **111**, 14 (1968).
52. H. Narita, T. Okimoto, and S. Machida, *J. Polym. Sci., Part A-1* **8**, 2725 (1971).
53. H. Narita, Y. Sakamoto, and S. Machida, *Makromol. Chem.* **143**, 279 (1971).
54. A. Nakano and Y. Minoura, *Kogyo Kagaku Zasshi* **71**, 732 (1968).
55. G. Talamini, A. Turrlia, and E. Vianello, *Chim. Ind. (Milan)* **45**, 335 (1963).
56. S. Saki, S. Fujii, M. Kitamura, and Y. Ishi, *J. Polym. Sci., Part B* **3**, 955 (1965).
57. M. Suzuki and E. Hirata, Japanese Patent 66/1797 (1966); *Chem. Abstr.* **65**, 12358 (1966).
58. H. G. Burrows and S. M. Todd, British Patent 1,024,172 (1966).
59. H. I. Patzelt, L. J. Connelly, E. G. Bellweber, D. B. Korzenski, and K. L. Slepicka, South African Patent 6,901,065 (1969); *Chem. Abstr.* **72**, 67748e (1970).
60. M. Santappa, V. Makadevan, and K. Jijie, *Proc. Indian Acad. Sci., Sect. A* **64** (3), 128 (1966).
61. J. P. Riggs and F. Rodriguez, *J. Polym. Sci., Part A-1* **5**, 3167 (1967).
62. N. P. Dymarchuk, G. S. Kachalova, and L. A. Lavrova, *Tr. Leningrad. Tekhnol. Inst. Tsellyul.-Bum. Prom.* **27**, 192 (1970).
63. G. S. Misra and C. V. Gupta, *Makromol. Chem.* **156**, 195 (1972).
64. J. S. Shukla and D. C. Misra, *Makromol. Chem.* **158**, 9 (1972).
65. J. A. Caskey and A. L. Fricke, *Makromol. Chem.* **158**, 27 (1972).
66. N. W. Johnston and N. J. McCarthy, Jr., *Polym. Prepr., Amer. Chem. Soc., Div. Polym. Chem.* **13** (2), 1278 (1972).

67. C. L. Arcus, *J. Chem. Soc., London* p. 2732 (1949).
68. H. C. Haas and R. L. MacDonald, *J. Polym. Sci., Part A-1* **9**, 3583 (1971).
69. G. D. Jones, U.S. Patent 2,533,166 (1950).
70. B. Antoine, *Science* **138**, 977 (1962).
71. M. L. Peterson, U.S. Patent 3,681,215 (1972).
72. H. F. Mengoli and A. L. Wayne, *Nature (London)* **212**, 481 (1966).
73. C. S. Hsia Chen, *J. Polym. Sci., Part B* **2**, 891 (1964).
74. C. S. Hsia Chen, *J. Polym. Sci., Part A* **3**, 1807 (1965).
75. C. S. Hsia Chen, *J. Polym. Sci., Part A* **3**, 1107, 1127, 1137, and 1155 (1965).
76. N.-L. Yang and G. Oster, *J. Phys. Chem.* **74**, 856 (1970).
77. H. Kothandaraman and M. Santappa, *J. Polym. Sci., Part A-1* **9**, 1351 (1971).
78. H. L. Needles, *Polym. Prepr.* **10** (1), 302 (1969).
79. K. Matsumae, S. Nakagawa, and A. Furahashi, Japanese Patent 72/08,716 (1972).
80. I. Kaetsu, S. Okamura, and K. Hayashi, Japanese Patent 72/23,174 (1972).
81. D. J. Currie, F. S. Dainton, and W. S. Watt, *Polymer* **6**, 451 (1965).
82. K. Venkatarao and M. Santappa, *J. Polym. Sci., Part A-1* **5**, 637 (1967).
83. D. J. Shields and H. W. Coover, Jr., *J. Polym. Sci.* **39**, 332 (1959).
84. K. Butler, P. R. Thomas, and G. J. Tyler, *J. Polym. Sci.* **48**, 357 (1960).
85. A. S. Matlack, U.S. Patent 2,672,480 (1954).
86. D. S. Breslow, G. E. Hulse, and A. S. Matlack, *J. Amer. Chem. Soc.* **79**, 3760 (1957).
87. J. P. Kennedy and T. Otsu, *J. Macromol. Sci., Part C* **6**, 237 (1972).
88. K. Yamaguchi and Y. Minoura, *J. Polym. Sci., Part A-1* **10**, 1217 (1972).
89. D. S. Breslow, G. E. Hulse, and A. S. Matlack, *Macromol. Syn.* **2**, 12 (1966).
90. K. Kojima, N. Yoda, and C. S. Marvel, *J. Polym. Sci., Part A-1* **4**, 1121 (1966).
91. S. Nogichi and Y. Tamura, *J. Polym. Sci., Part A-1* **5**, 2911 (1967).
92. T. Asahara and N. Yoda, *J. Polym. Sci., Part A-1* **6**, 2477 (1968).
93. T. Asahara, K. Ikeda, and N. Yoda, *J. Polym. Sci., Part A-1* **6**, 2489 (1968).
93a. K. Yamaguchi, O. Sonoda, and Y. Minoura, *J. Polym. Sci., Part A-1* **10**, 63 (1972).
94. J. F. Bork, D. P. Wyman, and L. E. Coleman, *J. Appl. Polym. Sci.* **7**, 451 (1963).
95. L. E. Coleman, J. F. Bork, and D. P. Wyman, *Polym. Prepr., Amer. Chem. Soc., Div. Polym. Chem.* **5** (1), 250 (1964).
96. L. E. Coleman, *Encycl. Polym. Sci. Technol.* **15**, 353 (1971).
96a. "Diacetone Acrylamide," Tech. Bull. Lubrizol Corporation, Cleveland, Ohio, 1972.
97. A. D. Abkin, P. M. Khomikovskii, V. F. Gromov, Z. Y. Bezzubik, I. A. Yakubovich, P. I. Paradnya, N. P. Pashkin, and M. P. Vilyanskii, U.S.S.R. Patent 189,579 (1966).
98. D. E. Nagy, L. L. Williams, and A. T. Coscia, U.S. Patent 3,488,720 (1970).
99. N. Yamashita, M. Yoshihara, and T. Maeshima, *J. Polym. Sci., Part B* **10**, 643 (1972).
100. G. Mino and S. Kaizerman, *Macromol. Syn.* **2**, 84 (1966).
101. G. Mino, U.S. Patent 2,861,982 (1958).
102. T. Hori and H. Kodama, U.S. Patent 3,764,501 (1973); J. P. Communal, J. Fritz, and B. Papillon, German Patent 2,248,715 (1973); L. Perce, *J. Polym. Sci., Part B* **11**, 267 (1973).

Chapter 13

ORGANOPHOSPHORUS POLYMERS

I. INTRODUCTION

Organophosphorus polymers and copolymers [1,1a] are gaining in interest as a result of their effectiveness as flame retardants [2]. Most definitive reviews [1,3,3a–3d] or monographs [4,4a,4b] lack preparative details for the preparation of organophosphorus monomers and polymers, and few scientific papers on the polymer chemistry of phosphorus compounds are evident in the polymer journals [5,5a,5b]. However, much of the application technology appears in an ever-increasing volume of patents, as can be seen by a quick review of *Chemical Abstracts* under the topics of "Fire retardants" or "Phosphorus compounds—polymers."

Phosphorus polymers are also of interest in other areas besides fire retardants since the P—O—C group has a plasticizing effect on polymers resulting in a reduction of the melting point or softening point. The P—OH and $R_3P^+X^-$ groups improve water solubility, and in the case of the former the solubility in base is markedly improved. Phosphorus compounds are also of interest as fungicides, bactericides, pesticides, and insecticides.

The most stable phosphorus compounds are the phosphine oxides, followed by the phosphonates with a P—C bond. The P—O- and P—N-bonded compounds are hydrolytically unstable. The phosphonates are usually the best practical compromise.

The unique chemistry of phosphorus can be better understood by examining the electronic structure of the orbitals and the valency possibilities.

$$P^{15} = 1s^2 2s^2 2p^6 3s^2 3p^3$$

The bonding for phosphorus includes either $3p$ orbitals, or various $3S$–$3P$ hybrids. In addition *Spd* hybrids are possible because of the readily available *d* orbitals. Some of the bonding possibilities are shown for the following compounds [4a,6,7].

Compound	Bond hybridization	Directional characteristics
PCl_5	sp^3d	Trigonal bipyramidal
PCl_6^-	sp^3d^2	Tetragonal bipyramidal
PCl_3	p^3 and sp^3	Trigonal pyramidal
PO_4^{2-}	sp^3	Tetrahedral
$POCl_3$	sp^3	Tetrahedral

The P—O bond usually involves localized π-bonding as in $Cl_3P{=}O$, which has been confirmed by X-ray data. On the other hand, the π-bond character of the P—O bond is spread over all the P—O bonds in $(HO)_3P{=}O$ [4a].

The weak to moderate $d\pi$–$p\pi$ bonding is associated with the rearrangement in Eq. (1) and the existence of the phosphonitrilic compounds $(PNCl_2)$. The vacant *d* orbitals are also responsible for the multiple-bond character of the

TABLE I

NOMENCLATURE AND OXIDATION STATES OF SOME REPRESENTATIVE PHOSPHORUS COMPOUNDS

Compound	Oxidation state	Name
$(HO)_3PO$	+5	Phosphoric acid
$(RO)_3PO$		Trialkyl phosphate
Cl_3PO		Phosphorus oxychloride
Cl_5P		Phosphorus pentachloride
$(RO)_2P(=O)$—Cl		Dialkyl chlorophosphate
$(HO)_2P(=O)$—H	+3	Phosphorous acid
PCl_3		Phosphorus trichloride
$(RO)_3P$		Trialkyl phosphite
$RP(=O)$—$(OH)_2$		Alkylphosphonic acid
$(RO)_2PCl$		Dialkyl chlorophosphite
$(RNH)_3P$		Trialkyl phosphorous triamide
$H_2P(=O)$—OH	+1	Hypophosphorous acid
$R_2P(=O)$—OH		Dialkylphosphinic acid
$RP(=O)(H)$—OH		Alkylphosphonous acid
$(C_2H_5)_2P(=O)$—OR		Alkyl diethylphosphinate
R_3PO	−1	Trialkyl phosphine oxide
R_2POH		Dialkylphosphinous acid
H_3P	−3	Phosphine
R_3P		Trialkylphosphine
$R_4P^+Cl^-$		Tetraalkylphosphonium chloride

oxygen atom shown in Eq. (2). Phosphorus can also undergo valency

$$\gt P{-}OH \rightleftharpoons H{-}P{=}O \qquad (1)$$

$$R_3P^+{-}O^- \longleftrightarrow R_3P{=}O \qquad (2)$$

expansion, as is evidenced in its having coordination numbers of 5 or 6.

In Table I is a brief outline of the nomenclature and the oxidation states of some typical phosphorus compounds.

Polymers containing phosphorus can be obtained by the polymerization of unsaturated phosphorus monomers, by the condensation of reactive phosphorus compounds with polyols, carboxylic acids, amines, olefins and diolefins, isocyanates, and many other functional groups, or by self-polymerization (including ring opening). Phosphonitrilic polymers are also an important class and have recently been reviewed in a monograph [8]. Some of the many polymer-forming reactions are briefly outlined in Scheme 1.

The reactions of carbon–phosphorus heterocycles [30] to give polymers have received little attention. Polyesters should be possible using the following heterocyclic phosphorus dicarboxylic acid.

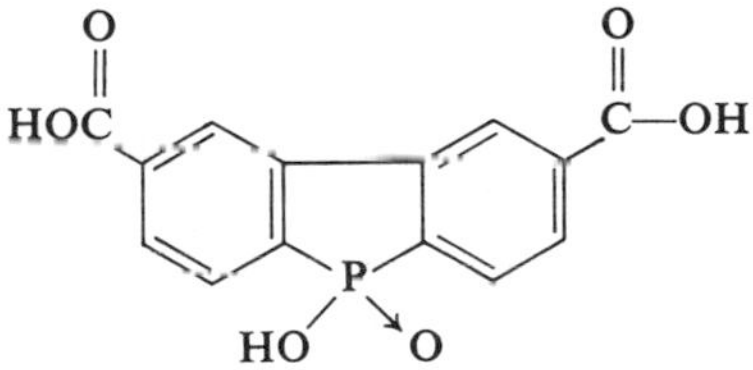

HOC(=O)–[dibenzophosphole, P(OH)→O]–C(=O)–OH

Some representative methods for preparing the starting phosphorus compounds are shown in Scheme 2.

The Diels-Alder reaction is also useful in preparing unsaturated cyclic monomers, as shown for the vinylphosphonates in Eqs. (3) and (4) [62].

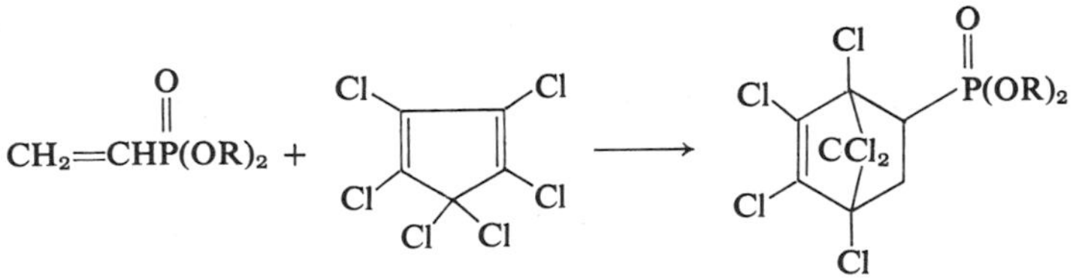

$$CH_2{=}CH\overset{O}{\overset{\|}{P}}(OR)_2 + C_5Cl_6 \longrightarrow \text{Cl}_6\text{-bicyclo adduct with } CCl_2 \text{ bridge}{-}\overset{O}{\overset{\|}{P}}(OR)_2 \qquad (3)$$

$$CH_2{=}CH\overset{O}{\overset{\|}{P}}(OR)_2 + CH_3CH{=}CH{-}CH{=}CHCH_3 \longrightarrow \text{3,6-dimethylcyclohexene}{-}\overset{O}{\overset{\|}{P}}(OR)_2 \qquad (4)$$

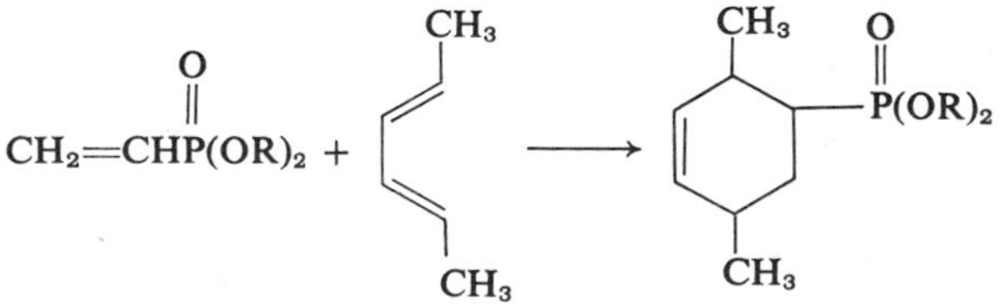

SCHEME 1

Polymer-Forming Reactions Utilizing Phosphorus Compounds

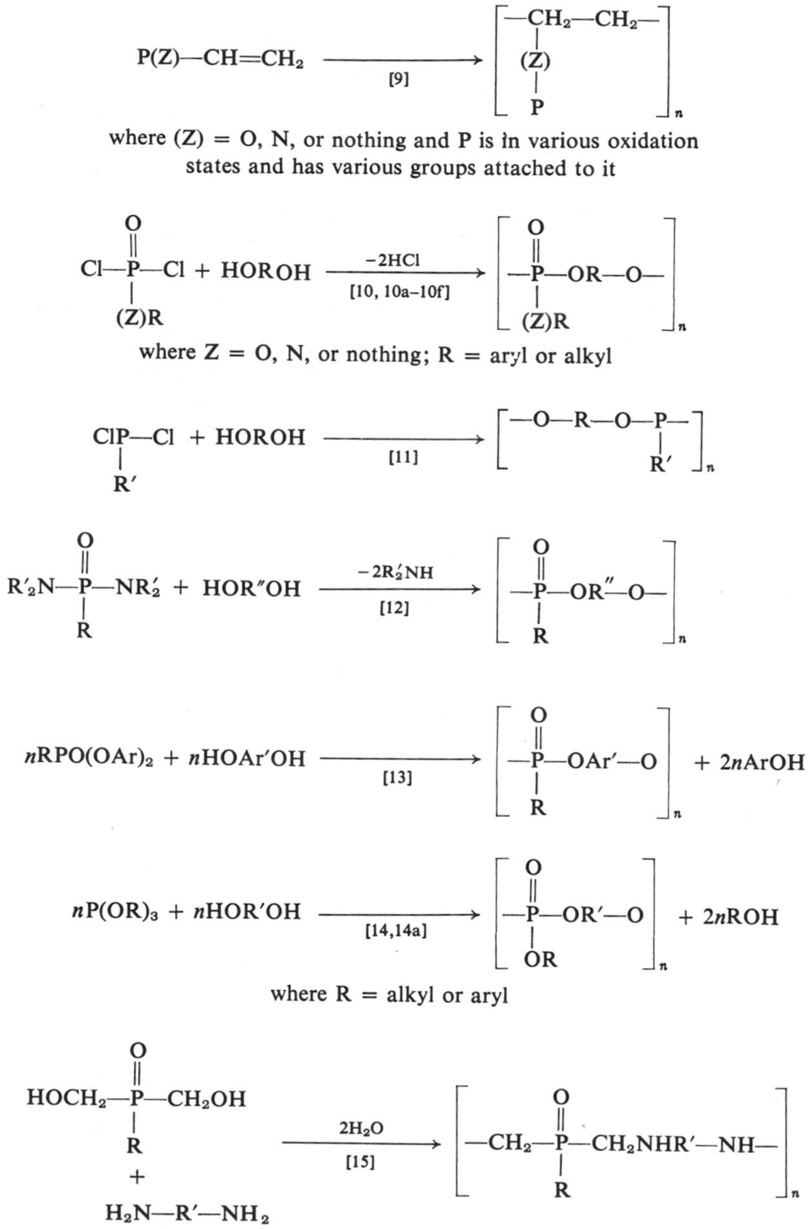

$$\mathrm{P(Z){-}CH{=}CH_2} \xrightarrow[\text{[9]}]{} \left[\begin{array}{c}\mathrm{{-}CH_2{-}CH_2{-}} \\ | \\ \mathrm{(Z)} \\ | \\ \mathrm{P}\end{array}\right]_n$$

where (Z) = O, N, or nothing and P is in various oxidation states and has various groups attached to it

$$\mathrm{Cl{-}\underset{\underset{(Z)R}{|}}{\overset{\overset{O}{\|}}{P}}{-}Cl} + \mathrm{HOROH} \xrightarrow[\text{[10, 10a–10f]}]{-2\mathrm{HCl}} \left[\mathrm{{-}\underset{\underset{(Z)R}{|}}{\overset{\overset{O}{\|}}{P}}{-}OR{-}O{-}}\right]_n$$

where Z = O, N, or nothing; R = aryl or alkyl

$$\mathrm{Cl\underset{\underset{R'}{|}}{P}{-}Cl} + \mathrm{HOROH} \xrightarrow[\text{[11]}]{} \left[\mathrm{{-}O{-}R{-}O{-}\underset{\underset{R'}{|}}{P}{-}}\right]_n$$

$$\mathrm{R'_2N{-}\underset{\underset{R}{|}}{\overset{\overset{O}{\|}}{P}}{-}NR'_2} + \mathrm{HOR''OH} \xrightarrow[\text{[12]}]{-2\mathrm{R'_2NH}} \left[\mathrm{{-}\underset{\underset{R}{|}}{\overset{\overset{O}{\|}}{P}}{-}OR''{-}O{-}}\right]_n$$

$$n\mathrm{RPO(OAr)_2} + n\mathrm{HOAr'OH} \xrightarrow[\text{[13]}]{} \left[\mathrm{{-}\underset{\underset{R}{|}}{\overset{\overset{O}{\|}}{P}}{-}OAr'{-}O}\right]_n + 2n\mathrm{ArOH}$$

$$n\mathrm{P(OR)_3} + n\mathrm{HOR'OH} \xrightarrow[\text{[14,14a]}]{} \left[\mathrm{{-}\underset{\underset{OR}{|}}{\overset{\overset{O}{\|}}{P}}{-}OR'{-}O}\right]_n + 2n\mathrm{ROH}$$

where R = alkyl or aryl

$$\mathrm{HOCH_2{-}\underset{\underset{R}{|}}{\overset{\overset{O}{\|}}{P}}{-}CH_2OH} + \mathrm{H_2N{-}R'{-}NH_2} \xrightarrow[\text{[15]}]{2\mathrm{H_2O}} \left[\mathrm{{-}CH_2{-}\underset{\underset{R}{|}}{\overset{\overset{O}{\|}}{P}}{-}CH_2NHR'{-}NH{-}}\right]_n$$

$$\text{(RO)P—OROP(OR)} + \text{XR'X} \xrightarrow[\text{[16,16a]}]{-2\text{RX}} \left[-\overset{\overset{\text{O}}{\|}}{\underset{\underset{\text{OR}}{|}}{\text{P}}}-\text{OR}-\text{O}\overset{\overset{\text{O}}{\|}}{\underset{\underset{\text{OR}}{|}}{\text{P}}}-\text{R'}- \right]_n$$

Michaelis-Arbuzov reaction

$$\text{RO—P}\langle{}^{\text{O}}_{\text{O}}\rangle\text{R'} \xrightarrow[\text{[17,17a]}]{\text{Arbuzov rearrangement}} \left[-\overset{\overset{\text{O}}{\|}}{\underset{\underset{\text{OR}}{|}}{\text{P}}}-\text{OR'}- \right]_n + \left[-\overset{\overset{\text{O}}{\|}}{\underset{\underset{\text{R'}}{|}}{\text{P}}}-\text{ORO}- \right]_n$$

$$\text{(RO)}_2\text{P—CH}_2\text{Cl} \xrightarrow[\text{[18]}]{\text{Arbuzov reaction}} \left[-\overset{\overset{\text{O}}{\|}}{\underset{\underset{\text{OR}}{|}}{\text{P}}}-\text{CH}_2- \right]_n$$

$$\text{R(RO)P—CH}_2\text{Cl} \xrightarrow[\text{[18]}]{\text{Arbuzov reaction}} \left[-\overset{\overset{\text{O}}{\|}}{\underset{\underset{\text{R}}{|}}{\text{P}}}-\text{CH}_2- \right]_n$$

$$\text{(ClCH}_2\text{CH}_2\text{O)}_2\text{P—C}_6\text{H}_4\text{CH}_2\text{Cl} \xrightarrow[\text{[19]}]{\text{Arbuzov reaction}} \left[-\overset{\overset{\text{O}}{\|}}{\underset{\underset{\text{OCH}_2\text{CH}_2\text{Cl}}{|}}{\text{P}}}-\text{C}_6\text{H}_4\text{CH}_2- \right]_n$$

$$\text{(ClCH}_2\text{CH}_2\text{O)}\overset{\overset{\text{O}}{\|}}{\text{P}}\text{—R} \xrightarrow[\substack{200°\text{–}250°\text{C} \\ \text{[20]}}]{-\text{ClCH}_2\text{CH}_2\text{Cl}} \left[-\overset{\overset{\text{O}}{\|}}{\underset{\underset{\text{R}}{|}}{\text{P}}}-\text{OCH}_2\text{CH}_2\text{O}- \right]_n$$

$$\begin{matrix}\text{H}_2\text{C—O} \\ | \quad\quad \rangle\overset{\text{O}}{\underset{\text{R}}{\text{P}}} \\ \text{H}_2\text{C—O}\end{matrix} \xrightarrow{\text{[21]}} \text{(same polymer as above)}$$

$$\text{(Z)R—}\overset{\overset{\text{O}}{\|}}{\underset{\underset{\text{R}}{|}}{\text{P}}}\text{—R(Z)} \xrightarrow[\text{[22,23]}]{\text{XRX}} \text{Polymer}$$

where Z = $—\text{C}_6\text{H}_4\text{COOH}$, $—\text{C}_6\text{H}_4\text{NH}_2$, $—\text{CH}_2\text{CH}_2\text{COOH}$, $—\text{CH}_2\text{CH}_2\text{CH}_2\text{NH}_2$, $—(\text{CH}_2)_n—\text{OH}$

X = H, COOH, NH_2, OH, $\overset{\overset{\text{O}}{\|}}{\text{C}}\text{—Cl}$, NCO, etc.

R = aryl or alkyl

$$R'PH_2 + CH_2{=}CH{-}R{-}CH{=}CH_2 \xrightarrow[\text{[16a]}]{} \left[\begin{array}{l} {-}\underset{\displaystyle R'}{\underset{|}{P}}{-}CH_2CH_2{-}R{-}CH_2CH_2{-} \end{array} \right]$$

$$\underset{\text{(small amounts)}}{R{-}P{-}Cl_2} + \underset{\displaystyle R'}{\underset{|}{CH_2{=}CH}} \xrightarrow[\text{[24]}]{} \left[{-}\underset{\displaystyle R'}{\underset{|}{CH}}{-}CH_2{-}\overset{Cl\ \ Cl}{\overset{\diagdown\ \diagup}{\underset{\displaystyle R}{\underset{|}{P}}}}{-} \right]_n \xrightarrow{2nCH_3OH} \left[{-}\underset{\displaystyle R'}{\underset{|}{CH}}{-}CH_2{-}\overset{O}{\overset{\|}{\underset{\displaystyle R}{\underset{|}{P}}}}{-} \right]_n$$

($R' = CN, C_6H_5, OAc, CH{=}CH_2$, etc.)

$$RPH_2 + R'(NCO)_2 \xrightarrow[\text{[25]}]{} \left[{-}\underset{\displaystyle R}{\underset{|}{P}}{-}CONHR'{-}NHCO{-} \right]_n$$

$$PCl_5 + NH_4Cl \xrightarrow[\text{[8,26]}]{} \frac{1}{n} \left[{-}\overset{Cl}{\overset{|}{\underset{\displaystyle Cl}{\underset{|}{P}}}}{=}N{-} \right]_n + 4HCl$$

$$\xrightarrow[\text{[27]}]{Z{-}R{-}Z} \left[{-}\overset{|}{\underset{|}{P}}{=}N{-} \right]_n$$

($Z = OH, NH_2$, etc.)

$$H_2N{-}\overset{O}{\overset{\|}{\underset{\displaystyle R}{\underset{|}{P}}}}{-}NH_2 \xrightarrow[\text{[28]}]{-NH_3} \left[{-}\overset{O}{\overset{\|}{\underset{\displaystyle R}{\underset{|}{P}}}}{-}NH{-} \right]_n$$

$$R_2PH + B_2H_6 \xrightarrow[\text{[29]}]{-H_2} \left[{-}\overset{R}{\overset{|}{\underset{\displaystyle R}{\underset{|}{P}}}}{-}BH_2{-} \right]_n$$

SCHEME 2

Preparation of Starting Phosphorus Monomers and Key Intermediates

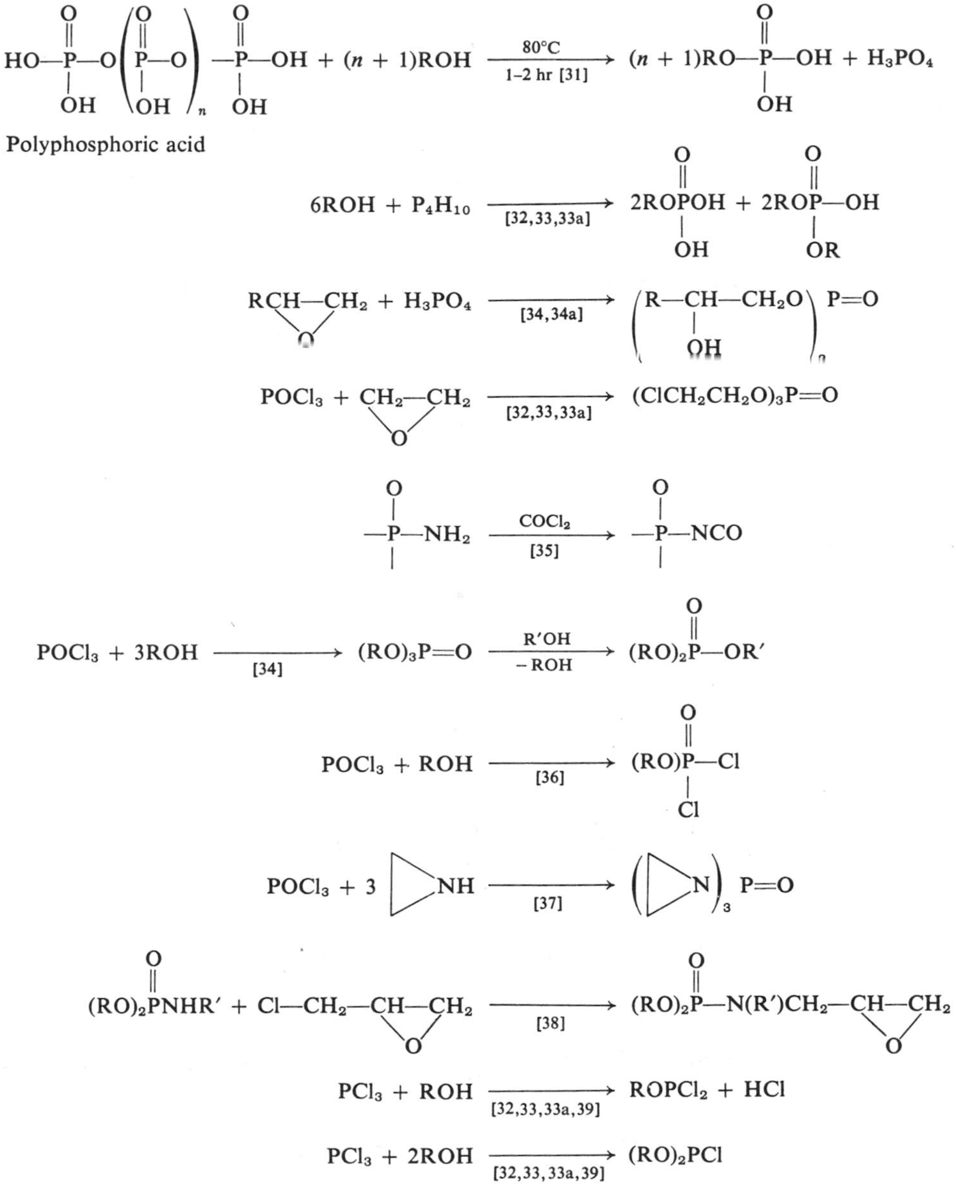

$$HO{-}\overset{O}{\overset{\|}{P}}(OH){-}O\left(\overset{O}{\overset{\|}{P}}(OH){-}O\right)_n{-}\overset{O}{\overset{\|}{P}}(OH){-}OH + (n+1)ROH \xrightarrow[1\text{–}2\ hr\ [31]]{80^\circ C} (n+1)RO{-}\overset{O}{\overset{\|}{P}}(OH){-}OH + H_3PO_4$$

Polyphosphoric acid

$$6ROH + P_4H_{10} \xrightarrow[[32,33,33a]]{} 2RO\overset{O}{\overset{\|}{P}}(OH)OH + 2RO\overset{O}{\overset{\|}{P}}(OR){-}OH$$

$$R\underset{\diagdown O \diagup}{CH{-}CH_2} + H_3PO_4 \xrightarrow[[34,34a]]{} \left(R{-}\underset{OH}{\underset{|}{CH}}{-}CH_2O\right)_n P{=}O$$

$$POCl_3 + \underset{\diagdown O \diagup}{CH_2{-}CH_2} \xrightarrow[[32,33,33a]]{} (ClCH_2CH_2O)_3P{=}O$$

$$-\overset{O}{\overset{|}{\underset{|}{P}}}{-}NH_2 \xrightarrow[[35]]{COCl_2} -\overset{O}{\overset{|}{\underset{|}{P}}}{-}NCO$$

$$POCl_3 + 3ROH \xrightarrow[[34]]{} (RO)_3P{=}O \xrightarrow[-ROH]{R'OH} (RO)_2\overset{O}{\overset{\|}{P}}{-}OR'$$

$$POCl_3 + ROH \xrightarrow[[36]]{} (RO)\overset{O}{\overset{\|}{\underset{Cl}{\underset{|}{P}}}}{-}Cl$$

$$POCl_3 + 3\ \triangleright NH \xrightarrow[[37]]{} (\triangleright N)_3 P{=}O$$

$$(RO)_2\overset{O}{\overset{\|}{P}}NHR' + Cl{-}CH_2{-}\underset{\diagdown O \diagup}{CH{-}CH_2} \xrightarrow[[38]]{} (RO)_2\overset{O}{\overset{\|}{P}}{-}N(R')CH_2{-}\underset{\diagdown O \diagup}{CH{-}CH_2}$$

$$PCl_3 + ROH \xrightarrow[[32,33,33a,39]]{} ROPCl_2 + HCl$$

$$PCl_3 + 2ROH \xrightarrow[[32,33,33a,39]]{} (RO)_2PCl$$

$$PCl_3 + 3ROH \xrightarrow[[32,33,33a,39]]{\text{no base}} (RO)_2\overset{\overset{\displaystyle O}{\|}}{P}H + RCl + 2HCl$$

$$PCl_3 + 3ROH \xrightarrow[[39]]{R_3N} (RO)_3P$$

$$PCl_3 + 3ArOH \longrightarrow (ArO)_3P + 3HCl$$

$$PCl_3 + 3RX \xrightarrow[[40]]{6Na} R_3P \xrightarrow[[41]]{\frac{1}{2}O_2} R_3PO$$

$$PCl_3 + HO{-}R{-}OH \xrightarrow[[42]]{} \underset{\lfloor\_\_\_\_\_\_\_\_O}{RO{-}{-}PCl} + 2HCl$$

$$PCl_3 + 3\underset{\diagdown O \diagup}{CH_2{-}CH_2} \xrightarrow[[35]]{-3HCl} (ClCH_2CH_2O)_3P \xrightarrow[\text{reaction}]{\text{Michaelis-Arbuzov}} (ClCH_2CH_2O)_2\overset{\overset{\displaystyle O}{\|}}{P}CH_2CH_2Cl$$

$$(RO)_3P + R'X \xrightarrow[[43,43a]]{\text{Arbuzov reaction}} (RO)_2{-}\overset{\overset{\displaystyle O}{\|}}{P}R' + RX$$

$$(C_2H_5O)_3P + BrCH_2{-}\overset{\diagup O \diagdown}{CH{-}CH_2} \xrightarrow[[44]]{-C_2H_5Br} (C_2H_5O)_2\overset{\overset{\displaystyle O}{\|}}{P}{-}CH_2\underset{\diagdown O \diagup}{CH{-}CH_2}$$

$$(RO)_2POH \xrightarrow{KOH} (RO)_2POK \xrightarrow[[45]]{R'X} (RO)_2{-}\overset{\overset{\displaystyle O}{\|}}{P}R' + KX$$

$$R'\overset{\overset{\displaystyle O}{\|}}{P}Cl_2 + 2ROH \xrightarrow[[46]]{} R'\overset{\overset{\displaystyle O}{\|}}{P}(OR)_2$$

$$(RO)_3P + HOR'OH \xrightarrow[[14,14a,47]]{} R'\langle\begin{smallmatrix}O\\O\end{smallmatrix}\rangle P{-}OR'{-}OH + ROH$$

$$PCl_3 + 3NaOCN \xrightarrow[[48]]{} P(NCO)_3$$

$$RPCl_2 + 2NaOCN \xrightarrow[[48]]{} RP(NCO)_2$$

$$PCl_3 + RCH{=}O \xrightarrow{[49]} Cl{-}CH(R){-}P(=O){-}(OH)_2$$

$$PCl_3 + RCl \xrightarrow[{[34a,50]}]{AlCl_3} RP(=O){-}Cl_2$$

$$HP(=O)(OC_2H_5){-}OC_2H_5 + RCH{=}CH_2 \xrightarrow{[51,52]} (C_2H_5O)_2P(=O){-}CH_2CH_2R$$

$$(CH_3O)_2P{-}Z + \underset{\diagdown O \diagup}{CH_2{-}CH_2} \xrightarrow[{[53]}]{BF_3} (CH_3O)_2P(=O){-}CH_2CH_2Z$$

(Z = Cl or OH)

$$NaPH_2 + RX \longrightarrow RPH_2 + NaX \xrightarrow[{[54]}]{X_2} RPX_2 + 2HX$$

$$R_2POR' + R''X \xrightarrow[{[55,56]}]{\text{Arbuzov reaction}} RR''PO + R'X$$

$$PH_3 + \underset{\diagdown O \diagup}{CH_2{-}CH_2} \xrightarrow{[57]} (HOCH_2CH_2){-}PH_2 \longrightarrow (HOCH_2CH_2)_3P$$

$$PH_3 + CH_2{=}O \xrightarrow[{[58]}]{H_2O,\ HgCl_2} (HOCH_2)_4P^+ + OH^- \longrightarrow (HOCH_2)_3P$$

$$(HOCH_2)_4P^+ \longrightarrow (HOCH_2)_3P{=}O$$

$$PCl_5 + NH_4Cl \xrightarrow{[59]} \frac{1}{n}(NPCl_2)_n + 4HCl$$

$$PCl_5 + NH_2C(=O)NH_2 \xrightarrow{[60]} (Cl_3P{=}N)_2C{=}O$$

$$CH_2{=}C(CH_3){-}COOCH_2CH_2OH + P_2O_5 \xrightarrow{[61]} \left(CH_2{=}C(CH_3){-}COOCH_2CH_2O\right)_3 P{=}O$$

Some important reactions of phosphites which can lead to important monomer starting materials are shown in Eqs. (5)–(28).

$$(RO)_2P(O)H + CH_2{=}CHR \xrightarrow[\text{peroxides [63, 63a]}]{\text{base or}} (RO)_2P(O)CH_2CH_2R \quad (5)$$

$$(RO)_2P(O)H + R'CH{=}O \xrightarrow[\text{[64,65]}]{} (RO)_2P(O)CH(OH){-}R' \quad (6)$$

$$(RO)_2P(O)H + (CH_2O)_x + (C_2H_5)_2NH \xrightarrow[\text{[63a]}]{} (RO)_2P(O)CH_2N(C_2H_5)_2 \quad (7)$$

$$(RO)_2P(O)H + R'NCO \xrightarrow[\text{[66]}]{} (RO)_2P(O)CONHR' \quad (8)$$

$$(RO)_2P(O)H + R'NCS \xrightarrow[\text{[67]}]{} (RO)_2P(O)CSNHR' \quad (9)$$

$$(RO)_2P(O)H + Cl_2 \xrightarrow[\text{[68]}]{} (RO)_2P(O)Cl \quad (10)$$

$$(RO)_2P(O)H + SOCl_2 \xrightarrow[\text{[69]}]{} (RO)_2P(O)Cl \quad (11)$$

$$2(RO)_2P(O)H + 2ClSCCl_3 \xrightarrow[\text{[70]}]{} (RO)_2P(O)Cl \quad (12)$$

$$(C_2H_5O)_2P(O)H + HOCH_2CH_2OH \xrightarrow[\text{[71]}]{} \begin{matrix} CH_2{-}O \\ | \\ CH_2{-}O \end{matrix}\!\!>\!P(O)H + 2C_2H_5OH \quad (13)$$

$$(C_2H_5O)_2P(O)H + 2C_8H_{17}OH \xrightarrow[\text{[72]}]{\text{H+ or base}} (C_8H_{17}O)_2P(O)H + 2C_2H_5OH \quad (14)$$

$$(RO)_3P + CH_2{=}CHR \xrightarrow[\text{[73]}]{} (RO)_2P(O)CH_2CH_2R \quad (15)$$

$$(RO)_3P + R'CH{=}O \xrightarrow[\text{[74]}]{} (RO)_2P(O)CH(R')OR \quad (16)$$

$$(RO)_3P + Cl_2 \xrightarrow[\text{[75,76]}]{} (RO)_2P(O)Cl + RCl \quad (17)$$

$$(RO)_3P + PCl_3 \xrightarrow[\text{[69]}]{} (RO)_2PCl + ROPCl_2 \quad (18)$$

$$(RO)_3P + I_2 \xrightarrow[\text{[77]}]{} (RO)_2P(O)I + RI \quad (19)$$

$$(RO)_3P + O_2 \xrightarrow[\text{[78]}]{} (RO)_3PO \quad (20)$$

$$(RO)_3P + HgO \xrightarrow[\text{[79]}]{} (RO)_3PO + Hg \quad (21)$$

$$(RO)_3P + KMnO_4 \xrightarrow[\text{[80]}]{} (RO)_3PO \quad (22)$$

$$(RO)_3P + H_2O \xrightarrow[{[81]}]{} (RO)_2P(O)H + ROH \tag{23}$$

$$(RO)_3P + HCl \xrightarrow[{[82]}]{} (RO)_2P(O)H + RCl \tag{24}$$

$$(RO)_3P + HOCH_2CH_2OH \xrightarrow[{[83]}]{} \begin{matrix} CH_2{-}O \\ | \\ CH_2{-}O \end{matrix}\!\!\!> P{-}OR + 2ROH \tag{25}$$

$$(C_2H_5O)_3P + 2\text{-}i\text{-}AmOH \xrightarrow[{[84]}]{} (i\text{-}AmO)_3P + 3C_2H_5OH \tag{26}$$

$$(ClCH_2CH_2O)_3P \xrightarrow[{[85]}]{} ClCH_2CH_2P(O)(OCH_2CH_2Cl)_2 \xrightarrow[{[85]}]{KOH} (CH_2{=}CH)P(O)(OCH_2CH_2Cl)_2 \tag{27}$$

$$(C_2H_5O)_3P + C_6H_5{-}\overset{\overset{O}{\|}}{C}{-}\underset{\underset{X}{|}}{C}(CH_3)_2 \xrightarrow[{[86]}]{\text{Perkow reaction}} C_6H_5\overset{\overset{\overset{O}{\|}}{OP(OC_2H_5)_2}}{\overset{|}{C}}{=}C(CH_3)_2 + C_2H_5X \tag{28}$$

$$(X = Br \text{ or } Cl)$$

The infrared spectra of various phosphorus compounds have been reported and the review is worthwhile consulting [87].

This chapter is divided into sections describing polymers prepared from either phosphate, phosphonate, or phosphite monomers. In addition, the phosphonitrile rubbers and phosphonamide polymers are also described.

2. ORGANOPHOSPHATE POLYMERS

A. Condensation Reactions

As described in Scheme 1, one of the main reactions used to afford phosphate polymers is that involving the reaction of alkyl dichlorophosphates with either alcohols or phenols. Examples of this and other methods are given in Table II.

The aromatic polyphosphates were first developed in the 1950s by several companies [10d–10f,88,88a]. The commercial application of the phosphate polyesters has been hindered by the fact that they are mostly too brittle and lack resistance to hydrolysis [89].

The polyphosphate esters produced by the reaction in Eq. (29) can be chain-extended with typical polyester reactants [90]. For example, bromoethylphosphoric acid dichloride can be condensed with a 5–8 *M* excess of diol

TABLE II

PREPARATION OF POLYPHOSPHATES BY VARIOUS CONDENSATION REACTIONS INVOLVING HYDROXYL COMPOUNDS

Phosphorus reagent	Hydroxyl compound	Solvent	Catalyst	Reaction conditions		Polymer properties	Ref.
				Temp. (°C)	Time (hr)		
$POCl_3$	Polyglycol mol. wt. 700	C_6H_6	—	80	$1\frac{1}{3}$	Waxy	*a*
$POCl_3$	Hydroquinone	—	—	—	—	Elastic and soluble in organic solvent	*b*
$POCl_3$	Bisphenol A	Xylene	—	—	—	Elastic and soluble in organic solvent	*a,c*
$C_6H_5OPOCl_2$	4,4′-Dihydroxy-biphenyl	—	BF_3 diacetate	170–220	20–26	Transparent, brown hard resin s.p. = 125°C	*d*
$4\text{-}ClC_6H_4OPOCl_2$	Resorcinol	—	—	165–195	20–26	Transparent, light rubbery s.p. = 50°C	*d,e*
$4\text{-}CH_3OC_6H_4OPOCl_2$	Hydroquinone	—	Tin	140–200	16	Transparent, brittle s.p. = 35°–38°C	*f*
$4\text{-}C_6H_5C_6H_4OPOCl_2$	Hydroquinone	—	—	150 180–185	7	Transparent, light colored s.p. = 115°C	*g*
$POCl_3$	Poly(vinyl alcohol)	Dioxane	—	—	—	Gray powder	*h*
$(RO)_2P(=O)\text{—}Cl$	Poly(vinyl alcohol)	—	—	—	—	—	*i*
$(RO)_3P{=}O$	Polyester diols	—	—	230–300	10–24	—	*j*
$(BrCH_2CH_2O)_3P{=}O$	Glycols and diols	—	—	—	—	—	*k*

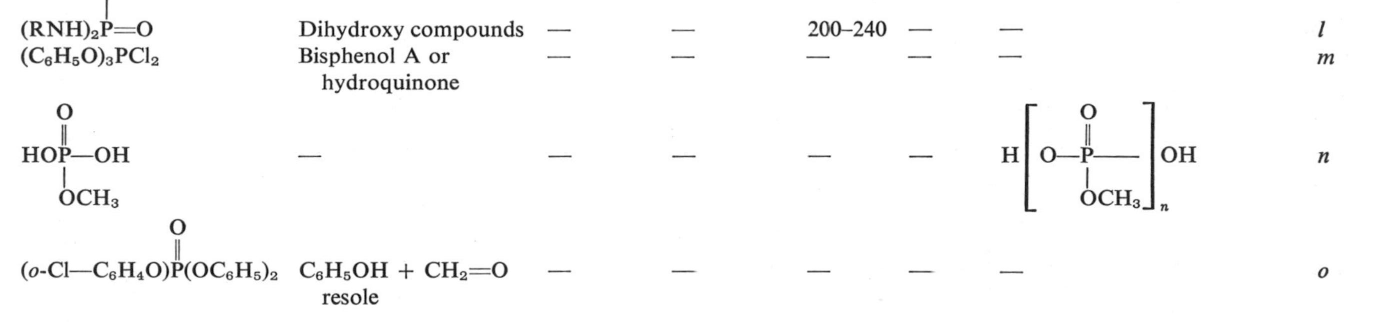

$(RNH)_2P(OC_2H_5)=O$	Dihydroxy compounds	—	—	200–240	—	—		*l*
$(C_6H_5O)_3PCl_2$	Bisphenol A or hydroquinone	—	—	—	—	—		*m*
$HOP(=O)(OCH_3)—OH$	—	—	—	—	—		$H[O—P(=O)(OCH_3)—]_nOH$	*n*
$(o\text{-}Cl—C_6H_4O)P(=O)(OC_6H_5)_2$	$C_6H_5OH + CH_2=O$ resole	—	—	—	—	—		*o*

[a] A. G. Metallgesellschaft, British Patent 706,410 (1954).
[b] B. Hilferich and H. G. Schmidt, German Patent 905,318 (1954).
[c] J. A. Arvin, U.S. Patent 2,058,394 (1937).
[d] H. Zenftman and A. McClean, U.S. Patent 2,636,876 (1953).
[e] W. E. Cass, U.S. Patent 2,616,873 (1952).
[f] V. V. Korshak, I. A. Gribova, and M. A. Andreeva, *Izv. Akad. Nauk SSSR, Otd. Khim. Nauk* p. 880 (1958).
[g] H. Zenftman, British Patent 679,834 (1952).
[h] G. C. Danl, J. D. Reid, and R. M. Reinhardt, *Ind. Eng. Chem.* **46**, 1042 (1954).
[i] D. E. Kvalnes and N. O. Brace, U.S. Patent 2,691,567 (1954).
[j] Kurashiki Rayon Co. Ltd., British Patent 1,060,401 (1967); *Chem. Abstr.* **67**, 12440 (1967).
[k] Nederlandse Organisalie voor Toeglpast-Natuurwetenschappelijk underzoek ten Behoeve van Nijverheid, Handel an Verkeen, Belgian Patent 619,217 (1962).
[l] K. A. Petrov, V. P. Evdakov, L. I. Mizrakh, and V. A. Kravchenko, U.S.S.R. Patent 152,572 (1963).
[m] F. Yaku and I. Yamashita, *Osaka Kogyo Gijutsu Shikensho Kiho* **18**, 117 (1967).
[n] B. Topley, *Quart. Rev., Chem. Soc.* **3**, 345 (1949).
[o] R. Dahms, U.S. Patents 3,549,479 and 3,549,480 (1970).

to give a condensate which can be further reacted with maleic or phthalic anhydride [91]. The application of these polyphosphates for flame retardants has recently been discussed [92].

$$\mathrm{Cl{-}\overset{\overset{\displaystyle O}{\|}}{\underset{\underset{\displaystyle OR}{|}}{P}}{-}Cl + HO{-}R'{-}OH \longrightarrow \left[{-}\overset{\overset{\displaystyle O}{\|}}{\underset{\underset{\displaystyle OR}{|}}{P}}{-}O{-}R'{-}O{-} \right]_n + 2HCl} \qquad (29)$$

Another method involves the condensation of 3 moles of a glycol with $POCl_3$ and subsequent reaction of the OH groups with a dicarboxylic acid or its monoester (Eqs. 30 and 31). In addition, OH-terminated polyesters can also react with $POCl_3$ or alkyl dichlorophosphates [93]. Some typical examples are given in Table II.

$$\mathrm{3HO{-}R{-}OH + POCl_3 \longrightarrow (HO{-}R{-}O)_3P{=}O} \qquad (30)$$

$$\mathrm{(HO{-}R{-}O)_3P{=}O \xrightarrow{R(COOH)_2} phosphate\ polyester}$$

$$\mathrm{HO{-}R{-}OH + R(COOH)_2 \longrightarrow polyester \xrightarrow{POCl_3} phosphate\ polyester} \qquad (31)$$

(R = alkylene or arylene)

The phosphate polyesters are also obtained by the ring-opening polymerization of cyclic phosphates [94].

$$\mathrm{\begin{matrix} CH_2{-}O \\ | \\ CH_2{-}O \end{matrix} \!\!>\! P(=O)OR \longrightarrow \left[{-}\overset{\overset{\displaystyle O}{\|}}{\underset{\underset{\displaystyle OR}{|}}{P}}{-}OCH_2CH_2{-}O{-} \right]_n} \qquad (32)$$

Dialkyl hydrogen phosphates react with epoxides to give polyphosphates [95–97].

$$\mathrm{R'CH{-}CH_2\ (epoxide,\ \underbrace{\ }_{O}) + RO{-}\overset{\overset{\displaystyle O}{\|}}{\underset{\underset{\displaystyle OH}{|}}{P}}{-}OR \xrightarrow{heat} \left[{-}\overset{\overset{\displaystyle O}{\|}}{\underset{\underset{\displaystyle OCH_2{-}CH(OH){-}R'}{|}}{P}}{-}O{-} \right]_n + 2ROH} \qquad (33)$$

Related polyphosphates are also produced by heating trialkyl phosphates with P_2O_5 to 130°C [98].

The Friedel-Crafts reaction has also been used to prepare polyphosphates by the reaction of bischloromethyl compounds as shown in Eq. (34) [99].

$$ClCH_2-C_6H_4-O-\overset{\displaystyle O}{\overset{\|}{\underset{\underset{\displaystyle OAr}{|}}{P}}}-O-C_6H_4-CH_2Cl + Ar'H \xrightarrow[\substack{150°-200°C \\ -2HCl}]{FeCl_3} \left[-CH_2-C_6H_4-O-\overset{\displaystyle O}{\overset{\|}{\underset{\underset{\displaystyle OAr}{|}}{P}}}-O-C_6H_4-CH_2-Ar'-\right]_n \quad (34)$$

As mentioned earlier, phosphate esters are hydrolytically not as stable as the phosphonate esters, as shown in Table III. This has been a limitation for some commercial applications. On the other hand, phosphonate esters appear to be more unstable in the presence of base.

TABLE III

FIRST-ORDER RATE CONSTANTS FOR THE NUCLEOPHILIC REACTION OF PHOSPHATE ESTERS AND PHOSPHONATE ESTERS AT 80°C[a]

Compound	Nucleophilic[b] reagent	k^c (min^{-1})
$(CH_3O)_3PO$	H_2O	2.0×10^{-2}
$(CH_3O)_2(C_2H_5)PO$	H_2O	1.5×10^{-3}
$(C_6H_5CH_2O)_2(CH_3)PO$	H_2O	6.4×10^{-2}
$(CH_3O)_3PO$	Cl^-	0.49×10^{-2}
$(CH_3O)_2(C_2H_5)PO$	Cl^-	0.30×10^{-3}
$(C_6H_5CH_2O)_2(CH_3)PO$	Cl^-	8.3×10^{-2}
$(CH_3O)_3PO$	OH^-	1.67
$(CH_3O)_2(C_2H_5)PO$	OH^-	4.70
$(C_6H_5CH_2O)_2(CH_3)PO$	OH^-	59.0

[a] Data taken from R. F. Hudson and D. C. Harper, *J. Chem. Soc., London* p. 1356 (1958).

[b] 0.2 N anion in H_2O at 80°C.

[c] k is the first-order constant for the reaction with a particular anion.

The rate of hydrolysis of various diethyl vinyl phosphates with 0.1 N HCl–40% ethanol at 85°C is compared to that of triethyl phosphate and shown in Table IV. For other data on the nucleophilic substitution in phosphate esters see a review by Cox and Ramsay [100].

TABLE IV
RATE OF HYDROLYSIS OF DIETHYL VINYL PHOSPHATES IN 0.1 *N* HYDROCHLORIC ACID–40% ETHANOL AT 85°C[a]

Diethyl vinyl phosphate	$K \times 10^3$ [b] (min^{-1})	$t_{1/2}$ (hr)
$(C_2H_5O)_2P(\rightarrow O)OC(OC_2H_5)=CHCOOC_2H_5$	130	0.1
$(C_2H_5O)_2P(\rightarrow O)OC(CH_3)=CH_2$	14.88	0.78
$(C_2H_5O)_2P(\rightarrow O)OC(C_6H_5)=CH_2$	6.46	1.79
$(C_2H_5O)_2P(\rightarrow O)OC(OC_2H_5)=CCl_2$	3.77	3.06
$(C_2H_5O)_2P(\rightarrow O)OC(COOC_2H_5)=CH_2$	2.38	4.86
$(C_2H_5O)_2P(\rightarrow O)OCH=CCl_2$	1.34	8.56
$(C_2H_5O)_2P(\rightarrow O)OC(CH_3)=CHCOOC_2H_5$	0.99	11.71
$(C_2H_5O)_2P(\rightarrow O)OCH=CH_2$	0.46	25.21
$(C_2H_5O)_2P(\rightarrow O)OCH=CH(CH_2)_3CH_2$ (ring: C and CH_2 bonded)	0.37	31.59
$(C_2H_5O)_2P(\rightarrow O)OC_2H_5$	0.037	308.70

[a] Reprinted from F. W. Lichtenthaler, *Chem. Rev.* **61**, 607 (1961). Copyright 1961 by the American Chemical Society. Reprinted by permission of the copyright owner.

[b] *K* represents the unimolecular reaction constants, calculated with the formula

$$K = \frac{2.303}{t_2 - t_1} \log_{10} \frac{100}{100 - x}$$

x being the percentage of hydrolyzed vinyl phosphate. The molecularity of the reaction has not been established.

The synthesis of phosphorus-containing polyesters and organo polyphosphates has also been reviewed [101].

2-1. *Preparation of a Polyphosphate by the Reaction of Phenyl Phosphoryl Chloride with Hydroquinone* [10d]

$$C_6H_5O\overset{O}{\overset{\|}{P}}Cl_2 + HO{-}C_6H_4{-}OH \longrightarrow H\left[{-}O{-}C_6H_4O\overset{O}{\overset{\|}{P}}\underset{\underset{C_6H_5}{O}}{|}{-}\right]_n Cl \qquad (35)$$

To a resin kettle equipped with a mechanical stirrer, gas inlet tube, thermometer, and reflux condenser with a drying tube is added 211 gm (1.0 mole) of phenyl phosphoryl chloride and 115 gm (1.05 mole) of hydroquinone. While passing a slow stream of dry nitrogen over the reaction mixture, the flask is stirred vigorously for 14 hr at 168°–197°C. The hydrogen chloride gas is evolved during the reaction. After this time the reaction mixture is heated in an oil bath at 169°–184°C while applying vacuum of 2.5–5.0 mm Hg until bubbling practically ceases. The resulting product is a clear, light brown resin which is plastic and nontacky at room temperature. Additional heating to 300°C on a hot plate converts the polymer to a solid, flexible, rubbery material.

Similar polymers are prepared using resorcinol or 4,4′-dihydroxybiphenyl [88a] in place of hydroquinone.

2-2. *Preparation of Poly(vinyl phosphate)* [102]

$$\left[{-}CH_2{-}\underset{OH}{\underset{|}{CH}}{-}\right]_n + POCl_3 \xrightarrow{-HCl} \left[{-}CH_2{-}\underset{\underset{O{=}PCl_2}{|}}{\underset{O}{\underset{|}{CH}}}{-}\right]_n \xrightarrow{H_2O} \left[{-}CH_2{-}\underset{\underset{O{=}P{-}(OH)_2}{|}}{\underset{O}{\underset{|}{CH}}}{-}\right]_n \qquad (36)$$

To a resin kettle containing 4.5 gm (0.10 unit mole) poly(vinyl alcohol) is added dropwise a solution of 16 gm (0.10 mole) of phosphorus oxychloride in 50 ml of dioxane. The evolved hydrogen chloride is removed with the aid of slight vacuum to finally give a gray powder. The powder is insoluble in organic solvents and contains chlorine. When water is added to the poly(vinyl phosphoryl dichloride) the polymer turns yellow-orange and swells but does not dissolve. The product is washed with water to give a chlorine-free product containing 14.72% phosphorus.

TABLE V
STABILITY OF POLYVINYL PHOSPHATE (SODIUM SALT)[a,b]

Solution	P Content after 2-hr treatment (%)	
	Room temperature	60°C
0.1 *N* sodium hydroxide	12.9	13.3
1 *N* sodium hydroxide	13.9	14.6
0.1 *N* sodium chloride	11.8	13.4
1 *N* sodium chloride	12.9	14.5
0.1 *N* hydrochloric acid	12.0	13.3
1 *N* hydrochloric acid	13.0	14.5

[a] Reprinted from G. C. Daral, J. D. Reid, and R. M. Reinhardt, *Ind. Eng. Chem.* **46,** 1042 (1954). Copyright 1954 by the American Chemical Society. Reprinted by permission of the copyright owner.
[b] Phosphorus content of untreated sample was 14.5%.

Heating of the free acid gives a cross-linked polymer. Poly(vinyl phosphate) is hydrolytically stable towards dilute acids or bases as shown in Table V.

2-3. Preparation of Poly(vinylammonium phosphate) [102]

$$\left[\begin{array}{c}-CH_2-\underset{\displaystyle OH}{\underset{|}{CH}}-\end{array}\right]_n + H_3PO_4 + 2NH_2-\overset{\displaystyle O}{\overset{\|}{C}}-NH_2 \longrightarrow$$

$$\left[\begin{array}{c}-CH_2-CH- \\ | \\ O \\ | \\ O{=}P(ONH_4)(OH)\end{array}\right]_n + H_2NCONHCONH_2 + H_2O \quad (37)$$

To a resin kettle is added 300 gm (2.6 moles) of 85% orthophosphoric acid and 175 gm (2.9 moles) of urea. The mixture is stirred until the urea dissolves and then 100 gm (2.3 unit mole) of low-viscosity grade polyvinyl alcohol dissolved in 300 ml of water is added. The mixture is stirred well and then the excess water is removed by heating the mixture in a pan at 110°C for 3 hr or more in a circulating air oven, with occasional stirring. The mass is then heated at 150°C for 15 min, at which time ammonia is evolved as urea decomposes and the white mass expands to several times its volume. The polymer is dissolved in 300 ml of water and then precipitated by pouring slowly into alcohol which is being agitated in a blender. The polymer is dried in a vacuum oven at 60°C, ground to a powder, and extracted for 4 hr with alcohol in a Soxhlet extractor to afford 304.6 gm (94%), % P = 19.6 and % N = 9.8 (calcd. for the monoammonium salt of polyvinyl phosphoric acid with a substitution of 3 phosphoric acid groups/4 vinyl groups % P = 19.9%, % N = 9.0). The free phosphoric acid derivative as prepared in Preparation 2-2 is obtained by acidification with hydrochloric acid.

2-4. Preparation of Poly(diisopropyl vinyl phosphate) [33a]

$$\left[\begin{array}{c}-CH_2-CH- \\ \quad\ | \\ \quad\ OH\end{array}\right]_n + (i\text{-}C_3H_7O)_2\overset{O}{\overset{\|}{P}}-Cl \longrightarrow \left[\begin{array}{c}-CH_2-CH- \\ | \\ O-P{=}O \\ | \\ (OC_3H_7\text{-}i)_2\end{array}\right]_n \qquad (38)$$

To a resin kettle equipped with a stirrer, condenser, and dropping funnel is added 22 gm (0.50 unit mole) of 90% hydrolyzed poly(vinyl alcohol) and 392 gm (5.0 moles) pyridine. Then 97 gm (0.50 mole) of diisopropyl chlorophosphate is added dropwise and the reaction mixture stirred for 18 hr at room temperature. Then sodium carbonate is added to neutralize the pyridine hydrochloride and the free pyridine is steam-distilled from the reaction mixture. The aqueous salt solution is separated from the viscous syrup of poly(diisopropyl vinyl phosphate) and the latter dried in a vacuum oven at 60°C. The product contains 10% phosphorus (approx. 30% esterification of the free hydroxyl groups).

The product is capable of further reaction with diisocyanates to give a solid polymer.

B. Free-Radical Addition Reactions

Vinyl phosphates polymerize in a manner similar to that of vinyl esters and the rates depend on the substituents present. Some typical examples are shown in Table VI.

The vinyl phosphates are synthesized by either the Perkow reaction [3] (Eq. 39) or by the condensation of dialkyl chlorophosphates with hydroxy derivatives of olefins or olefinic acids (Eq. 40).

$$(RO)_3P + XCH_2COCH_3 \longrightarrow (RO)_2\overset{O}{\overset{\|}{P}}-O\underset{CH_3}{\underset{|}{C}}{=}CH_2 + (RO)_2\overset{O}{\overset{\|}{P}}CH_2COCH_3 + RX \qquad (39)$$

$$(RO)_2\overset{O}{\overset{\|}{P}}-Cl + HOCH_2X-CH{=}CH_2 \longrightarrow (RO)_2\overset{O}{\overset{\|}{P}}-OCH_2X-CH{=}CH_2 \qquad (40)$$

$$(X = CH_2O\overset{O}{\overset{/\!/}{C}}-,\ CH_2-,\ \text{nothing, etc.})$$

TABLE VI

POLYMERIZATION AND COPOLYMERIZATION OF VINYL PHOSPHATES

$(RO)_nP(=O)—(X—OCR'=CH_2)_{n'}$							Reaction conditions			
R	n	X	R′	n'	Co-monomer	Catalyst (%)	Temp. (°C)	Time (hr)	Polymer properties	Ref.
						Benzoyl peroxide				
C_2H_5	3	—	H	1	—	?	?	2	Colorless visc. liq.	*a*
$CH_2=CH$	1	—	H	2	—	1.0	50–70	5	Yellow solid	*b*
C_2H_5	2	—	$COOC_2H_5$	1	—	1.5	25	5	Light yellow tacky solid	*c*

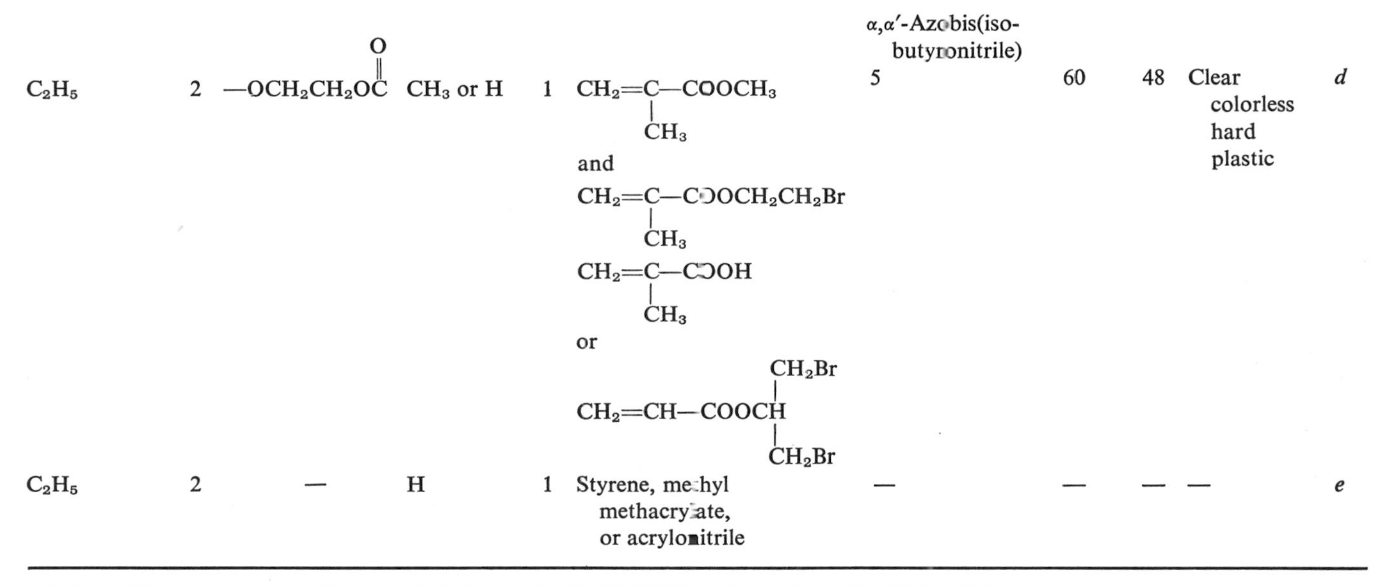

C_2H_5	2	$—OCH_2CH_2OC(=O)$	CH_3 or H	1	$CH_2{=}C(CH_3)—COOCH_3$ and $CH_2{=}C(CH_3)—COOCH_2CH_2Br$ $CH_2{=}C(CH_3)—COOH$ or $CH_2{=}CH—COOCH(CH_2Br)_2$	α,α′-Azobis(isobutyronitrile) 5	60	48	Clear colorless hard plastic	*d*
C_2H_5	2	—	H	1	Styrene, methyl methacrylate, or acrylonitrile	—	—	—	—	*e*

[a] R. R. Whetstone and D. Harman, U.S. Patent 2,765,331 (1957); *Chem. Abstr.* **51**, 5816 (1957).
[b] E. L. Gefter and M. I. Kabachnik, *Dokl. Akad. Nauk SSSR* **114**, 541 (1957); *Chem. Abstr.* **52**, 295 (1958).
[c] F. W. Lichtenthaler, Dissertation, University of Heidelberg, Germany (1959).
[d] J. L. I'Brian, U.S. Patent 2,993,033 (1961).
[e] R. W. Upson, *J. Amer. Chem. Soc.* **75**, 1763 (1953); U.S. Patent 2,557,805 (1951); *Chem. Abstr.* **45**, 8298 (1955).

The Perkow reaction is a modification of the Michaelis-Arbuzov reaction which occurs with haloaldehydes, lactones, some esters, nitroparaffins, and dicarbonyl compounds. The reaction will be described in further detail in the phosphonate section of this chapter (Section 3, Table VIII and Eq. 50).

Additional methods involve the dehydrohalogenation of 2-haloalkyl phosphates [103–105] (Eqs. 41 and 42)

$$(RO)_2P(=O)—OCH_2CH_2X \xrightarrow{B} (RO)_2P(=O)—OCH{=}CH_2 + BX \quad (41)$$

$$(RO)_2P(=O)—CH(OH)—CCl_3 \xrightarrow{B} (RO)_2P(=O)—OCH{=}CCl_2 \quad (42)$$

and the phosphorylation of enolates [106] (Eq. 43).

$$POCl_3 + 3CH_3CH{=}O + 3R_3N \xrightarrow{-3R_3NHCl} (CH_2{=}CH—O)_3P{=}O \quad (43)$$

Additional synthetic methods are described in Section 8.

Allyl phosphate monomers with and without bromine substituents can be polymerized to give products useful as flameproofing agents for paper and textiles. Walters and Hornstein [107] were the first to report that brominated poly(triallyl phosphate) gave a useful flameproofing compound. The preparation and polymerization of triallyl phosphate has been studied quite extensively [108,109]. The phosphate polymers suffer from the disadvantage that they are hydrolytically unstable [110]. Bromoform has been used as a chain-transfer agent in the polymerization of triallyl phosphate and thereby gave polymers containing bromine [111]. The latter polymers were also prepared in emulsion form using poly(vinyl alcohol) as a stabilizer [111]. Several other examples as well as the conditions of polymerization are given in Table VII.

2-5. Preparation of a Copolymer of Diethyl Vinyl Phosphate with Styrene [112]

$$(C_2H_5O)_2P(=O)OCH{=}CH_2 + C_6H_5CH{=}CH_2 \longrightarrow \left[—CH_2—CH(C_6H_5)—\right]_x \left[—CH_2—CH(—O—P(=O)(OC_2H_5)_2)—\right]_y \quad (44)$$

To a polymer tube is added 1.8 gm (0.011 mole) diethyl vinyl phosphate, 1.04 gm (0.01 mole) of styrene, and 0.15 gm of benzoyl peroxide. The tube is heated at 80°C under nitrogen at atmospheric pressure for 42 hr to give a

TABLE VII

POLYMERIZATION OF ALLYL PHOSPHATE MONOMERS

		Reaction conditions			
Phosphate monomer	Catalyst	Temp. (°C)	Time (hr)	Polymer properties	Ref.
Bis(β-bromoallyl) allyl phosphate	Benzoyl peroxide	100	6	Slight amount of polymerization—mainly decomposition	*a*
Diallyl *n*-butyl phosphate	—	110	2	Taffy-like polymer	*b*
Diallyl 2,3-dibromopropyl phosphate	Benzoyl peroxide	110	?	Solution polymerization. Decomposition. Very little polymer	*c* *c*
Diethyl allyl phosphate	Benzoyl peroxide	88	18	Does not polymerize in attempted bulk polymerization	*d*
Triallyl phosphate	Benzoyl peroxide	100	1	White polymer	*e*
Triallyl phosphate + chloroform	Benzoyl peroxide	85	1	White polymer—somewhat tacky	*f*
Triallyl phosphate + bromoform	$K_2S_2O_8$	80	1	Emulsion polymer	*g*
Triallyl phosphate + vinyl acetate	Benzoyl peroxide	40 65	48 240	Copolymer obtained which is flame-retardant	*h*

[a] R. Esteve and R. Laible, University of Rhode Island, QM Contract DA44-109-QM-421, Progress Report No. 7 (1952).

[b] R. Laible and R. Esteve, University of Rhode Island, QM Contract DA44-109-QM-421, Progress Report No. 3 (1951).

[c] R. Esteve and L. Laible, University of Rhode Island, QM Contract DA44-109-QM-421, Progress Report No. 2 (1951).

[d] A. D. F. Toy and R. Cooper, *J. Amer. Chem. Soc.* **76**, 2191 (1954).

[e] G. Griffin, R. Esteve, and R. Laible, University of Rhode Island, QM Contract DA44-109-QM-421, Progress Report No. 1 (1951).

[f] R. Esteve and R. Laible, "Polymerization of Triallyl Phosphate." University of Rhode Island, Kingston (unpublished work).

[g] J. Frick, J. Weaver, R. Arceneaux, and M. Stansbury, *J. Polym. Sci.* **20**, 307 (1956); J. Frick, J. Weaver, and J. A. Reid, *Text. Res. J.* **25**, 100 (1955).

[h] J. Haworth, British Patent 675,783 (1952); *Chem. Abstr.* **46**, 11778b (1953).

colorless viscous liquid. The polymer is isolated by precipitation in 50 ml of methanol to give 1.2 gm (43%) of a white, solid copolymer, softening point, 68°C; % P = 25.2.

Diethyl vinyl phosphate copolymerizes under similar conditions with methyl methacrylate and acrylonitrile but not with vinyl acetate. Furthermore, diethyl vinyl phosphate does not homopolymerize using free-radical catalysts or sodium in liquid ammonia.

A general procedure applicable to the polymerization of allyl phosphate esters is given in Preparation 3-6 and some further examples are given in Tables XIII–XV.

3. POLYPHOSPHONATES

Polyphosphonates are prepared either by the condensation polymerization of polyols with disubstituted phosphonates, ring-opening polymerizations of epoxides or cyclic phosphonates, the Arbuzov reaction, or the free-radical polymerization of vinyl phosphonates. Additional methods for preparing polyphosphonates are discussed where applicable.

A. Condensation Reactions

The interfacial condensation of alkyl dichlorophosphonates with aromatic diols gives polyphosphonates with molecular weights of about 25,000 [113–115]. These products are resistant to hydrolysis and have good mechanical properties.

$$X{-}\overset{\overset{\displaystyle O}{\|}}{\underset{\underset{\displaystyle R}{|}}{P}}{-}X + HO{-}R'{-}OH \xrightarrow{-2HCl} \left[{-}\overset{\overset{\displaystyle O}{\|}}{\underset{\underset{\displaystyle R}{|}}{P}}{-}OR'{-}O{-} \right]_n \qquad (45)$$

R′ = Ar [113,114], —Ar—C(CH_3)$_2$—Ar [10,10a,10b]
R = alkyl or alkenyl [10,10a,10b,113,114]
X = Cl [10,10a,10b,113,114] or NR''_2 [114]

The ring-opening polymerization of cyclic phosphonates also gives polymeric phosphonates.

$$\begin{matrix} CH_2{-}O & & O \\ | & P & \\ CH_2{-}O & & R \end{matrix} \longrightarrow \left[{-}\overset{\overset{\displaystyle O}{\|}}{\underset{\underset{\displaystyle R}{|}}{P}}{-}OCH_2{-}CH_2{-}O{-} \right]_n \qquad (46)$$

(R = alkyl [116], alkoxy [117], vinyl [118], or aryl [116])

The ring-opening polymerization of cyclic phosphites gives polyphosphonates as shown in Eq. (47) [17,17a].

Polyphosphonates are also prepared by typical condensation reactions

$$\text{RO—P}\underset{\text{O}}{\overset{\text{O}}{\diamond}}\text{R}' \longrightarrow \left[\begin{array}{c}\text{O}\\ \| \\ \text{—P—OR}'\text{—}\\ | \\ \text{OR}\end{array}\right]_n + \left[\begin{array}{c}\text{O}\\ \| \\ \text{—P—OR}'\text{—O—}\\ | \\ \text{R}\end{array}\right]_n \qquad (47)$$

involving carboxylic acid, alcohol, amine, and isocyanate derivatives, such as that in Eq. (48).

$$(\text{RO})_2\overset{\text{O}}{\overset{\|}{\text{P}}}\text{—CH}_2\text{N(CH}_2\text{CH}_2\text{OH)}_2 + \text{R(COOH)}_2 \xrightarrow{-2\text{H}_2\text{O}} \text{polyester} \qquad (48)$$

Epoxy phosphonates can also be self-polymerized via the epoxy group with catalysts or reacted with anhydrides, amines, etc. The epoxy phosphonates [119] are prepared by the Arbuzov reaction as shown in Eq. (49). Some typical examples for the preparation of a variety of monomers by the Arbuzov reaction are shown in Table VIII.

$$(\text{C}_2\text{H}_5\text{O})_3\text{P} + \text{Br—CH}_2\text{—}\underbrace{\text{CH—CH}_2}_{\text{O}} \xrightarrow{\text{reflux, 140°C}}$$

$$(\text{C}_2\text{H}_5\text{O})_2\overset{\text{O}}{\overset{\|}{\text{P}}}\text{—CH}_2\text{—}\underbrace{\text{CH—CH}_2}_{\text{O}} \xrightarrow{\text{catalyst}} \left[\begin{array}{l}\text{—O—CH}_2\text{—CH—}\\ \qquad\qquad\quad | \\ \qquad\qquad\text{CH}_2\text{—P(OC}_2\text{H}_5)_2\\ \qquad\qquad\qquad\quad \| \\ \qquad\qquad\qquad\quad \text{O}\end{array}\right]_n \qquad (49)$$

The Arbuzov reaction [120] can be used to modify existing halogenated polymers to afford phosphonate side groups as shown in Eq. (50) [43a,121, 122].

$$\left[\begin{array}{l}\text{—CH}_2\text{—CH—}\\ \qquad\quad | \\ \qquad\text{OC—CH}_2\text{Cl}\\ \qquad\ \| \\ \qquad\ \text{O}\end{array}\right]_n + (\text{C}_2\text{H}_5\text{O})_3\text{P} \xrightarrow{-\text{C}_2\text{H}_5\text{—Cl}}$$

$$\left[\begin{array}{l}\text{—CH}_2\text{—CH—}\quad\ \text{O}\\ \qquad\quad | \qquad\qquad \| \\ \qquad\text{OC—CH}_2\text{P—(OC}_2\text{H}_5)_2\\ \qquad\ \| \\ \qquad\ \text{O}\end{array}\right]_n \qquad (50)$$

The self-condensation of bis(2-chloroethyl) phosphonate eliminates 1,2-dichloroethane to give a low molecular weight polyphosphonate as shown in Eq. (51).

$$\text{ClCH}_2\text{CH}_2\text{O—}\overset{\text{O}}{\overset{\|}{\underset{\text{R}}{\underset{|}{\text{P}}}}}\text{—O—CH}_2\text{CH}_2\text{Cl} \xrightarrow[\text{200°–250°C}]{\text{Arbuzov reaction}} \left[\text{—}\overset{\text{O}}{\overset{\|}{\underset{\text{R}}{\underset{|}{\text{P}}}}}\text{—OCH}_2\text{CH}_2\text{O—}\right]_n + \text{ClCH}_2\text{CH}_2\text{Cl} \qquad (51)$$

(R = alkyl [123]; vinyl [124]; H [125]; or Cl [17a])

Additional examples of condensation polymer preparations are shown in Table IX.

TABLE VIII

PREPARATION OF MONOMERS UTILIZING THE ARBUZOV REACTION

$(RO)_3P$ R =	R′—X R′ =	Reaction conditions Temp. (°C)	Time (hr)	$(RO)_2P(=O)$—R′ structure	Properties B.p., °C (mm Hg)	η_D (°C)	Ref.
C_2H_5	$—CH_2CH—CH_2$ (epoxide, –O–) (X = Br)	140	—	$(C_2H_5O)_2P(=O)—CH_2—CH—CH_2$ (epoxide, –O–)	101 (1.5)	—	*a*
HOR—	CCl_3 (X = Cl)	Reflux	7	$(HORO)_2P(=O)CCl_3$	—	—	*b*
$CH_2{=}CHCH_2$	$ClC_6H_4CH_2—$ (X = Cl)	90	10	$(CH_2{=}CH—CH_2)_2P(=O)—CH_2C_6H_4Cl$	—	—	*c*
$CH_2{=}CHCH_2$	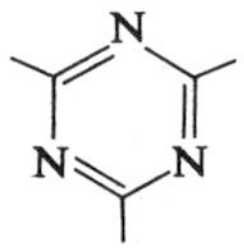 $(CH_3)_2C_3N_3—$ (triazinyl) (X = Cl)	85–90	—	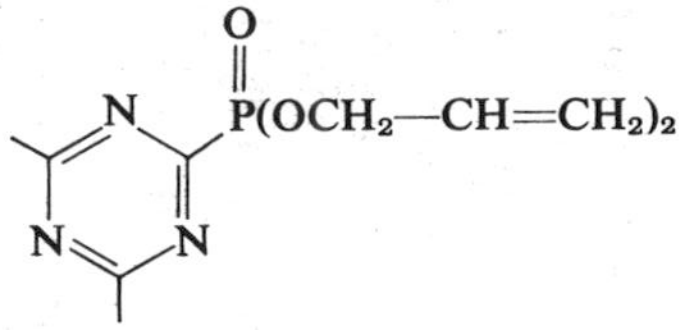 $(CH_3)_2C_3N_3—P(=O)(OCH_2—CH{=}CH_2)_2$	—	—	*d*
$XCH_2CH_2—$ where (X = Cl,Br)	—	(1) 160 (2) NaOH 50	— 20	$(XCH_2CH_2O)_2P(=O)—CH{=}CH_2$	137–139 (4)	—	*e*

$(X—CH_2)_2CH—$	—	105	—	$[(XCH_2)_2CHO]\overset{O}{\overset{\Vert}{P}}CH(CH_2X)_2$	—	—	*f*
C_2H_5	$ClCH_2C_6H_4CH_2CH_2Cl$	90	20	$CH_2{=}CHC_6H_4CH_2—\overset{O}{\overset{\Vert}{P}}(OC_2H_5)_2$	120 (2.0)	—	*g*
		80	$KOH—C_2H_5OH$, 6	$CH_2{=}CHC_6H_4CH_2—\overset{O}{\overset{\Vert}{P}}(OC_2H_5)_2$	120 (2.0)	—	*g*
CH_3	$BrCH_2—CH{=}CH_2$	126–157	—	$(CH_3O)_2\overset{O}{\overset{\Vert}{P}}—CH_2CH{=}CH_2$	—	—	*h*
C_2H_5	$ClCH_2COOCH{=}CH_2$	—	1½	$(C_2H_5O)_2\overset{O}{\overset{\Vert}{P}}—CH_2COOCH{=}CH_2$	115–123 (1.2)	1.4431	*i*
C_2H_5	$CH_2{=}C(X)COOCH_3$	25	½	$(C_2H_5O)_2\overset{O}{\overset{\Vert}{P}}—CH{=}CHCOOCH_3$	109–110	1.4483 (20)	*j*
C_2H_5	$CH_2{=}C(CH_3)—COCl$	40–50	½	Polymer	—	—	*k*

[a] E. E. Hardy and T. Reetz, U.S. Patent 2,770,620 (1956).
[b] Pure Chemicals, Ltd., British Patent 1,012,630 (1965).
[c] G. F. D'Alelio, U.S. Patent 3,325,569 (1967).
[d] A. D. F. Toy and R. S. Cooper, *J. Amer. Chem. Soc.* **76**, 2191 (1954).
[e] A. D. B. Graham, U.S. Patent 3,255,145 (1966); Stauffer Chem. Co., Belgian Patent 7[illegible]3,066 (1968).
[f] B. S. Taylor and M. R. Lutz, French Patent 1,372,907 (1964); G. E. Schroll, U.S. Patent 3,250,827 (1966).
[g] J. G. Abramo, A. Y. Garner, and E. C. Chapin, U.S. Patents 3,051,740 (1962) and 3,161,667 (1964).
[h] A. E. Arbuzov and A. J. Razumov, *Bull. Acad. Sci. USSR* p. 714 (1951).
[i] R. H. Wiley, U.S. Patent 2,478,441 (1949).
[j] H. W. Coover, Jr., M. A. McCall, and J. B. Dickey, *J. Amer. Chem. Soc.* **79**, 1963 (1957).
[k] M. L. Ernsberger, U.S. Patent 2,491,920 (1949).

TABLE IX

PREPARATION OF POLYPHOSPHONATES BY CONDENSATION POLYMERIZATION METHODS

Reactants				Reaction conditions		
Phosphorus compounds	Non-P compounds	Catalyst	Structure	Temp. (°C)	Time (hr)	Ref.
$RPOCl_2$	HOR′OH	—	$\left[-P(=O)(R)-OR'-O-\right]_n$ where R = C_6H_5 and R′ = tetrachlorophenyl then product m.p. = 241°C	175–200	4–8	*a*
2,4-$Br_2C_6H_3POCl_2$	HOC_6H_4OH	—	—	95–200	13	*b*
$(C_6H_5O)_2P(=O)H$	$HOC_6H_4C(CH_3)_2C_6H_4OH$	$MgCl_2$	Polymeric, waxy solid, visc. oil	190–240	1–2	*c*
$CH_3P(=O)Cl_2$	—	$NaAlO_2$	$\left[-P(=O)(CH_3)-O-\right]$	25–60	—	*d*
$C_2H_5P(=O)(OC_2H_5)_2$	—	P_4O_{10}	$\left[-P(=O)(C_2H_5)-O-P(=O)(OC_2H_5)-O-\right]_n$	90	6	*e*

CH_2-O $\vert\quad\quad\quad>P-Cl$ CH_2-O	$(CH_3)_2C{=}O$	$CHCl_3, H_2O$	$\left[-O-\underset{CH_3}{\overset{CH_3}{C}}-\underset{OCH_2CH_2Cl}{\overset{O}{\overset{\Vert}{P}}}-\right]_n$ Solid	56	20–40	*f*
CH_2-O $\vert\quad\quad\quad>P-OC_6H_5$ CH_2-O	CH_2-O $\vert\quad\quad\vert$ $CH_2-C{=}O$	—	$\left[-\underset{OC_6H_5}{\overset{O}{\overset{\Vert}{P}}}-CH_2CH_2COOCH_2CH_2O-\right]_n$	150–180	$\frac{1}{6}$	*g*
			Viscous oil, n_D^{20} 1.5233	10–65	5	*g*
$(C_2H_5O)_3P$	CH_2-O $\vert\quad\quad\vert$ $CH_2-O{=}O$	—	$(C_2H_5O)_2\overset{O}{\overset{\Vert}{P}}-[CH_2CH_2COO]_n\cdot$ $CH_2CH_2COOC_2H_5$ Viscous oil, n_D^{20} 1.4570, mol. wt. 490	—	—	*g*
$R\overset{O}{\overset{\Vert}{P}}(OAr)_2$	HOArOH	—	$\left[-\underset{R}{\overset{O}{\overset{\Vert}{P}}}-OAr-O-\right]_n$	—	—	*h*
$CH_3PO(OCH_2CH_2Cl)_2$	—	—	—	230	10	*i*
$ClCH_2PO(OCH_2CH_2Cl)_2$	—	—	—	230	10	*i*
$C_6H_5PO(OCH_2CH_2Cl)_2$	—	—	—	250	10	*i*

[a] A. D. F. Toy, U.S. Patents 2,435,252 (1948) and 2,572,076 (1951).

[b] H. Zenftman and A. McLean, U.S. Patent 2,636,876 (1953).

[c] H. W. Coover, Jr. and R. L. McConnell, U.S. Patent 3,271,329 (1966).

[d] D. Grant, J. R. Van Wazer, and C. H. Dungan, *J. Polym. Sci., Part A-1* **5**, 57 (1967); A. N Pudovik, A. A. Muratova, F. F. Sushentsova, and M. M. Zoreva, *Vysokomol. Soedin.* **6**, 258 (1964).

[e] Farb. Hoechst, A.-G., Belgian Patent 671,561 (1965).

[f] A. Carson, W. E. Feely, and M. J. Hurwitz, U.S. Patent 3,371,131 (1968).

[g] R. L. McConnell and H. W. Coover, Jr., U.S. Patent 3,062,788 (1962); *J. Amer. Chem. Soc.* **78**, 4453 (1956).

[h] H. W. Coover and M. A. McCall, U.S. Patent 2,682,522 (1954).

[i] V. V. Korshak, I. A. Gribova, and V. K. Shitikov, *Izv. Akad. Nauk SSSR, Otd. Khim Nauk* p. 210 (1958).

3-1. *Preparation of Poly(ethylene methylphosphonate) by the Ring-Opening Polymerization of Ethylene Methyl Phosphite* [17a]

$$\begin{array}{l} CH_2-O \\ | \qquad\quad \backslash \\ | \qquad\quad\; P-OCH_3 \\ | \qquad\quad / \\ CH_2-O \end{array} \xrightarrow{AlCl_3} \left[-CH_2CH_2O\overset{\overset{O}{\|}}{\underset{\underset{CH_3}{|}}{P}}-O- \right]_n \qquad (52)$$

To a thick-walled test tube is added 1.0 gm (0.008 mole) of dry ethylene methyl phosphite and 0.04 gm of aluminum chloride catalyst. The tube is cooled, sealed under a nitrogen stream, and placed in a 150°C temperature bath for 8 hr to give a polymer in quantitative yield, $\eta_{inh} = 1.01$ (DMF at 9°C); IR, 1300 cm^{-1} [P—(C_2H_5)], 1225 cm^{-1} (P═O), and 1255 cm^{-1} (P═O).

3-2. *Condensation of Triethyl Phosphite with β-Propiolactone to Give Ethyl O-(3-Diethylphosphonopropionyl)polyhydracrylate* [126]

$$(C_2H_5O)_3P + \underbrace{CH_2-CH_2-C{=}O}_{\llcorner\;O\;\lrcorner} \longrightarrow (C_2H_5O)_3P^+CH_2CH_2COO^-$$

$$\downarrow \; \underbrace{CH_2CH_2CH{=}O}_{\llcorner\;O\;\lrcorner}$$

$$(C_2H_5O)_2\overset{\overset{O}{\|}}{P}CH_2CH_2CO[OCH_2CH_2CO]_nOC_2H_5 \qquad (53)$$

To a 100 ml flask equipped with a stirrer, condenser, and dropping funnel is added 16.6 gm (0.1 mole) of triethyl phosphite and 28.8 gm (0.4 mole) of β-propiolactone. The mixture is stirred well while 5 ml of triethylamine is added dropwise. The temperature rises to 40°C and remains there for $\frac{1}{2}$ hr. The reaction mixture is stirred for 4 hr and then allowed to stand overnight. Then the starting materials are removed by distilling the volatiles off at 65°C (3.0 mm Hg) to give a light yellow viscous residue, n_D^{20} 1.4570, mol. wt. 490 (ebulliometric determination in benzene solution).

B. Free-Radical Addition Reactions

Some vinyl and allyl phosphonate monomers can be free-radically polymerized to give polyphosphonates.

The Michaelis-Arbuzov reaction [3c] is the most important method for the preparation of vinyl phosphonates by the action of halides with the corresponding phosphite [127]. Some examples of this method have already been given in Table VIII.

$$(RO)_3P + R'X \longrightarrow (RO)_2\overset{\overset{O}{\|}}{P}—R' + RX \quad (54)$$

Another method of preparing phosphonates involves the reaction shown in Eq. (55) [3c].

$$(RO)_2POM + R'X \longrightarrow (RO)_2\overset{\overset{O}{\|}}{P}—R' + MX \quad (55)$$

Examples of the homopolymerization and copolymerization of vinyl and allyl phosphonates are shown in Table X.

3-3. *Bulk Polymerization of Bis(2-chloroethyl) Vinylphosphonate* [128]

$$CH_2{=}CH—\overset{\overset{O}{\|}}{P}(OCH_2CH_2Cl)_2 \longrightarrow \left[—CH_2—\underset{\underset{\underset{(OCH_2CH_2Cl)_2}{|}}{\overset{|}{P=O}}}{CH}— \right]_n \quad (56)$$

To a test tube is added 20.0 gm (0.09 mole) of bis(2-chloroethyl) vinylphosphonate. The tube is warmed to 51°C and 0.4 gm of α,α'-azobis(α,γ-dimethylvaleronitrile) (Du Pont "Vazo 52") is added. The tube is agitated under an inert atmosphere for 25 hr while heating at 51°C. The polymer is obtained in quantitative yield as a pale yellow viscous liquid.

Other polymers and copolymers of bis(2-chloroethyl) vinylphosphonate are described in Table XI.

3-4. *Preparation of a Copolymer of Lauryl Methacrylate and Diallyl Phenylphosphonate by Solution Polymerization* [5]

$$CH_2{=}\underset{\underset{CH_3}{|}}{C}—COOC_{12}H_{25} + (CH_2{=}CH—CH_2O)_2\overset{\overset{O}{\|}}{P}C_6H_5 \xrightarrow{80°C} \text{Copolymer} \quad (57)$$

To a polymer tube with a constricted neck is added 5.84 gm (0.023 mole) of lauryl methacrylate (r_1), 9.27 gm (0.039 mole) of diallyl phenylphosphonate (r_2), 5 ml of pure benzene, and 0.20 mole % of benzoyl peroxide. The monomer mixture is outgassed and sealed under vacuum. The tube is placed in a constant temperature bath at 80° ± 1°C and agitated for 2 hr. The tube is then opened and the contents added to 500 ml of ethanol containing 0.5 gm of hydroquinone. The polymer is separated from the supernatant liquid in a centrifuge and then reprecipitated from benzene twice to afford 2.6 gm (17%), % P = 1.24. Calcd. $r_1 = 19.5$, $r_2 = 0.072$ (reactivity ratios).

TABLE X

PREPARATION OF PHOSPHONATE POLYMERS FROM THE CORRESPONDING VINYL AND ALLYL MONOMERS

Phosphonate	Initiator	Conditions: Temp. (°C)	Conditions: Time (hr)	Yield (%)	Ref.
$CH_2{=}CH{-}R{-}P(O)(OR)_2$	—	—	—	—	*a*
$CH_2{=}CHP(O)(OC_2H_5)_2$ + styrene	$(CH_3)_3COOH$	116	11	85	*b*
$CH_2{=}CH{-}P(O)(OC_2H_5)_2$	Bz_2O_2	—	—	—	*c*
$CH_2{=}C(OCOCH_3)P(O)(OC_2H_5)_2$	Bz_2O_2	—	—	—	*d*
$CH_2{=}C(CH_3)P(O)(OCH_3)_2$ + butadiene (15:85)	$K_2S_2O_8$	50	18	45	*e*
$CH_3OC(O){-}CH{=}CH{-}P(O)(OR)_2$ + styrene or other monomer	Bz_2O_2	—	—	—	*f*
$CH_2{=}C(R){-}CH_2P(O)(OR')_2$ + styrene or other monomers	—	—	—	—	*g*
$CH_2{=}CR{-}C(O){-}P(O)(OR')_2$ + methyl methacrylate or by itself	Bz_2O_2	12	—	—	*h*

$CH_2{=}CHOC({=}O){-}CH_2P({=}O)(OC_2H_5)_2$	Bz_2O_2	60–70	200	—	*i*
$CH_2{=}CHOC({=}O){-}CH_2P({=}O)(OC_2H_5)_2$ + vinyl acetate or other monomers	Bz_2O_2	65	10	—	*j*
$(CH_2{=}C(CH_3){-}COOCH_2CH_2O)_2P({=}O)R$ alone and with co-monomers	Bz_2O_2	70	—	—	*k*
$CH_2{=}CH{-}CH_2OC{-}CH_2P(OR)_2$ alone or with co-monomer	Bz_2O_2	60–70	200	—	*l*
$CH_3{-}P({=}O)(OCH{=}CH_2)(OR)$	Bz_2O_2	50–70	150	—	*m*
$R{-}P({=}O)(OCH{=}CH_2)_2$	Bz_2O_2	50	50	—	*l,m,n*
$CH_3{-}P({=}O)(OCH_2CH{=}CH_2)_2$	Bz_2O_2	87–88	18	—	*n*
$Cl{-}CH_2P({=}O)(OCH_2CR{=}CH_2)_2$ alone or with co-monomers	Bz_2O_2	85	—	—	*o*
$CCl_3P({=}O){-}(OCH_2CH{=}CH_2)_2$	Bz_2O_2	60–70	80–90	—	*p*
$C_2H_5P({=}O)(OCH_2CH{=}CH_2)_2$	Bz_2O_2	87–88	18	—	*n*
$CH_3C({=}O){-}P({=}O)(OCH_2CH{=}CH_2)_2$ alone or with co-monomers	Bz_2O_2	—	—	—	*q*
$C_6H_5C({=}O){-}P({=}O)(OCH_2CH{=}CH_2)_2$	Bz_2O_2	—	—	—	*q*

(*cont.*)

TABLE X (*cont.*)

Phosphonate	Initiator	Conditions: Temp. (°C)	Conditions: Time (hr)	Yield (%)	Ref.
$[(CH_2{=}C(R)CH_2O)_2P(=O)-CH_2-]_2O$	$[(CH_3)_3CO]_2$	110	20	—	*r*
$R-C_6H_4P(=O)-(OCH_2CH{=}CH_2)_2$ alone and with co-monomers	Bz_2O_2	100	2	—	*s*
$(CH_3)_2C{=}CHP(=O)(OCH_2CR{=}CH_2)_2$ with co-monomers	Bz_2O_2	70	24	—	*t*
$ROC(=O)-P(=O)(OCH_2CH{=}CH_2)_2$ alone or with co-monomers	Bz_2O_2	70	70–100	—	*u*
$CH_2{=}CH-P(=O)(OCH{=}CH_2)_2$	Bz_2O_2	70	30	—	*v*
$(CH_3)_3C-CH_2C(CH_3){=}CHP(=O)(OCH_2CH{=}CH_2)_2$	Bz_2O_2	87–88	18	—	*n*
$C_6H_5CH{=}CH-P(=O)(OCH_2C(CH_3){=}CH_2)_2$	Bz_2O_2	87–88	18	—	*n,w*
$CH_2{=}CH-CH_2-P(=O)(OCH_2-CH{=}CH_2)_2$	Bz_2O_2	98–100	6	—	*x*

$CH_2{=}CH{-}\overset{O}{\overset{\|}{C}}{-}\overset{O}{\overset{\|}{P}}(OCH_2CH{=}CH_2)_2$	—	—	—	—	*q*
$CH_2{=}CHO\overset{O}{\overset{\|}{C}}{-}CH_2\overset{O}{\overset{\|}{P}}(OCH_2CH{=}CH_2)_2$	Bz_2O_2	70	6	No polymer	*u*
$(CH_3)(CRCH_2CH_2O)\overset{O}{\overset{\|}{P}}OCH{=}CH_2$	Bz_2O_2	50–70	200	Soft yellow	*l*
$R{-}\overset{O}{\overset{\|}{P}}(OCH_2{-}CH{=}CH_2)_2$ + lauryl methacrylate where R = C_6H_5, C_4H_9, and $(C_2H_5)_2P{-}CH_2{-}CH{=}CH_2$ in place of $R\overset{O}{\overset{\|}{P}}(OCH_2CH{=}CH_2)_2$	Bz_2O_2	80	2	—	*y*
$CH_2{=}CH{-}\overset{O}{\overset{\|}{P}}(OCH_2CH_2Cl)_2$ alone and with co-monomers	Bz_2O_2	70	120	94.1	*z*

[a] A. D. F. Toy and R. S. Cooper, *J. Amer. Chem. Soc.* **76**, 2191 (1954).
[b] C. L. Arcus and R. J. S. Matthews, *J. Chem. Soc., London* p. 4607 (1956).
[c] M. I. Kabachnik, *Izv. Akad. Nauk SSSR, Otd. Khim. Nauk* p. 233 (1947); E. K. Fields, U.S. Patent 2,579,810 (1951).
[d] J. Kennedy and G. M. Meaburn, *Chem. Ind. (London)* p. 930 (1956).
[e] C. S. Marvel and J. C. Wright, *J. Polym. Sci.* **8**, 255 (1952); R. V. Lindsey, Jr., U.S. Patent 2,439,214 (1948).
[f] J. B. Dickey and H. W. Coover, U.S. Patent 2,559,854 (1951).
[g] H. W. Coover and J. B. Dickey, U.S. Patent 2,636,027 (1953); J. B. Dickey and H. W. Coover, U.S. Patents 2,721,876 (1956) and 2,780,616 (1957).
[h] M. L. Ernsberger, U.S. Patent 2,491,920 (1949); G. Kamai and V. A. Kukhtin, *Tr. Kazan. Khim.-Tekhnol. Inst.* **16**, 29 (1952).
[i] G. Kamai and V. A. Kikhtin, *Zh. Obshch. Khim.* **24**, 1855 (1954); R. H. Wiley, U.S. Patent 2,478,441 (1949).
[j] R. H. Wiley, U.S. Patent 2,478,441 (1949).
[k] A. A. Berlin, L. P. Raskina, L. A. Zhil'tsova, and B. E. El'tsefon, *Vysokomol. Sayer A.B.* **1**, 174 (1971).
[l] G. Kamai and V. A. Kukhtin, *Zh. Obshch. Khim.* **24**, 1855 (1954); *Tr. Kazan. Khim.-Tekhnol.* **16**, 29 (1952).
[m] E. L. Gefter and M. I. Kabachnik, *Dokl. Akad. Nauk SSSR* **114**, 541 (1957); *Chem. Abstr.* **52**, 295 (1958); A. J. Castro and W. E. Elwell, *J. Amer. Chem. Soc.* **72**, 2275 (1950).

References to Table X (cont.)

[n] A. D. F. Toy and R. S. Cooper, *J. Amer. Chem. Soc.* **76**, 2191 (1954).

[o] A. D. F. Toy and K. H. Rattenbury, U.S. Patent 2,714,100 (1955); *Chem. Abstr.* **49**, 14380c (1955); U.S. Patent 2,735,789 (1956).

[p] G. Kamai and V. A. Kukhtin, *Dokl. Akad. Nauk SSSR* **89**, 309 (1953); J. Kennedy, E. S. Lane, and B. K. Robinson, *J. Appl. Chem.* **8**, 459 (1918).

[q] G. Kamai and V. A. Kykhtin, *Tr. Kazan. Khim.-Tekhnol.* **16**, 29 (1952).

[r] D. Harman and A. R. Stiles, U.S. Patent 2,632,756 (1953).

[s] A. D. F. Toy and L. V. Brown, U.S. Patent 2,586,885 (1952); *Ind. Eng. Chem.* **40**, 2276 (1948); A. D. F. Toy, U.S. Patent 2,538,810 (1951); *J. Amer. Chem. Soc.* **70**, 185 (1948).

[t] A. D. F. Toy, U.S. Patent 2,485,677 (1949).

[u] G. Kamai and V. A. Kukhtin, *Zh. Obshch. Khim.* **24**, 1855 (1954).

[v] E. L. Gefter, *Dokl. Akad. Nauk SSSR* **114**, 541 (1957).

[w] A. D. F. Toy, U.S. Patent 2,497,638 (1950).

[x] J. Kennedy, E. S. Lane, and B. K. Robinson, *J. Appl. Chem.* **8**, 459 (1958).

[y] K. I. Beynon, *J. Polym. Sci., Part A* **1**, 3343 (1963).

[z] Stauffer Chemical Company, Technical Brochure on Fyrol Bis-Beta[Bis(beta-chloroethyl)vinylphosphonate] No. 1567A-11/70.

TABLE XI

POLYMERIZATION AND COPOLYMERIZATION OF BIS(2-CHLOROETHYL) VINYLPHOSPHONATE

Bis(2-chloroethyl) vinyl-phosphonate (r_1) (moles)	Co-monomer (r_2) (moles)	Solvent (ml)	Catalyst (gm)	Polymerization conditions		Yield (%)	Properties			Reactivity ratios		
				Temp. (°C)	Time (hr)		η_{inh}	mol. wt.	Ref.	r_1	r_2	Ref.
4.06 gm	—	—	Benzoyl peroxide (0.0405)	70	120	94.1	0.045 (acetone) 25°C	—	*a*	—	—	—
1.0	—	Hexane	Diisopropyl peroxycarbonate (3%)	70	—	50	—	2700	*b*	—	—	—
1.0	Methyl methacrylate (3.3)	Hexane	Diisopropyl peroxycarbonate (3%)	70	—	99	—	3000	*b*	0.11	7.89	*a*
170 gm	Acrylonitrile (68 gm)	N_2O (1150 + 5.5 gm Na dodecyl sulfate)	$K_2S_2O_8$ (24) + $NaHSO_3$ (24)	40	17	87	—	—	*d*	0.25	3.38	*a*
										0.40	0.93	*c*
26 gm	—	Toluene (60)	BuMgCl (5 ml, 3 *N*)	0	5	21 gm	—	—	*e*	—	—	—
1.5	Vinyl acetate (1.0)	—	Benzoyl peroxide or $K_2S_2O_8$	60–95	—	—	—	—	*f*	—	—	—

[a] S. Konya and M. Yokayama, *Kogyu Kagaku Zasshi* **68**, 1080 (1965).
[b] B. J. Muray, *J. Polym. Sci., Part C* **16**, 1869 (1967).
[c] S. Fujii, *J. Sci. Hiroshima Univ., Ser. A-2* **31**, 89 (1967).
[d] V. E. Shashoua, U.S. Patent 2,888,434 (1959); *Chem. Abstr.* **53**, 16554c (1959).
[e] F. J. Welch, U.S. Patent 3,312,674 (1967).
[f] Farbwerke Hoechst A.G., German Patent 1,077,215 (1967).

3-5. Preparation of Diethyl-1-keto-2-methyl-2-propenephosphonate Polymer [129]

$$CH_2{=}\underset{\displaystyle CH_3}{\underset{|}{C}}{-}COCl + (C_2H_5O)_3P \longrightarrow (C_2H_5O)_2\overset{\displaystyle O}{\overset{\|}{P}}{-}\overset{\displaystyle O}{\overset{\|}{C}}{-}\underset{\displaystyle CH_3}{\underset{|}{C}}{=}CH_2 + C_2H_5Cl \quad (58)$$

$$\downarrow \text{Polymer}$$

To a flask equipped with a thermometer, stirrer, and condenser is added 40 gm (0.37 mole) of methacrylyl chloride (CAUTION: toxic, lachrymator) (b.p. 97.5°–98.5°C) which is stabilized by hydroquinone. Then 55 gm (0.33 mole) of triethyl phosphite is added dropwise over a $\frac{1}{2}$ hr period while the temperature of the reaction is kept at 40°–55°C with cooling. HCl is evolved during the reaction and after the addition the temperature is raised to 85°–90°C for 20 min. The temperature is raised to 200°C while reducing the pressure to 5 mm Hg. The residue remaining appears polymeric and is isolated in 59% (56 gm) yield; mol. wt., 480,450 (ebullioscopic determination in C_6H_6). The polymer is viscous, water-soluble, and decolorizes bromine in carbon tetrachloride.

The structure of this polymer was not determined and still awaits analysis.

3-6. General Procedure for the Polymerization of Allyl Esters of Phosphonic Acids [130]

To a polymer tube is added 5.0 gm of the allyl ester along with 3% of benzoyl peroxide catalyst. The sample is heated for 18 hr under a nitrogen atmosphere at 87°–88°C. Diallyl phosphite requires only 2% benzoyl peroxide, since 3% or more causes a violent reaction. The results of the polymerization of several compounds by this method are shown in Tables XIII–XV, grouped according to the properties of the polymers isolated.

3-7. Polymerization of Diallyl Cyclohexanephosphonate [131]

$$C_6H_{11}{-}\overset{\displaystyle Cl}{\underset{\displaystyle Cl}{P}}{=}O + 2CH_2{=}CH{-}CH_2OH \longrightarrow C_6H_{11}{-}P(OCH_2CH{=}CH_2)_2$$

$$\xrightarrow[\text{70°–115°C}]{Bz_2O_2}$$

$$\left[-CH_2-\overset{|}{C}HCH_2O-\underset{\displaystyle C_6H_{11}}{\underset{|}{\overset{\displaystyle O}{\overset{\|}{P}}}}-OCH_2-\overset{|}{C}H-CH_2- \right]_n \quad (59)$$

TABLE XII

COMPOUNDS YIELDING HARD GLASSY SOLIDS ON POLYMERIZATION[a]

Compounds	Yield (%)	B.p. (°C)	(mm)	d_{25}^{25}	n_D^{25}	Analyses, P, %	
						Calc.	Found
$HP(O)(OCH_2CH{=}CH_2)_2$	—	58–61	(1)	1.07[illegible]	1.4459	19.1	19.0
$(CH_3)_2C{=}CH{-}P(O)(OCH_2CH{=}CH_2)_2$	69	87–9[illegible]	(0.5)	1.0[illegible]85	1.4670	14.3	14.0
$(CH_3)_2C{=}CH{-}P(O)(OCH_2C(CH_3){=}CH_2)_2$[b]	64.3	108–115	(2)	1.0[illegible]91	1.4668	12.7	12.6
$C_6H_5P(O)(OCH_2CH{=}CH_2)_2$	—	—	—	—	—	—	
$C_6H_5OP(O)(OCH_2CH{=}CH_2)_2$[c]	64	102	(0.5)	1.[illegible]42	1.4957	12.2	12
$C_6H_5CH_2P(O)(OCH_2CH{=}CH_2)_2$	79.4	141–143	(1.5)	1.[illegible]99	1.5113	12.3	12.3
$C_6H_5CH_2P(O)(OCH_2C(CH_3){=}CH_2)_2$	—	158–163	(2)	1.[illegible]630	1.5053	11.06	11.0

[a] Reprinted from A. D. F. Toy and R. S. Cooper, *J. Amer. Chem. Soc.* **76**, 2191 (1954). Copyright 1954 by the American Chemical Society. Reprinted by permission of the copyright owner.

[b] This polymer is a hard, but opaque solid.

[c] Phosphate ester given here for comparison.

TABLE XIII

COMPOUNDS POLYMERIZED INTO MODERATELY HARD AND FLEXIBLE-TO-SOFT SOLIDS OR RUBBERY GELS

Compounds	Yield (%)	B.p., °C (mm)	d_{25}^{25}	n_D^{25}	Analyses, P, %	
					Calc.	Found
$Cyclo—C_6H_{11}—P(O)(OCH_2C(CH_3)=CH_2)_2$	73	94–98 (0.5)	1.027	1.4738	11.4	11.4
$Cyclo—C_6H_{11}—P(O)(OCH_2CH=CH_2)_2$	74.6	100–103 (0.5)	1.0485	1.4734	12.7	12.8
$CH_3—P(O)(OCH_2C(CH_3)=CH_2)_2$	69.5	95 (1)	1.024	1.4491	15.2	15.1
$CH_3P(O)(OCH_2CH=CH_2)_2$	47	77–85 (0.5)	1.044	1.4468	17.6	17.5
$C_2H_5P(O)(OCH_2C(CH_3)=CH_2)_2$	76.8	93–95 (1)	1.006	1.4502	14.2	14.2
$C_2H_5P(O)(OCH_2CH=CH_2)_2$	65.8	73–78 (0.5)	0.9963	1.4470	16.3	16.3
iso-$C_3H_7P(O)(OCH_2C(CH_3)=CH_2)_2$	60	106–108 (2)	0.990	1.4490	13.4	13.5
iso-$C_3H_7P(O)(OCH_2CH=CH_2)_2$	60.8	52 (0.5)	1.0166	1.4459	15.2	15.4
$nC_3H_7P(O)(OCH_2C(CH_3)=CH_2)_2$	84	107–112 (1–2)	0.9909	1.4512	13.4	13.2

$nC_3H_7\overset{O}{\overset{\|}{P}}(OCH_2CH{=}CH_2)_2$	63.3	91 (1)	1.021	1.4472	15.2	15.1
$nC_4H_9\overset{O}{\overset{\|}{P}}(OCH_2\overset{CH_3}{\overset{\|}{C}}{=}CH_2)_2$	85.4	10[illegible]–106 (1)	0.9789	1.4519	12.6	12.8
$nC_4H_9\overset{O}{\overset{\|}{P}}(OCH_2CH{=}CH_2)_2$	74	75–81 (0.5)	1.002	1.4478	14.2	14.2
iso-$C_4H_9\overset{O}{\overset{\|}{P}}(OCH_2CH{=}CH_2)_2$	79.2	65–67 (0.5)	1.005	1.4456	14.2	14.0
2-$C_8H_{17}\overset{O}{\overset{\|}{P}}(OCH_2\overset{CH_3}{\overset{\|}{C}}{=}CH_2)_2$	59.5	125 (0.5)	0.947	1.4543	10.3	10.4
$C_6H_5CH{=}CH{-}\overset{O}{\overset{\|}{P}}(OCH_2\overset{CH_3}{\overset{\|}{C}}{=}CH_2)_2$	77.4	*a*	1.0692	1.5360	10.6	10.4
$C_6H_5CH{=}CH\overset{O}{\overset{\|}{P}}(OCH_2CH{=}CH_2)_2$	70.0	*b*	1.0996	1.5445	11.7	11.4
$(CH_3)_3CCH_2\overset{CH_3}{\overset{\|}{C}H}{-}CH_2\overset{O}{\overset{\|}{P}}(OCH_2CH{=}CH_2)_2$	84	92 (0.5)	0.9641	1.4489	11.3	11.3

[a] Purified in a Hickman still; bath temperature 135°–142°C (0.003 mm); rate, drop/2–3 sec.

[b] Purified in a Hickman still; bath temperature 130°–140°C (0.007 mm); rate, drop/2–4 sec.

[c] These compounds are quite soluble in water but not in saturated sodium chloride solution.

TABLE XIV

COMPOUNDS REMAINING AS LIQUIDS AFTER THE POLYMERIZATION TEST[a]

Compounds	Yield (%)	B.p., °C (mm)	d_{25}^{25}	n_D^{25}	Analyses, % Calc.	Found
$(CH_3)_3CCH_2C(CH_3)=CH-P(=O)(OCH_2C(CH_3)=CH_2)_2$	82.8	121–125 (1)	0.9638	1.4668	P, 10.3	P, 10.4
$(CH_3)_3CCH_2C(CH_3)=CH-P(=O)(OCH_2CH=CH_2)_2$	87.5	124–125 (2)	0.9795	1.4660	P, 11.4	P, 11.4
$C_6H_5P(=S)(OCH_2CH=CH_2)_2$	64.5	126–129 (1)	1.115	1.5508	P, 12.2; S, 12.6	P, 12.4; S, 12.9
$C_6H_5P(=O)(NHCH_2C(CH_3)=CH_2)_2$	55.8	m.p. 88–89	—	—	P, 11.7; N, 10.6	P, 11.7; N, 10.3
$(CH_3)_2N-P(=O)(OCH_2CH=CH_2)_2$[b]	71.4	75 (0.5)	1.0561	1.4465	P, 15.2; N, 6.9	P, 15.2; N, 6.8
$(C_2H_5O)_2P(=O)N(CH_2CH=CH_2)_2$[c]	72.0	77–81	1.013	1.4430	P, 13.3; N, 6.0	P, 13.1; N, 5.8
$(C_2H_5O)_2P(=O)NH-CH_2C(CH_3)=CH_2$	81.3	96 (0.5)	1.124	1.4412	P, 15.0; N, 6.8	P, 15.0; N, 6.6
$(C_2H_5O)_2P(=O)OCH_2CH=CH_2$	72.5	63 (0.5)	1.073	1.4216	P, 16.0	P, 16.1

[a] Reprinted from A. D. F. Toy and R. S. Cooper, *J. Amer. Chem. Soc.* **76**, 2191 (1954). Copyright 1954 by the American Chemical Society. Reprinted by permission of the copyright owner.

[b] Original viscosity = 2.3 centipoises. Polymerized sample, viscosity = 26 centipoises. 55.8% of the original monomer recovered unchanged by distillation at 0.5 mm.

[c] Original viscosity = 2.8 centipoises. Polymerized sample, viscosity = 5.2 centipoises; 84% of the original monomer was recovered unchanged from the polymerized sample by distillation at 0.5 mm.

TABLE XV
POLYMERIZATION OF DIALLYL AND DIMETHALLYL ISOOCTENYLPHOSPHONATE AT 87°–88°C[a]

	Viscosity in centipoises			
	Diallyl ester		Dimethallyl ester	
Bz_2O_2 (%)	24 hr	48 hr	24 hr	48 hr
0	5	5	6	7
2	10	10	52	54
4	18	18	340	390
6	30	31	2,260	2,770
8	50	51	14,650	25,300
10	79	88	145,000	215,000

[a] Reprinted from A. D. F. Toy and R. S. Cooper, *J. Amer. Chem. Soc.* **76**, 2191 (1954). Copyright 1954 by the American Chemical Society. Reprinted by permission of the copyright owner.

A stoppered test tube containing 3.0 gm (0.012 mole) of diallyl cyclohexanephosphonate and 0.15 gm of benzoyl peroxide is placed in an oil bath at 70°C while the temperature is raised to 115°C over a 7 hr period. After this time the liquid turns to a clear, colorless moderately hard solid.

4. POLYPHOSPHITES

Polyphosphites are not as commonly available as polyphosphates or polyphosphonates. They are prepared on a limited basis usually at low molecular weights. Some are quite unstable and rearrange to polyphosphonates. Others can be oxidized to polyphosphates. Low molecular weight polyphosphites can be chain-extended with anhydrides, diisocyanates, or other reactive difunctional reagents. The condensation reactions involve phosphorus trichloride and derivatives, while the free-radical reactions involve the vinyl diolefin allyl phosphite derivatives.

A. Condensation Reactions

Dialkyl or diacyl phosphites are prepared in good yield by the reaction of 2 moles of alcohol (or phenol) per mole of phosphorus trichloride in the presence of 2 moles of a tertiary amine base (Eq. 60).

$$2ROH + PCl_3 + 2R_3N \longrightarrow (RO)_2PCl + 2R_3NHCl \qquad (60)$$

The monoalkyl ester is formed using only 1 mole of alcohol in the absence of base (Eq. 61).

$$ROH + PCl_3 \longrightarrow ROPCl_2 + HCl \tag{61}$$

The latter compounds can be reacted with glycols to give polyphosphites (Eq. 62) which can be chain-extended by reacting with dianhydrides [132].

$$ROPCl_2 + HOR'OH \longrightarrow H\left[-O-R'-O-\underset{\underset{OR}{|}}{P}-\right]OH \tag{62}$$

Along with the polyphosphites there are also formed varying amounts of cyclic esters (Eq. 63) [133–135].

$$ROPCl_2 + HOR'OH \longrightarrow R-\underset{\underset{O}{\|}}{P}\langle{}^{O}_{O}\rangle R' \tag{63}$$

The reaction of 3 moles of an alkanol with phosphorus trichloride in the absence of base affords the diester [136]. However, in the presence of base the triester is obtained (Eq. 64).

$$3ROH + PCl_3 \longrightarrow (RO)_2\overset{\overset{O}{\|}}{P}H + RCl + 2HCl \tag{64}$$

$$3ROH + PCl_3 \xrightarrow[0°C]{3R_3N} (RO)_3P$$

The reaction of 3 moles of phenol with phosphorus trichloride in the absence of base affords triaryl phosphite (Eq. 65).

$$3ArOH + PCl_3 \longrightarrow (ArO)_3P + 3HCl \tag{65}$$

The reaction of glycols with phosphorus trichloride affords cyclic phosphites (Eq. 66). Trialkyl phosphites also react with glycols to give cyclic phosphites (Eq. 67) [47].

$$HOCH_2CH_2OH + PCl_3 \longrightarrow \begin{matrix} CH_2-O \\ | \\ CH_2-O \end{matrix}\rangle P-Cl + 2HCl \tag{66}$$

$$HOR'OH + (RO)_3P \longrightarrow R'\langle{}^{O}_{O}\rangle P-OR'OH \tag{67}$$

Triaryl phosphites react with diols to give cyclic phosphites as well as polymeric phosphates [14a].

Polyolefin glycols afford polyolefin glycol polyphosphites [137].

Dialkyl phosphites in the presence of basic catalysts are readily transesterified to produce phosphite-containing polyols [138,139].

The latter reactions are transesterification reactions and have been reported to occur rapidly. For example, triethyl phosphite exchanges ethoxy groups for methoxy groups when exposed to CH_3OH in the presence of catalysts as determined by PMR studies [140].

Poly(hexamethylene phosphite) and poly(octamethylene phosphite) of molecular weights 50,000–70,000 were prepared by the reaction of diethyl phosphite with a moderate excess of the diol [141]. Reaction of these products with alcohols or phenols in the presence of tertiary amine gave alkyl(aryl) alkylene phosphates [141]. Other methods of converting the polyphosphites to polyphosphates have been described [142].

Also important is the reaction of phosphite diols with various reagents (Eq. 68) [143].

$$\text{(HORO)POR}' + \text{X}_3\text{C}_6\text{H}_2\text{OH (2,4,6-trihalophenol)} \xrightarrow[\text{4 mm Hg}]{\text{NaOAr, 130°C}} \text{(HORO)}_2\text{PO}-\text{C}_6\text{H}_2\text{X}_3 \quad (68)$$

$\text{(HORO)}_2\text{PO}-\text{C}_6\text{H}_2\text{X}_3 \longrightarrow$ Various condensation reaction polyphosphites

$\text{(HORO)}_2\text{PO}-\text{C}_6\text{H}_2\text{X}_3 \xrightarrow{\text{R(NCO)}_2}$ Polyurethane phosphites

(X = Br or Cl)

4-1. Preparation of a Polyphosphite by the Transesterification Reaction of Hexane-1,6-diol with Triphenyl Phosphite [14a]

$$\text{HO(CH}_2)_6\text{OH} + (\text{C}_6\text{H}_5\text{O})_3\text{P} \xrightarrow{\text{Na}} \left[-\text{O(CH}_2)_6-\text{O}-\underset{\text{OC}_6\text{H}_5}{\underset{|}{\text{P}}}- \right]_n + 2\text{C}_6\text{H}_5\text{OH} \quad (69)$$

To a flask equipped with a mechanical stirrer, condenser, dropping funnel, and thermometer and containing 38.8 gm (0.125 mole) of triphenyl phosphite is added dropwise over a 75 min period at 90°C, 14.8 gm (0.125 mole) of hexane-1,6-diol containing 0.2 gm (0.01 gm atom) of dissolved sodium metal. During the addition a temporary clouding of the mixture clears up quickly. After standing overnight the product is a rubberlike mass, insoluble in ether and dioxane, which dissolves slowly in boiling chloroform. Concentration of the chloroform affords 12 gm (51%) phenol at a bath temperature of 130°C

TABLE XVI

TRANSESTERIFICATION OF DIOLS AND GLYCOLS WITH TRIPHENYL PHOSPHITE TO GIVE CYCLIC PHOSPHITES AND POLYMERIC PHOSPHITES[a]

Triphenyl phosphite (mole)	Diol or glycol (moles)	Na catalyst (gm)	Reaction conditions		Cyclic phosphite			Polymeric phosphite		
			Temp. (°C)	Time (hr)	%	B.p., °C (mm Hg)	n_D(°C)	%	B.p., °C (mm Hg)	n_D(°C)
0.5	Ethylene glycol (0.5)	—	100	1	45	59–61 (0.09)	1.5356 (20)	45	61–76 (0.1–0.3)	1.5070 to 1.5827 (20)
0.5	(0.5)	0.01	100	1½	60	70–76 (0.2)	1.5370 (20)	—	—	—
0.5	Propylene (0.5)	0.01	100	2	52	81–87 (0.6)	1.5155 (20)	—	—	—
0.33	Trimethylene glycol (0.33)	—	100	6	27	69–75 (0.07)	1.526 (20)	58	165–173 (0.05 to 0.1)	1.5576 (20)
0.5	(0.5)	0.01	100	2	53	89 (0.4) m.p. 38°–42°	—	—	—	—
0.2	Butane-1,4-diol (0.2)	0.01	100	1½	20	91–95 (0.3)	1.5359 (22)	—	—	—

0.24	Pentane-1,5-diol (0.24)	0.01	100	$1\frac{1}{2}$	13	98–104 (0.0015)	1.5268 to 1.5346 (18)	—	—	—
0.125	Hexane-1,6-diol (0.125)	0.01	90	$1\frac{1}{4}$	—	—	—	? Polymer	—	—
0.1	*cis*-Cyclopentane-1,2-diol (0.1)	0.01	—	1	49	72–75 (0.002)	1.5266 (23)	16	120–130 (0.0015)	1.5780 (23)
0.1	*cis*-Cyclohexane-1,2-diol (0.1)	0.01	—	1	37	96–112 (0.004)	—	26	120–133 (0.003)	—
0.1	*trans*-Cyclohexane-1,2-diol (0.1)	0.01	—	1	29	82–100 (0.0045)	1.5320–1.5360 (37)	9	—	—
0.5	Glycerol (0.5)	0.01	100	5	—	—	—	100	—	—

[a] Data taken from D. C. Ayres and H. N. Rydon, *J. Chem. Soc., London* p. 1109 (1957).

(0.3 mm Hg). Raising the bath temperature to 180°C at 0.004 mm Hg gives a little more phenol but shows no sign of boiling.

The reaction of other diols and glycols with triphenyl phosphite by a procedure similar to 4-1 is shown in Table XVI.

B. Free-Radical Addition Reactions

Diallyl [130] and triallyl [109] phosphites have been polymerized with the aid of benzoyl peroxide at 88°–100°C and the data are summarized in Table XVII. The procedure for this polymerization is described in Preparation 3-6.

TABLE XVII
POLYMERIZATION OF DIALLYL AND TRIALLYL PHOSPHITES

Phosphite	Initiator	Solvent	Temp. (°C)	Time (hr)	% Conversion	Comments
Diallyl[a]	Bz_2O_2	None	88	18	—	Hard glassy polymer
Triallyl	Bz_2O_2	—	100	—	60	Taffy-like polymer

[a] B.p. 58°–61°C (1.0 mm Hg), n_D^{25} 1.4459.

5. POLYPHOSPHORYLAMIDES, POLYPHOSPHONAMIDES, AND POLYPHOSPHINAMIDES

A. Condensation Reactions

Phosphonic dihalides have been known to react with amines [144] but it was not until 1959 [145] that the reaction with diamines was reported to yield a polyphosphonamide.

The relationship between the various phosphorus amides is shown in Eqs. (70)–(73). The polymer applications of this class are few and further research remains to find different areas of application.

Phosphorylamides

$$(HO)_3P{=}O \longrightarrow (HO)_2P(NHR){=}O \longrightarrow HO{-}P(NHR)_2{=}O \longrightarrow (RNH)_3P{=}O \xleftarrow{3RNH_2} POCl_3 \quad (70)$$

Phosphonamides

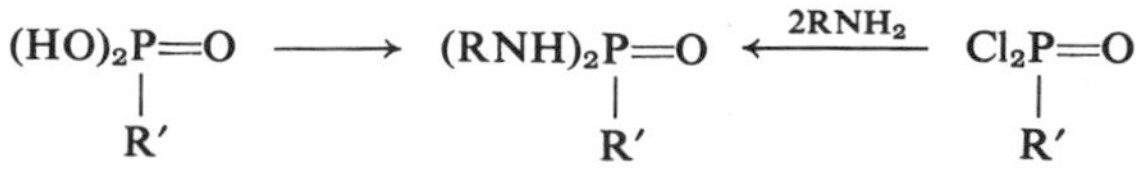

$$\underset{\displaystyle R'}{(HO)_2P{=}O} \longrightarrow \underset{\displaystyle R'}{(RNH)_2P{=}O} \xleftarrow{2RNH_2} \underset{\displaystyle R'}{Cl_2P{=}O} \tag{71}$$

Phosphinamides

$$\underset{\displaystyle OH}{R'_2P{=}O} \longrightarrow \underset{\displaystyle NHR}{R'_2P{=}O} \xleftarrow{RNH_2} \underset{\displaystyle R'_2}{Cl{-}P{=}O} \tag{72}$$

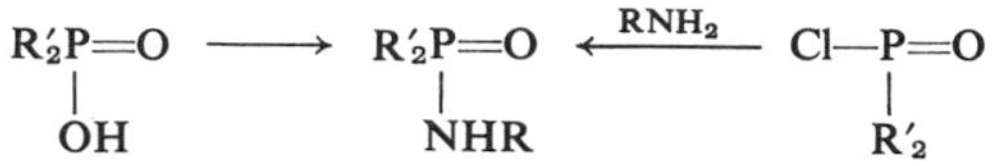

(R = H, alkyl, or aryl, and R′ = alkyl or aryl)

Some examples of several types of phosphorus amides and their preparative details are shown in Table XVIII.

5-1. *Preparation of Poly(hexamethylenephenylphosphondiamide)* [145]

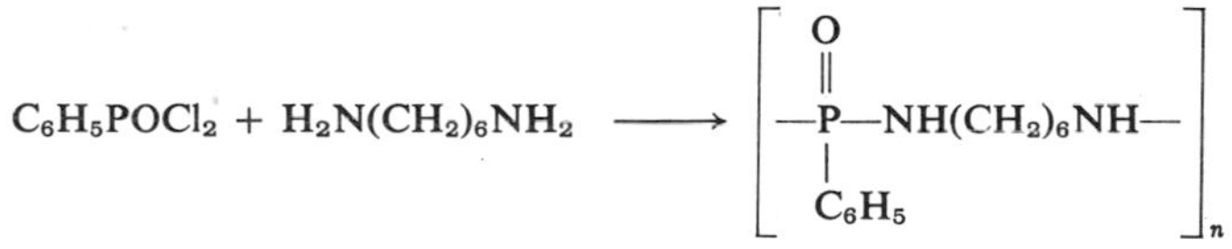

$$C_6H_5POCl_2 + H_2N(CH_2)_6NH_2 \longrightarrow \left[\begin{array}{c} O \\ \| \\ -P-NH(CH_2)_6NH- \\ | \\ C_6H_5 \end{array} \right]_n \tag{73}$$

(n = 5–20)

To a flask containing 195.0 gm (1.0 mole) of phenylphosphonic dichloride in 500 ml of chloroform is added with vigorous stirring an aqueous solution of 116.0 gm (1.0 mole) of hexamethylenediamine containing 82 gm of NaOH. The polymer that forms is soluble in the chloroform layer and is isolated by precipitation in hexane or by evaporation. The isolated polymer has an n value of 5–20 with an empirical formula of approximately $C_{12}H_{19}N_2OP$.

Other strong amines such as Bisaniline A can also be used but weak amines such as *N,N′*-diethylethylenediamine or diamine give no polymers.

B. Free-Radical Reactions

Olefinic phosphorus amides can be polymerized or copolymerized to give polymers with phosphonamide side groups [146–148].

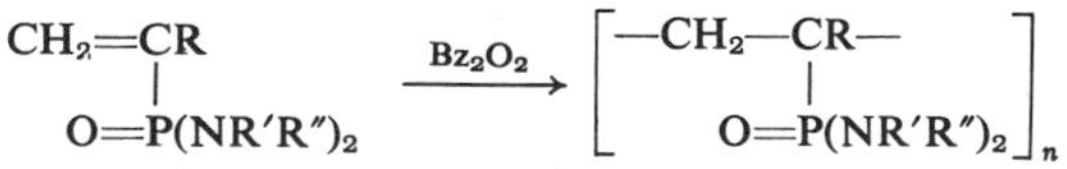

$$\begin{array}{c} CH_2{=}CR \\ | \\ O{=}P(NR'R'')_2 \end{array} \xrightarrow{Bz_2O_2} \left[\begin{array}{c} -CH_2-CR- \\ | \\ O{=}P(NR'R'')_2 \end{array} \right]_n \tag{74}$$

TABLE XVIII

PREPARATION OF PHOSPHORUS AMIDES

Phosphorus compound	Amine compound	Diol	Solvent	Reaction temp. (°C)	Structure of polymer product	S.p. (°C)	Ref.
$(Ph)_2NPOCl_2$	—	$HO—C_6H_4OH$	—	205	$[—O—P(=O)(N(C_6H_5)_2)—O—C_6H_4—]_n$	—	*a*
$POCl_3$	NH_3	—	Kerosene	200	$[—P(=O)(NH_2)—NH—]_n$	dec.	*b*
$(C_2H_5)_2NP(=O)Cl_2$	NH_3	—	Benzene	20	$[—P(=O)(N(C_2H_5)_2)—NH—]_n$	dec.	*c*
$C_6H_5—P(=O)Cl_2$	$NH_2(CH_2)_6NH_2$	—	H_2O, NaOH	20	$[—P(=O)(C_6H_5)—NH(CH_2)_6NH—]$	—	*d*
$R'P(=O)(NHR)_2$	—	—	—	200–300	$[—P(=O)(R')—N(R)—]_n$	> 200	*e*
$R'P(=O)(NHR)_2$	$H_2NC(=O)NH_2$	—	—	200–300	$[—P(=O)(R)—NHCONH—]_n$	—	*f*

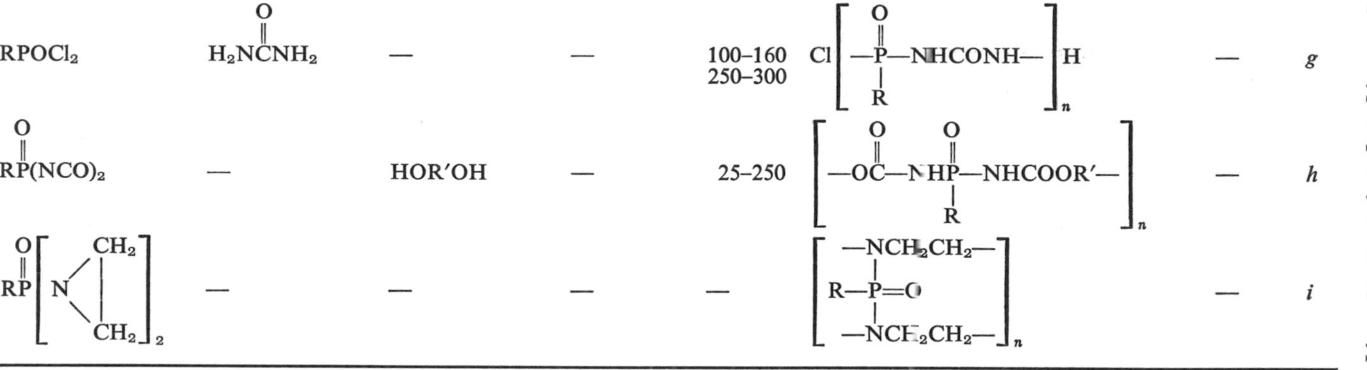

$RPOCl_2$	$H_2NC(=O)NH_2$	—	—	100–160 250–300	$Cl[-P(=O)(R)-NHCONH-]_nH$	—	*g*
$RP(=O)(NCO)_2$	—	HOR′OH	—	25–250	$[-OC(=O)-NHP(=O)(R)-NHCOOR'-]_n$	—	*h*
$RP(=O)[N(CH_2)_2]_2$	—	—	—	—	$[R-P(=O)(-NCH_2CH_2-)(-NCH_2CH_2-)]_n$	—	*i*

[a] H. W. Coover, U.S. Patent 2,682,521 (1954).

[b] J. E. Malowan and F. R. Hurley, U.S. Patent 2,596,935 (1952); J. E. Malowan, U.S. Patent 2,749,233 (1956); E. H. Rossin and M. J. Scott, U.S. Patent 2,816,004 (1957).

[c] Farb. Bayer A.-G., British Patent 830,800 (1961).

[d] D. M. Harris, R. L. Jenkins, and M. L. Nielsen, *J. Polym. Sci.* **35**, 540 (1959); J. Frago, U.S. Patent 3,116,268 (1963).

[e] J. B. Dickey and H. W. Coover, U.S. Patent 2,666,750 (1954); H. W. Coover, U.S. Patent 2,642,413 (1953); H. W. Coover, R. L. McConnell, and N. H. Shearer, Jr., *Ind. Eng. Chem.* **52**, 412 (1960).

[f] H. W. Coover, U.S. Patent 2,642,413 (1953).

[g] A. C. Haven, Jr., U.S. Patent 2,716,639 (1955).

[h] A. C. Haven, Jr., *J. Amer. Chem. Soc.* **78**, 842 (1956).

[i] J. Heyna and W. Noll, German Patent 854,651 (1952); R. P. Parker, D. R. Seeger, and E. Kuh, U.S. Patent 2,606,901 (1952); H. Bestian, *Justus Liebigs Ann. Chem.* **566**, 210 (1950); R. P. Parker, D. R. Seeger, and E. Kuh U.S. Patents 2,606,900 and 2,606,902 (1952); E. Kuh and D. R. Seeger, U.S. Patent 2,670,347 (1954); H. Z. Lecher and E. Kuh, U.S. Patent 2,654,738 (1953).

TABLE XIX
Condensation Polymerizations of Substituted Phosphine and Phosphine Oxide Derivatives

Phosphorus compound	Coreactant	Solvent	Catalyst	Reaction conditions: Temp. (°C)	Reaction conditions: Time (hr)	Structure of polymer	Ref.
$C_6H_5PH_2$	$CH{=}CCH_2OH$	C_6H_6	Peroxide	100	40	$[-CH(CH_2CH_2OH)-P(=O)(C_6H_5)-]_n$	*a*
$RP(OH)_2$	$R'(NCO)_2$	C_6H_6	Amine	80	5	$-N<[R'NHCOP(=O)R]_n$	*b*
$C_6H_5P(=O)Cl_2$	LiC_6H_4Li	THF	—	−40	—	$[-P(=O)(C_6H_5)-C_6H_4-]_n$	*c*
$CH_3P(=O)(C_6H_4NH_2)_2$	$R(COCl)_2$	—	—	—	—	$[-C_6H_4-P(=O)(CH_3)-C_6H_4NHC(=O)RC(=O)NH-]_n$	*d*

$R'PH_2$	$R(CH{=}CH_2)_2$	—	—	—	—	$[-P(R')-CH_2CH_2RCH_2CH_2-]_n$	*e*
$RP(=O)(CH_2OH)_2$	$R'(NH_2)_2$	—	—	—	—	$[-CH_2-P(=O)(R)-CH_2NHR'NH-]_n$	*f*
RPH_2	$R(NCO)_2$	—	—	—	—	$[-P(R)-CONHRNHCO]_n$	*g*

[a] E. C. Chapin and A. Y. Garner, U.S. Patent 3,158,642 (1964); A. Y. Garner, U.S. Patent 3,235,536 (1966).
[b] S. A. Buckler and M. Epstein, U.S. Patent 3,213,042 (1965).
[c] P. R. Bloomfield, U.S. Patent 3,044,984 (1962).
[d] T. Y. Medved, T. M. Frunze, C. M. Hu, V. V. Kurashev, V. V. Korshak, and M. I. Kabochnik, *Vysokomol. Soedin.* **5**, 1309 (1963).
[e] E. Steininger and M. Sander, *Kunststoffe* **54**, 507 (1964).
[f] K. A. Petrov, V. A. Parshina, and G. M. Tsypina, *Plast. Massy* No. 11, p. 11 (1963); *Chem. Abstr.* **60**, 5651 (1964).
[g] S. A. Buckler and M. Epstein, British Patent 926,268 (1963).

6. POLYPHOSPHINES AND DERIVATIVES

A. Condensation Reactions

The reaction of arylphosphonous dihalides [3a] with olefins yields a polymer which upon reaction with methanol affords the corresponding polymer with phosphine oxide groups [149–152].

$$R'PCl_2 + nRCH{=}CH_2 \longrightarrow \left[-\underset{\substack{|\\ R}}{CH}-CH_2-\underset{\substack{|\\ R'}}{\overset{Cl\ \ Cl}{\overset{\diagdown\,\diagup}{P}}}- \right]_n \xrightarrow[\substack{-H_2O\\ -2nCH_3Cl}]{2nCH_3OH} \left[-\underset{\substack{|\\ R}}{CH}-CH_2\underset{\substack{|\\ R'}}{\overset{\overset{O}{\|}}{P}}- \right]_n \qquad (75)$$

$R' = C_2H_5$ or C_6H_5

$R = CH_2{=}CH$, $CH_2{=}\underset{\substack{|\\ CH_3}}{C}H-$, $CH_2{=}\underset{\substack{|\\ Cl}}{C}$, $CH_2{=}\underset{\substack{|\\ Br}}{C}-$, CN, CH_3OCH_2-, C_6H_5

The Friedel-Crafts condensation of arylphosphonous dichloride with 1,2-diphenylethane affords a polyphosphine [153].

$$ArPCl_2 + C_6H_5-CH_2-CH_2-C_6H_5 \xrightarrow{AlCl_3}$$

$$H\left[-C_6H_4-CH_2-CH_2-\underset{\substack{|\\ Ar}}{P}- \right]_n -C_6H_4-CH_2-CH_2-C_6H_5 \qquad (76)$$

The condensation of dibutyl phosphite with pentamethylenebis(magnesium bromide) gives a poly(phosphine oxide) [154]

$$(C_4H_9O)_2P(O)H + BrMg(CH_2)_5MgBr \longrightarrow \left[-(CH_2)_5-\underset{H}{\overset{\overset{O}{\|}}{P}}- \right]_n \qquad (77)$$

The reaction of tetrakis(hydroxymethyl)phosphonium chloride with melamine [155], aziridine [156], phenol [157,158], or urea [159] affords phosphonium chloride cross-linked polymers.

Some additional polyphosphines and poly(phosphine) oxides are summarized in Table XIX.

$$(HOCH_2)_4P^+Cl^- \xrightarrow{\text{melamine } (H_2N\text{-}C_3N_3(NH_2)_2)} \text{polymers} \quad (78)$$

$(HOCH_2)_4P^+Cl^-$ with ethylenimine (CH_2—CH_2—NH) → polymers; with C_6H_5OH → polymers; with $H_2NC(=O)NH_2$ → polymers

B. Free-Radical Reactions

Some typical free-radical polymerizations in which olefinic phosphines or phosphine oxides are used to give polymeric derivatives are shown in Table XX.

6-1. Homopolymerization of Diphenyl-p-styrylphosphine Oxide [160]

$$CH_2{=}CH{-}C_6H_4{-}P(O)(C_6H_5)_2 \xrightarrow{AIBN} \left[-CH_2-CH(C_6H_4P(=O)(C_6H_5)_2)- \right]_n \quad (79)$$

To a heavy-walled glass tube is added a 50% benzene solution of 5 gm (0.016 mole) of diphenyl-*p*-styrylphosphine oxide. Then 0.025 gm of azobis-(isobutyronitrile) (AIBN) is added. The tube is frozen, sealed under vacuum, and warmed at 60°C for 20 hr to afford a 96% yield of polymer; [] = 0.89 dl/gm (30°C, $CHCl_3$).

The monomer from Preparation 6-1 copolymerizes well with styrene or methyl methacrylate.

7. PHOSPHONITRILIC POLYMERS AND RELATED COMPOUNDS

Liebig [161] and also Rose [162] have reported that the reaction of ammonia with phosphorus pentachloride yields two products as shown in

TABLE XX

FREE-RADICAL POLYMERIZATIONS OF OLEFINIC PHOSPHINE AND PHOSPHINE OXIDE DERIVATIVES

Phosphorus compound	Coreactant	Solvent	Catalyst (gm/gm monomer)	Reaction conditions: Temp. (°C)	Reaction conditions: Time (hr)	Structure of polymer	Ref.
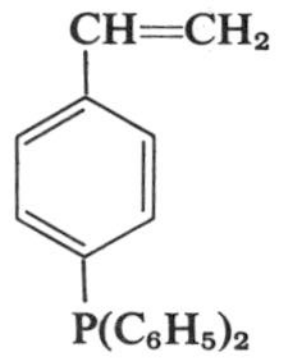 $CH{=}CH_2$–C_6H_4–$P(C_6H_5)_2$ (para)	—	C_6H_6	AIBN (0.0005)	60	50	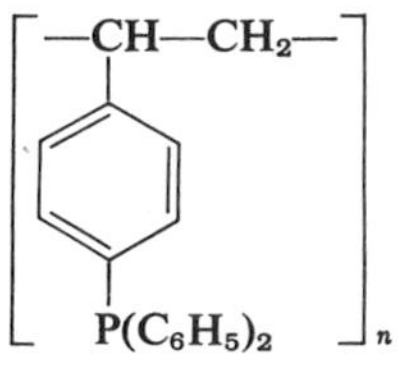 $[-CH(C_6H_4P(C_6H_5)_2)-CH_2-]_n$ (para); 97% yield	*a*

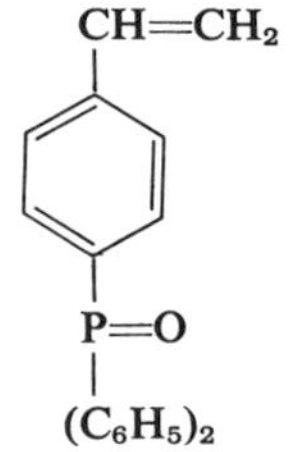 $CH{=}CH_2$–C_6H_4–$P(=O)(C_6H_5)_2$	—	C_6H_6	AIBN (0.005)	60	20	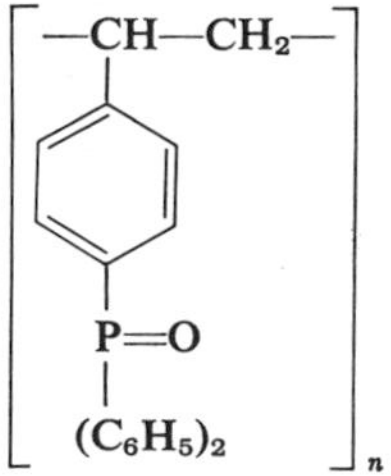 $[-CH(C_6H_4P(=O)(C_6H_5)_2)-CH_2-]_n$ yield 96%	*a*
$(CH_2{=}CH{-}CH_2)_2P(=O)C_6H_5$	Lauryl methacrylate	—	—	—	—		*b*
$R_2P(=O)CH{=}CH_2$	—	—	Peroxide and transition metal salt	100	—	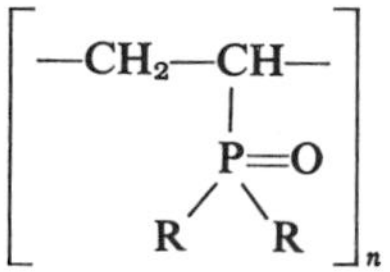 $[-CH_2-CH(P(=O)R_2)-]_n$	*c*

[a] R. Rabinowitz, R. Marcus, and J. Pellon, *J. Polym. Sci., Part A* **2**, 1241 (1964).
[b] K. I. Geynon, *J. Polym. Sci., Part A* **1**, 3357 (1963).
[c] F. J. Welch and H. J. Paxton, *J. Polym. Sci., Part A* **3**, 3427 and 3439 (1965).

Eq. (80). The chlorophosphazenes were able to be steam-distilled and were

$$NH_3 + PCl_5 \longrightarrow \underset{\text{Phospham}}{(NPNH)_n} + \underset{\text{Chlorophosphazenes}}{(NPCl_2)_n} \quad (80)$$

not decomposed by hot acid. The chlorophosphazenes were later analyzed by Gerhardt [163] and Laurent [164], who determined the empirical formula as $NPCl_2$. Gladstone and Holmes [165] and also Wichelhaus [166] used vapor density measurements to show that the molecular formula was $(NPCl_2)_3$. Later workers [167], particularly Stokes, reported on the hydrolysis substitution and polymerization reactions of the phosphazenes. Stokes [168] also reported the properties of the series of higher cyclic homologs $(NPCl_2)_{4-7}$ [b.p. 127°C to 289°–294°C (13 mm Hg)]. The compound $(NPCl_2)_{11}$ was reported to be an oil. Stokes [168] also reported on the preparation of "inorganic rubber" by the thermal polymerization of the chlorophosphazenes and the depolymerization at elevated temperatures under reduced pressure. Recent work had shown indirect evidence for $(NPCl_2)_{9-17}$ [169,169a].

Stokes was the first to suggest that hexachlorotriphosphazene was cyclic and his work laid the foundation for phosphazene chemistry. The name "phosphonitrilic chloride" was suggested by Stokes for the $(NPCl_2)$ trimer and the name is still used for this compound and its polymers. The structure of the trimer and the tetramer are shown below:

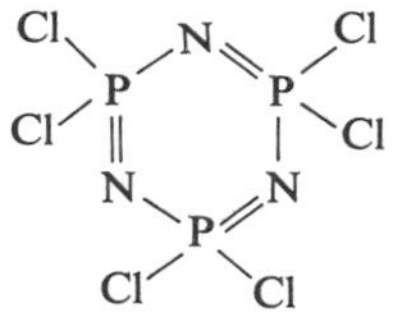

Trimer
(Hexachlorocyclotriphosphazene)

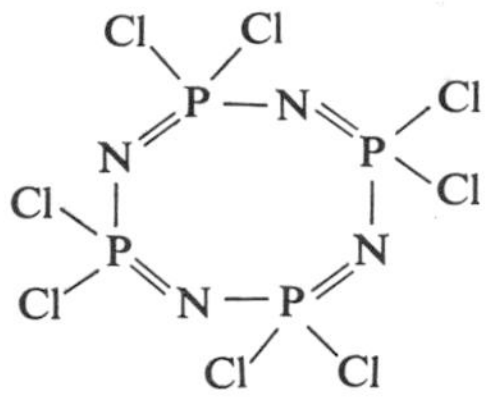

Tetramer
(Octachlorocyclotetraphosphazene)

The stereochemistry of these rings and their substituted derivatives has been described [170].

The present nomenclature uses the term phosphazene for structure (I) and phosphazanes for structures (II) and (III) or their polymers.

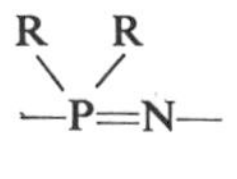

(I) 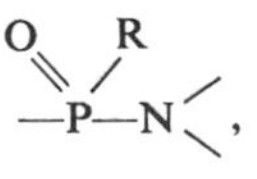 (II) 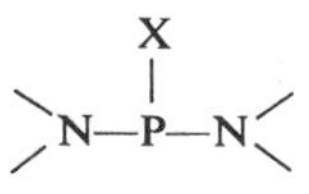 (III)

A. Condensation Reactions

The Stokes procedure for the preparation of the chlorophosphazenes was hazardous and difficult to carry out. This procedure was improved by Schenk and Römer [171] in 1924 and is the basis of the present commercial process (e.g., Eq. 81). This reaction was only recently investigated in greater detail

$$nNH_4Cl + nPCl_5 \xrightarrow[120°C]{\text{solvent}} (NPCl_2)_n + 4nHCl \tag{81}$$

[169,172]. Fluorophosphazenes [173], homophosphazenes [174], and the direct synthesis of several organophosphazenes, $(NPR_2)_n$, were recently described [175]. The study of the polymers has been accelerated by the demand of the aerospace industry for high-temperature polymers [176–178].

Bromo, iodo, fluoro, and mixed halogen derivatives are prepared by related methods. The physical properties of some of these derivatives are shown in Table XXI [170].

TABLE XXI
HALOGENOCYCLOTRIPHOSPHAZENES AND HALOGENOCYCLOTETRAPHOSPHAZENES[a]

Compound	M.p. (°C)	B.p. (°C)
$N_3P_3F_6$	27.8	50.9
$N_3P_3F_4Cl_2$	—	114.7
$N_3P_3F_4Cl_2$	—	84.4
$N_3P_3F_3Cl_3$	—	150.1
$N_3P_3F_2Cl_4$	—	181.6
$N_3P_3F_2Cl_4$	—	140–142
$N_3P_3Cl_4$	114	256.5
$N_3P_3Cl_5Br$	122.5–123.5	—
$N_3P_3Cl_4Br_2$	134.5–136.5	—
$N_3P_3Cl_2Br_4$	167.5–169	—
$N_3P_3Br_6$	192	—
$N_4P_4F_8$	30.5	89.7
$N_4P_4F_6Cl_2$	−12.4 to −12.1	105.8[b]
$N_4P_4F_4Cl_4$	−25.2 to −24.9	130.5[b]
$N_4P_4Cl_8$	124	328.5
$N_4P_4Br_8$	202	—

[a] Reprinted from R. A. Shaw, B. W. Fitzsimmons, and B. C. Smith, *Chem. Rev.* **62**, 247 (1962). Copyright 1962 by the American Chemical Society. Reprinted by permission of the copyright owner.

[b] Estimated from vapor pressure data.

The reaction of halophosphazenes with several nucleophilic reagents and Friedel-Crafts catalysts has been studied in detail. Several reviews [3d,170,179] and a recent monograph [8] are worthwhile consulting.

7-1. Preparation of Hexachlorocyclotriphosphazene (Phosphonitrilic Chloride Trimer) [180]

$$PCl_5 + NH_4Cl \longrightarrow \tfrac{1}{3}(PNCl_2)_3 + 4HCl \qquad (82)$$

(*a*) *Without solvent method* [180]. To a 50 cm test tube made from 35–50 mm Pyrex tubing is added 52.1 gm (0.25 mole) of phosphorus pentachloride and 25.0 gm (0.48 mole) of ammonium chloride. The tube is immersed in an oil bath to such a level that the ammonium chloride solid is above the liquid level. The tube outlet is connected to a sulfuric acid trap and the tube is heated at 145°–160°C for 4–6 hr, at which time the hydrogen chloride evolution ceases. The reaction mixture is cooled, extracted with low-boiling (50°–70°C) petroleum ether, and then concentrated to afford 11 to 12.5 gm (38–43%) of the trimer–tetramer mixture. The pure trimer is obtained by fractional distillation under reduced pressure to give a 37% yield of trimer at b.p. 115°–130°C (10 mm Hg), m.p. 114°C.

(*b*) *Solvent method* [181,182]. To a flask equipped with a condenser, an exit tube to an HCl trap, and a dropping funnel is added 62.5 gm (0.3 mole) of phosphorus pentachloride, 1.0 liter of *sym*-tetrachloroethane, and 176.5 gm (3.3 moles) of ammonium chloride. The mixture is refluxed and during this time a solution of 563 gm (2.7 moles) of phosphorus pentachloride in 1.0 liter of *sym*-tetrachloroethane is added over an 8 hr period. During a total reaction time of $10\frac{1}{2}$ hr, 11.7 moles (98%) of hydrogen chloride is evolved. The cooled reaction mixture is filtered of the 20.7 gm of excess ammonium chloride and the solution distilled under reduced pressure to afford 237 gm (68%), b.p. 115°–130°C (10 mm Hg), m.p. 114°C.

If the product is not distilled, the crystalline-oily residue is filtered by suction on a sintered-glass funnel and washed quickly with some cold ethanol to afford a white powder, m.p. 108°–114°C. Low yields of product may be caused if excess ammonium chloride reacts with the chloride to give phospham [183].

Phosphonitrilic chloride trimer (cyclic) is thermally polymerized at 250°–300°C in the absence of solvents to give a rubbery polymer. The polymer is

$$(Cl_2PN)_3 \xrightarrow[\text{48 hr, vacuum}]{250^\circ\text{–}300^\circ C} \left| \begin{array}{c} Cl\quad Cl \\ -P{=}N- \end{array} \right|_n \xrightarrow{250^\circ\text{–}300^\circ C} \text{Cross-linked polymer} \qquad (83)$$

Poly(dichlorophosphazene)

probably mainly un-cross-linked since it is soluble in benzene, toluene, or THF [5b,183,184]. The un-cross-linked polymers are in the range of $n = 5{,}000$–$10{,}000$ or molecular weights up to approx. 2,000,000. The polymers are usually isolated in various ranges of molecular weights and the final determined value is reported as an average value.

If the polymerization is carried out in the *sym*-tetrachloroethane or chlorobenzene then oily low molecular weight polymers are obtained in 10–25% yields with the remainder being cyclic species [171,185].

$$NH_4Cl + PCl_5 \longrightarrow [Cl_3P{=}N{-}(PCl_2{=}N)_n{-}PCl_3]^+PCl_6^- + [Cl_3P{=}N{-}(PCl_2{=}N)_2PCl_3]^+Cl^- \quad (84)$$

If the reaction is carried out in the presence of metal halides, polymers of similar molecular weight are obtained [186].

Prolonged heating of the rubbery polymer gives a cross-linked polymer which is tough and insoluble in organic solvents [184,187]. The cross-linked polymer shows properties similar to elastomers [188].

Most information about molecular weights has been obtained from light-scattering data [189] and from the organo-substituted derivative properties [184].

The unstretched pure polymer is amorphous but the stretched polymer shows crystallinity in the X-ray fiber diffraction pattern [190].

7-2. *Preparation of Poly(dichlorophosphazene)* [184]

$$(NPCl_2)_3 \xrightarrow[48\text{ hr}]{250°C} -\overset{Cl}{\underset{Cl}{|}}{P}{=}N-\left[\overset{Cl}{\underset{Cl}{|}}{P}{=}N\right]_n-\overset{Cl}{\underset{Cl}{|}}{P}{=}N- \quad (85)$$

Hexachlorocyclotriphosphazene (1000 gm) is recrystallized from *n*-heptane (1500 ml) at a temperature below 75°C. The hot solution is decolorized with activated carbon and cooled to afford 621 gm of a first crop of crystals. This sample is dried under reduced pressure to afford trimer, m.p. 108°–113°C.

To a 32.5 × 3.5 cm constricted Pyrex tube is added 200 gm of the above recrystallized hexachlorocyclotriphosphazene. The tube is evacuated for ½ hr at 0.1 mm Hg, and the trimer melted and allowed to solidify again. The tube is evacuated for an additional ½ hr at 0.1 mm and sealed. The tube is immersed in a 250°C constant-temperature bath for 48 hr. The tube is cooled, opened, and the polymer cut into small pieces in a stream of nitrogen and dissolved in 1000–3000 ml of dry benzene to afford a colorless viscous solution after 24–48 hr of continuous agitation. Precipitation of the polymer into heptane indicates 70% of un-cross-linked polymer. The un-cross-linked polymer is a colorless transparent elastomer which dissolves slowly in

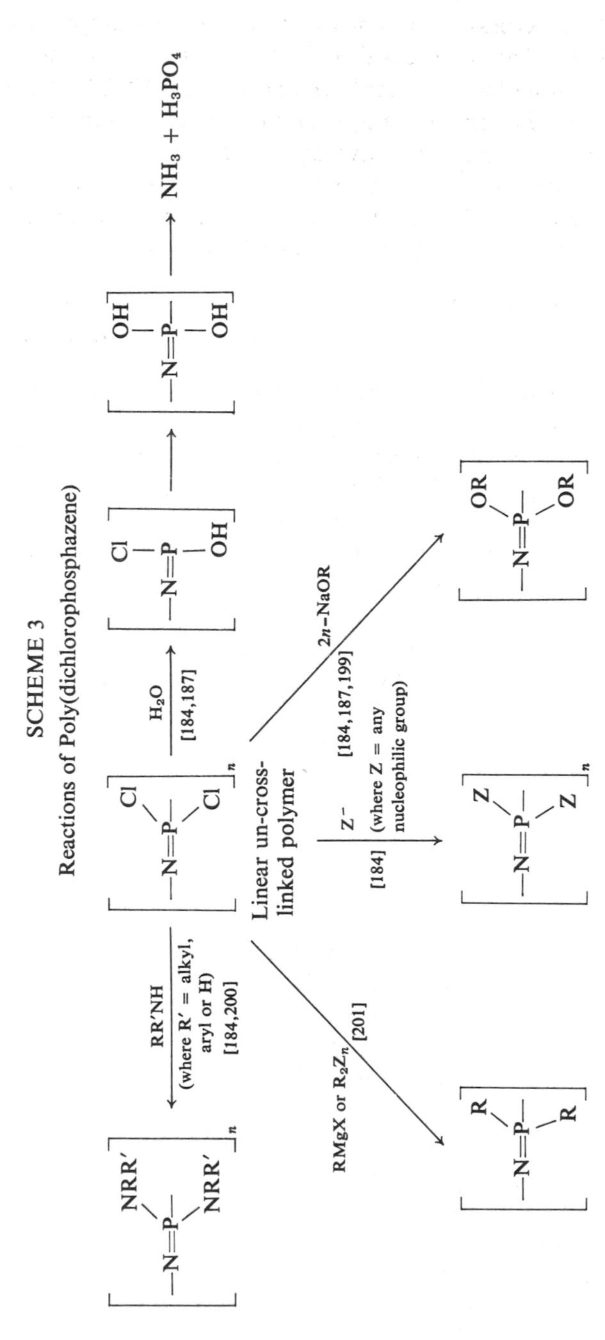

Where R = CH_2CF_3 then a very stable and mechanically stable polymer is obtained which can be molded into films and fibers.

benzene, toluene, tetrahydrofuran, or xylene to give viscous solutions. Solutions of poly(dichlorophosphazene) in dry benzene undergo a slow viscosity increase and gel after 6 days at room temperature. Gelation is speeded up by the addition of over 0.6 mole of $SnCl_4/NPCl_2$ unit but unaffected by UV or visible light irradiation. Rigorous removal of water also did not retard gelation. The molecular weights of the un-cross-linked polymer average about 1,500,000.

Polymers prepared using ammonium persulfate catalysts (ionic mechanism) is reported to give lighter-colored products [191]. Konecny and Douglas reported that benzoic acid is also an effective catalyst [192]. MacCallum and Werninck also reported that benzoic acid and sodium benzoate are equally effective polymerization catalysts at 235°C [5b].

Gimblett [193] suggested that the thermal and benzoic acid-catalyzed polymerization of the phosphonitrilic chloride trimer proceeds by an anionic mechanism. Completely substituted phosphonitrilic trimers which do not contain halogen are incapable of undergoing ionization and consequent polymerization. For example, hexaphenyl, hexamethyl, or hexaethyl phosphonitrilic trimers do not polymerize at elevated temperatures but decompose [194–196]. Other mechanisms [5b,197] have been suggested but more data are necessary to draw more definite conclusions.

The poly(dichlorophosphazene) can be reacted with a variety of reagents (see Table XXII). Reaction with water gives hydrolysis products with the evolution of HCl, NH_4Cl, $(NH_4)_3PO_4$, or H_3PO_4 [198]. The reaction with various nucleophilic reagents removes the chlorine atoms and gives a hydrolytically stable polymer showing good high-temperature properties. Some important examples are shown in Scheme 3 but others are possible and further research into this area should be quite fruitful.

The reactions in Scheme 3 can also be carried out effectively with hexachlorocyclotriphosphazene (cyclic phosphonitrilic chloride trimer). For examples see Eq. (86).

$$(PNCl_2)_3 \xrightarrow[\text{[202]}]{H_2NRNH_2} [\text{—}(P_3N_3)\text{—}NHR\text{—}NH\text{—}]_n \tag{86}$$

$$(PNCl_2)_3 \xrightarrow[\text{[204]}]{R(OH)_n} \text{Polymers}$$

$$(PNCl_2)_3 \xrightarrow[\text{[178, 203]}]{C_6H_3(OH)_3} \left[\text{—}P_3N_3\text{—}O\text{—}C_6H_3\begin{matrix} O^- \\ \\ O^- \end{matrix}\right]_n$$

TABLE XXII

POLYMERS OBTAINED BY THE REACTION OF POLYPHOSPHONITRILIC CHLORIDE WITH VARIOUS REAGENTS

Poly-(phosphonitrilic chloride) (unit moles)	Reagent (moles)	Solvent (ml)	Reaction conditions		Yield (%)	Polymer properties			Ref.
			Temp. (°C)	Time (hr)		T_g (°C)	[η] (dl/gm)	Mol. wt.	
1.0	$NaOCH_3$ (2.5 in 400 ml CH_3OH)	C_6H_6 (600)	60–64	12	12	—	1.22 ($CHCl_3$)	640,000	*a*
1.575	$NaOC_2H_5$ (4.0)	Ethanol (2000)	25	120	45	—	1.9 ($CHCl_3$)	—	*a*
1.58	$NaOCH_2CF_3$ (5.22 in 600 ml THF)	C_6H_6 (1500)	66	16	27	—	2.7 (acetone)	1.7×10^6	*a*
1.0	$NaOC_6H_5$ (3.5 in 700 ml THF)	C_6H_6 (500)	74 25	4 60	—	—	2.7 (C_6H_6)	3.7×10^6	*a*
1.0	$C_6H_5NH_2$ (8.6)	C_6H_6 (600)	66 25	48 168	15	—	1.44 (THF)	1.65×10^6	*b*
1.0	$C_2H_5NH_2$ (6.0)	Toluene (1500)	25	192		30	0.47 (1.0 *M* NaCl–0.1 *N* HCl)	—	*b*

1.73	$(CH_3)_2NH$ (5.4)	C_6H_6 (2500)	0–5 25	2 25	50 gm 230	—	0.89 (0.1 *M* NaCl–0.1 *N* NaCl)	363,000	*b*
0.5	Piperidine (2.2)	C_6H_6 (1000)	80	48	—	19	1.33 (C_6H_6)	1.58×10^6	*b*
1.73	NH_3 (25.0)	C_6H_6 (2000)	25	48	93/133	—	—	—	*b*
1.0	CH_3NH_2 (5.0 in 1500 ml THF)	Toluene (1500)	25	48	—		Cross-linked polymer		*b*
1.0	$(C_2H_5)_2NN$ or $(C_6H_5)_2NH$ or $CH_3(C_6H_5)NH$ (2.0+)	C_6H_6 —	80	95	—		Practically no reaction		*b*
0.43	$NaOCH_2CF_3$ (1.0) + $NaOCH_2CF_2CF_2CF_3$ (1.0 in 250 ml THF)	C_6H_6 (300+) + *n*-Bu_2O (300)	108	18	31	−77	1.46 (89/11CF_3CCl_3/acetone)		*c*

[a] H. R. Allcock, R. L. Kugel, and K. J. Valan, **5**, 1709 (1966); H. R. Allcock and E. L. Kugel, U.S. Patent 3,370,020 (1968); H. R. Allcock, *Abstr., 164th Nat. Meet., Amer. Chem. Soc., New York* Polymer No. 27 (1972).

[b] H. R. Allcock and R. L. Kugel, *J. Inorg. Chem.* **5**, 1716 (1966); U.S. Patent 3,364,189 (1968).

[c] S. H. Rose, *J. Polym. Sci.* **6**, 837 (1968); S. H. Rose and K. A. Reynard, *Abstr., 164th Nat. Meet., Amer. Chem. Soc., New York* Polymer No. 28 (1972).

7-3. Preparation of Polybis(trifluoroethoxy)phosphazene [184,205]

$$(NPCl_2)_n + 2nNaOCH_2CF_3 \longrightarrow [NP(OCH_2CF_3)_2]_n \quad (87)$$

To a flask equipped with a condenser and dropping funnel is added 600 ml of tetrahydrofuran and 600 gm (6.0 moles) of trifluoroethanol. Sodium metal is added portionwise until 120 gm (5.22 gm-atoms) had been reacted. Then a solution of 300 gm (2.58 moles) of polydichlorophosphazene in 1500 ml of benzene is added dropwise over a 3 hr period. The reaction is exothermic and causes the solvent to reflux. Then the reaction mixture is refluxed for 16 hr, cooled to room temperature, and acidified with conc. hydrochloric acid. The reaction mixture is filtered, the solid polymer washed with water, and then with aqueous 95% ethanol. The polymer is dissolved in 3 liters of acetone, filtered under pressure, and then reprecipitated in water to remove sodium salts. The polymer is air-dried, dissolved in acetone, precipitated into benzene, and the process repeated again to remove oligomers. The isolated polymer, which weighs 170 gm (27%), is white and fibrous, m.p. 242°–243°C, $\eta_i = 2.7$ dl/gm (acetone). A film cast from the polymer is flexible, does not burn, and is water-repellent.

The infrared spectrum shows a P=N stretching peak as part of a multiple band at 1270 cm^{-1} and a ^{31}P NMR spectrum shows a single peak at 47.5 ppm [187]. The polymer is soluble in acetone, tetrahydrofuran, methyl ethyl ketone, ethylene glycol, and dimethyl ether [187].

The poly(organophosphazenes) have properties useful for flame retardants, water repellents, for stable visible and UV film, adhesives, and heat-stable (up to 200°C) polymer applications.

Cyclolinear polymers can also be prepared utilizing the phosphonitrilic chloride cyclic trimer (hexachlorocyclotriphosphazene) or derivatives and reaction with various difunctional reagents: alkyl or aryl diols, diamines, diolefins, epoxides, phenol-formaldehyde combinations, siloxanes, etc. [190,206].

8. MISCELLANEOUS METHODS

1. Preparation of polyphosphate esters by the reaction of alkyl dichlorophosphate with quinones [207].

$$Cl-\overset{\overset{O}{\|}}{\underset{\underset{OR}{|}}{P}}-Cl + O=C_6H_4=O \longrightarrow \left[-\overset{\overset{O}{\|}}{\underset{\underset{OR}{|}}{P}}-O-C_6H_4Cl_2-O-\right]_n \quad (83)$$

2. Preparations of esters of polyphosphoric acid by use of phosphorus oxychloride [208] or other reagents [209].

$$4POCl_3 + 9ROH \longrightarrow RO\left[\begin{array}{c} O \\ \| \\ -P-O- \\ | \\ OR \end{array}\right]_n R + 3RCl + 4HCl \quad (89)$$

3. Preparation of vinyl phosphates by the mercuric ion-catalyzed addition of dialkyl hydrogen phosphates to acetylenes [210,210a].

4. Preparation of vinyl phosphoric acid esters by the mercuric ion-catalyzed acid-exchange reaction between vinyl acetate and phosphoric acid [211].

5. Chlorination of diethyl vinyl phosphates [212].

6. Hydrogenation of vinyl phosphates [213].

7. Monodialkylation of dialkyl vinyl phosphates by heating with alkali halides in a ketone solvent [214].

8. Transesterification of dialkyl vinyl phosphates [215].

9. Substitution of the halogen group in the vinyl ester portion of the phosphate ester [210,216].

10. The use of the Diels-Alder reaction to prepare substituted phosphate esters [217].

11. Reaction of ethylene [218] and tetrafluoroethylene [218a] with dialkylphosphonates under pressure using free-radical catalysts affords polyalkylene phosphonate [218,218a].

12. Reaction of trialkyl phosphites with methacrylic [219] acid or lactones [220].

13. Aluminum chloride-catalyzed reaction between *tert*-butylbenzene and phosphorus trichloride [221].

14. Phosphorylation of olefins [222].

15. Preparation of 1-thiovinyl phosphates [223].

16. Halogenation of trialkyl phosphates [224].

17. Preparation of vinyl-type polymers containing phosphorus side chains by reaction of carbonyl-containing polymers with phosphorus trichloride [225].

18. Reaction of phosphorus pentachloride with either SBR, natural rubber, or butadiene polymers to give phosphorus-containing rubbers [226].

19. Reduction of phosphinates and phosphonates with bis(2-methoxyethoxy)aluminum hydride to give phosphine oxides [227].

20. A new cyclotriphosphazene by a ring closure reaction of a phosphinimine and $P[N(CH_3)_2]_3$ [228].

21. Preparation of metal complexes of *P,P'*-dialkyl-*N,N,N',N'*-tetramethylpyrophosphoramide [229].

22. Transesterification of dialkyl phosphites with polyols [230].

23. Preparation of fire-retardant epoxy compounds [231].

24. Oxyphosphorylation of propylene [232].

25. The use of 2-dimethylphosphoryl ethyl methacrylate and 2-diethylphosphoryl ethyl methacrylate to copolymerize with methyl methacrylate [233].

26. Reaction of phosphorus with KOH, H_2O, and acrylamide to give

$$(H_2N\overset{\overset{O}{\|}}{C}-CH_2CH_2)_3PO$$

Methylolation of this compound gives a flame-retardant fabric treatment material [234].

27. Phosphorylation of cellulose [235].

28. Phosphoramidates for wood treating [236].

29. Preparation of self-extinguishing polyester resins [237].

30. The preparation of phosphorus-containing Mannich bases [238].

31. Preparation of dihydroxy phosphonates [239].

32. Lithium aluminum hydride reduction of alkyl alkylene diphosphonates to give alkylene diphosphines [240].

33. Preparation of (dialkoxyphosphinyl)ethylene glycols [241].

34. Preparation of organophosphonium halides [242].

35. Tris(1-aziridinyl)phosphine oxide to give durable fire-retardant finishes for cellulosic textile [243].

36. Radiation-induced copolymerization of vinyl phosphonyl dichloride with acrylonitrile [244].

37. Phosphorylation of poly(allyl alcohol) with dimethyl phosphite [245].

38. Preparation and use of 3-phosphonopropionamides [246].

39. Phosphine oxide copolymers [247].

40. Preparation of vinyl esters containing phosphorus [248].

41. Interaction of diaryl chlorophosphites with saturated ketones [249].

42. Reaction of P_2O_5 with hydroxyethyl acrylate to give organophosphorus thermosetting monomers [250].

43. Preparation of vinyl phosphates by the Perkow reaction [251].

44. Polymerization of metal carbonyls with phosphinic acids [252].

45. Preparation of poly dichlorophosphoric acids [253].

46. Preparation of poly arylphosphonates [254].

47. Polymerization of 1,3-dienephosphonates with styrene and methylmethacrylate [255].

48. Methylol phosphorus derivatives [256].

49. Preparation of epoxyalkyldimethylphosphine oxides [257].

50. Preparation of cyclic phosphoramidates [258].

51. Preparation of phosphorus containing *s*-triazines [259].

52. Preparation of polyphosphonates [260].

53. Preparation of cyclic phosphates of bis(hydroxymethyl)carboranes [261].

54. Preparation of hydrocarbyl carbamoyl phosphonates [262].

55. Preparation of polydiphenyl and diisobutylvinylphosphine oxide [263].

56. Preparation of tetraphosphonates of diallyl chlorendate [264].

57. Preparation and polymerization of allylaminophosphonitrile compounds [265].

58. Preparation and use of alkoxyphosphazenes [266].

59. Preparation and polymerization of esters of diolefinic phosphonic acids [267].

60. Polymerization of dimethyl 1-propene-2-phosphonate [147].

61. Ring-forming polymerizations of nitrogen–phosphorus compounds [268].

62. Reaction of phosphoro or phosphonohalides with an aldehyde and trivalent phosphorous ester to give a polymer [269].

REFERENCES

1. E. L. Gefter, "Organophosphorus Monomers and Polymers" (transl. by G. M. Kosolapoff; L. Jacolev, ed.). Associated Technical Services, Inc., Glen Ridge, New Jersey, 1962.

1a. R. C. Laible, *Chem. Rev.* **58**, 807 (1958).

2. J. W. Lyons, "The Chemistry & Uses of Fire Retardants." Wiley (Interscience), New York, 1970.

3. F. W. Lichtenthaler, *Chem. Rev.* **61**, 607 (1961).

3a. A. W. Frank, *Chem. Rev.* **61**, 389 (1961).

3b. R. A. Shaw, B. W. Fitzsimmons, and B. C. Smith, *Chem. Rev.* **62**, 247 (1963); J. R. Cox, Jr. and O. B. Ramsay, *ibid.* **64**, 317 (1964); A. N. Pudovik and V. K. Khairullen, *Russ. Chem. Rev.* **37**, 317 (1968); V. K. Promovienkov and S. Z. Ivin, *ibid.*, p. 670; K. D. Berlin and G. B. Butler, *Chem. Rev.* **60**, 243 (1960); J. G. Cadogan, *Quart. Rev., Chem. Soc.* **16**, 208 (1962); P. C. Crofts, *ibid.* **12**, 34 (1958); G. O. Doak and L. D. Freedman, *Chem. Rev.* **61**, 31 (1961); M. Yokoyama, *Yuki Gosei Kagaku Kyokai Shi* **28**, 443 (1970); M. Sander and E. Steininger, *J. Macromol. Sci., Rev. Macromol. Chem.* **2**, 1, 33, and 57 (1968).

3c. G. M. Kosolapoff, *Org. React.* **6**, 273 (1951).

3d. N. L. Paddock, *Quart. Rev., Chem. Soc.* **18**, 168 (1964).

4. G. M. Kosolapoff, "Organophosphorus Compounds." Wiley, New York, 1950; A. J. Kirby and S. G. Warren, "The Organic Chemistry of Phosphorus." Amer. Elsevier, New York, 1967.

4a. J. R. Van Wazer, "Phosphorus and its Compounds," Vol. 1. Wiley (Interscience), New York, 1958.

4b. J. R. Van Wazer, ed., "Phosphorus and its Compounds," Vol. 2. Wiley (Interscience), New York, 1961.

5. K. I. Baynon, *J. Polym. Sci., Part A* **1**, 3343 (1963).

5a. A. A. Berlin, L. P. Raskina, L. A. Zhil'tsova, and B. S. El'tsefon, *Vysokomol. Soedin., Ser. A* **13**, 174 (1971); J. R. Soulen and M. S. Silverman, *J. Polym. Sci., Part A* **1**, 823 (1963).
5b. J. R. MacCallum and A. Werninck, *J. Polym. Sci., Part A-1* **5**, 3061 (1967).
6. R. F. Hudson, "Structure and Mechanism in Organo-Phosphorus Chemistry." Academic Press, New York, 1966.
7. F. A. Cotton and G. Wilkinson, "Advanced Inorganic Chemistry, A Comprehensive Text," Chapter 20. Wiley (Interscience), New York, 1967.
8. H. R. Allcock, "Phosphorus–Nitrogen Compounds: Cyclic, Linear, and High Polymeric Systems." Academic Press, New York, 1971.
9. A. D. F. Toy and R. S. Cooper, *J. Amer. Chem. Soc.* **76**, 219 (1954); I. A. Krivosheeva, A. I. Razumov, and G. S. Kolesnikov, *Vysokomol. Soedin.* **3**, 1247 (1961); R. W. Hiley, U.S. Patent 2,478,441 (1949); G. Kamai and V. A. Kukhtin, *Zh. Obshc. Khim.* **24**, 1855 (1954).
10. D. Israelov and L. A. Rodivilova, *Polym. Sci. USSR* **8**, 1557 (1966).
10a. H. W. Coover, Jr., U.S. Patent 2,743,258 (1956).
10b. W. B. McCormack and H. E. Schroeder, U.S. Patent 2,891,915 (1959).
10c. V. V. Korshak, I. A. Gribova, and M. A. Andreeva, *Izv. Akad. Nauk SSSR, Otd. Khim. Nauk.*, p. 880 (1958); H. Zenftman and A. McLean, U.S. Patent 2,636,876 (1953); British Patent 644,468 (1958); H. Zenftman and H. R. Wright, *Brit. Plast.* **25**, 274 (1952); E. V. Kuznetsov, I. M. Shermerhorn and V. A. Belyaeva, U.S.S.R. Patent 137,673 (1961); A. J. Conix, German Patent 1,199,500 (1965).
10d. W. E. Cass, U.S. Patent 2,616,873 (1952).
10e. H. W. Coover, Jr., R. L. McConnell, and M. A. McCall, *Ind. Eng. Chem.* **52**, 409 (1960).
10f. B. Helferich and K. G. Schmidt, *Chem. Ber.* **92**, 2051 (1959).
11. A. D. F. Toy, U.S. Patent 2,435,252 (1948).
12. K. A. Petrov, V. P. Evdakov, L. I. Mizrakh, and V. A. Kravchenko, U.S.S.R. Patent 152,572 (1963).
13. H. W. Coover, Jr. and M. A. McCall, U.S. Patent 2,682,522 (1954).
14. Chemische Werke Hüls G.m.b.H., British Patent 772,486 (1957).
14a. D. C. Ayres and H. N. Rydon, *J. Chem. Soc., London*, p. 1109 (1957).
15. K. A. Petrov, V. A. Parshina, and G. M. Tsypina, *Plast. Massy* No. 11, p. 11 (1963).
16. M. Sander, *Makromol. Chem.* **55**, 191 (1962).
16a. E. Steininger and M. Sander, *Kunststoffe* **54**, 507 (1964).
17. R. J. McManimie, U.S. Patent 2,893,961 (1959).
17a. T. Shimidzu, T. Hakozaki, T. Kagiya, and K. Fukui, *J. Polym. Sci., Part B* **3**, 871 (1956).
18. A. Y. Garner, U.S. Patents 3,161,607 and 3,161,687 (1964).
19. M. I. Kabachnik and E. N. Tsvetkov, *Zh. Obshch. Khim.* **31**, 684 (1961).
20. V. V. Korshak, I. A. Gribova, and V. K. Shitikov, *Izv. Akad. Nauk SSSR, Otd. Khim. Nauk*, p. 210 (1958); E. L. Gefter and A. Yuldashev, *Plast. Massy* No. 2, p. 49 (1962); S. M. Shner, I. K. Rubtoova, and E. L. Gefter, *Vysokomol. Soedin.* **7**, 1684 (1965).
21. V. V. Korshak, I. A. Gribova, and M. A. Andreeva, *Bull. Acad. Sci. USSR, Div. Chem. Sci.* p. 641 (1957); V. V. Korshak, I. A. Gribova, M. A. Andreeva and T. Ya. Medved, *Vysokomol. Soedin.* **6**, 117 (1964); K. A. Petrov, E. E. Nifanteev, and L. V. Fedorchuk, *ibid.* **2**, 417 (1960).
22. V. V. Korshak, *J. Polym. Sci.* **31**, 319 (1958); J. Pellon, *J. Polym. Sci., Part A* **1**, 3561 (1963).

23. T. M. Frunze, V. V. Korshak, A. A. Izyneev, and V. V. Kurashev, *Vysokomol. Soedin.* **7**, 285 (1965).
24. W. E. McCormack, U.S. Patents 2,671,077–2,671,080 (1954) A. Borniec; and B. Laskiewicz, *J. Polym. Sci., Part A* **1**, 1963 (1963); L. A. Errede, *J. Polym. Sci.* **49**, 253 (1961); L. A. Errede and N. Knoll, *ibid.* **60**, 33 (1962); Minnesota Mining & Manufacturing Co., German Patent 1,145,799 (1963).
25. S. A. Buckler and M. Epstein, British Patent 926,268 (1963).
26. H. R. Allcock and R. L. Kugel, *J. Amer. Chem. Soc.* **87**, 4216 (1965); D. L. Herring, *Chem. Ind.* (*London*) p. 717 (1960); G. Tesi, C. P. Haber, and C. H. Douglas, *Proc. Chem. Soc., London* p. 219 (1960).
27. *Chem. Eng. News* **39**, 63 (1961); **42**, 54 (1964); R. A. Shaw and T. Ogawa, *J. Polym. Sci., Part A* **3**, 3343 (1965).
28. L. Parts, M. L. Nielsen, and J. T. Miller, *Inorg. Chem.* **3**, 1261 (1964).
29. A. L. McCloskey, *in* "Inorganic Polymers" (F. G. A. Stone and W. A. G. Graham, eds.), p. 171. Academic Press, New York, 1962; J. M. C. Thompson, *in* "Developments in Inorganic Polymer Chemistry" (M. F. Lappert and G. J. Leigh, eds.), p. 76. Amer. Elsevier, New York, 1962; V. A. Zamjatina and N. I. Bekasova, *Usp. Khim.* **30**, 48 (1961); **33**, 1216 (1964).
30. K. D. Berlin and D. M. Hellwege, *Top. Phosphorus Chem.* **6**, 1–186 (1969).
31. F. B. Clarke and J. W. Lyons, *J. Amer. Chem. Soc.* **88**, 4401 (1966).
32. C. A. Gleason and G. H. Slack, U.S. Patent 3,328,492 (1967); H. Grabhofer, H. Muller, R.-F. Posse, and H. Ulrich, U.S. Patent 3,342,903 (1967).
33. H. R. Guest and B. W. Kiff, U.S. Patent 2,957,856 (1960).
33a. D. E. Kvalnes and N. O. Brace, U.S. Patent 2,691,567 (1954).
34 T F. Ronay and R. D. Dexheimer, German Patent 1,251,450 (1967); A. L. Austin, R. J. Hartman, and J. T. Patton, Jr., U.S. Patent 3,439,067 (1969).
34a. V. Trescher, G. Braun, and H. Nordt, Canadian Patent 759,469 (1967).
35. H. P. Latscha and P. B. Hormuth, *Z. Anorg. Allg. Chem.* **359**, 78 (1968); G. Tomaschewski and B. Breittfeld, *J. Prakt. Chem.* [4] **311**, 256 (1969).
36. G. M. Kosolapoff, "Organophosphorus Compounds," Chapter 9. Wiley, New York, 1950; D. H. Chadwick and R. S. Watt, *in* "Phosphorus and its Compounds" (J. R. Van Wazer, ed.), Vol. 2, Chapter 19. Wiley (Interscience), New York, 1961.
37. R. A. Pizzarello and A. F. Schneid, U.S. Patent 3,335,131 (1967); M. Lidaks, J. Hillers, and A. Medne, *Latv. PSR Zinat. Akad. Vestis* p. 87 (1959).
38. J. C. Frick, Jr. and R. L. Arceneaux, U.S. Patent 2,939,849 (1960).
39. C. E. Adams and B. H. Shoemaker, U.S. Patent 2,372,244 (1945); E. Bergmann and A. Bondi, *Ber. Deut. Chem. Ges.* **63**, 1158 (1930).
40. G. M. Kosolapoff, "Organophosphorus Compounds," Chapters 2–6. Wiley, New York, 1950.
41. I. K. Jackson and W. J. Jones, *J. Chem. Soc., London* p. 575 (1931).
42. H. E. Sorstokke, Canadian Patent 787,946 (1968).
43. A. E. Arbuzov and I. Arbuzova, *J. Russ. Phys. Chem. Soc.* **62**, 1533 (1930); B. Ackerman, R. M. Chladek, and D. Swern, *J. Amer. Chem. Soc.* **79**, 6524 (1957); G. M. Kosolapoff, *ibid.* **66**, 109 and 1511 (1944); R. Sasin, R. M. Nauman, and D. Swern, *ibid.* **80**, 6336 (1958); G. Kamai and V. A. Kakhtin, *Dokl. Akad. Nauk SSSR* **89**, 309 (1953).
43a. R. H. Wiley, U.S. Patent 2,478,441 (1949).
44. E. E. Hardy and T. Reetz, U.S. Patent 2,770,620 (1956).
45. G. Reuter and E. Wolf, German Patent 1,046,047 (1958); E. Klanke, E. Kühle, I. Hammann, and W. Lorenz, U.S. Patent 3,407,248 (1968); A. Y. Garner, U.S. Patent 2,916,510 (1959).

46. S. C. Ternin, U.S. Patents 3,065,183 (1960) and 3,179,533 (1965).
47. L. Friedman, U.S. Patent 3,433,856 (1969).
48. J. J. Pitts, M. A. Robison, and S. I. Trotz, *Inorg. Nucl. Chem. Lett.* **4**, 483 (1968); Imperial Chem. Ind. Inc., British Patent 907,029 (1962); *Chem. Abstr.* **58**, 279 (1963).
49. H. Coates and J. D. Collins, Canadian Patent 794,378 (1968); A. D. F. Toy and K. H. Rattenbury, U.S. Patent 2,841,604 (1958); German Patent 1,041,251 (1958).
50. F. E. King, V. F. G. Cooke, and J. Lincoln, British Patent 872,206 (1961); H. Feinholz and F. Winder, U.S. Patent 3,378,502 (1968).
51. N. V. deBataafsche Petroleum Maatschappij, Netherlands Patent 69,357 (1952); F. Johnston, U.S. Patent 2,754,319 (1956); Victor Chem. Works, British Patent 766,722 (1957); K. Schimmelschmidt and H. J. Kleiner, Canadian Patent 804,618 (1969).
52. S. A. Zahir, U.S. Patent 3,374,292 (1968).
53. A. N. Pudovik and B. E. Ivanov, *Izv. Akad. Nauk SSSR Otd. Khim. Nauk* p. 947 (1952).
54. C. Walling, U.S. Patents 2,437,796 and 2,437,798 (1948).
55. A. E. Arbuzov, *J. Russ. Phys.-Chem. Soc.* **42**, 395 and 549 (1910); A. E. Arbuzov and K. V. Nikonorov, *Zh. Obshch. Khim.* **18**, 2008 (1948).
56. A. Michaelis and W. LaCoste, *Ber. Deut. Chem. Ges.* **18**, 2109 (1885).
57. I. L. Knunyantz and R. N. Sterlin, *Dokl. Akad. Nauk SSSR* **56**, 49 (1947); W. J. Vullo, U.S. Patent 3,452,098 (1969).
58. M. Reuter and F. Jacob, German Patent 1,040,549 (1958); M. Reuter and L. Orthner, German Patent 1,035,135 (1959).
59. M. Becke-Goehring, German Patent 1,059,186 (1959); C. A. Redfarn and H. Bates, British Patent 788,735 (1958).
60. A. V. Kirsanov, *J. Gen. Chem. USSR* **33**, 1545 (1963).
61. Dow Chem. Co., Netherland Patent 7,008,737 (1972).
62. A. N. Pudovik and M. G. Imaev, *Izv. Akad. Nauk SSSR* p. 916 (1952); *Chem. Abstr.* **47**, 10463 (1953); W. M. Daniewski and C. E. Griffin, *J. Org. Chem.* **31**, 3236 (1966); J. B. Dickey, H. W. Coover, Jr., and H. N. Shearer, Jr., U.S. Patent 2,550,651 (1951); *Chem. Abstr.* **45**, 8029b (1951); E. C. Ladd, U.S. Patent 2,622,096 (1952); *Chem. Abstr.* **47**, 9344c (1953).
63. V. Jagodic, *Chem. Ber.* **93**, 2308 (1960); R. L. McConnell and H. W. Coover, Jr., *J. Amer. Chem. Soc.* **79**, 1961 (1957); S. H. Metzer and A. F. Isbell, *J. Org. Chem.* **29**, 623 (1964); A. N. Pudovik, *Dokl. Akad. Nauk SSSR* **125**, 826 (1959); *Zh. Obshch. Khim.* **29**, 3338 and 3342 (1959); A. R. Stiles, W. E. Vaughan, and F. F. Rust, *J. Amer. Chem. Soc.* **80**, 714 (1958).

63a. A. N. Pudovik, *Zh. Obshch. Khim.* **23**, 263 (1953).

64. V. S. Abramov, *Tr. Kazan. Khim.-Tekhnol. Inst.* **23**, 102 (1957); M. Kirilov, *C. R. Acad. Bulg. Sci.* **10**, 209 (1957); T. Y. Medved, *Izv. Akad. Nauk SSSR, Otd. Khim. Nauk* p. 1357 (1957); K. V. Nikonorov, *Khim. Primen. Fosfororg. Soedin., Tr. Konf., 1st, 1955* p. 223 (1957).
65. P. I. Sanin, *Dokl. Akad. Nauk SSSR* **132**, 145 (1960).
66. R. B. Fox and D. L. Venezky, *J. Amer. Chem. Soc.* **78**, 1661 (1956).
67. K. A. Petrov, *Zh. Obshch. Khim.* **29**, 1819 (1959).
68. H. D. Orloff, *J. Amer. Chem. Soc.* **80**, 727 (1958).
69. K. A. Petrov, *Zh. Obshch. Khim.* **29**, 1486 (1959).
70. V. Ettel, *Chem. Listy* **50**, 1261 (1956).
71. A. A. Oswald, *Can. J. Chem.* **37**, 1498 (1959).

72. Unpublished data from Mobil Chemical Company mentioned in Mobil Chemical Technical Bulletin on Dialkyl Hydrogen Phosphites and Trialkyl Phosphites (1972).
73. G. Kamai, *Zh. Obshch. Khim.* **27**, 2372 (1956); V. A. Kikhtin, *ibid.* **28**, 2790 (1958).
74. V S. Abramov, *Dokl. Akad. Nauk SSSR* **95**, 991 (1954).
75. E. E. Hardy and G. M. Kosolapoff, U.S. Patent 2,409,039 (1946); H. McCombie, B. C. Saunders, and G. J. Stacey, *J. Chem. Soc., London* p. 380 (1945).
76. R. L. Jenkins, U.S. Patent 2,426,691 (1947).
77. Unpublished data from Mobil Chemical Company mentioned in Mobil Chemical Company Technical Bulletin on Dialkyl and Trialkyl Phosphites (1972).
78. A. E. Arbuzov, B. A. Arbuzov, and B. P. Ludovkin, *Bull. Acad. Sci. USSR, Cl. Sci. Chem.* p. 535 (1947); *Chem. Abstr.* **42**, 1886 (1949); E. V. Kuznetsov, *Zh. Obshch. Khim.* **29**, 2017 (1959).
79. E. M. Crook, *J. Chem. Soc., London* p. 710 (1961).
80. K. Dimroth, *Chem. Ber.* **90**, 801 (1957).
81. A. E. Arbuzov, *J. Russ. Phys.-Chem. Soc.* **46**, 291 (1914); *Dokl. Akad. Nauk SSSR* **112**, 856 (1957).
82. B. R. Carrell and W. Gerrard, *Chem. Ind. (London)* p. 1289 (1958).
83. A. A. Oswald, *Can. J. Chem.* **37**, 1498 (1959).
84. M. G. Imaev, *Zh. Obshch. Khim.* **31**, 1770 (1961).
85. M. I. Kabachnik, *Izv. Akad. Nauk SSSR, Otd. Khim. Nauk* p. 2135 (1959).
86. I. J. Borowitz, S. Firstenberg, G. B. Borowity, and D. Schuessler, *J. Amer. Chem. Soc.* **94**, 1623 (1972); W. Perkow, K. Ullerich, and F. Meyer, *Naturwissenschaften* **39**, 353 (1952); W. Perkow, *Chem. Ber.* **87**, 755 (1954); W. Perkow, E. W. Krockow, and K. Knoevenagel, *ibid.* **88**, 662 (1955).
87. D. E. C. Corbridge, *Top. Phosphorus Chem.* **6**, 235–265 (1969).
88. H. Zenftman and A. McLean, British Patent 644,468 (1950); B. Helferich and K. G. Schmidt, German Patent 843,753 (1952).
88a. H. Zenftman and A. McLean, U.S. Patent 2,636,876 (1953).
89. H. Zenftman and H. R. Wright, *Brit. Plast.* **25**, 374 (1952).
90. D. Heuck, F. Rochlitz, H. Schmidt, H. Vilczek, and J. Winter, German Patent 1,117,305 (1958).
91. A. Schors and A. Heslinga, German Patent 1,203,464 (1965).
92. R. C. Nametz, *Ind. Eng. Chem.* **59**, 99 (1967).
93. R. C. Nametz, *Ind. Eng. Chem.* **59**, 99 (1967).
94. K. A. Petrov, E. E. Nifanteev and L. V. Fedorchuk, *Vysokomol. Soedin.* **2**, 417 (1960); *Chem. Abstr.* **54**, 24372 (1960).
95. W. M. Lanham, British Patent 812,390 (1959).
96. Hercules, Inc., Belgian Patent 725,477 (1968).
97. M. Umemura and A. Hatano, Japanese Patent 17,597 (1968).
98. H. Feinholz and F. Winder, U.S. Patent 3,378,502 (1968).
99. British Thompson-Houston Co., Ltd., British Patent 524,510 (1940).
100. J. R. Cox, Jr. and O. B. Ramsay, *Chem. Rev.* **64**, 317 (1964).
101. A. Munoz, J. Novech, and J. P. Vives, *Colloq. Nat. Cent. Nat. Rech. Sci. (Fr.)* 253 253 (1965).
102. G. C. Daul, J. D. Reid, and R. M. Reinhardt, *Ind. Eng. Chem.* **46**, 1042 (1954).
103. A. H. Ford-Moore and J. H. Williams, *J. Chem. Soc., London* p. 1465 (1947).
104. G. M. Kosolapoff, *J. Amer. Chem. Soc.* **70**, 1971 (1948).
105. E. O. Leopold and H. Zorn, German Patent 1,006,414 (1959).
106. E. L. Gefter and M. I. Kabachnik, *Dokl. Akad. Nauk SSSR* **114**, 541 (1957).
107. G. Walters and I. Hornstein, U.S. Patents 2,660,542 and 2,660,543 (1953).

108. G. Griffin, R. Esteve, and R. C. Laible, University of Rhode Island, QM Contract DA 44-109-QM-421, Progress Report No. 1 (1951); A. J. McQuade, *Amer. Dyest. Rep.* **44**, 749 (1955).
109. L. Whitehill and R. Barker, U.S. Patent 2,394,829 (1946).
110. R. C. Laible and R. Esteve, University of Rhode Island, QM Contract DA 44-109-QM-421, Progress Report No. 3 (1951).
111. J. C. Frick, Jr., J. Weaver, R. L. Arceneaux, and M. Stansbury, *J. Polym. Sci.* **20**, 307 (1956).
112. R. W. Upson, *J. Amer. Chem. Soc.* **75**, 1763 (1953).
113. E. V. Kuznetsov, I. M. Shermerhorn, and V. A. Belyaeva, U.S.S.R. Patent 137,673 (1961); *Chem. Abstr.* **55**, 29790 (1961).
114. A. J. Conix, German Patent 1,199,500 (1965).
115. K. A. Petrov, V. P. Evdakov, L. I. Mizrakh, and V. A. Kravchenko, U.S.S.R. Patent 152,572 (1963); *Chem. Abstr.* **59**, 1804 (1963).
116. V. V. Korshak, I. A. Gribova, and M. A. Andreeva, *Izv. Akad. Nauk SSSR, Otd. Khim. Nauk* p. 210 (1958).
117. K. A. Petrov, E. E. Nifanteev, and L. V. Fedorchuk, *Vysokomol. Soedin.* **2**, 417 (1960); *Chem. Abstr.* **54**, 24372 (1960).
118. V. V. Korshak, I. A. Gribova, M. A. Andreeva, and T. Y. Medved, *Vysokomol. Soedin.* **6**, 117 (1964); *Chem. Abstr.* **61**, 8416 (1964).
119. E. E. Hardy and T. Reetz, U.S. Patent 2,770,620 (1956).
120. R. T. Harvey and E. R. DeSombre, *Top. Phosphorus Chem.* **1**, 57–111 (1964).
121. I. Heckenbleiker and K. R. Mott, Canadian Patent 804,619 (1969).
122. E. N. Rostovskii, O. V. Shehelkunova, and N. S. Bondareva, *Vysokomol. Soedin., Khim. Svoistra Modif. Polim.* p. 151 (1964).
123. V. V. Korshak, I. A. Gribova, and V. K. Shitikov, *Izv. Akad. Nauk SSSR, Otd. Khim. Nauk* p. 210 (1958); *Chem. Abstr.* **52**, 12804 (1958).
124. E. L. Gefter and A. Yuldashev, *Plast. Massy*, No. 2, p. 49 (1962); *Chem. Abstr.* **57**, 1039 (1962).
125. S. M. Shner, I. K. Rubtoova, and E. L. Gefter, *Vysokomol. Soedin.* **7**, 1684 (1965).
126. R. L. McConnell and H. W. Coover, Jr., *J. Amer. Chem. Soc.* **78**, 4453 (1956).
127. R. T. Harvey and E. R. DeSombre, *Top. Phosphorus Chem.* **1**, 58 (1964).
128. S. Konya and M. Yokoyama, *Kogyu Kagaku Zasshi* **68**, 1080 (1965); W. Karo, Final Report, Material R and D. Commission, Natick, Mass., May 30, 1960, Contract No. DA-19-129-QM-841, QM Project No. 7-93-15-004.
129. M. L. Ernsberger, U.S. Patent 2,491,920 (1949).
130. A. D. F. Toy and R. S. Cooper, *J. Amer. Chem. Soc.* **76**, 2191 (1954).
131. A. J. Castro and W. E. Elwell, *J. Amer. Chem. Soc.* **72**, 2275 (1950).
132. C. Heuck, F. Rochlitz, H. Schmidt, H. Vilczek, and J. Winter, German Patent 1,117,315 (1958).
133. H. J. Lucas, F. W. Mitchell, and C. N. Scully, *J. Amer. Chem. Soc.* **72**, 5791 (1950).
134. B. A. Arbuzov, K. V. Nikonorov, and Z. G. Shishova, *Izv. Akad. Nauk SSSR, Otd. Khim. Nauk* p. 823 (1954).
135. V. V. Korshak, I. A. Gribova, and V. K. Shitikov, *Izv. Akad. Nauk SSSR, Otd. Khim. Nauk* p. 631 (1957).
136. H. E. Sorstokke, Canadian Patent 787,946 (1968).
137. L. Friedman, U.S. Patent 3,330,888 (1967).
138. V. I. Kirilovich, I. K. Rubtoova, and E. L. Gefter, *Plast. Massy* p. 20 (1963).
139. G. Nishk, H. Holtschmidt, and I. Ugi, German Patent 1,190,186 (1965).
140. D. H. Gerlach, W. G. Peet, and E. L. Muetterties, *J. Amer. Chem. Soc.* **94**, 4545 (1972).

141. K. A. Petrov, E. E. Nifanteev, R. G. Gol'tsova, and S. M. Korneev, *Vysokomol. Soedin.* **6**, 929 (1964).
142. K. A. Petrov, E. E. Nifanteev, and R. G. Gol'tsova, *Vysokomol. Soedin.* **6**, 1545 (1964).
143. M. S. Larrison, U.S. Patent 3,333,026 and 3,333,027 (1967).
144. G. M. Kosolapoff, "Organophosphorus Compounds," p. 279. Wiley, New York, 1950.
145. D. M. Harris, R. L. Jenkins, and M. L. Nielsen, *J. Polym. Sci.* **35**, 540 (1959).
146. H. W. Coover, Jr. and J. B. Dickey, U.S. Patent 2,725,371 (1955).
147. R. V. Lindsey, Jr., U.S. Patent 2,439,214 (1948).
148. J. B. Dickey and H. W. Coover, Jr., U.S. Patent 2,666,750 (1954).
149. W. B. McCormack, U.S. Patent 2,671,077–2,671,078 (1954).
150. W. B. McCormack, U.S. Patent 2,671,079 (1954).
151. W. E. McCormack, U.S. Patent 2,663,737 (1953).
152. W. B. McCormack, U.S. Patent 2,663,738 (1953).
153. V. V. Korshak, G. S. Kolesnikov, and B. A. Zhubanuv, *Izv. Akad. Nauk SSSR, Otd. Khim. Nauk* p. 618 (1958).
154. G. M. Kosolapoff, *J. Amer. Chem. Soc.* **77**, 6658 (1955).
155. Albright & Wilson, Ltd., British Patent 740,269 (1955).
156. Albright & Wilson, Ltd., British Patent 764,313 (1956).
157. W. A. Reeves and J. D. Guthrie, U.S. Patent 2,846,413 (1958).
158. L. M. Kindley, H. E. Podall, and N. Filipescu, *SPE* (*Soc. Plast. Eng.*) *Trans.* **2**, 122 (1962).
159. W. A. Reeves and J. D. Guthrie, *Ind. Eng. Chem.* **48**, 64 (1956); U.S. Patent 2,809,701 (1957).
160. R. Rabinowitz, R. Marcas, and J. Pellon, *J. Polym. Sci., Part A* **2**, 1241 (1964).
161. J. Liebig, *Ann. Pharm.* **11**, 14–17 and 139 (1834).
162. H. Rose, *Ann. Pharm.* **11**, 131 (1834).
163. C. Gerhardt, *Ann. Chim. Phys.* [3] **18**, 188 (1846); *C. R. Acad. Sci.* **22**, 858 (1846).
164. A. Laurent, *C. R. Acad. Sci.* **31**, 356 (1850).
165. J. H. Gladstone and J. D. Holmes, *J. Chem. Soc., London* **17**, 225 (1864); *Ann. Chim. Phys.* [4] **3**, 465 (1864); *Bull. Soc. Chim. Fr.* [2] **3**, 113 (1865).
166. H. Wichelhaus, *Ber. Deut. Chem. Ges.* **3**, 163 (1870).
167. J. H. Gladstone, *Ann. Chem. Pharm.* **76**, 74 (1850); *J. Chem. Soc., London* **2**, 121 (1850); **3**, 135 and 353 (1851); *Ann. Chem. Pharm.* **77**, 314 (1851); A. Besson, *C. R. Acad. Sci.* **111**, 972 (1890); **114**, 1264 and 1479 (1892); A. Besson and G. Rosset, *ibid.* **143**, 37 (1906); **146**, 1149 (1908); W. Couldridge, *J. Chem. Soc., London* **53**, 398 (1888); *Bull. Soc. Chim. Fr.* [2] **50**, 535 (1888).
168. H. N. Stokes, *Ber. Deut. Chem. Ges.* **28**, 437 (1895); *Amer. Chem. J.* **17**, 275 (1895); **18**, 629 and 780 (1896); **19**, 782 (1897); **20**, 740 (1898); *Z. Anorg. Chem.* **19**, 36 (1899).
169. L. G. Lund, N. L. Paddock, J. E. Proctor, and H. T. Searle, *J. Chem. Soc., London* p. 2542 (1960).
169a. M. Becke-Goehring and G. Koch, *Chem. Ber.* **91**, 1188 (1959).
170. R. A. Shaw, B. W. Fitzsimmons, and B. C. Smith, *Chem. Rev.* **62**, 247 (1962).
171. R. Schenck and G. Römer, *Ber. Deut. Chem. Ges. B* **57**, 1343 (1924).
172. M. Becke-Goehring and E. Fluck, *Angew. Chem.* **74**, 382 (1962); M. Becke-Goehring and W. Lehr, *Z. Anorg. Allg. Chem.* **327**, 128 (1964).
173. F. Seel and J. Langer, *Angew. Chem.* **68**, 461 (1956).
174. K. John and T. Moeller, *J. Amer. Chem. Soc.* **82**, 2647 (1960).
175. C. P. Haber, D. L. Herring, and E. A. Lawton, *J. Amer. Chem. Soc.* **80**, 2116 (1958); R. A. Shaw and C. Stratton, *Chem. Ind.* (*London*) p. 52 (1959); D. L. Herring, *ibid.*,

p. 717 (1960); G. Tesi, C. P. Haber, and C. H. Douglas, *Proc. Chem. Soc., London* p. 219 (1960); I. I. Berzman and J. H. Smalley, *Chem. Ind.* (*London*) p. 1839 (196); A. J. Bilbo, *Z. Naturforsch. B* **15**, 330 (1960); V. V. Korshak, I. A. Gribova, T. V. Artamonova, and A. N. Bushmarina, *Vysokomol. Soedin.* **2**, 377 (1960).

176. F. Goldschmidt and B. Dishon, *J. Polym. Sci.* **3**, 481 (1948); J. O. Konecny, C. H. Douglas, and M. Y. Gray, *ibid.* **42**, 383 (1960).
177. H. R. Allcock, *J. Macromol. Sci., Rev. Macromol. Chem.* **4**, 3 (1970); R. G. Rice, B. H. Geib, and L. A. Kaplan, U.S. Patent 3,121,704 (1964).
178. C. A. Redfarn, U.S. Patent 2,866,733 (1858).
179. M. Becke-Goehring, *Quart. Rev., Chem. Soc.* **10**, 437 (1956).
180. R. Steinman, F. B. Schirmer, Jr., and L. F. Andrieth, *J. Amer. Chem. Soc.* **64**, 2377 (1942).
181. L. G. Lund, N. L. Paddock, J. E. Proctor, and H. T. Searle, *J. Chem. Soc., London* p. 2542 (1960).
182. M. L. Nielsen, U.S. Patent 3,347,643 (1967).
183. C. J. Brown, *J. Polym. Sci.* **5**, 465 (1950).
184. H. R. Allcock, R. L. Kugel, and K. J. Valan, *Inorg. Chem.* **5**, 1709 (1966).
185. J. E. Proctor, N. L. Paddock, and H. T. Searle, Australian Patent 233,600 (1959); N. L. Paddock, U.S. Patent 3,026,174 (1962); M. Becke-Goehring and G. Koch, *Chem. Ber.* **92**, 1188 (1959); E. F. Moran, *J. Inorg. Nucl. Chem.* **30**, 1405 (1968).
186. G. M. Nichols, *Rep. Conf. High Temp. Polym. Fluid Res., 1962* ASD-TDR-62-372 (1962).
187. H. R. Allcock and R. L. Kugel, *J. Amer. Chem. Soc.* **87**, 4216 (1965).
188. H. Spencer, *Angew. Chem.* **65**, 299 (1953).
189. R. Knoesel, J. Parrod, and H. Benoit, *C. R. Acad. Sci.* **251**, 2994 (1960).
190. H. R. Allcock, "Phosphorus Nitrogen Compounds: Cyclic, Linear, and High Polymeric Systems," Chapter 16. Academic Press, New York, 1971.
191. R. L. Vale, *Macromol. Syn.* **2**, 91 (1966).
192. J. O. Konecny and C. H. Douglas, *J. Polym. Sci.* **36**, 195 (1959).
193. F. G. R. Gimblett, *J. Polym. Sci.* **60**, S29 (1963).
194. R. A. Shaw, *J. Polym. Sci.* **50**, 21 (1961).
195. H. T. Searle, *Proc. Chem. Soc., London* p. 7 (1959).
196. R. A. Shaw, *Chem. Ind.* (*London*) p. 412 (1959).
197. F. Patat and F. Kollinsky, *Makromol. Chem.* **6**, 292 (1951); F. Patat and K. Frömbling, *Monatsh. Chem.* **86**, 718 (1955); F. Patat and P. Derst, *Angew. Chem.* **71**, 105 (1959); J. R. MacCallum and J. Tanner, *J. Polym. Sci., Part B* **7**, 743 (1969).
198. F. G. R. Gimblett, *Trans. Faraday Soc.* **56**, 528 (1960); S. M. Zhivukhin, V. B. Tolstoguzov, and Y. V. Meitn, *Vysokomol. Soedin.* **3**, 414 (1961).
199. H. R. Allcock, *Abstr., 164th Nat. Meet., Amer. Chem. Soc., New York*, Polymer No. 27 (1972); S. H. Rose, *J. Polym. Sci., Part B* **6**, 837 (1968); *Abstr., 164th Nat. Meet., Chem. Soc., New York*, Polymer No. 28 (1972); G. L. Hagnauer, N. Schneider, and R. E. Singler, *161st Meet., Amer. Chem. Soc., Los Angeles, Paper 82, Polymer Div.* (1971); *Polym. Prepr., Amer. Chem. Soc., Div. Polym. Chem.* **12**, 525 (1971); G. Allen, C. J. Lewis, and S. M. Todd, *Polymer* **11**, 31 and 44 (1970); H. R. Allcock and E. J. Walsh, *J. Amer. Chem. Soc.* **94**, 4538 (1972); F. Goldschmidt and B. Dishon, *J. Polym. Sci.* **3**, 481 (1948).
200. H. R. Allcock and D. P. Mack, *Chem. Commun.* p. 685 (1970).
201. C. F. Liu and R. L. Evans, U.S. Patent 3,169,933 (1965); J. R. MacCallum and J. Tanner, *J. Polym. Sci., Part A-1* **6**, 3163 (1968).
202. M. Becke-Goehring and D. Neubauer, German Patent 1,143,027 (1963).

203. H. C. Wu and M. C. Kao, *Kao Fen Tzu T'ung Hsun* 7, 229 (1965); *Chem. Abstr.* **64,** 9897b (1966); R. G. Rice and M. V. Ernest, Canadian Patent 781,991 (1968); K. Nakamura, Japanese Patent 22,224 (1971).
204. U. Sallberger and R. Wolf, German Patent 1,951,082 (1970).
205. H. R. Allcock and R. L. Kugel, U.S. Patent 3,370,020 (1968).
206. H. Allcock and R. L. Kugel, U.S. Patent 3,299,128 (1967); H. Allcock, U.S. Patents 3,294,872 (1966) and 3,356,769 (1967); H. Allcock and W. M. Thomas, U.S. Patent 3,329,663 (1967); H. Allcock and L. A. Siegel, U.S. Patent 3,356,768 (1967).
207. R. Wegler, E. Kühle, and W. Schäfer, *Angew. Chem.* **70**, 350 (1958).
208. E. P. Plueddemann, U.S. Patent 2,558,380 (1951).
209. G. M. Kosolapoff, *Science* **108**, 485 (1948); L. P. Kyrides, U.S. Patent 2,510,033 (1950); D. C. Hull and J. R. Snodgrass, U.S. Patent 2,492,153 (1949); A. Closse, *Chem.-Ztg.* **81**, 72–75, 103–105, and 141–144 (1957).
210. W. R. Diveley, U.S. Patent 2,864,741 (1959).
210a. R. L. McConnell and H. W. Coover, Jr., U.S. Patent 2,912,450 (1960); R. L. McConnell and J. B. Dickey, U.S. Patent 2,861,093 (1960); R. C. Morris and J. L. Van Winkle, U.S. Patent 2,744,128 (1958); H. H. Wassermann and D. Cohen, *J. Amer. Chem. Soc.* **82**, 4435 (1960).
211. E. Baer, L. J. Ciplÿauskas, and T. Visser, *J. Biol. Chem.* **234**, 1 (1959).
212. Food Machinery and Chem. Corp., British Patent 783,697 (1958); J. F. Allen, S. K. Reed, O. H. Johnson, and N. J. Brunsvold, *J. Amer. Chem. Soc.* **78**, 3715 (1956).
213. W. Kiessling, *Chem. Ber.* **68**, 597 (1935); H. I. Jacobson, M. J. Griffin, S. Preis, and E. V. Jensen, *J. Amer. Chem. Soc.* **79**, 2608 (1957).
214. F. Cramer and D. Voges, *Chem. Ber.* **92**, 952 (1959); E. Y. Spencer, A. R. Todd, and R. F. Webb, *J. Chem. Soc., London* p. 2968 (1958).
215. Y. Nishizawa, *Bull. Agr. Chem. Soc. Jap.* **25**, 61 (1961).
216. Y. Nishizawa, *Bull. Agr. Chem. Soc. Jap.* **24**, 261 (1960).
217. R. R. Whetstone, *Proc. Int. Congr. Crop Prot., 4th, 1957*, Vol. 2, p. 1157 (1960).
218. W. E. Hanford and R. M. Joyce, U.S. Patent 2,478,390 (1949).
218a. J. A. Bittles and R. M. Joyce, U.S. Patent 2,559,754 (1951).
219. V. A. Kukhtin, G. Kamai, and L. A. Sinchenko, *Dokl. Akad. Nauk SSSR* **118**, 505 (1958); V. A. Kukhtin and G. Kamai, *Zh. Obshch. Khim.* **28**, 1196 (1958).
220. R. L. McConnell and H. W. Coover, Jr., *J. Amer. Chem. Soc.* **78**, 4453 (1956).
221. R. Brooks and C. A. Bunton, *J. Org. Chem.* **35**, 2642 (1970).
222. E. Jungermann, J. J. McBride, Jr., R. Clutter, and A. Mais, *J. Org. Chem.* **27**, 606 (1962).
223. L. F. Ward, Jr., R. R. Whetstone, G. E. Pollard, and D. D. Phillips, *J. Org. Chem.* **33**, 4470 (1968).
224. A. W. Frank and C. F. Baranaukas, *J. Org. Chem.* **31**, 872 (1966).
225. C. S. Marvel and J. C. Wright, *J. Polym. Sci.* **8**, 495 (1952).
226. J. P. Schroeder, *J. Polym. Sci.* **46**, 547 (1960); G. D. Martin, U.S. Patent 2,386,968 (1945); R. L. Sibley, *India Rubber World* **106**, 246 (1942).
227. R. W. Wetzel and G. L. Kenyon, *J. Amer. Chem. Soc.* **94**, 1774 (1972).
228. M. Bermann and J. R. Van Wazer, *J. Inorg. Chem.* **11**, 209 (1972).
229. M. D. Joesten and Y. T. Chen, *J. Inorg. Chem.* **11**, 429 (1972).
230. V. I. Kirilovich, I. K. Rubtoova, and E. L. Gefter, *Blast. Massy* p. 20 (1962); G. Nishk, H. Holtschmidt, and I. Ugi, German Patent 1,190,186 (1965).
231. Chemische Fabrik Kalk G.m.b.H., Belgian Patent 673,400 (1965).
232. D. Bellus, Z. Manasek, and M. Lazar, *Vysokomol. Soedin.* **5**, 145 (1963).

233. J. L. O'Brian and C. A. Lane, U.S. Patent 3,030,347 (1962); C. A. Lane, U.S. Patent 2,934,554 (1960); J. L. O'Brian, U.S. Patent 2,993,033 (1961).
234. L. H. Chance, W. A. Reeves, and G. L. Drake, Jr., *Text Res. J.* **35**, 291 (1965).
235. M. L. Nielsen *Text Res. J.* **27**, 603 (1957).
236. P. C. Arni and E. Jones, *J. Appl. Chem.* **14**, 221 (1964).
237. R. C. Nametz, *Ind. Eng. Chem.* **59**, 99 (1967).
238. L. Maier, U.S. Patent 3,359,266 (1967); German Patent 1,292,654 (1969).
239. M. Ingenuin and K. R. Molt, U.S. Patent 3,382,301 (1968).
240. Monsanto Co., British Patent 1,130,487 (1968).
241. E. Eimers, German Patent 1,920,293 (1970).
242. M. Grayson, P. T. Keough, and M. M. Rauchut, U.S. Patent 3,332,962 (1967).
243. R. G. Weyker and G. L. Mina, U.S. Patent 3,650,819 (1972).
244. S. Fujii and K. Maeda, *Nippon Kagaku Zasshi* p. 215 (1972).
245. L. M. Kolesova, E. E. Nifanteev, and V. P. Zubov, *Vysokomol. Soedin., Ser. A* **14**, 304 (1972).
246. H. Nachbur and A. Maeder, German Patent 2,139,180 (1970).
247. A. F. Kerst, U.S. Patents 3,645,918 and 3,645,919 (1972).
248. B. A. Trofimov, V. M. Nikitin, A. S. Stavin, and M. Y. Khil'ko, *Zh. Obshch. Khim.* **24**, 342 (1972).
249. K. S. Nurtdinov, R. S. Khairallin, V. S. Tsivunin, T. V. Zykova, and G. Kamai, *Zh. Obshch. Khim.* **24**, 123 (1972).
250. S. R. Hargis, Jr., German Patent 2,028,492 (1971).
251. I. J. Borowitz, S. Firstenberg, E. W. R. Casper, and B. K. Crouch, *J. Org. Chem.* **36**, 3282 (1971).
252. H. E. Podall and T. L. Iapalucci, *J. Polym. Sci., Part B* **1**, 457 (1963).
253. H. Grunze, *Z. Chem.* **6**, 266 (1966); J. Damelson and S. E. Ramussen, *Acta Chem. Scand.* **17**, 1971 (1963); C. E. Wilkes and R. A. Jacobson, *Inorg. Chem.* **4**, 99 (1965).
254. Y. Masei, X. Kato, K. Murayama, and N. Fukui, German Patent 2,111,202 (1971).
255. K. A. Makarov, L. N. Mashlyakovskii, S. D. Shenkov, A. F. Nikolaev, and I. S. Okhrimenko, *Vysokomol. Soedin., Ser. B* **13**, 675 (1971).
256. M. I. Bakhitov, E. V. Kuznetsov, and M. Y. Obryadina, *Sin. Fiz.-Khim. Polim.* **7**, 56 (1970).
257. A. Ohorodnik, U. Dettmeier, K. Sennewald, and H. J. Berns, German Patent 1,943,712 (1971).
258. B. N. Wilson and R. R. Hindersinn, U.S. Patent 3,597,503 (1971).
259. J. P. Moreau and L. H. Chance, *Amer. Dyest. Rep.* **59**, 37 (1970).
260. S. Nogami, M. Ishigami, and J. Kuritz, German Patent 1,809,861 (1969).
261. J. Green and A. P. Kotloby, U.S. Patent 3,450,798 (1969).
262. A. A. R. Sayigh and J. N. Tilley, U.S. Patent 3,458,605 (1969).
263. R. Rabinowitz, German Patent 1,241,995 (1967).
264. E. Kuehn, U.S. Patent 3,639,535 (1972).
265. H. R. Allcock, P. S. Forgione, and K. J. Valan, *J. Org. Chem.* **30**, 947 (1965).
266. L. E. A. Godfrey and J. W. Schappel, *Ind. Eng. Chem., Prod. Res. Develop.* **9**, 426 (1970).
267. G. M. Kosolapoff, U.S. Patent 2,389,576 (1945).
268. R. J. Cotter and M. Matzner, "Ring-Forming Polymerizations," Vol. B,1 and B,2. Academic Press, New York, 1972.
269. G. H. Birum, U.S. Patents 3,014,956 (1961), 3,042,701 (1962), 3,058,941 (1962), and 3,192,242 (1965).

Chapter 14

FREE-RADICAL INITIATORS: DIACYL PEROXIDES

I. INTRODUCTION

A. Free-Radical Initiators

The polymerization of vinyl monomers may usually be initiated by the homolytic scission of a variety of compounds such as diacyl peroxides, organic peroxides, hydroperoxides, and aliphatic azo compounds. Such reagents are usually used in bulk, suspension, or in organic solution polymerizations. In aqueous systems such as aqueous solution or emulsion polymerizations, water-soluble reagents are commonly used, although some organic peroxides have been used in certain redox systems.

Factors which determine the choice of the initiator are many. Among these are the safety in storage and handling of the material, its solubility in the monomer system, the temperature at which the initiator is effective, the nature of fragments derived from the initiator, the extent to which the monomer undergoes induced decomposition, its chain-transfer activity, its oxidizing characteristics, its ability to be activated by ultraviolet light or by aromatic amines both with and without metallic "dryers," the absence of impurities, side reactions of the initiator, etc.

The decomposition of benzoyl peroxides, for example, generates free radicals, as is well known. Some of these free radicals may decarboxylate to produce phenyl radicals. The phenyl radicals may react with other free radicals to form phenyl benzoate. Clearly free radicals which go to form phenyl benzoate (or any of a number of other organic compounds) cannot be effective in initiating the polymerization of a monomer. Therefore, there is an efficiency factor associated with initiators, i.e., the fraction of the generated radicals which actually initiate polymerization.

The efficiency factor may be even more complex in the case of the decomposition of hydroperoxides, where two dissimilar radicals form on thermal scission:

$$\text{R—OOH} \longrightarrow \text{R—O}\cdot + \text{HO}\cdot$$

It would not be surprising to expect different initiating activities of these two radicals. Furthermore, chain transfer with undecomposed hydroperoxide may also enter the picture.

From this brief discussion, it is evident that it is desirable to have a selection of initiators available for study in various polymerizing systems. While manufacturers of initiators supply a modest range of peroxides in several forms and degrees of purity, many other initiators may be of interest. For this reason, it was decided to include in this book chapters on the methods of synthesis of the diacyl peroxides and of the organic hydroperoxides. These two functional groups were selected to present the preparation of symmetrically substituted diacyl peroxides and the highly unsymmetrically substituted hydroperoxides. Furthermore, the hydroperoxides are of particular interest because of their exceptional temperature stability while, by appropriate selection of the starting acids, diacyl peroxides covering a wide range of thermal stabilities may be obtained.

B. Diacyl Peroxides

The term "organic peroxide" has been widely used to denote derivatives of hydrogen peroxide in which one or both of the hydrogens are replaced by organic radicals. Thus among the classes of peroxides are the following.

Alkyl hydroperoxides	$R{-}OOH$
Peroxy acids	$R{-}C({=}O){-}OOH$
Dialkyl peroxides	$R{-}OO{-}R$
Alkyl peroxy esters	$R{-}C({=}O){-}OOR$
Ketone peroxides	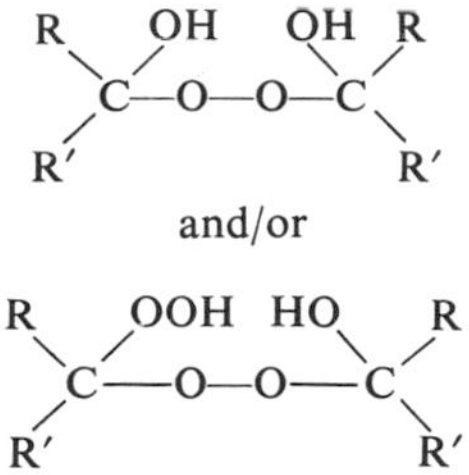 $RR'C(OH){-}O{-}O{-}C(OH)RR'$ and/or $RR'C(OOH){-}O{-}O{-}C(OH)RR'$
Diacyl peroxides	$R{-}C({=}O){-}O{-}O{-}C({=}O){-}R$

Among the diacyl peroxides, the dialkyl peroxydicarbonates,

$$R{-}O{-}C({=}O){-}O{-}O{-}C({=}O){-}O{-}R$$

may also be included.

Frequently the term "peroxide" is used rather loosely to refer to the diacyl peroxides. Furthermore, since relatively few unsymmetrically substituted diacyl peroxides have been prepared, the term "acyl peroxide" has been used for symmetrically substituted derivatives.

The primary interest in the peroxides is in their application as initiators for free-radical addition polymerization. Consequently a large number of monographs and review articles exist and References [1]–[11] are a representative selection of these. Much useful information is also available from the manufacturers of peroxides.

In the manufacture of addition polymers, the concern is with the extent of conversion of the monomer to polymer, the molecular weight distribution of the product, the degree of cross-linking or branching of the polymer chains, the reaction rates, the exothermic nature of the reaction, inhibition, etc. Many of these and related factors can be controlled by the judicious selection of the initiator. In effect, it is the thermal decomposition characteristics of a peroxide which influence the choice of the diacyl peroxides which are to be used in a polymerization process.

It is beyond the scope of this chapter to enter into a detailed discussion of the thermal decomposition of diacyl peroxides, since many summaries of this topic exist (see References [1]–[11], for example). Briefly stated, in nonpolar solvents such as benzene, the overall kinetics of the thermal decomposition of diacyl peroxides appears to be first order. Since there is a definite concentration dependence of the overall first-order rate constants, the reaction mechanism is presumed to be more complex and is said to involve induced decompositions of the peroxide by free radicals generated during the reaction [12,13].

Either by mathematical analysis of the data [12,13] or by use of "inhibitors" [14,15] the effect of the induced decomposition may be obviated and the first-order reaction constant may be studied further. By this means it was found that para and meta substitution of dibenzoyl peroxides affects the first-order decomposition rates. These rates may be correlated to the Hammett σ values. Electron-donating substituents such as methoxy groups tend to stabilize the peroxide bond, while electron-withdrawing groups, such as nitro groups, tend to reduce the stability of the peroxide bond [13–15].

For purposes of selecting an acyl peroxide for polymerizations, the half-life of the peroxide at a given temperature is used. The comparison is usually carried out on data which have been gathered under standard conditions of solvent (usually benzene), concentration, decomposition, and analytical procedures. Thus only the effect of temperature on the overall half-life is determined [16,17].

In Fig. 1, some of the half-life data gathered from a variety of sources are presented in the form of a nomograph.

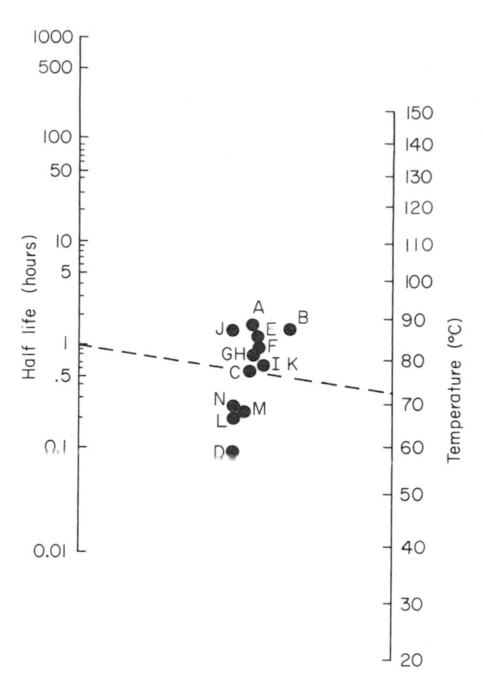

FIG. 1. Nomograph: Half-life of diacyl peroxides vs. temperature. See Table I on p. 450 for key to symbols.

To find the half-life of a diacyl peroxide at a given temperature by use of this nomograph, a line is drawn from the preselected temperature through the point representing the peroxide to the half-life scale. The dashed line in Fig. 1, for example, shows that the half-life of bis(2,4-dichlorobenzoyl) peroxide is 1 hr at 73°C. Table I gives the code letters and half-life data used in the construction of this nomograph.

If half-life data are available for two temperatures it may be entered into this nomograph by finding the intersection of the two lines connecting each half-life with its temperature. The nomograph is based on the well-known relationship of the logarithm of the half-life and the reciprocal of the absolute temperature.

In connection with work on the curing of resins, as well as the polymerization of monomers, it is frequently desirable to know whether an effective concentration of undecomposed peroxide is still present in the composition at a particular temperature. As a rule-of-thumb, we have frequently used the assumption that at a given temperature, the effective concentration of a peroxide approaches zero at 5 to 6 times the half-life of the peroxide at this temperature. Since the nature of the monomer and of the reaction medium does affect the half-life of the initiator, and the viscosity of the medium may

TABLE I
HALF-LIFE DATA FOR THE SERIES OF DIACYL PEROXIDES PRESENTED IN FIG. 1

Nomograph code	Diacyl peroxide	Half-life (hr)	Temp. (°C)	Half-life (hr)	Temp. (°C)
A	Benzoyl	1	91	10	72
B	4-Chlorobenzoyl	1	90	10	78
C	2,4-Dichlorobenzoyl	1	73	10	54
D	Isobutyroyl	1	38	10	21
E	Acetyl	1	84	10	69
F	Propionyl	1	82	10	64
G	Lauroyl	1	80	10	62
H	Decanoyl	1	80	10	61
I	Pelargonoyl	0.5	80	10	62
K	"Isononanoyl"	1	77	10	59
L	Diisopropyl peroxydicarbonate	0.33	60	10	35
M	Di-*n*-propyl peroxydicarbonate	0.55	60	6.6	40
N	Di-*sec*-butyl peroxydicarbonate	0.53	60	5.8	40

affect the internal temperature of the polymerizing system ("gel-effect" or "Trommsdorff effect"), this rule must be used with considerable caution.

The primary method of preparing symmetrically substituted diacyl peroxides consists of the reaction of acyl halides with solutions of sodium peroxide or hydrogen peroxide in the presence of bases (Scheme 1).

SCHEME 1

$$2\mathrm{R{-}\overset{\overset{\displaystyle O}{\|}}{C}{-}Cl} + \mathrm{Na_2O_2} \longrightarrow \mathrm{R{-}\overset{\overset{\displaystyle O}{\|}}{C}{-}O{-}O{-}\overset{\overset{\displaystyle O}{\|}}{C}{-}R} + 2\mathrm{NaCl}$$

(R = alkyl, aryl, or RO)

Other methods of preparation are less well known and consequently may have only limited application. These include the acylation of peroxy acids or their salts, the oxidation of compositions of aldehydes and acid anhydrides, the reaction of peroxyacids and imidazolides, and the reaction of acids with hydrogen peroxide in the presence of *N,N'*-dicyclohexylcarbodiimide. Some of these methods are particularly useful for the preparation of the less well-known unsymmetrically substituted diacyl peroxides.

C. Safety Notes

In considering the safety aspects of all of the peroxides, some distinction must be drawn between the more or less pure compounds and the variety of solutions, pastes, and other mixtures or compositions into which they may enter. Furthermore, the actual quantity of the peroxide at hand may influence the hazards involved. For example, a very small quantity of a diacyl peroxide in an open container may decompose violently but without serious damage to personnel or property. On the other hand, due to local overheating, the decomposition of a similar quantity of a diacyl peroxide may initiate a violent explosion if it is part of a large quantity of material.

If incorrectly handled, the organic peroxides can be hazardous. Consequently only very small quantities of materials should be prepared in the laboratory with due precautions against explosive hazards. Generally they should be protected from all sources of heat and radiation, friction, shock, and contamination from acids, bases, amines, oxidizing and reducing agents, heavy metals, and heavy metal compounds. In connection with the hazards associated with friction, diacyl peroxides should not be stored in screwcap bottles. All of the aromatic diacyl peroxides should be considered hazardous. Of the aliphatic diacyl peroxides, bis(lauroyl peroxide) and bis(decanoyl peroxide) are said to be somewhat less hazardous while bis(acetyl peroxide) and bis(propionyl peroxide) are extremely shock- and friction-sensitive. The latter two peroxides are supplied commercially only in 25% solutions and must be stored at recommended temperatures to prevent crystallization from solution while preventing decomposition at too high a temperature [18]. Since the dialkyl peroxydicarbonates decompose at even lower temperatures, handling of these compounds should follow manufacturers' recommendations [19,20]. Further information on the safety problems associated with peroxides will also be found in References [9], [21], and [22] and in bulletins issued by various agencies concerned with safety, transportation of hazardous materials, insurance, etc. While emphasis is on commercially available peroxides, the data that are available should be used as a guide to handling related compounds.

One aspect which must not be overlooked is the safe disposal of spills, excess material, and solutions used in the preparation of peroxides. Liquids or solutions must never be dumped into a sewer. Usually liquids or solutions may be absorbed on a noncombustible material such as Vermiculite for ultimate disposal by burning in an isolated outdoor location (ignition being carried out with a torch) [18]. The diacyl peroxides themselves usually burn too rapidly to be disposed of by this method. These compounds may be decomposed by cautious hydrolysis with an excess of cold 10% sodium hydroxide solution [18].

The above discussion should not be considered as including definitive recommendations for all situations. It is presented to alert the reader to the problem so that precautions will be taken.

2. CONDENSATION REACTIONS

A. Acylation of Sodium Peroxide

The classical method for the preparation of diacyl peroxides consists of the treatment of an aqueous solution of sodium peroxide with a solution of an acyl chloride [23]. With minor variations, this still represents the most general procedure for the synthesis of the symmetrically substituted compounds.

The method has been well described [24], except that the method of purification, recrystallization from warm solvents, was subsequently considered sufficiently hazardous that a warning concerning this procedure was issued in a later volume of *Organic Syntheses.*

The example given here illustrates several points of technique not generally mentioned in the literature. In particular, it was found that the most reliable results were obtainable only when acyl chlorides which had been fractionally distilled under reduced pressure at least twice were used.

2-1. Preparation of Bis(p-methylbenzoyl) Peroxide [25]

$$2CH_3\text{—}C_6H_4\text{—}\overset{\overset{\large O}{\|}}{C}\text{—}Cl + Na_2O_2 \longrightarrow CH_3\text{—}C_6H_4\text{—}\overset{\overset{\large O}{\|}}{C}\text{—}O\text{—}O\text{—}\overset{\overset{\large O}{\|}}{C}\text{—}C_6H_4\text{—}CH_3 + 2NaCl \quad (1)$$

In a well-ventilated hood, behind a safety shield, a solution of 3.9 gm (0.05 mole) of fresh sodium peroxide in 50 ml of distilled water, which is rapidly stirred with an all-glass stirrer, is cooled with an ice-salt bath to 0°C. While maintaining the temperature at 0°C, to this solution is added slowly a solution of 16.6 gm (0.0974 mole) of doubly distilled *p*-toluoyl chloride in 50 ml of dry toluene. After the addition has been completed, stirring at 0°C is continued for a total period of 4 hr from the beginning of the addition.

The reaction mixture is filtered through a sintered glass funnel. The product is preserved in a refrigerator while the filtrate is stirred for an additional 12 hr. The solid which separates from the filtrate is collected and combined with the first crop of the product and carefully but cautiously dried between filter papers. (Metal spatulas should not be used in transferring the product.)

The combined product is dissolved in a minimal quantity of chloroform. The solution is filtered, if necessary, and treated with absolute ethanol to reprecipitate the product. The product is filtered off and air-dried. Yield: 12.1 gm (92%); m.p. (on a Dennis-Parr bar) 143°C (dec.), purity 96% or better.

Table II lists a number of substituted benzoyl peroxides prepared by similar reaction procedures, although the purification procedures may have varied

TABLE II
PROPERTIES OF BIS(SUBSTITUTED BENZOYL) PEROXIDES PREPARED BY ACYLATION OF SODIUM PEROXIDE

Substituent	M.p., °C (dec.) (% purity)	Ref.
(H)	107° (98), 106–107° (98.5), 105° (100 ± 1)	13,15,26
p-Chloro	139° (95), 140° (100), 141° (100 ± 1)	13,15,26
p-Methyl	143° (96), 136° (100), 137° (100)	13,15,26
p-Methoxy	127° (98), 128° (99.2), 126°–127°	13,15,24
p-Nitro	158° (92), 157°–158° (99), 156°, 158° (100 ± 1)	13,15,24,26
m-Nitro	136°–137°, 137° (98)	15,24
p-Bromo	144°	21
2,4,5-Tribromo	186°	24
m-Methoxy	82°–83° (99.2)	15
m-Chloro	123° (100)	15
m-Methyl	54° (99.1)	15
3,5-Dinitro	161°–162° (100)	15
o-Chloro	95° (99.9)	15
o-Nitro	145° (99.8)	15
o-Methyl	52.5°–53.5° (99.5)	15
o-Methoxy	85°–86° (98.7)	15
p-Phenoxy	66°–67° (97.1)	15
p-Cyano	178° (99.9)	15

[13,15,24,26]. It should be mentioned that bis(acetylsalicylyl) peroxide could not be prepared satisfactorily by this general method [24] and that the ortho-substituted diacyl peroxides tend to decompose on standing at room temperature, although this decomposition was not explosive in nature [15].

By this general procedure even complex diacyl peroxides such as bis(4-phenylazobenzoyl) peroxide,

$$C_6H_5-N{=}N-C_6H_4-\overset{O}{\overset{\|}{C}}-O-O-\overset{O}{\overset{\|}{C}}-C_6H_4-N{=}N-C_6H_5$$

(decomposition temperature, 148°–150°C)

and its 3-isomer (decomposition temperature 102°–103°C) have been prepared [27].

In the determination of decomposition points of the diacyl peroxides, we recommend the use of an electrically heated metal bar such as the Dennis-Parr bar. Results are quite reproducible and the hazard of damaging explosions is reduced. Furthermore, the rate of heating does not affect the decomposition point as materially as might be the case when a capillary melting point is determined. We are aware that capillary melting procedures have been used in these determinations; the end point is readily noted by an explosion. Obviously this is an extremely hazardous approach and must not be condoned.

Aliphatic diacyl peroxides have also been prepared by the treatment of sodium peroxide with acyl halides. In this connection it must again be pointed out that bis(acetyl) peroxide ("acetyl peroxide") and bis(propionyl) peroxide are extremely dangerous compounds. They are sensitive to friction and shock. These compounds are only handled in solution at concentrations no higher than 25% by weight. Since bis(lauroyl) peroxide is relatively stable, no definite statement can be made as to the minimum number of carbon atoms in an aliphatic diacyl peroxide which is required to render the molecule reasonably stable. We suggest that, because of the serious explosion hazards, certainly work with bis(acetyl) peroxide, or with bis(propionyl) peroxide, be avoided, if at all possible. If this cannot be done, only dilute solutions should be used in small quantities and with extreme protective measures. Examples of the preparation of the higher aliphatic diacyl peroxides are given here. While no difficulties have been reported with these, precautions still should be taken.

It will be noted that in some of the preparations given in this chapter, ether solutions are used in the presence of organic peroxides, hydrogen peroxide, and/or sodium peroxide. We consider this an unsafe procedure since the possibility of the formation of the explosive hydroperoxide of diethyl ether exists, particularly if prolonged reaction times are used. It is suggested that the use of alternative solvents be explored.

2-2. Preparation of Bis(n-heptanoyl) Peroxide [28]

$$2CH_3(CH_2)_5\overset{\overset{\displaystyle O}{\|}}{C}—Cl + Na_2O_2 \longrightarrow CH_3(CH_2)_5—\overset{\overset{\displaystyle O}{\|}}{C}—O—O—\overset{\overset{\displaystyle O}{\|}}{C}—(CH_2)_5CH_3 + 2NaCl \quad (2)$$

SAFETY NOTE: In regard to the use of ether in the presence of oxidizing agents the reader is referred to the paragraph immediately preceding Preparation 2-2.

In a hood, behind a safety shield, to a stirred solution of 113 gm (1.4 moles) of sodium peroxide in 1 liter of ice-water, maintained between 0° and 2°C, is added over a 45 min period a solution of 104 gm (0.70 mole) of redistilled *n*-heptanoyl chloride in 300 ml of anhydrous ether (see Note). After the addition has been completed, stirring is continued at 3°–5°C for 1 hr. Then the ether layer is separated and washed in turn with cold water, a cold solution of sodium bicarbonate, and again with cold water. After drying over magnesium sulfate, the ether is removed under reduced pressure and finally in a slow air stream. Yield: 75 gm (80%), purity 97.6 ± 0.5%; n_D^{25} 1.4340 (96% pure) [26].

NOTE: As described earlier it is safer to use an alternate solvent that does not form peroxides or hydroperoxides.

Acyl halides have also been added without prior solution to sodium peroxide to produce diacyl peroxides. Thus, for example bis(*endo*-norbornene-5-carbonyl) peroxide [m.p. 45°–47°C (dec.)] has been prepared by the addition of the neat acyl chloride to a slurry of sodium peroxide in ether at 1°–2°C with occasional additions of small amounts of water [29]. Preparation 2-3 is another example of this approach which is particularly interesting since it indicates the relative reactivities of two different kinds of bromine substituents.

2-3. Preparation of Bis(bromoacetyl) Peroxide [30]

$$2BrCH_2\overset{O}{\overset{\|}{C}}—Br + Na_2O_2 \longrightarrow BrCH_2\overset{O}{\overset{\|}{C}}—O—O—\overset{O}{\overset{\|}{C}}—CH_2Br + 2NaBr \quad (3)$$

SAFETY NOTE: See Preparation 2-2.

In a hood, behind a safety shield, to 10 ml of a stirred aqueous solution containing 0.5 gm (0.064 mole) of sodium peroxide maintained at −5°C is added, over a 10 min period, 2.3 gm (0.011 mole) of bromoacetyl bromide. After the addition has been completed, stirring is continued for 10 min. The product is filtered off. Yield: 0.46 gm (30%), m.p. 47.5°C (dec.) (99% purity). IR spectrum, 1818 and 1786 cm^{-1} (C=O); 877 and 862 cm^{-1} (—O—O—).

Since low yields are sometimes associated with impurities in the acyl halides and with the use of excessive amounts of sodium peroxide, it is, therefore suggested that these two points be reevaluated in this particular type of synthesis.

Attention has been focused on the importance of the pH of the reaction. Thus, for example, mention is made that the cyclic phthaloyl peroxide has been prepared from phthaloyl chloride in chloroform solution with aqueous sodium peroxide in the presence of a phosphate buffer [31].

B. Acylation of Hydrogen Peroxide in Basic Media

With the increased availability of hydrogen peroxide solutions in higher concentrations, a greater interest in their use in the preparation of diacyl peroxides has arisen. This reagent adds flexibility to the procedures by permitting control of the base concentration independently of the peroxide concentration, by making the use of organic bases possible, and by allowing reactions to be carried out in certain organic solvent systems.

In an aqueous system, treatment of hydrogen peroxide with sodium hydroxide may constitute an *in situ* preparation of aqueous sodium peroxide.

Bis(acetylsalicylyl) peroxide, which was difficult to prepare from the acyl chloride with aqueous sodium peroxide, could be prepared in acetone solution with hydrogen peroxide, for example [24].

A variety of the unstable esters of peroxydicarbonic acid,

$$R{-}O{-}\overset{\overset{\displaystyle O}{\|}}{C}{-}O{-}O{-}\overset{\overset{\displaystyle O}{\|}}{C}{-}OR$$

have been prepared from the corresponding alkyl chloroformates. Some of these esters are currently available commercially because of their importance in the polymerization of vinyl chloride.

Purchase of these materials from their suppliers is a fascinating example of industrial logistics. After assurances to the supplier that safe and adequate refrigerated storage facilities are at hand, a purchased quantity is delivered from a special refrigerated truck which has detailed delivery routes to a series of customers in a given part of the country. After prior notice of the exact date and time of delivery, the truck arrives quite reasonably on schedule. Since any one location can obviously only be serviced on one particular day each week, the flexibility required for research is lost. While the suppliers are generally very accommodating and helpful, to set this elaborate machinery in motion for the modest and intermittent requirements of a laboratory is a problem.

Preparation 2-4 gives a laboratory procedure for the preparation of diisopropyl peroxydicarbonate (IPP). For a discussion of safety and health aspects of this compound, Reference [20] should be consulted.

2-4. Preparation of Diisopropyl Peroxydicarbonate [32]

$$2(CH_3)_2CHO{-}\overset{\overset{\displaystyle O}{\|}}{C}{-}Cl + H_2O_2 \xrightarrow{2NaOH} (CH_3)_2CHO{-}\overset{\overset{\displaystyle O}{\|}}{C}{-}O{-}O{-}\overset{\overset{\displaystyle O}{\|}}{C}{-}OCH(CH_3)_2 + 2NaCl + 2H_2O \quad (4)$$

SAFETY NOTE: The product melts between 9° and 10°C when pure and decomposes with a steady evolution of gases at 35°–38°C. Therefore low temperatures should be maintained throughout the preparation and precautions against explosions should be taken. The manufacturers' recommendations on handling hydrogen peroxide solutions should be observed.

(*a*) *Preparation of "sodium peroxide" solution.* To a cooled solution of 44.0 gm (1.10 moles) of sodium hydroxide in 300 ml of water is added 68.2 gm (0.55 mole) of 27.4% hydrogen peroxide (analyzed iodometrically just prior to use) while keeping the temperature between 10° and 15°C. The small quantity of solid which separates (probably sodium peroxide octahydrate) is maintained in suspension by stirring.

(*b*) *Acylation.* To 122.6 gm (1.0 mole) of vigorously stirred isopropyl chloroformate, cooled with an ice-salt mixture, the sodium peroxide mixture from (a) is added through a dropping funnel at such a rate that the internal temperature of the reaction mixture is maintained carefully between 6° and 10°C. After the addition has been completed, stirring is continued for 0.5 hr.

A sample of the oily product is withdrawn and tested for completeness of reaction by adding the sample to cold 20% aqueous pyridine followed by acidification with nitric acid and testing for chloride ions with silver nitrate solution. Only when the oil product is free from chloride ions is the separation of the product continued as follows.

The oil is separated, washed with cold distilled water (maintained between 5° and 10°C) until the water washes are free of chloride ions. The oil is then dried by adding cold anhydrous sodium sulfate and maintaining the temperature between 5° and 10°C. The oil is filtered and preserved in Dry Ice. Yield: 83–92 gm (81–89%).

In those cases where this type of reaction is to be carried out either with a solid chloroformate or to produce a solid product, a volatile solvent which may either be washed out with cold water or removed by evaporation under reduced pressure may be used, e.g., ether or ethyl acetate.

Table III gives the properties of a series of dialkyl peroxydicarbonates prepared by this method.

Using a more dilute mixture of sodium hydroxide and 30% hydrogen peroxide, heptane solutions of aliphatic and olefinic acyl chlorides have also been converted to the corresponding diacyl peroxides. Unfortunately no physical constants for these products were obtained. Thermal decomposition data for bis(nonanoyl), bis(2-nonenoyl), bis(4-ethyl-2-octenoyl), bis(3-nonenoyl), bis(2-ethyl-2-hexenoyl), bis(2-ethyl-4-methylpentenoyl), and bis(2-ethylhexanoyl) peroxides were presented [33].

A procedure in which solutions of long-chain acyl chlorides are simultaneously treated with sodium hydroxide solutions and hydrogen peroxide has been described. The method suffers from the formation of emulsions which are difficult to break and from low yields. Both difficulties are probably interrelated and caused by the formation of soaps from the long-chain acyl chlorides [34]. The method is given here since it may find use in the preparation of lower molecular weight diacyl peroxides, possibly by more careful control

TABLE III

PROPERTIES OF DIALKYL PEROXYDICARBONATES, $ROC(=O)—O—O—C(=O)OR$

R	Scale of preparation (moles)	Yield (%)	m.p. (°C)	Refractive index n_D (20°C)	Ref.
Methyl	0.1	6	Explosive compound	—	32
Ethyl	5.0	81	—	1.4017	32
n-Propyl	0.5	81	< -70	1.4106	19,32
Isopropyl	0.5–6.0	81–89	8–10	1.4034	19,32
Isobutyl	1.3	75	—	1.4148	32
sec-Butyl	—	—	< -80	1.4132	19
Neopentyl	0.08	67	45	—	32
Cyclohexyl	0.15	72	46	—	32
Allyl	0.05	46	Explosive compound	1.434	32
Tetrahydrofurfuryl	0.05	—	Decomposes between 10° and 20°C	—	32
Benzyl	0.03	90	101–102	—	32
β-Methoxyethyl	1	28	Explodes at 34°C	1.4250	32
α-Carbethoxyethyl	0.25	27	—	1.4266	32
2-Chloroethyl	0.1	79	—	1.4582	32
2-Carbamyloxyethyl	0.05	58	Solid at 25°C	—	32
2-Nitrobutyl	0.1	85	50	—	32
2-Nitro-2-methylpropyl	0.4	78	100.5–101	—	32

of the pH during the addition stage or by the use of a cationic surfactant to suppress emulsion formation by the anionic soaps.

2-5. *Preparation of Bis(palmitoyl) Peroxide by Simultaneous Addition of Aqueous Sodium Hydroxide and Hydrogen Peroxide* [34]

$$2CH_3(CH_2)_{14}C(=O)—Cl + H_2O_2 \xrightarrow{2NaOH} CH_3(CH_2)_{14}C(=O)—O—O—C(=O)(CH_2)_{14}CH_3 + 2NaCl + 2H_2O \quad (5)$$

With the safety precautions mentioned generally throughout this chapter, to a solution of 20 gm (0.073 mole) of palmitoyl chloride in 20 ml of chloroform cooled to 5°C is added simultaneously with vigorous stirring, 11.6 gm (0.0725 mole) of a 25% solution of aqueous sodium hydroxide and 2.5 gm (0.037 mole) of 50% aqueous hydrogen peroxide. The temperature is maintained below 15°C throughout the addition. After the addition has been completed, stirring is continued for 1 hr. The mixture is then poured into a mixture of chloroform and water and partially acidified to break the emulsion which forms.

The chloroform solution is separated, washed with water and dried over anhydrous sodium sulfate, filtered, and freed of solvent by evaporation under reduced pressure. The residue represents a 77% yield of 64% purity. The product is difficult to purify by direct crystallization.

A Polish report describes the preparation of bis(benzoyl) peroxide from benzoyl chloride, hydrogen peroxide, and sodium peroxide at 1°–6°C over a 2.5–3 hr period. In this paper the use of a 5% excess of sodium peroxide is recommended [35]. In a recent patent, benzoyl chloride is treated with hydrogen peroxide and aqueous ammonia while maintaining a pH between 8.0 and 9.5 at 5°–15°C followed by an adjustment of pH, after 2 hr, to 8.0–8.5 prior to isolation of the product [36].

A method of synthesis which is carried out in a substantially homogeneous reaction system makes use of the extensive solubility of 60% hydrogen peroxide in ether, particularly in the presence of pyridine. This procedure is said to give high yields of essentially pure diacyl peroxides. We have already commented on the problem associated with the use of ether in the presence of oxidizing agents and therefore suggest that the procedure be reexamined with other solvents which may also have a high solubility for hydrogen peroxides but a lower tendency to form dangerous hydroperoxides.

Care must also be taken in working up the product solution because of the solubility of amine salts in moist ether solutions.

2-6. Preparation of Bis(myristoyl) Peroxide [34]

$$2CH_3(CH_2)_{12}\overset{\overset{\displaystyle O}{\|}}{C}Cl + H_2O_2 \xrightarrow{\text{2 pyridine}} CH_3(CH_2)_{12}\overset{\overset{\displaystyle O}{\|}}{C}—O—O—\overset{\overset{\displaystyle O}{\|}}{C}C(CH_2)_{12}CH_3 + H_2O \quad (6)$$

With the safety precautions mentioned generally throughout this chapter, to a solution of 24.7 gm (0.10 mole) of myristoyl chloride in 175 ml of ether (see Note), cooled to 0°C, is added 4.25 gm (0.075 mole) of 60% hydrogen peroxide. To this stirred mixture is added dropwise, while maintaining the temperature between 0° and 5°C, 9.5 gm (0.12 mole) of pyridine. During this addition, the product precipitates as it forms. If necessary more ether is added to facilitate stirring of the slurry. After the addition has been completed, the cooling bath is removed and stirring is continued for a total period of 1 hr. By adding more ether, a homogeneous solution is formed at room temperature.

The ether solution is washed in turn with dilute hydrochloric acid, 5% potassium bicarbonate, and water. After drying with anhydrous sodium sulfate and filtration, the solvent may be evaporated under reduced pressure. The yield (by titration of the purified ether solution) is said to be 98%.

NOTE: Solvents other than ether are preferred for safety reasons as described earlier.

Recently this method was carried out in a cold room to prepare bicyclo-[2.2.2]octane-1-formyl peroxide [36a].

Variations of Procedure 2-6 include the following:

1. Addition of aliphatic acyl chloride at $-5°$ to $-10°C$ to a cold mixture of pyridine, ether, and 30% hydrogen peroxide [37].

2. Reaction of a mixture of an aliphatic acyl chloride with a mixture of 30% hydrogen peroxide in ether [38].

In both 1 and 2 the final ether solutions were concentrated on a rotary evaporator under reduced pressure.

3. The use of *n*-heptane or benzene as the reaction solvent for the preparation of aliphatic diacyl peroxides by reaction of acyl halides, hydrogen peroxide, and sodium hydroxide has been patented [39]. This procedure might be extended to reactions in which the neutralization is carried out with pyridine.

Table IV lists the properties of a series of aliphatic diacyl peroxides.

TABLE IV

PROPERTIES OF ALIPHATIC DIACYL PEROXIDES, $RC(=O)—O—O—C(=O)R$

R	M.p. (°C) (dec.)	Refractive index (n_D^{30})	Specific gravity (d_4^{30})	Ref.
CH_3	30	—	—	7
$CH_3(CH_2)_5$	—	n_D^{25} 1.4340	—	26,28
$CH_3(CH_2)_6$	21.8–22.4	1.4363	0.9275	34
$CH_3(CH_2)_7$	—	n_D^{25} 1.4410	—	30
	13.0–13.5	1.4388	0.9182	34
$CH_3(CH_2)_8$	40.5–41.0	—	—	34
$CH_3(CH_2)_{10}$	54.7–55.0	—	—	34
$CH_3(CH_2)_{11}$	48.3–48.8	—	—	34
$CH_3(CH_2)_{12}$	63.9–64.4	—	—	34
$CH_3(CH_2)_{14}$	71.4–71.9	—	—	34
$CH_3(CH_2)_{16}$	76.5–76.9	—	—	34
$BrCH_2$	47.5	—	—	30
$ClCH_2$	85	—	—	7

C. Acylation of Peroxy Acids and Their Salts

The acylation of sodium peroxide or hydrogen peroxide leads to symmetrically substituted diacyl peroxides. On the other hand, acylation of peroxy acids or their salts offers a method for the preparation of either symmetrically or unsymmetrically substituted diacyl peroxides. Thus, if lauroyl chloride is allowed to react with perlauric acid, the product is the symmetric bis(lauroyl) peroxide. If, on the other hand, *p*-chloroperbenzoic acid is acylated with bromoacetic anhydride, the product is the unsymmetrical bromoacetyl-*p*-chlorobenzoyl peroxide.

The limitation of this method is the availability of peroxy acids. The preparation of these acids is reviewed in Reference [1]. Perhaps the best method involves the acid-catalyzed reaction of carboxylic acids with hydrogen peroxide. Since the stability of organic peroxy acids is variable, due precautions must, of course, be taken in preparing and handling these compounds.

2-7. *Preparation of Bis(lauroyl) Peroxide* [34]

$$CH_3(CH_2)_{10}\overset{\overset{\displaystyle O}{\|}}{C}—O—O—H + CH_3(CH_2)_{10}\overset{\overset{\displaystyle O}{\|}}{C}—Cl \xrightarrow{\text{pyridine}}$$

$$CH_3(CH_2)_{10}\overset{\overset{\displaystyle O}{\|}}{C}—O—O—\overset{\overset{\displaystyle O}{\|}}{C}—(CH_2)_{10}CH_3 + C_5H_5N\cdot HCl \quad (7)$$

With all due precautions, to a solution of 4.80 gm (0.0222 mole) of perlauric acid in 54 ml of ether (see Note), cooled to 0°C is added 4.87 gm (0.0222 mole) of lauroyl chloride. While maintaining the reaction system between 0° and 5°C, with stirring, 2.1 gm (0.0266 mole) of pyridine is added dropwise. After completion of the addition, the ice bath is removed and stirring is continued for 40 min. The ether solution is washed with water and dried over anhydrous sodium sulfate. A peroxide content representing an 89.3% yield is found by analysis. The product may be isolated by evaporation of the solvent under reduced pressure. Recrystallization from a 1:1 mixture of ether–petroleum naphtha is mentioned in the literature.

NOTE: Solvents other than ether should be used, if possible, for safety reasons.

An example of the preparation of an unsymmetrically substituted diacyl peroxide is given in Preparation 2-8. In this case an anhydride is used instead of the more common acyl chloride.

2-8. *Preparation of Bromoacetyl p-Chlorobenzoyl Peroxide* [30]

With the usual precautions, a solution of 4.4 gm (0.025 mole) of *p*-chloroperbenzoic acid in 150 ml of methylene chloride is shaken with 100 ml of a

$$BrCH_2-\overset{O}{\overset{\|}{C}}-O-\overset{O}{\overset{\|}{C}}-CH_2Br + Cl-C_6H_4-\overset{O}{\overset{\|}{C}}-OONa \longrightarrow BrCH_2-\overset{O}{\overset{\|}{C}}-O-O-\overset{O}{\overset{\|}{C}}-C_6H_4-Cl + NaO\overset{O}{\overset{\|}{C}}CH_2Br \quad (8)$$

4% solution of aqueous sodium hydroxide. The aqueous solution is retained in a separatory funnel. To this aqueous solution is added a solution of 6.7 gm (0.025 mole) of α,α'-dibromoacetic anhydride in 75 ml of methylene chloride. The mixture is shaken. The organic layer is drawn off and dried over anhydrous magnesium sulfate. After filtration, the solvent is removed at room temperature under reduced pressure. The residual product weighs 2.73 gm (29% yield); m.p. 36°C (dec.) (97% purity); IR spectrum: 1793 and 1767 cm^{-1} (C=O); 855 and 848 cm^{-1} (—O—O—).

A somewhat larger quantity of an unsymmetrically substituted diacyl peroxide has been prepared by more conventional procedures, as in Preparation 2-9.

2-9. Preparation of 3-Thienoyl Benzoyl Peroxide [40]

$$C_4H_3S-\overset{O}{\overset{\|}{C}}-Cl + \left[C_6H_5-\overset{O}{\overset{\|}{C}}-O-O\right]^- Na^+ \longrightarrow C_4H_3S-\overset{O}{\overset{\|}{C}}-O-O-\overset{O}{\overset{\|}{C}}-C_6H_5 + NaCl \quad (9)$$

With the usual safety precautions, a solution of 13.8 gm (0.086 mole) of sodium perbenzoate in approximately 100 ml of water in a 500 ml flask is cooled with a Dry Ice–isopropanol bath to −5°C. While this solution is rapidly stirred and maintained at −5°C a solution of 12.7 gm (0.086 mole) of 3-thienoyl chloride in approximately 50 ml of cyclohexane (enough solvent is used to dissolve the acyl chloride completely) is added from a dropping funnel. After the addition has been completed, stirring is continued at −5°C for another hour. The white product is filtered off, washed repeatedly with ice cold water, and air-dried under reduced pressure. Yield: 14.3 gm (66%), m.p. 102°–102.3°C (dec.).

Table V lists a number of unsymmetrical diacyl peroxides prepared by this technique.

TABLE V
PROPERTIES OF UNSYMMETRICAL DIACYL PEROXIDES [40]

$$R-\overset{O}{\overset{\|}{C}}-O-O-\overset{O}{\overset{\|}{C}}-C_6H_5$$

$R-\overset{O}{\overset{\|}{C}}-$	Yield (%)	M.p. (°C) (dec.)
2-Furoyl	20	57.5–58
2-Thienoyl	83	92–92.5
3-Thienoyl	66	102–102.3
3-Methyl-2-thienoyl	53	49.5–50

Perbenzoyl phenyl carbonate has been prepared by the reaction of phenylchloroformate with barium perbenzoate in ether suspension at 0°C [m.p. 60°C (dec.)] [41].

$$C_6H_5-O-\overset{O}{\overset{\|}{C}}-O-O-\overset{O}{\overset{\|}{C}}-C_6H_5$$

Other perbenzoyl aryl carbonates were prepared by the method of Preparation 2-10.

2-10. Preparation of Perbenzoyl p-Tolyl Carbonate [41]

$$C_6H_5-\overset{O}{\overset{\|}{C}}-O-OH + CH_3-C_6H_4-O-\overset{O}{\overset{\|}{C}}-Cl \xrightarrow{\text{pyridine}}$$

$$CH_3-C_6H_4-O-\overset{O}{\overset{\|}{C}}-O-O-\overset{O}{\overset{\|}{C}}-C_6H_5 + C_5H_5N\cdot HCl \qquad (10)$$

With the safety precautions discussed above, to a solution of 4.8 gm (0.0348 mole) of perbenzoic acid in 75 ml of methylene chloride maintained at −20°C is added 5.5 gm (0.0322 mole) of *p*-tolyl chloroformate. To the stirred solution is added a solution of 4 ml of dry pyridine in 25 ml of methylene chloride over a 2 min period. The reaction mixture is washed in turn with 1% sulfuric acid and ice-water. The organic solution is dried with sodium sulfate and filtered. The solvent is evaporated under reduced pressure. The residue is recrystallized from light petroleum to afford 7.0 gm (80%), m.p. 84°C.

By a technique similar to Preparation 2-10, except that the reaction was carried out through the drying stage at $-35°C$, perbenzoyl 2,6-dimethylphenyl carbonate (m.p. 49°C) was also prepared.

D. Acylation of Peroxy Acids with Imidazolides and Related Preparations

In an extension of a study of derivatives of imidazoles interesting methods for the preparation of diacyl peroxides, both of the symmetrical and the unsymmetrical type, have been developed.

Basically the reaction involves the acylation of hydrogen peroxide with imidazolides in tetrahydrofuran (THF) at room temperature. The reaction sequence may be represented by Eqs. (11) and (12) [42].

$$\text{Im}\text{—}\overset{\text{O}}{\overset{\|}{\text{C}}}\text{—R} + \text{H}_2\text{O}_2 \longrightarrow \text{ImH} + \text{R—}\overset{\text{O}}{\overset{\|}{\text{C}}}\text{—OOH} \quad (11)$$

$$\text{R}\overset{\text{O}}{\overset{\|}{\text{C}}}\text{—OOH} + \text{Im}\text{—}\overset{\text{O}}{\overset{\|}{\text{C}}}\text{—R} \longrightarrow \text{ImH} + \text{R}\overset{\text{O}}{\overset{\|}{\text{C}}}\text{—OO—}\overset{\text{O}}{\overset{\|}{\text{C}}}\text{R} \quad (12)$$

Since the imidazolides may be prepared directly by reaction of a carboxylic acid and *N,N′*-dicarbonyl diimidazole, in principle a "single reactor" procedure for the preparation of symmetrical diacyl peroxide is possible (Eq. 13).

$$2\text{R—}\overset{\text{O}}{\overset{\|}{\text{C}}}\text{—OH} + \text{Im—}\overset{\text{O}}{\overset{\|}{\text{C}}}\text{—Im} \xrightarrow{\text{H}_2\text{O}_2} 2\,\text{ImH} + \text{R—}\overset{\text{O}}{\overset{\|}{\text{C}}}\text{—O—O—}\overset{\text{O}}{\overset{\|}{\text{C}}}\text{—R} + \text{CO}_2 + \text{H}_2\text{O} \quad (13)$$

Equation (12) is the basis for the preparation of unsymmetrical diacyl peroxides. However, it was discovered that the reaction of peroxy acids with imidazolides leads to a mixture of diacyl peroxides, as represented by Eq. (14).

$$\text{R—}\overset{\text{O}}{\overset{\|}{\text{C}}}\text{—OOH} + \text{Im—}\overset{\text{O}}{\overset{\|}{\text{C}}}\text{—R}' \longrightarrow \text{R}\overset{\text{O}}{\overset{\|}{\text{C}}}\text{—OO—}\overset{\text{O}}{\overset{\|}{\text{C}}}\text{—R}',\ \text{R}\overset{\text{O}}{\overset{\|}{\text{C}}}\text{—OO—}\overset{\text{O}}{\overset{\|}{\text{C}}}\text{—R},\ \text{R}'\text{—}\overset{\text{O}}{\overset{\|}{\text{C}}}\text{—O—O—}\overset{\text{O}}{\overset{\|}{\text{C}}}\text{—R}' \text{ and ImH} \quad (14)$$

(Im = 1-imidazolyl ring; ImH = imidazole, drawn as ring structures in the original.)

Further research showed that the imidazole which formed in the first step of the reaction catalyzed a transacylation reaction which resulted in the formation of mixed products [42,43]. To overcome this difficulty, a procedure was devised in which imidazole is precipitated with maleic acid as it forms.

2-11. *Preparation of Bis(benzoyl) Peroxide (Dibenzoyl Peroxide or Benzoyl Peroxide) from Benzoic Acid (Single Reactor Method)* [42]

$$2\,C_6H_5\text{—}\overset{O}{\overset{\|}{C}}\text{—OH} + (C_3H_3N_2)\text{—}\overset{O}{\overset{\|}{C}}\text{—}(C_3H_3N_2) \xrightarrow{H_2O_2} 2\,C_3H_3N_2H + C_6H_5\text{—}\overset{O}{\overset{\|}{C}}\text{—O—O—}\overset{O}{\overset{\|}{C}}\text{—}C_6H_5 + H_2O + CO_2 \quad (15)$$

With the usual precautions discussed above, a solution of 8.1 gm (50 mmoles) of *N*,*N'*-carbonyldiimidazole in 40 ml of tetrahydrofuran is slightly warmed with 6.1 gm (50 mmoles) of benzoic acid. After the evolution of carbon dioxide has subsided, the reaction mixture is cooled in running water. The solution is then shaken with 29.2 ml (25 mmoles) of an ethereal solution of hydrogen peroxide (prepared by shaking peroxide-free ether repeatedly with 30% hydrogen peroxide until the analysis of the $MgSO_4$-dried solution analyzes as 2.91 gm of hydrogen peroxide in 100 ml or solution)

The mixture is maintained at room temperature for 1 hr and then evaporated under reduced pressure. The residue is washed repeatedly with cold water and finally dissolved in a minimum quantity of chloroform. The solution is filtered and treated with methanol to precipitate the product. The crude yield prior to precipitation is 5.15 gm (m.p. 104°C). Since this is more than the theoretical yield, this product requires thorough purification.

By similar procedures, the following diacyl peroxides were prepared [42]: bis(*p*-methylbenzoyl) peroxide (89–91% yield), m.p. 136.5°C (dec.); bis(*p*-chlorobenzoyl) peroxide (79.5–85% yield), m.p. 139.5°–140°C (dec.); bis-(lauroyl) peroxide (64–88% yield), m.p. 55.8°–56.3° (dec.); bis(*p*-nitrobenzoyl) peroxide (87% yield), m.p. 153.5°C (dec.); bis(cinnamoyl) peroxide (63% yield), m.p. 134°C (dec.); bis(palmitoyl) peroxide, m.p. 70.5°–71°C (dec.).

2-12. *Preparation of Benzoyl p-Chlorobenzoyl Peroxide* [42]

$$C_6H_5\text{—}\overset{O}{\overset{\|}{C}}\text{—OOH} + (C_3H_3N_2)\text{—}\overset{O}{\overset{\|}{C}}\text{—}C_6H_4\text{—Cl} \xrightarrow[\text{THF}]{\text{maleic acid}} C_6H_5\text{—}\overset{O}{\overset{\|}{C}}\text{—O—O—}\overset{O}{\overset{\|}{C}}\text{—}C_6H_4\text{—Cl} + C_3H_3N_2H\cdot HOCOCH{=}CHCOOH \quad (16)$$

With the usual precautions, to a stirred solution of 4.12 gm (20 mmoles) of *N*-(*p*-chlorobenzoyl)imidazole in tetrahydrofuran is added dropwise, with stirring, at room temperature, a solution of 3.1 gm (20 mmoles) of 89% pure perbenzoic acid and 2.55 gm (22 mmoles) of maleic acid in 50 ml of THF. After 90 min, the imidazolium maleate is separated and the filtrate is evaporated to dryness under reduced pressure. The residue is repeatedly extracted with a saturated sodium bicarbonate solution. The crude residue weighs 4.16 gm (75%), m.p. 84.5°–85°C (dec.). On dissolving in chloroform and reprecipitating with methanol, the melting point is raised to 85°–85.5°C (dec.).

Other products prepared by this method were [42] benzoyl *p*-nitrobenzoyl peroxide, m.p. 113°–114°C (dec.); benzoyl palmitoyl peroxide, m.p. 46°–47°C (dec.); benzoyl *o*-methylbenzoyl peroxide, m.p. 34°–35.5°C (dec.); benzoyl *p*-methylbenzoyl peroxide, m.p. 89.5°–90.5°C (dec.); *p*-nitrobenzoyl *p*-methoxybenzoyl peroxide, m.p. 106°–107°C (dec.).

E. Acylation of Hydrogen Peroxide in the Presence of *N,N'*-Dicyclohexylcarbodiimide

Carboxylic acids may be condensed with hydrogen peroxide in the presence of *N,N'*-dicyclohexylcarbodiimide to produce symmetrically substituted diacyl peroxides. Unsymmetrically substituted diacyl peroxides are formed on treatment of mixtures of carboxylic acids and peroxy acids with *N,N'*-dicyclohexylcarbodiimide. This procedure may be carried out in nonaqueous media. It also does not require the presence of a base [44]. Equation (17) indicates the courses of the reactions involved.

$$2R{-}\overset{O}{\overset{\|}{C}}{-}OH + H_2O_2 + 2R'{-}N{=}C{=}N{-}R' \longrightarrow R{-}\overset{O}{\overset{\|}{C}}{-}O{-}O{-}\overset{O}{\overset{\|}{C}}{-}R + 2R'NH\overset{O}{\overset{\|}{C}}NHR'$$

$$R\overset{O}{\overset{\|}{C}}{-}OH + R\overset{O}{\overset{\|}{C}}{-}O{-}OH + R'{-}N{=}C{=}N{-}R' \longrightarrow R{-}\overset{O}{\overset{\|}{C}}{-}O{-}O{-}\overset{O}{\overset{\|}{C}}{-}R + 2R'NH\overset{O}{\overset{\|}{C}}NHR' \quad (17)$$

Since the reaction involves the use of mixtures of 90% hydrogen peroxide and ether, we prefer not to give detailed directions for this procedure on grounds of safety. Instead we suggest that this procedure be reexamined to establish whether safer concentrations of hydrogen peroxides may not be utilized, preferably in a solvent which does not hydroperoxidize as readily as ether.

Table VI gives the properties of a series of peroxides prepared by this general method.

TABLE VI

PREPARATION AND PROPERTIES OF DIACYL PEROXIDES BY USE OF *N,N'*-DICYCLOHEXYLCARBODIIMIDE[a]

Peroxide	Starting materials	Yield (%)	M.p. (°C) (dec.)
Bis(benzoyl)	Acid + H_2O_2	90	105–106
Bis(benzoyl)	Peracid	80	105.5–105.5
Bis(*p*-nitrobenzoyl	Acid + H_2O_2	85	154
Bis(*p*-nitrobenzoyl)	Peracid	70	155–156
Bis(*p*-bromobenzoyl)	Acid + H_2O_2	90	143
Mono-*p*-bromobenzoyl benzoyl	*p*-Bromoperbenzoic acid + benzoic acid	85	93.4–94.5
Bis(*p*-methoxybenzoyl)	Acid + H_2O_2	80	124.5–125
p-Methoxybenzoyl *p'*-nitrobenzoyl	*p*-Methoxybenzoic acid + *p*-nitroperbenzoic acid	70	107–108
Bis(palmitoyl)	Acid + H_2O_2	75	70.5–71.5
Bis(*trans-tert*-butylcyclohexanoyl)	Acid + H_2O_2	75	89–90
trans-tert-Butylcyclohexanoyl benzoyl	Perbenzoic acid + *trans-tert*-butylcyclohexanecarboxylic acid	40	81
3,4-Dichlorophenylacetylbenzoyl	Perbenzoic acid + 3,4-dichlorophenylacetic acid	65	65–66
Phthaloyl (monomeric)	Phthalic anhydride + H_2O_2	40	114

[a] From Reference [44].

3. OXIDATION REACTIONS

Toward the end of the nineteenth century, Nef prepared acetyl benzoyl peroxide by allowing a mixture of benzaldehyde and acetic anhydride to be oxidized by air. To increase the surface area exposed, the procedure involved mixing the reagents with a large excess of sand [45]. The same compound was prepared more recently by this general procedure. The melting point of the product was reported to be 35.5°C [46].

The concept of oxidizing mixtures of aldehydes and anhydrides has been explored further and developed into a method of preparing a range of unsymmetrically substituted diacyl peroxides [47]. In this procedure, the ratio of aldehyde to anhydride may be varied widely (1:1.2 to 1:4); either air or oxygen is circulated through the mixture at a rate of 0.25 to 1.5 liter/min. Since the reactants are volatile, a special closed recirculating system is used. The reaction system is kept between 20° and 25°C for reactions with aliphatic aldehydes and between 40° and 45°C for aromatic aldehyde and diffused light such as a 50–75 watt electric bulb is used for irradiation. Anhydrous sodium acetate or calcium carbonate serve as catalysts for the reaction.

While this procedure seems satisfactory for the preparation of unsymmetrically substituted aliphatic diacyl peroxides, only small amounts of bis(benzoyl) peroxide are formed from benzaldehyde and benzoic anhydride. Many aromatic-aliphatic diacyl peroxides may be prepared by this method. However, no peroxides were formed when the reaction was attempted with acetic anhydride and *m*-nitro-, *p*-nitro-, and *p*-dimethylamino-substituted benzaldehydes.

3-1. Generalized Preparation of Unsymmetrically Substituted Diacyl Peroxide by the Oxidation of Aldehyde and Anhydride Mixtures [47]

$$\mathrm{R{-}\overset{\overset{\displaystyle O}{\|}}{C}{-}H + R'{-}\overset{\overset{\displaystyle O}{\|}}{C}{-}O{-}\overset{\overset{\displaystyle O}{\|}}{C}{-}R' \xrightarrow{O_2} R{-}\overset{\overset{\displaystyle O}{\|}}{C}{-}O{-}O{-}\overset{\overset{\displaystyle O}{\|}}{C}{-}R' + R'{-}\overset{\overset{\displaystyle O}{\|}}{C}{-}OH} \quad (18)$$

In an apparatus which permits the recirculation of air or oxygen and of volatile organic components, a mixture of an aldehyde and an anhydride (mole ratio of 1:1.2 to 1:4) and 0.1 to 0.2% of anhydrous sodium acetate is treated with air at a rate of 0.25 to 1.5 liter/min until the absorption of oxygen stops (progress of the reaction may be followed iodometrically). The reaction mixture is poured into cold water and left in contact with water for 1 day.

The water layer is decanted and the residue is dissolved in ether. The ether solution is washed, in turn, with dilute aqueous sodium bicarbonate solution and with water. After drying over sodium sulfate and filtration, the ether is distilled off under reduced pressure. The residue is further purified.

Some of the crude products may be purified by washing with 50% nitric acid followed by treatment with cold water, extraction with ether, and treatment of the ether solution with 5% aqueous sodium hydroxide, followed by a water wash and drying over sodium sulfate. The product is then isolated by evaporation of the solvent under reduced pressure. Other products are crystallized from petroleum ether or by chilling solutions in petroleum ether–diethyl ether (1:1) at −70°C.

Table VII gives further details of the preparation of 14 diacyl peroxides by this method. More recently, this reaction was extended to the reaction of aldehyde, acyl chlorides, and oxygen in the presence of sodium hydroxide (Eq. 19).

TABLE VII

Preparation of Diacyl Peroxides Using the Method of Preparation 3-1 [Eq. (18)][a]

Aldehyde (moles)	Anhydride (moles)	Reaction conditions Temp. (°C)	Time (hr)	M.p. (°C)	B.p., °C (mm)	d_4^{20}	n_D^{20}	Yield (%)	Purity (%)
o-Methylbenz- (0.025)	Acetic (0.117)	45	3	—	—	1.1[illegible]20	1.5126	81.4	96.8
m-Methylbenz- (0.020)	(0.070)	50	4	32–32.5	—	—	—	78.0	99.6
p-Methylbenz- (0.017)	(0.039)	40	4	65–65.5	—	—	—	74.9	99.5
2,4-Dimethylbenz- (0.022)	(0.147)	40	5.5	—	—	1.1376	1.5210	73.6	92.6
p-Methoxylbenz- (0.068)	(0.210)	30	12	59.5	—	—	—	63.0	96.5
o-Chlorobenz- (0.036)	(0.106)	40	4	—	—	1.[illegible]589	1.5305	81.5	90.5
m-Chlorobenz- (0.054)	(0.210)	40	10	53–54	—	—	—	92.5	96.0
p-Chlorobenz- (0.018)	(0.150)	40	5	49.5	—	—	—	83.0	99.7
Benz- (0.056)	Propionic (0.110)	40	5	—	—	1.1530	1.5097	79.0	97.0
m-Chlorobenz- (0.060)	Propionic (0.120)	40	6	—	—	1.2222	1.5170	81.5	92.0
Benz- (0.034)	*n*-Butyric (0.070)	40	12	—	—	1.0671	1.5040	83.0	95.0
Benz- (0.040)	Monochloroacetic (0.090)	50	5	—	—	1.2386	1.5313	78.0	98.0
n-Butyr- (0.050)	Acetic (0.100)	20–25	4	—	37–37.5 (2)	1.0610	1.4123	81.1	98.2
Isovaler- (0.050)	Acetic (0.105)	20–25	3	—	40–41.5 (2)	1.0260	1.4145	75.5	96.0

[a] From Ol'dekop *et al.* [47].

$$\text{R—}\overset{\overset{\text{O}}{\|}}{\text{C}}\text{—H} + \text{R}'\text{—}\overset{\overset{\text{O}}{\|}}{\text{C}}\text{—Cl} + \text{NaOH} \xrightarrow{\text{O}_2} \text{R—}\overset{\overset{\text{O}}{\|}}{\text{C}}\text{—O—O—}\overset{\overset{\text{O}}{\|}}{\text{C}}\text{—R}' + \text{NaCl} + \text{H}_2\text{O} \quad (19)$$

Two procedures were developed. In one, air is percolated through an aqueous solution of sodium hydroxide and a benzene solution of the aldehyde and acyl chloride at 20°–40°C for 3–8 hr; in the other an acetone solution of the reagents is treated with ozone at 0°–5°C. Yields were said to be in the 70–75% range. The mole ratio of acyl halide to aldehyde may be varied from 2:1 to 1:1. By this method *n*-valeryl benzoyl peroxide (n_D^{20} 1.4991, d^{20} 1.0981) and *o*-bromobenzoyl benzoyl peroxide (m.p. 81°C) were prepared [48].

A recent patent claims the use of lithium benzoate as a catalyst for the oxidation of mixtures of aldehydes and anhydrides in acetophenone solution [49].

One aspect of the chemistry of unsymmetrically substituted diacyl peroxides which has been discovered recently is their tendency, in the presence of alkaline reagents such as sodium carbonate, to symmetrize. Aliphatic aromatic diacyl peroxides are converted to bis(aroyl) peroxides as shown in Eq. (20) [50].

$$2\text{Ar}\overset{\overset{\text{O}}{\|}}{\text{C}}\text{—O—O—}\overset{\overset{\text{O}}{\|}}{\text{C}}\text{—R} \xrightarrow{\text{OH}^-} \text{Ar}\overset{\overset{\text{O}}{\|}}{\text{C}}\text{—O—O—}\overset{\overset{\text{O}}{\|}}{\text{C}}\text{—Ar} \quad (20)$$

4. MISCELLANEOUS METHODS

A. Diacyl Peroxides

1. Condensation of peroxy acids and anhydrides [47].
2. Acylation of barium peroxide with anhydrides in ether [46].
3. Preparation from cyclic diketones with *p*-$(OCH_3)_3$ and ozone [51].
4. Acylation of hydrogen peroxide–urea or cyclohexylamine–hydrogen peroxide complexes with acyl chlorides in nonaqueous systems [52].
5. Acylation of 98% hydrogen peroxide with di-*n*-butylmalonic acid in methanesulfonic acid to produce 4,4-di-*n*-butyl-1,2-dioxolane-3,5-dione, a cyclic acyl peroxide [53].

B. Preparation of Peroxy Acids

Detailed directions for the preparation of a variety of per-acids are given by Swern [1, p. 476ff]. A general discussion will be found in the same reference

[1, p. 315ff]. A few of these synthetic methods given are outlined here with leading references to the literature.

1. Acid-catalyzed equilibrium reaction between carboxylic acids and hydrogen peroxide [54].
2. Oxidation of cation-exchange resins with hydrogen peroxide in presence of *p*-toluenesulfonic acid to produce a peroxy acid-type ion-exchange resin [55].
3. Oxidation of aldehydes with oxygen after initiation with ozone [56,57].
4. Oxidation of aldehydes in the presence of selenium or manganese catalysts [58].
5. Hydrolysis of diacyl peroxides [59].
6. Perhydrolysis of anhydrides [60].
7. Perhydrolysis of imidazolides in alkaline media [61].
8. Perhydrolysis of diacyl diethyl phosphates [62].

C. Organometallic and Organometalloid Peroxides

1. Review of organometallic and organometalloid peroxides [63].

REFERENCES

1. D. Swern, ed., "Organic Peroxides," Vol. 1. Wiley (Interscience), New York, 1970.
2. D. Swern, ed., "Organic Peroxides," Vol. 2. Wiley (Interscience), New York, 1971.
3. A. G. Davies, "Organic Peroxides." Butterworth, London, 1961.
4. E. G. E. Hawkins, "Organic Peroxides." Van Nostrand-Reinhold, Princeton, New Jersey, 1961.
5. A. V. Tobolsky and R. B. Mesrobian, "Organic Peroxides." Wiley (Interscience), New York, 1954.
6. O. L. Mageli and J. R. Kolczynski, *Mod. Plast. Encycl.* p. 482 (1967).
7. O. L. Mageli and C. S. Sheppard, *Encycl. Chem. Technol.* **14**, 766 (1967).
8. J. O. Edwards, "Peroxide Reaction Mechanisms." Wiley (Interscience), New York, 1962.
9. R. Criegee, *in* "Methoden der organischen Chemie," Vol. 8 (Houben-Weyl, ed.), 4th ed., p. 1. Thieme, Stuttgart, 1952.
10. V. Karnojitzki, "Les peroxyde organiques." Hermann, Paris, 1958.
11. W. Eggerlüss, "Organische Peroxyde." Verlag Chemie, Weinheim, 1951.
12. K. Nozaki and P. D. Bartlett, *J. Amer. Chem. Soc.* **68**, 1686 (1946).
13. W. Karo, Thesis, Cornell University, Ithaca, New York (1949).
14. C. G. Swain, W. H. Stockmayer, and J. T. Clarke, *J. Amer. Chem. Soc.* **72**, 5426 (1950).
15. A. T. Blomquist and A. J. Buselli, *J. Amer. Chem. Soc.* **73**, 3883 (1951).
16. D. F. Doehnert and O. L. Mageli, *Prepr., 13th Annu. Meet. Reinforced Plast. Div., Soc. Plast. Ind.* Sect. 1-B, p. 1 (1961).
17. O. L. Mageli and J. R. Kolczynski, *Ind. Eng. Chem.* **58**, 25 (1966).

18. Lucidol Division, Pennwalt Corporation, Reprints 30.40, 30.42, and 30.43.
19. P.P.G. Industries, Inc., Chemical Division, Bulletins 80A, 80B, and 80C.
20. W. A. Strong, *Ind. Eng. Chem.* **56**, 33 (1964).
21. D. C. Noller, S. J. Mazurowski, G. F. Linden, F. J. G. De-Leeuw, and O. L. Mageli, *Ind. Eng. Chem.* **56**, 18 (1964).
22. J. B. Armitage and H. W. Strauss, *Ind. Eng. Chem.* **56**, 28 (1964).
23. H. von Pechmann and L. Vanino, *Ber. Deut. Chem. Ges.* **27**, 1510 (1894).
24. C. C. Price and E. Krebs, *Org. Syn., Collect. Vol.* **3**, 649 (1955); C. C. Price, R. W. Kell, and E. Krebs, *J. Amer. Chem. Soc.* **64**, 1103 (1942).
25. W. Karo, unpublished data.
26. J. G. Cadogan, D. H. Hey, and P. G. Hibbert, *J. Chem. Soc., London* p. 3939 (1965).
27. H. Kämmerer, K.-G. Steinfort, and F. Rocaboy, *Makromol. Chem.* **63**, 214 (1963).
28. D. F. DeTar and D. V. Wells, *J. Amer. Chem. Soc.* **82**, 5839 (1960).
29. H. Hart and F. J. Chloupek, *J. Amer. Chem. Soc.* **85**, 1155 (1963).
30. J. G. Cadogan, D. H. Hey, and P. G. Hibbert, *J. Chem. Soc., London* p. 3950 (1965).
31. M. Schulz and K. Kirschke, *Advan. Heterocycl. Chem.* **8**, 165 (1967).
32. F. Strain, W. E. Bissinger, W. R. Dial, H. Rudoff, B. J. DeWitt, H. C. Stevens, and J. H. Langstron, *J. Amer. Chem. Soc.* **72**, 1254 (1950).
33. J. E. Guillet, T. R. Walker, M. F. Meyer, J. P. Hawk, and E. B. Towne, *Ind. Eng. Chem., Prod. Res. Develop.* **3**, 257 (1964).
34. L. S. Silbert and D. Swern, *J. Amer. Chem. Soc.* **81**, 2364 (1959).
35. F. Gregor and E. Pavlacha, *Chem. Prum.* **14**, 299 (1964); *Chem. Abstr.* **61**, 10613f (1964).
36. S. Jablonski, F. Ksiezak, E. Blaszcyk, E. Brylka, E. Rosinski, and L. Glygiel, Polish Patent 53,450 (1967); *Chem. Abstr.* **69**, P76955k (1968).
36a. J. E. Leffler and A. A. Moore, *J. Amer. Chem. Soc.* **94**, 2483 (1972).
37. R. A. Sheldon and J. K. Kochi, *J. Amer. Chem. Soc.* **92**, 4395 (1970).
38. C. L. Jenkins and J. K. Kochi, *J. Org. Chem.* **36**, 3095 (1971).
39. W. Walczyk, J. Stepowski, P. Papageorgios, and H. Eisermann, Polish Patent 56,517 (1968); *Chem. Abstr.* **71**, 2986r (1969).
40. D. R. Byrne, F. M. Gruen, D. Priddy, and R. D. Schuetz, *J. Heterocycl. Chem.* **3**, 369 (1966).
41. V. A. Dodonov and W. A. Waters, *J. Chem. Soc., London* p. 2459 (1965).
42. H. Staab, W. Rohr, and F. Graf, *Chem. Ber.* **98**, 1122 (1965).
43. H. Staab, F. Graf, and W. Rohr, *Chem. Ber.* **98**, 1128 (1965).
44. F. D. Greene and J. Kazan, *J. Org. Chem.* **28**, 2168 (1963).
45. J. U. Nef, *Justus Liebigs Ann. Chem.* **298**, 282 (1897).
46. Y. Ogata, Y. Furuya, and K. Aoki, *Bull. Chem. Soc. Jap.* **38**, 838 (1965).
47. Yu. A. Ol'dekop, A. V. Sevchenka, I. P. Zyat'kov, G. S. Bylina, and A. P. El'nitskii, *J. Gen. Chem. USSR* **31**, 2706 (1961).
48. Yu. A. Ol'dekop, G. S. Bylina, and S. F. Petrashkevich, *Probl. Poluch. Poluprod. Prom. Org. Sin.* p. 152 (1967); *Chem. Abstr.* **68**, 39292a (1968).
49. H. R. Appell, U.S. Patent 3,397,245 (1968); *Chem. Abstr.* **69**; P76954j (1968).
50. Yu. A. Ol'dekop and A. P. El'nitskii, *Zh. Obshch. Khim.* **34**, 3478 (1964); *Chem. Abstr.* **62**, 3971e (1965).
51. F. Ramirez, N. B. Desai, and R. B. Mitra, *J. Amer. Chem. Soc.* **83**, 492 (1961).
52. D. F. DeTar and L. A. Carpino, *J. Amer. Chem. Soc.* **77**, 6370 (1955).
53. W. Adam and R. Rucktäschel, *J. Amer. Chem. Soc.* **93**, 557 (1971).

54. B. Phillips, P. S. Starcher, and B. D. Ash, *J. Org. Chem.* **23**, 1823 (1958); L. S. Silbert, E. Siegel, and D. Swern, *Org. Syn.* **43**, 93 (1963).
55. K. Koyama, M. Nishimura, and I. Hashida, *J. Polym. Sci., Part A-1* **9**, 2439 (1971).
56. C. R. Dick and R. F. Hanna, *J. Org. Chem.* **29**, 1218 (1964).
57. S. Miyajima, I. Koga, H. Harada, and T. Hirai, German Patent 2,061,456 (1971); *Chem. Abstr.* **75**, 109856h (1971).
58. N. Ota, T. Imamura, and T. Matsuzaki, Japanese Patent 70/21,807 (1970); *Chem. Abstr.* **73**, 76670w (1970).
59. G. Braun, *Org. Syn., Collect. Vol.* **1**, 431 (1941).
60. G. B. Payne, *J. Org. Chem.* **24**, 1354 (1959); H. Böhme, *Org. Syn., Collect. Vol.* **3**, 619 (1955).
61. U. Folli and D. Iarossi, *Boll. Sci. Fac. Chim. Ind. Bologna* **26**, 61 (1968).
62. D. A. Konen and L. S. Silbert, *J. Org. Chem.* **36**, 2162 (1971).
63. G. Sosnovsky and J. H. Brown, *Chem. Rev.* **66**, 529 (1966).

Chapter 15

FREE-RADICAL INITIATORS: HYDROPEROXIDES

I. INTRODUCTION

Compounds of the general formula R—O—O—H are termed organic hydroperoxides [1]. Their chemistry has been extensively reviewed, references [2]–[12] being a selection of such reviews. Of these, attention should be directed to the review of Sosnovsky and Brown [8], which deals with the chemistry of organometallic and organometalloid peroxides, a topic which is beyond the scope of this chapter.

The hydroperoxides are not only of interest because they frequently are intermediates in the oxidation of organic molecules, but also because of their use in polymer chemistry. With the exception of the very lowest molecular weight members of the series, most are remarkably stable thermally. Consequently they find application in those areas of polymer chemistry in which polymerization is to be initiated at relatively high temperatures, such as in certain unsaturated polyester resin systems and in the rubber and silicone field. The fact that most hydroperoxides are liquids is a great convenience in preparing compositions of these materials with various monomers and viscous resin mixtures.

Despite their thermal stability, in the presence of certain reducing agents, metal "dryers," or other accelerators, the hydroperoxides may initiate free-radical polymerizations at room temperature and below. Indeed many of the so-called "cold recipes" for the emulsion copolymerization of butadiene with styrene involve hydroperoxides as the oxidizing agent of the redox system. Many of the anaerobic polymerization compositions consist of solutions of dimethacrylates and hydroperoxides. Presumably, because of the thermal stability of the hydroperoxides, such compositions are stable at room temperature for prolonged periods of time, at least as long as oxygen is permitted to diffuse through the liquid phase [13].

Table I lists the half-lives of several hydroperoxides at various temperatures [14–16]. The first six compounds listed in Table I are of some interest commercially. In addition, diisopropylbenzene hydroperoxide is also available. However, the exact composition of the commercial product is still in doubt. The pure hydroperoxide (i.e., a product substantially free of diisopropylbenzene) is not available.

In the preparation of hydroperoxides, particularly by oxidation procedures, it is frequently the practice to carry the reactions out to relatively low conversions. The separation of the hydroperoxide from unreacted starting material is frequently more easily accomplished than the separation of the product from decomposition products which may form if the reaction is carried to high conversions. Many of the commercially available hydroperoxides are supplied as solutions in the unreacted starting material. The half-life of such

TABLE I

HALF-LIFE VS. TEMPERATURE DATA FOR HYDROPEROXIDES

Hydroperoxide	Temp. (°C)	Half-life (hr)	Temp. (°C)	Half-life (hr)	Ref.
tert-Butyl	130	520	160	29	14
1,1,3,3-Tetramethylbutyl	130	11	160	0.68	14
2,5-Dihydroperoxy-2,5-dimethylhexane	130	67	160	6.1	14
Cumene	130	110	145	29	14
p-Menthane	130	12.5	160	0.93	14
Pinane	130	27	160	2	15
1-Phenyl-2-methylpropyl-1	133.8	60.5	163.7	4.76	16
1-Phenyl-2-methylpropyl-2	144.2	38.2	165.6	6.6	16

solutions may be shorter at any given temperature than that of the pure compound.

Although the higher hydroperoxides are fairly stable thermally when pure, impure materials may be hazardous. The lower hydroperoxides are said to be explosive. In general, safety precautions discussed in the diacyl peroxide chapter should be followed in the preparation and handling of hydroperoxide, even though any specific compound may be substantially more stable than the diacyl peroxides.

The nomenclature of hydroperoxides is generally quite straightforward. It is the usual practice to designate these compounds as "alkyl hydroperoxides." For example,

$$CH_3-\overset{\displaystyle CH_3}{\underset{\displaystyle CH_3}{\overset{|}{\underset{|}{C}}}}-OOH$$

is termed *tert*-butyl hydroperoxide.

Since many hydroperoxides are prepared by oxidation of complex hydrocarbons to mixtures of products whose structures are not precisely known and may even be supplied as solutions in the parent hydrocarbon, they are also called "alkane hydroperoxides," e.g., diisopropylbenzene hydroperoxide. In Table VI, we have adopted this latter method of nomenclature to indicate the hydrocarbons from which the compounds are derived. This leads to a few difficulties, e.g., benzyl hydroperoxide is simply listed in the left-hand column as "benzyl." A designation such as "2-phenylethyl 2-" is used to indicate that the hydroperoxide radical is attached to the number 2 carbon of the two-carbon chain. Alternatively, "2-pentane" also implies that the hydroperoxide moiety is attached at the number 2 carbon of a normal five-carbon chain.

The major methods of preparing hydroperoxides may be divided into two groups: oxidation with oxygen and reactions with hydrogen peroxide.

The oxygen-oxidations are often termed autoxidations, which implies that a free-radical chain mechanism is operative. The autoxidation of alkanes and olefins may be carried out without any catalyst or with typical free-radical initiators such as 2,2′-azobis(isobutyronitrile), the product hydroperoxide, other organic peroxides, heavy metal salts such as the "metal dryers," ultraviolet radiation (with or without photosensitizers), and ozone. Singlet oxygen and ozone may be used to form hydroperoxides. Organometallic compounds, such as Grignard reagents and trialkylboranes, may be autoxidized with oxygen.

Reactions involving hydrogen peroxide include the reaction with olefins and a displacement reaction which may be represented most simply by Scheme 1.

SCHEME 1

Preparation of Hydroperoxides by the Reaction of Various Functional Groups with Hydrogen Peroxide

$$R{-}X + H_2O_2 \longrightarrow R{-}OOH + HX$$

$$X = OH,\ {-}O\overset{\overset{\displaystyle O}{\|}}{C}{-}R,\ {-}OR',\ R_2B{-},\ \text{halogen},\ {-}OSO_3H,\ {-}OSO_2OR,\ \text{and}\ {-}OSO_2R'$$

In general, all of the methods for the preparation of hydroperoxides proceed most satisfactorily if a tertiary hydroperoxide is to be formed. Secondary and primary hydroperoxides are less readily produced.

The best methods for the preparation of secondary and primary hydroperoxides are the oxidation of Grignard reagents and the reaction of hydrogen peroxide with alkyl methanesulfonates. The oxidation of trialkylboranes also may have merit for the preparation of all types of hydroperoxides. However, the method does not lend itself to the synthesis of large quantities of material.

The hydrolysis of peresters is primarily of importance as an analytical tool for the identification of the structure of peresters. Since peresters cannot be formed by the esterification of peracids, the usual preparation being the reaction of hydroperoxides with acids, obviously this approach has little preparative value. Recently, in the field of fluoro-organic chemistry, a novel method of establishing the perester bond has been developed in which no hydroperoxide is needed as a starting material. Naturally, in this situation, hydrolysis of the perester has some value for the isolation of trifluoromethyl hydroperoxide.

2. OXIDATION REACTIONS WITH OXYGEN

A. Uncatalyzed Autoxidation of Alkanes

Hydroperoxides have been prepared from such hydrocarbons as alkanes, aralkanes, and olefins, as well as the hydrocarbon moieties of more complex molecules such as lactones and ethers, by the action of molecular oxygen. Since the reaction is generally considered to be a free-radical chain process, it is designated autoxidation. The mechanism of the reaction has been suggested to be as given in Eqs. (1)–(8) [3].

Initiation

$$RH \longrightarrow R\cdot + [H\cdot] \tag{1}$$

Propagation

$$R\cdot + O_2 \longrightarrow RO_2\cdot \tag{2}$$

$$RO_2\cdot + RH \longrightarrow RO_2H + R\cdot \tag{3}$$

Termination

$$R\cdot + R\cdot \longrightarrow RH + R(^{-}H) \tag{4}$$

$$R\cdot + R\cdot \longrightarrow R{-}R \tag{5}$$

$$RO_2\cdot + R\cdot \longrightarrow R{-}O{-}O{-}R \tag{6}$$

$$RO_2\cdot + OH\cdot \longrightarrow ROH + O_2 \tag{7}$$

$$RO_2\cdot + RO_2\cdot \longrightarrow O_2 + R{-}O{-}O{-}R \text{ (or carbonyl and other)} \tag{8}$$

etc.

There is said to be a marked difference in the course of the reaction, depending on the reaction temperature. Thus, for example, in the gas phase, at temperatures above 200°C, linear paraffinic hydrocarbons are more readily oxidized than branched molecules. At lower temperatures, or in the liquid phase, the reaction is slower and more selective. Tertiary carbons are first attacked, then secondary carbons, and lastly the primary carbons are attacked. Olefins are attacked not at the double bond but on the adjacent activated methylene group, i.e., the allylic carbon [2,3].

The free-radical reactions are subject to inhibition by typical inhibitors of such processes, e.g., phenols. The reactions are initiated by free-radical sources such as 2,2′-azobis(isobutyronitrile), hydroperoxides, and by ultraviolet radiation. As is common to many free radicals, they are subject to chain-transfer reactions. Since the hydroperoxides formed in the reaction may also act as initiators, studies of the kinetics of the reaction are complex. The preparation of hydroperoxides in the presence of free-radical initiators is discussed in the next section of this chapter.

The decomposition of hydroperoxide is of importance not only in the application of these compounds but also in the selection of appropriate conditions for their preparation and isolation.

The free-radical decomposition of these compounds is said to involve a step such as shown in Eq (9).

$$\text{R—O—O—H} \longrightarrow \text{RO}\cdot + \cdot\text{OH} \tag{9}$$

Acid-catalyzed decompositions may proceed as in Eq. (10).

$$\text{R—O—O—H} \xrightarrow{\text{H}^+} \text{RO}^+ + \text{OH}^- \tag{10}$$

Base-catalyzed decompositions involve rupture of the oxygen hydrogen bond (Eq. 11).

$$\text{R—O—O—H} \xrightarrow{\text{OH}^-} \text{R—O—O}^- + \text{H}^+ \tag{11}$$

The catalysis of hydroperoxide decompositions by metals is represented as involving the oxidation of a metal ion from one oxidation state to another (Eq. 12) [3].

$$\text{R—O—O—H} \xrightarrow{\text{M}^+} \text{RO}^- + \cdot\text{OH}\ (+\text{M}^{2+}) \tag{12}$$

It is an unfortunate fact that the literature refers to such decompositions as metal-catalyzed when they are more precisely metal-ion catalyzed. This matter may be of some significance in the cases of anaerobic polymerizations, which usually are brought about on metal surfaces.

In the oxidation of hydrocarbons with oxygen, it is to be expected that acidic by-products are formed. It is, therefore, customary to remove acids completely from the reaction mixture prior to final isolation of the hydroperoxide [17]. On the other hand, base-catalyzed decompositions seem to present no serious problem, since the common method of separating hydroperoxides from the reaction mixture involves their conversion to their sodium or potassium salts followed by neutralization with a weak acid [18–20]. As to the stability of hydroperoxides towards metals, it should be noted that at least arylalkane hydroperoxides are formed in higher yields in an apparatus made from copper or silver than in glass or stainless steel equipment [21]. Preparation 2-1 is based on the patented process of N. V. DeBataafsche Petroleum Maatschappij, Den Haag, the Netherlands [21].

2-1. Preparation of Cumene Hydroperoxide by Autoxidation [21]

$$\text{C}_6\text{H}_5\text{—CH(CH}_3\text{)—CH}_3 + \text{O}_2 \longrightarrow \text{C}_6\text{H}_5\text{—C(CH}_3\text{)}_2\text{—OOH} \tag{13}$$

In a 500 ml copper flask fitted with a gas inlet tube leading to the bottom of the flask, a thermometer, a copper stirrer with six blades, and a reflux condenser is placed 225 ml (194 gm, 1.61 moles) of carefully purified cumene (b.p. 150°–151°C, n_D^{20} 1.4912–1.4913). The gas inlet tube is connected through a flowmeter to a source of technically pure oxygen while the reflux condenser is connected through a second flowmeter to a vent. During the reaction the oxygen uptake is calculated from the flowmeter readings.

The flask is placed in an electrically heated oil bath maintained at 120°C, the stirrer is maintained at 880 rpm, and oxygen is introduced at a rate of 25 liter/hr. By iodometric titration, after 3 hr, the reaction flask contains 34.7% by weight of cumene hydroperoxide. This represents a 96% yield based on oxygen actually absorbed. Table II shows results of similar preparations under different conditions of time, temperature, and reactor material.

TABLE II

OXIDATION OF CUMENE WITH OXYGEN; EFFECT OF TIME, TEMPERATURE, AND REACTOR MATERIAL[a,b]

Material of reactor and stirrer	Reaction temp. (°C)	Reaction time (hr)	Hydroperoxide content of product (% by weight)	Yield, based on oxygen taken up (%)
Copper	120	1	11.0	95
Copper	120	2	22.5	93
Copper	120	3	34.7	96
Copper	130	1	25.0	81
Silver	120	1	15.9	—
Stainless steel	120	1	2.1	—

[a] Oxygen flow rate: 25 liter/hr, stirring rate: 880 rpm.
[b] From Reference [21].

By similar procedures, diisopropylbenzene and *p*-cymene (1-isopropyl-4-methylbenzene) were converted to the corresponding hydroperoxides. It may be anticipated that in the case of the oxidation of diisopropylbenzene, a mixture of the mono- and dihydroperoxides as well as some α-hydroxy-hydroperoxide is formed [20].

It will be noted that in Procedure 2-1, the flow rate of oxygen, 25 liter/hr for approximately 200 gm of cumene, is rather high. This supply of oxidizing agent may very well overwhelm the decomposition of the hydroperoxide on the metal surfaces of the apparatus. An earlier report of the preparation of cumene hydroperoxide, which we judge, on the basis of the equipment pictured, to involve a much lower flow of oxygen, states that the addition of copper or lead to the reaction system, after an initial acceleration of the rate of oxidation, ultimately leads to a rapid decline of the rate of oxidation [22]. The resolution of these diametrically opposed points of view is difficult, particularly in view of the recommendation by Fortuin and Waterman [21] that a copper vessel is particularly useful for the storage of cumene hydroperoxide.

A generalized description of the procedure of Armstrong *et al.* [22] is given in Procedure 2-2 since the apparatus appears generally useful for oxidative preparations of hydroperoxides and because it describes methods of isolating the product from the unreacted starting materials.

Procedure 2-2 advocates the use of an emulsion of the hydrocarbon in an aqueous system containing an anionic surfactant such as sodium stearate or Teepol (an aqueous solution of mixed sodium higher-alkyl sulfates) and sodium carbonate (to buffer the reaction system between pH 8.5 and 10.5 since the rate of oxidation is substantially reduced at low pH values and the reaction efficiency is substantially reduced at a higher pH). Presumably the surfactant assists in achieving better contact between oxygen and the hydrocarbon reactant than may be accomplished by merely bubbling oxygen through a reaction system. From the practical standpoint, of course, care must be exercised to prevent excessive foam formation.

The apparatus used in this preparation is shown in Fig. 1. In this diagram, *A* is the oxygen supply cylinder; *B* is a control valve operating at constant pressure controlled by pressure regulator *G*; *C* and *F* are two wet-test meters of $\frac{1}{2}$ liter range; *L* is a 1 liter Pyrex flask; *S* is a stirrer which permits oxygen to enter its hollow stem through an orifice and ejects the oxygen stream by centrifugal forces through the hollow tube *T* at the bottom. The stirrer is

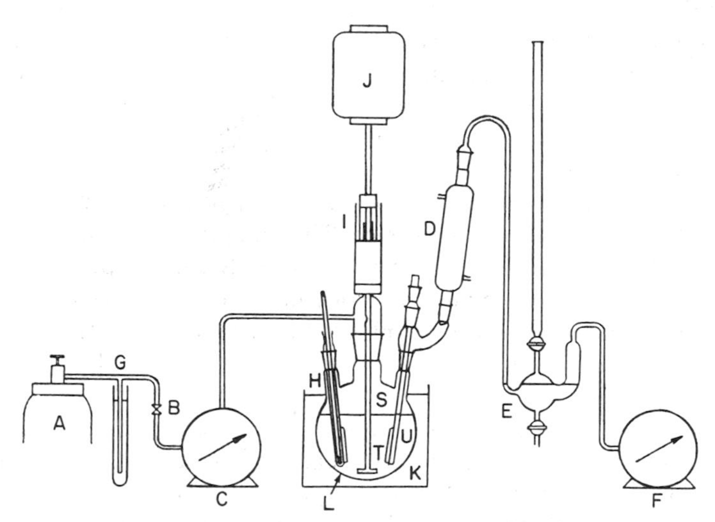

FIG. 1. Apparatus for oxidative preparation of hydroperoxides used in Procedure 2-2. See text for explanation. [Reprinted from G. P. Armstrong, R. H. Hall, and D. C. Quin, *J. Chem. Soc., London* p. 666 (1950). Copyright 1950 by the Chemical Society. Reprinted by permission of the copyright owner.]

connected to an explosion-proof motor *J* by means of the mercury seal (*I*) which, nowadays, may well be replaced by a ground glass stirrer bearing. The flask *L* is fitted with a thermometer well *H*, and a liquid sampling tube *U*, both of which have glass strips sealed to them to act as baffles which prevent swirling of the liquid. The flask is immersed in a thermostatically controlled oil bath *K*. The reflux condenser *D* is connected to a bubbler *E* which contains aqueous phenolphthalein solution and is connected to a buret containing 0.2 *N* aqueous sodium hydroxide which permits an estimate of the formation of carbon dioxide during the oxidation.

2-2. *Generalized Procedure for the Preparation of Cumene Hydroperoxide in an Emulsion System* [22]

$$C_6H_5{-}CH(CH_3){-}CH_3 + O_2 \longrightarrow C_6H_5{-}C(CH_3)_2{-}OOH \qquad (14)$$

Using the apparatus described above and shown in Fig. 1, the flask is heated to 85°C, the stirrer and the oxygen flow are started to replace the air in the equipment with oxygen. Two hundred milliliters of water containing 0.6 gm of sodium stearate and 2.6 gm of sodium carbonate are introduced in the flask and allowed to come to the reaction temperature. Then 100 ml of highly purified cumene is added to the contents of the flask. The course of the reaction is followed by periodic withdrawal of samples which are iodometrically analyzed and by recording the readings of the two wet-test meters.

When the hydroperoxide concentration in the flask is judged to have reached a desired level, the reaction product is cooled to room temperature and saturated with carbon dioxide. The oily layer is separated, cooled, with stirring, to below room temperature, and cautiously heated with a slight excess of cold 25% aqueous sodium hydroxide. The precipitated sodium salt of the hydroperoxide is filtered off and set aside. The filtrate is repeatedly heated with cold 25% aqueous sodium hydroxide until no further precipitation of the sodium salt takes place. The samples of the sodium salt of cumene hydroperoxide are combined, washed with petroleum ether (b.p. 40°–60°C), and dried under reduced pressure.

A suspension of the sodium salt in water is covered with a layer of petroleum ether (b.p. 40°–60°C) and saturated with carbon dioxide. The upper, organic layer is separated, washed with a small quantity of cold water, dried with anhydrous sodium sulfate, and freed of solvent under reduced pressure. The residue is distilled at low pressure, b.p. 53°C (0.0005 mm) (pump pressure), n_D^{20} 1.5243, d_D^{20} 1.062 {lit. b.p. 53°C (0.1 mm); 60°C (0.2 mm) [22]}.

Alternatively, the oil isolated from the reaction may be distilled directly under reduced pressure. Unreacted cumene is recovered at 41°–46°C/13 mm followed by traces of α-methylstyrene and acetophenone. The residue is then distilled at the low pressure indicated above.

The technique of separating a hydroperoxide from the unreacted starting material by means of its sodium salt has been described repeatedly [18,23–27]. By control of the concentrations of the sodium hydroxide solutions and the addition of organic solvent, the dihydroperoxides of the diisopropylbenzenes may be separated from their monohydroperoxides and the α-hydroxyhydroperoxides [20,24,26].

Since some of the sodium salts of the hydroperoxides are extremely hygroscopic, the potassium salts have been prepared in nonaqueous systems by treating *tert*-butyl hydroperoxide, for example, with ethanolic potassium hydroxide and precipitating the salt at 0°C with ice-cold acetone. The product may crystallize with *tert*-butanol of crystallization [19].

The preparation of 1-tetralin hydroperoxide (1,2,3,4-tetrahydronaphthyl 1-hydroperoxide) by oxygen oxidation of tetralin has been well described [28]. This product may be separated from the reaction mixture by distilling the unreacted tetralin off, dissolving the still-residue in toluene, and precipitating the product by cooling to −50°C. The crude product is then recrystallized from fresh toluene at −30°C.

Some hydrocarbons, such as the saturated cyclic terpenes, have been oxidized with oxygen by initiating the reaction at a relatively high temperature, followed by further oxidation at a more moderate temperature to reduce the formation of decomposition products [23].

While oxygen is frequently used to prepare hydroperoxides, air has also been used [21,29,30]. The work of van Lier and Kan [20] is of interest since it involves the autoxidation of cholesterol at 70°C for 3 months in the dark. By careful chromatography, hydroperoxides in the tertiary 20 and 25 positions, in the secondary 24 position and in the 26 position were isolated. The latter compound, 3β-hydroxy-5-cholestene 26-hydroperoxide, was present in the oxidate in a quantity of 140 mg/kilogram of starting material (m.p. 153°–155°C).

The oxidation of molecules with two tertiary carbon atoms such as dimethylalkanes of the type

$$
\begin{array}{c}
CH_3CH_3 \\
|| \\
CH_3—CH—(CH_2)_x—CH—CH_3
\end{array}
$$

where $x = 0$ to 3, indicates that intermediately formed monohydroperoxides or peroxy radicals assist in the oxidation of tertiary carbon atoms located in the β- or γ-position to the intermediate. 2,4-Dimethylpentane and 2,5-dimethylhexane are readily converted to the 2,4- or 2,5-diphydroperoxy derivatives, respectively. On the other hand, the hydroperoxidation of 2,3-dimethylpentane and 2,6-dimethylheptane does not seem to be catalyzed by intermediate monohydroperoxides [31]. An extention of this work to the 2,2'-azobis(isobutyronitrile)-initiated oxidation of syndiotactic polypropylene affords support to the mechanism which postulates hydrogen abstraction by peroxy radicals in the β-position [32].

Olefinic hydrocarbons may be oxidized to hydroperoxides, although catalyzed oxidations are more common [33]. In a recent report, 1-olefins in the range of C_{15} to C_{18} were autoxidized to alkenyl hydroperoxides. These products were used without further isolation as polymerization initiators [34].

B. Catalyzed Autoxidation of Alkanes

The autoxidation of alkanes described in the previous section is frequently slow or subject to an inhibition period. Since the reaction is considered to be of a free-radical nature, it is reasonable to induce the autoxidation with reagents which generate free radicals readily in the system. A variety of other catalyzing agents have also been proposed.

Naturally, the addition of such reaction initiators may create problems of separation of the reaction products derived from these compounds. However, a number of approaches are possible which minimize such difficulties.

At high concentrations of ozone in an air stream, the decomposition of hydroperoxide is accelerated. At low levels, however, ozone initiates the formation of hydroperoxides. Thus in a typical autoxidation of cumene with air in the presence of sodium carbonate, 0.5 to 1.5 volume % of ozone have been added to the gas stream during the first hour (reaction at 120°C). Then air oxidation is continued for another 8.5 hr. The product was found to contain 70% of cumene hydroperoxide. In a control preparation, when an air autoxidation was carried out without ozone initiator, only 5% of cumene hydroperoxide was present in the reactant [35].

The addition of hydroperoxide to a hydrocarbon prior to autoxidation with oxygen has been suggested as an improved method of synthesis [21,26]. This method has the advantage that a small amount of the expected product of the reaction may be added as an initiator, thus eliminating the problem of separating extraneous materials. In Preparation 2-3, cumene hydroperoxide is used as the initiator for the autoxidation of cumene.

2-3. Preparation of Cumene Hydroperoxide (Initiated Procedure) [21]

$$C_6H_5-CH(CH_3)CH_3 + O_2 \xrightarrow{\text{cumene hydroperoxide}} C_6H_5-C(CH_3)_2-OOH \qquad (15)$$

In the all-copper apparatus described in Preparation 2-1 is placed 250 ml (216 gm, 1.79 moles) of carefully purified cumene and 7.8 gm of cumene hydroperoxide. The reactor is immersed in the oil bath which is maintained at 120°C and, while stirring at 880 rpm, oxygen is introduced at a rate of 25 liter/hr. By iodometric titration, after 3 hr, the reaction flask contains 44.3% by weight of cumene hydroperoxide. This represents an 83.3% yield based on oxygen actually absorbed.

Mention has already been made of the use of 2,2′-azobis(isobutyronitrile) (AIBN) as the initiator of the autoxidation of a solution of syndiotactic polypropylene [32]. Aralkyl hydrocarbons have also been autoxidized in the presence of this initiator [36–38]. Some of this work attempted high conversions of the hydrocarbons with the result that undesirable extraneous products were also formed. For example, in the oxidation of cumene in chlorobenzene in the presence of AIBN, the product composition consisted of 60–75% of cumene hydroperoxide, 12.5–18.5% of dicumene peroxide, and 9.4–11.3% of acetophenone [38].

Other initiating systems for the autoxidation of hydrocarbons include cumene hydroperoxide in the presence of manganese resinate [39]; hydroperoxides in the presence of salts of copper, mercury, tin, lead, bismuth, iron, cobalt, or nickel [40]; cobalt acetylacetonate [41]; heavy metal salts [42]; hydroperoxides in the presence of copper phthalocyanine and sodium ethylenediaminetetraacetate or polyphosphate [43]; tertiary alkyl primary amines (Primene JMT and Primene 81-R) [44]; 2,4-diaminoazobenzene [45]; barium sulfate [46] or cupric sulfate [47]; hydroperoxides in the presence of potassium permanganate [42]; β-diketones in the presence of sodium bicarbonate [48]; malonic esters in the presence of sodium bicarbonate [49]; α-aryl ketones in the presence of sodium bicarbonate [50]; at a pH between 3 and 7 maintained with sodium carbonate [51] or sodium hydroxide [52]; and platinum [53,54].

In a flow system, gaseous branched alkanes such as isobutane have been oxidized with oxygen in the presence of a small amount of hydrogen bromide. In the case of isobutane, the major products are *tert*-butyl peroxide, *tert*-butyl alcohol, and alkyl bromides [55]. In this preparation it was found that the reaction was sensitive to the nature of the reactor surfaces. In a Pyrex glass apparatus, the oxidation rate rapidly decelerates. When boric acid was used to coat the glass surfaces of the reactor, improved reactions were observed; silica gel coatings tended to inhibit the reaction almost completely.

C. Autoxidation of Olefins

The autoxidation of olefins, as mentioned above, may be carried out under conditions which are similar to those used for the reaction of alkanes and aralkanes. It is usual to use catalysts in these reactions. Ultraviolet radiation is frequently used to induce the free-radical reaction. In this connection, photosensitizers have also been incorporated in the reaction system.

In the absence of photosensitizers, the attack is at an allylic carbon–hydrogen bond.

2-4. Preparation of Cyclohexene 3-Hydroperoxide [56]

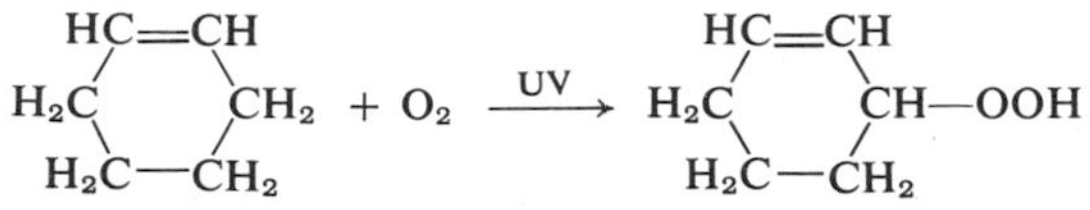

$$\text{cyclo-}[HC{=}CH{-}CH_2{-}CH_2{-}CH_2{-}CH_2] + O_2 \xrightarrow{UV} \text{cyclo-}[HC{=}CH{-}CH(OOH){-}CH_2{-}CH_2{-}CH_2] \quad (16)$$

In a 500 ml quartz flask connected to a gas buret containing oxygen, 169 gm (2.06 moles) of freshly distilled cyclohexene is shaken at 30°–35°C for 14 hr while exposing it to a mercury lamp. Then shaking is continued without exposure for 10 hr. In this 24 hr period a total of 8 liters of oxygen is taken up.

The reaction mixture is distilled under reduced pressure to remove unreacted cyclohexene. The residue weighs 32.3 gm (yield 79% based on oxygen consumption). The crude product may be distilled at 50°–52°C (0.3 mm). A yellow residue of 4 gm remains behind.

In the case of an olefin such as 4-vinylcyclohexene, autoxidation produces a mixture of olefinic hydroperoxides since there are several available allylic positions. The major reaction product is shown in Eq. (17) [57].

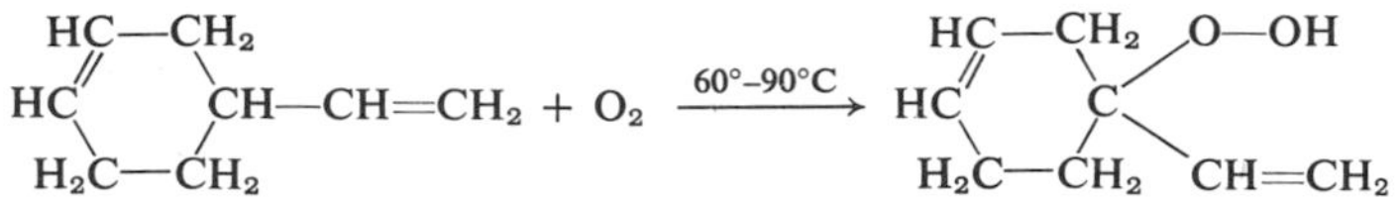

$$\text{cyclo-}[HC{=}HC{-}CH_2{-}CH(CH{=}CH_2){-}CH_2{-}H_2C] + O_2 \xrightarrow{60°\text{–}90°C} \text{cyclo-}[HC{=}HC{-}CH_2{-}C(O{-}OH)(CH{=}CH_2){-}CH_2{-}H_2C] \quad (17)$$

Similar results were obtained when the oxidation was carried out with *tert*-butyl hydroperoxide in the presence of cobalt naphthenate or with *tert*-butyl peroxybenzoate in the presence of cuprous bromide [57].

In the presence of hydrogen bromide, olefins are autoxidized with irradiation by ultraviolet radiation to produce hydroperoxides to which the elements of hydrogen bromide had also been added [Eq. (18)] [58]. To be noted is that

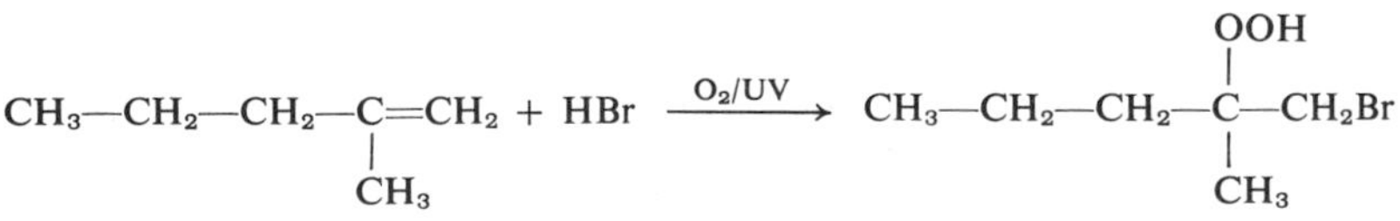

$$CH_3{-}CH_2{-}CH_2{-}C(CH_3){=}CH_2 + HBr \xrightarrow{O_2/UV} CH_3{-}CH_2{-}CH_2{-}C(OOH)(CH_3){-}CH_2Br \quad (18)$$

this reaction may be carried out with such olefinic compounds as allyl bromide, methallyl chloride, and allyl chloride [58].

The autoxidation of olefins has also been carried out in the presence of photosensitizers. Among these agents were rose bengale, chlorophyll [59], methylene blue, eosin Y, erythrosin B, and hematoporphyrin [60]. With olefins which have isolated double bonds and an available allylic hydrogen, a concerted reaction is said to take place with a shift of the double bond as shown in Eq. (19) [61].

$$\underset{1}{-\overset{|}{C}}=\underset{2}{C}-\underset{\underset{H}{|}}{\overset{|}{C_3}}- \xrightarrow[UV]{O_2/\text{photosensitizer}} \underset{\underset{OOH}{|}}{-\overset{|}{C_1}}-\underset{2}{C}=\underset{3}{C}- \tag{19}$$

While this general rule (sometimes called the Schenck rule) for olefins with isolated double bonds is said to have no exceptions, other systems are oxidized by different routes. For example, cyclohexadiene, on photosensitized autoxidation, produces an endoperoxide.

The Schenck rule is illustrated in Preparation 2-5. In this preparation we follow the nomenclature of the original literature although the product name does not clearly indicate the shift in the position of the double bond.

2-5. *Preparation of Δ^9-Octaline Hydroperoxide* [59]

$$\Delta^9\text{-octaline} \xrightarrow[\text{UV/rose bengale}]{O_2} \text{hydroperoxide (OOH)} \tag{20}$$

In a suitable apparatus, connected to a gas buret containing oxygen, 13.6 gm (0.1 mole) of Δ^9-octaline dissolved in 100 ml of isopropanol and containing 60 mg of rose bengale (see Note) is irradiated by a submerged high-pressure mercury lamp at 18°–20°C while oxygen flows through the system. Over a 90 min period, 2080 ml of oxygen is absorbed. The flow of gas is stopped at this point, the solvent is evaporated at 30°C (bath temperature) (12 mm). The residue is sublimed at 50°–55°C (bath temperature) (10^{-3} mm) to afford colorless needles. Yield: 14.8 gm (84% based on Δ^9-octaline), m.p. 59°–60°C.

NOTE: In our own experience in another connection, we have found that rose bengale (also called rose bengal) varies considerably in composition and ultraviolet absorption characteristics. The source of supply and the specific batch of a given supplier may be critical.

The reactive intermediate in the photosensitized autoxidation has been considered as electrophilic and the intermediate formation of singlet oxygen has been postulated [60].

The rate of methylene blue-sensitized autoxidation of olefins increases in the order: 1-nonene < 4-methylcyclohexene < cyclohexene < 2,3-dimethyl-1-butene < 2-hexene < cyclopentene < 1-methylcyclohexene < 2,3-dimethylcyclohexene < 1-methylcyclopentene < 1,2-dimethylcyclohexene < 2,3-dimethyl-2-butene. The rate of oxidation of the last compound in this sequence is 5500 times as fast as that of cyclohexene [60].

D. Oxidation with Singlet Oxygen

In the previous section, mention was made of the postulation of the intermediate formation of singlet oxygen in the photosensitized autoxidation of olefins [60].

Reactions of the type illustrated in Eq. (21) have been considered as diagnostic of the participation of singlet oxygen [62].

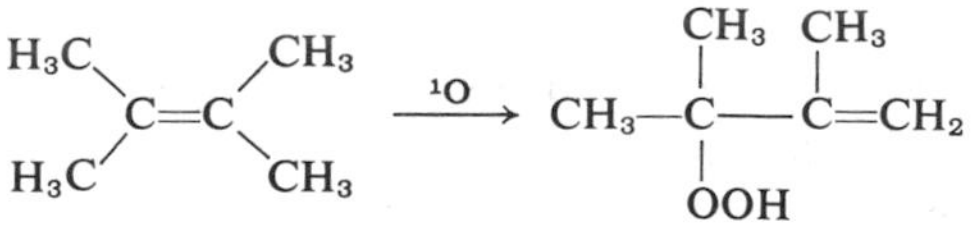

$$\begin{matrix} H_3C & & CH_3 \\ & C{=}C & \\ H_3C & & CH_3 \end{matrix} \xrightarrow{^1O} CH_3{-}\underset{OOH}{\overset{CH_3}{C}}{-}\overset{CH_3}{C}{=}CH_2 \qquad (21)$$

Thus, 9,10-diphenylanthracene peroxide

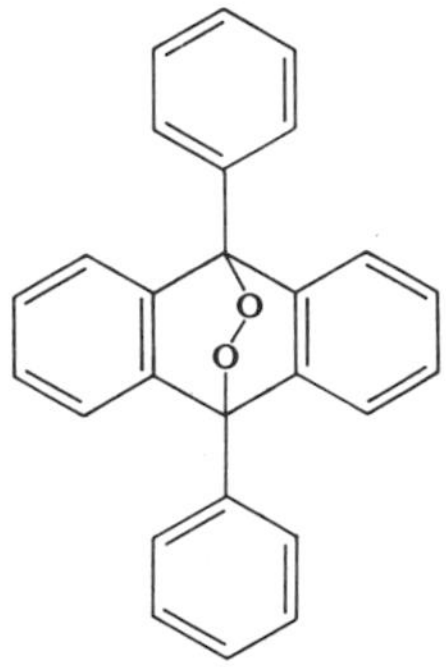

reacts with 1,1,2,2-tetramethylethylene to produce 2,3-dimethyl-1-butene 3-hydroperoxide. Consequently the reaction is considered to involve singlet oxygen, ^1O [63].

At $-15°C$, triphenyl phosphite ozonide seems to undergo decomposition to produce singlet oxygen. There is some question whether singlet oxygen really forms at appreciable rates at low temperatures. The observed fact is that at $-70°C$, triphenyl phosphite ozonide reacts with tetramethylethylene, as shown in Eq. (21), to produce the hydroperoxide in 81% yield within 90 min [62].

Singlet oxygen (^1O) has also been invoked as the agent produced by the microwave discharge of ground-state oxygen which, when passed over a film of *cis*-polybutadiene, causes the formation of hydroperoxides on the polymer surface [64].

E. Oxidation with Ozone

The use of ozone as an initiator for the autoxidation of hydrocarbons has been discussed above. More recently, ozonized air, per se, has been suggested as the oxidizing agent of paraffins to prepare hydroperoxides [65]. We presume that the mechanism of this procedure is closely related to the one previously mentioned [35]. The exact level of ozone used is expected to be quite critical in this procedure.

Generally, the cleavage of primary ozonides derived from olefins in methanol is expected to produce a methoxyhydroperoxide as indicated in Eq. (22).

$$>C=C< + O_3 \longrightarrow >C(O_3)C< \longrightarrow -C^+(-O-O^-)- + >C=O \xrightarrow{CH_3OH} -C(-O-OH)(-OCH_3)- \qquad (22)$$

$$(C_6H_5)_2C=C(C_6H_5)_2 \xrightarrow[CH_3OH/CHCl_3]{O_2/O_3} (C_6H_5)_2C(O_3)C(C_6H_5)_2 \longrightarrow (C_6H_5)_2C^+-O-O^- + (C_6H_5)_2C=O$$

A Criegee Zwitterion

$$\xrightarrow{H_2O_2} (C_6H_5)_2C(O-OH)_2 \qquad (23)$$

If the initial olefin is symmetrically substituted about the double bond, the structures of resultant products are readily predicted. However, if the substitution is unsymmetrical, the products of the reaction depend on the direction of the cleavage of the ozonide [66].

Recently an anomalous product was discovered upon ozonolysis of tetraphenylethylene. In a straightforward reaction of tetraphenylethylene with 1% of ozone in oxygen, the isolated products were the expected benzophenones (in somewhat more than the theoretically expected quantity) and a 50% yield of the unexpected diphenylmethyl bis(hydroperoxide).

It has been postulated that the intermediate Criegee zwitterion reacts with hydrogen peroxide, which forms in a side reaction, to produce the bis(hydroperoxide) (Eq. 23) [67].

Since the handling of ozone is hazardous, such reactions are usually carried out only on a very limited scale. Their value is primarily in the analytical field.

F. Autoxidation of Organometallic Compounds

A method of preparing hydroperoxides, which is not only suitable for the preparation of *tert*-alkyl hydroperoxides but also for primary and secondary alkyl hydroperoxides, was published almost simultaneously by two separate groups of investigators [68–71]. The method permits the synthesis of a wide variety of products. Even benzyl hydroperoxide, which had not been made by other methods, could be isolated using this procedure. Aromatic hydroperoxides, like phenyl hydroperoxides, however, could not be prepared.

The method involves the addition of a Grignard reagent solution to ether saturated with oxygen at temperatures in the range of −70°C. The Grignard reagent used in this reaction must be an alkylmagnesium chloride rather than the bromide.

2-6. Preparation of tert-Butyl Hydroperoxide by Inverse Oxidation of a Grignard Reagent [68]

$$(CH_3)_3C\text{—}MgCl \xrightarrow[2.\ H_2O]{1.\ O_2} (CH_3)_3C\text{—}O\text{—}OH + Mg(OH)Cl \tag{24}$$

To a 1 liter, four-necked flask fitted with an explosion-proof stirrer, sintered glass oxygen inlet tube, a graduated dropping funnel, and a condenser protected with a calcium chloride tube, cooled to −75°C, is added 400 ml of anhydrous ether. The ether is saturated with oxygen. With vigorous stirring and while maintaining the flow of oxygen, 300 ml of a 1.74 *N* solution of *tert*-butylmagnesium chloride in ether is added under the surface of the liquid over a 2.75 hr period.

After the addition has been completed, stirring is continued for 15 min while the flow of oxygen is continued. The reaction mixture is then allowed to warm to 0°C and poured over 200 gm of ice. The mixture is then slowly acidified with 6 *N* hydrogen chloride. The ether layer is separated and reserved. The aqueous layer is extracted three times with 90 ml portions of ether. The ether solutions and extracts are combined and dried with 100 gm of calcium chloride overnight. Titration of an aliquot of the solution indicates a 91.3% yield at this point.

The ether is evaporated under slightly reduced pressure. The residue is distilled through a 20 cm Vigreux column, b.p. 31°–33°C (17 mm). Yield: 38.7 gm (82.4%). On further, careful fractional distillation, the boiling point is raised to 34°–35°C (20 mm); purity 98.6%.

While this reaction has been used for the preparation of ethyl and *n*-butyl hydroperoxides, these products were not isolated. Particularly ethyl hydroperoxide was found to be quite explosive.

The preparation of benzyl hydroperoxide by this method required a minor modification. In this case, the Grignard reagent had to be introduced just above the surface of the ether solution at −75°C. When the more usual addition under the surface was attempted, the Grignard solution was found to freeze and yields were low. When an attempt was made to isolate the product by separating it from the by-products as the sodium salt, further losses were observed. Among the by-products were found benzyl alcohol and small amounts of benzaldehyde, compounds which may be expected from the decomposition of hydroperoxide by base [68].

Alkyllithium and dialkylzinc were also subjected to the inverse oxidation at low temperatures. Since the yields were somewhat lower than when the corresponding Grignard reagents were used, no special advantage could be claimed for the use of such starting materials [68].

A subsequent publication by a third group of investigators claims priority to the discovery of the above procedure [72,73]. To attempt to adjudicate this matter is outside the scope of the present work. A discussion of those aspects of the research which extend the application of the reaction is, however, quite appropriate.

It was found that the autoxidation of Grignard reagents is extremely rapid even at −150°C (using propane or low-boiling petroleum ether as solvents or suspending agents). Yields are generally improved at higher dilution of the reagents, at lower reaction temperatures, with increasing molecular weight of the organic portion of the organometallic compound, and with increasing branching of the carbon atom bearing the metallic atom [72].

While, as stated before, no particular advantage is gained by substituting dialkylzinc or -cadmium for a conventional Grignard reagent, a simplification of the reaction procedure is possible when a Grignard reagent is converted to alkylzinc (or cadmium) chloride prior to autoxidation. This modification permits the preparation of the reactant in the reaction vessel, thus eliminating the troublesome inverse addition. Also, at least for small quantities of reagents, substantially higher reaction temperatures are possible.

TABLE III
AUTOXIDATION OF ORGANOMETALLIC COMPOUNDS[a,b]

Structure of organometallic compound	Conc. (moles/liter)	Temp. (°C)	Yield of hydroperoxide (%)
C_4H_9Li	0.74	−75	31
Indenyl-3-Li[c]	0.3[d]	−75	—
C_4H_9MgCl	1.15	−75	67
$C_4H_9MgC_4H_9$	0.9	−75	59
$C_6H_{11}MgCl$[c]	1.75[e]	−75	65
$C_6H_{11}MgCl$[c]	1.25[f]	−110	79
1,6-$ClMg(CH_2)_6MgCl$[c]	0.2[g]	−130 to −110	—
C_4H_9ZnCl	0.96	−75 to −25	90
C_4H_9ZnCl	0.96	−20	93
C_4H_9ZnCl	0.96	0	92
C_4H_9ZnCl	0.96	+15	78
$C_4H_9ZnC_4H_9$	0.5	0	88
C_4H_9CdCl	0.5	+10	94
$C_4H_9CdC_4H_9$	0.5	0	95
n-$C_8H_{17}CdCl$	0.5	0	90
$C_6H_5CH_2CdCl$	0.5	−5	86
C_4H_9BFCl	0.5	+10	50
$C_4H_9AlCl_2$	1.0	−50	10
$C_4H_9AlCl_2$	1.0	0	9

[a] From Hock and Ernst [72].

[b] Reaction conditions: Addition of 50 ml of an ether solution of the organometallic compounds to 200 ml of anhydrous ether saturated with oxygen over a 45 min period.

[c] Hock and Ernst [73].

[d] 600 ml of an ethereal solution was autoxidized over a 4 hr period. The crude product was distilled in a molecular still at 10^{-3} to 10^{-4} mm.

[e] 400 ml of Grignard reagent added to 150 ml of ether.

[f] 810 ml of Grignard reagent was directly autoxidized.

[g] 1.7 liters of Grignard reagent was added to 220 ml of tetrahydrofuran saturated with oxygen. The autoxidation was continued for 4 hr. The product mixture is handled in several portions for safety reasons. The product mixture contained hexanediol, the hydroxy-hydroperoxide. The bishydroperoxide was purified by chromatographic separation on silica gel.

Table III summarizes the results of the autoxidation of a variety of organometallic compounds under various conditions of concentration and temperature using the inverse addition technique. The high yield of butyl hydroperoxide which was formed from butyl zinc chloride even at +15°C is to be noted.

Traditionally, the mechanism of the autoxidation of a Grignard reagent was considered to be an addition of the reagent to oxygen [68]. A more recent proposal is that radical intermediates from a one-electron transfer process are involved [74].

Trialkylboranes, prepared by the reaction of olefins with *N*-triethylborazone, are oxidized in two steps to an alkyl hydroperoxide as shown in Eqs. (25)–(28) [75].

$$3R{-}CH{=}CH_2 + BH_3 \cdot N(C_2H_5)_3 \xrightarrow{100°-120°C} B(CH_2CH_2R)_3 + N(C_2H_5)_3 \quad (25)$$

$$B(CH_2CH_2R)_3 \xrightarrow[0°C]{O_2} RCH_2CH_2B(OOCH_2CH_2R)_2 \quad (26)$$

$$RCH_2CH_2B(OOCH_2CH_2R)_2 \xrightarrow{peracid} RCH_2CH_2{-}O{-}B(OOCH_2CH_2R)_2 \quad (27)$$

$$RCH_2CH_2{-}O{-}B(OOCH_2CH_2R)_2 \xrightarrow{H_2O} 2RCH_2CH_2OOH + ROH + B(OH)_3 \quad (28)$$

The method is said to be particularly suitable for the preparation of hydroperoxides of medium-sized ring compounds where the Hock-Pritzkow-Walling method (usually called the Walling and Buckler method in the literature; see Refs. [68], [69], [72]) is unsatisfactory. It is also suitable for the preparation of aralkane and long-chain alkyl hydroperoxides [75]. A more satisfactory method of preparing and oxidizing trialkylboranes is discussed in Section 3 on the reactions of hydrogen peroxide.

Table VI at the end of this chapter gives the physical properties of a selection of hydroperoxides.

3. REACTIONS WITH HYDROGEN PEROXIDE

SAFETY NOTES: 1. The concentrated solutions of hydrogen peroxide are potentially hazardous not only because of their oxidizing action but also because of potential explosive hazards. For proper safety precautions, the manufacturers' literature should be consulted and followed.

2. In the reactions of alcohols with hydrogen peroxide in the presence of acids, it is believed that the correct, safe procedure consists of acidifying the hydrogen peroxide prior to the reaction with alcohols. Violent explosions are possible when acid is added to mixtures of alcohol and hydrogen peroxide [76].

By extension, we presume, quite generally only preacidified hydrogen peroxide should be used in other reactions which involve acidification of a system containing hydrogen peroxide. Extreme safety precautions must be taken in all cases.

Hydrogen peroxide and its anion are nucleophiles. Consequently they are capable of reaction with olefinic compounds and a large variety of alkyl compounds of the type RX where the leaving group X is electronegative, or where R can form a stable carbonium ion.

The reaction with olefins is indicated by Eq. (29).

$$\gt C{=}C\lt + HOOH \longrightarrow H{-}\overset{|}{\underset{|}{C}}{-}\overset{|}{\underset{|}{C}}{-}OOH \tag{29}$$

Equation (30) shows the general reaction of alkyl compounds with hydrogen peroxide.

$$RX + HOOH \longrightarrow ROOH + HX \tag{30}$$

$$X = -OH, -BR_2, -O-SO_2OH, -O-SO_2OR, -OSO_2R, -Cl-Br, -OR, -O\overset{O}{\overset{\|}{C}}-R$$

A. Reaction of Olefins

The addition of hydrogen peroxide to olefins in the presence of a trace of acid has been pictured as a reaction involving protonation of the olefin as the first step (Eq. 31) [77].

$$CH_3-\overset{CH_3}{\overset{|}{C}}{=}CH-CH_3 + H^+ \longrightarrow CH_3-\overset{CH_3}{\overset{|}{\underset{+}{C}}}-CH_2CH_3$$

$$CH_3-\overset{CH_3}{\overset{|}{\underset{+}{C}}}-CH_2CH_2 + HOOH \longrightarrow CH_3-\overset{CH_3}{\overset{|}{\underset{\underset{OOH}{|}}{C}}}-CH_2CH_3 + H^+ \tag{31}$$

It will be noted that, in this reaction, the proton adds to the least-substituted part of the double bond to produce a *tert*-hydroperoxide in the second step of the reaction. Cyclohexene appears to be unreactive in this system [78,79].

3-1. Preparation of tert-Pentyl Hydroperoxide [78]

With proper precautions for the handling of concentrated hydrogen peroxide, to 5 ml (1.33 gm, 0.19 mole) of 2-methyl-2-butene is added a cold

$$(H_3C)_2C{=}CH{-}CH_3 + H_2O_2 \xrightarrow{H_2SO_4} CH_3{-}CH_2{-}C(CH_3)_2{-}OOH \qquad (32)$$

mixture of 5 ml of 85–90% hydrogen peroxide and 0.02 ml of concentrated sulfuric acid. The mixture is stirred for 6 hr and then stored for 16 hr. After dilution with water, the product is extracted with petroleum ether. After the evaporation of the solvent, the residue is cautiously distilled, under reduced pressure. Yield: 1.29 gm (65%); b.p. 63°–64°C (37.5 mm); n_D^{20} 1.4158.

A procedure for the preparation of 2,4,4-trimethylpentyl 2-hydroperoxide from diisobutylene in a solution which is 12.5% in hydrogen peroxide and 40% in sulfuric acid has been described [80].

β-Halohydroperoxides have been prepared by the addition of hydrogen peroxide to olefins in the presence of *p*-toluenesulfonic acid and *N*-chloroacetamide or 1,3-dibromo-5,5-dimethylhydantoin as halogenating agents [81].

The controlled addition of acidified hydrogen peroxide to vinyl ethers has been used to prepare α-hydroperoxy ethers [82].

In view of the fact that olefins and, as will be discussed below, alcohols and esters of sulfuric acid, may be converted to hydroperoxides by similar procedures, it is interesting to speculate on the interrelationship of these syntheses.

B. Reaction of Alcohols

While ordinary primary and secondary alcohols do not usually react with hydrogen peroxide, alcohols, which can form carbonium ions under the reaction conditions, have been converted to hydroperoxides [77–79,83].

It is interesting to note that when some optically active secondary alcohols are subjected to the reaction, the configuration was substantially retained although the sign and magnitude of the optical solution changed during the conversion. Also, the source of both oxygens of the hydroperoxide is the hydrogen peroxide.

3-2. Preparation of 1-Phenylethyl Hydroperoxide [83]

$$C_6H_5{-}CH(CH_3){-}OH + H_2O_2 \xrightarrow{H+} C_6H_5{-}CH(CH_3){-}OOH + H_2O \qquad (33)$$

To minimize the explosion hazards and to facilitate handling, the following preparation is carried out on a relatively small scale with due precautions.

At 0°C, a mixture of 30 gm (0.0246 mole) of optically active 1-phenylethanol and 12 ml of 90% hydrogen peroxide which has been acidified with 0.1 gm of concentrated sulfuric acid is stirred for 1 hr. The temperature is then raised to room temperature and stirring is continued for 5 hr (with precautions against explosions). The reaction mixture is then diluted with 50 ml of water and the product is extracted with three 5 ml portions of ether. The ether extracts are combined, and washed in turn with saturated aqueous sodium bicarbonate and with water. After drying with anhydrous sodium sulfate, the ether is evaporated off at reduced pressure. The residue is distilled to produce optically active 1-phenylethyl hydroperoxide. Yield: 2.76 gm (81%), b.p. 66°C (0.5 mm).

By similar procedures, *tert*-butyl hydroperoxide, *tert*-butyl hydroperoxide, 1-isobutyl-1-methylpropyl hydroperoxide, 1-tetralin hydroperoxide [78], 1-ethyl-1-methylbutyl hydroperoxide, 4-phenyldiphenylmethyl hydroperoxide, 1-methyl-1-phenylpropyl hydroperoxide [79], and 2-phenyl-2-propyl hydroperoxide [84] have been synthesized.

1,1-Diphenylethyl and 1,1-diphenyl-1-propyl hydroperoxides have been prepared with acidified 50% hydrogen peroxide in acetic acid [85]. Chloro-*tert*-butyl hydroperoxide and the analogous bromo compound have been prepared by adding a chloroform solution of 1 mole of the appropriate halo alcohol to a solution of 4 moles of hydrogen peroxide (as a 30% solution) in 9.31 moles of 95% sulfuric acid [86].

An unsaturated tertiary alcohol such as 1,1-dimethylpent-2-yne-4-ene-1-ol has been converted to the corresponding ene–yne hydroperoxide in a solution of hydrogen peroxide in sulfuric acid [87].

C. Reaction of Esters

Esters which readily undergo alkyl–oxygen fission, such as hydrogen phthalate esters, react with concentrated hydrogen peroxide to produce hydroperoxides [77–79].

The reaction is usually carried out in a sodium bicarbonate buffer.

3-3. Preparation of 1-Methyl-1-phenylpropyl Hydroperoxide [79]

With all due precautions, a homogeneous solution of 2.0 gm (0.0067 mole) of 1-methyl-1-phenylpropyl hydrogen phthalate, 10 ml of 85–90% hydrogen peroxide, and 1.5 gm (0.018 mole) of sodium bicarbonate at 0°C is stirred for 2 hr. At about this time, an oil begins to separate. Stirring is continued for 20 hr, then 25 ml of water is added and the product is extracted with low-boiling petroleum ether. On evaporation of the solvent, 0.59 gm of a residual oil is obtained (53% yield). The product distills at 72°–73°C (0.02 mm); n_D^{20} 1.5224.

$$CH_3-CH_2-C(CH_3)(C_6H_5)-O-C(=O)-C_6H_4-COOH \xrightarrow[NaHCO_3]{H_2O_2} CH_3-CH_2-C(CH_3)(C_6H_5)-OOH + C_6H_4(CO_2H)_2 \quad (34)$$

D. Reaction of Ethers

The reaction of vinyl ethers with hydrogen peroxide has already been mentioned in the section dealing with the reactions of olefins (Section 3, A). Addition to the double bond is observed. In the case of alkyl ethers derived from branched alkyl moieties, cleavage of the ether takes place and, where possible, *tert*-hydroperoxides form. The reaction evidently can only take place when an acid is present during the reaction [88]. It is doubtful whether this procedure has any real value as a laboratory synthesis since ethers usually would have to be prepared first and the starting materials for the preparation of ethers may indeed be suitable materials for the direct preparation of hydroperoxides. However, the method has some value in mechanism studies.

3-4. Preparation of 1-Phenylethyl Hydroperoxide [88]

$$C_2H_5-O-CH(C_6H_5)-CH_3 \xrightarrow[H^+]{H_2O_2} CH_3-CH(C_6H_5)-OOH \quad (35)$$

With all due precautions, to a solution of 8 ml of 90% hydrogen peroxide and 0.1 gm of concentrated sulfuric acid cooled to 0°C is added with stirring 0.5 gm of ethyl 1-phenylethyl ether. The mixture is allowed to warm to room temperature. Stirring is continued for a total of 6 hr. Then 25 ml of water is added and the organic layer is extracted with ether. The ether extracts are combined, washed in turn with aqueous sodium bicarbonate and with water, and dried over anhydrous sodium sulfate. After evaporation of the ether, 0.1 gm of an oil identified as 1-phenylethyl hydroperoxide remains.

By similar procedures, ethyl triphenylmethyl ether and bis(diphenylmethyl) ether produced the expected hydroperoxides. On the other hand, bis(1-phenylethyl) ether was recovered essentially unchanged [88].

E. Reaction of Trialkylboranes

The reaction of trialkylboranes with gaseous oxygen has been mentioned in the section dealing with autoxidation. It will be recalled that the autoxidation itself led to a compound of the type $(RO_2)_2BR$ which required further oxidation with a peracid to form $(RO_2)_2BOR$, which, on hydrolysis, produced a hydroperoxide and an alcohol. Similar results are obtained when the oxidation is carried out with hydrogen peroxide instead of oxygen [89].

With the new hydroboration technology developed by H. C. Brown, the trialkylboranes are much more readily accessible than by previous techniques. By adopting the H. C. Brown hydrogenator (see the first volume of our series, S. R. Sandler and W. Karo, "Organic Functional Group Preparations," Academic Press, New York, 1968, pp. 3–8) for oxygenation, a convenient method of converting olefins by way of the boranes to hydroperoxides has been developed [90]. The procedure involves autoxidation of a trialkylborane with oxygen followed by a treatment with hydrogen peroxide to liberate the alkyl hydroperoxide.

3-5. Preparation of Cyclohexyl Hydroperoxide [90]

$$\text{Cyclohexene } (H_2C{-}CH_2{-}CH{=}CH{-}CH_2{-}CH_2 \text{ ring}) \xrightarrow[\text{THF}]{\text{Borane}} (C_6H_{10})_3B \tag{36}$$

$$(C_6H_{10})_3B \xrightarrow{O_2} C_6H_{10}B(OC_6H_{10})_2 \tag{37}$$

$$C_6H_{10}B(OC_6H_{10})_2 + H_2O_2 + 2H_2O \longrightarrow C_6H_{11}OH + 2C_6H_{11}OOH + B(OH)_3 \tag{38}$$

In a dry 200-ml flask fitted with a septum inlet and a Teflon-covered magnetic stirrer, flushed with nitrogen, to 75 ml of dry tetrahydrofuran is added 12.3 gm (0.15 mole) of cyclohexene. After cooling the solution to 0°C, the olefin is hydroborated by the dropwise addition of 16.3 ml of a 307 *M* solution of borane in tetrahydrofuran. Since this particular olefin reacts sluggishly, the reaction mixture is heated to 50°C for 3 hr. Then the solution is cooled to −78°C and attached to an H. C. Brown hydrogenator converted to act as an automatic oxygenator, which has been flushed with oxygen (by injecting 15 ml of 30% hydrogen peroxide into the apparatus with an empty 100 ml flask in place of the reaction flask). After the equipment is flushed further by injecting 5 ml of 30% hydrogen peroxide into the generator, with stirring at −78°C, the oxygen absorption is followed by reading the buret which has been filled with 3% aqueous hydrogen peroxide. When the absorption of oxygen has stopped, the reaction solution is warmed with stirring to 0°C. After a total of 2 moles of oxygen has been absorbed for each mole of borane, 16.5 ml of 30% aqueous hydrogen peroxide is added dropwise and stirring is continued for ½ hr at 0°C. Then 50 ml of hexane is added and the solution is washed with 25 ml of water. The hydroperoxide is separated by extraction of the reaction mixture with four 25-ml portions of 40% potassium hydroxide. The aqueous extracts are combined, washed with 50 ml of hexane, and then neutralized at 0°C with concentrated hydrochloric acid. The product is separated from the aqueous mixture by extraction with 50 ml of hexane. The hexane extract is dried with magnesium sulfate, filtered, and distilled. After evaporation of the hexane at reduced pressure, the product is distilled at 39°–40°C (0.8 mm). Yield: 9.5 gm (82%), n_D^{20} 1.4645.

In the reaction of tri-*n*-octylborane, methanol is added to the reaction system to facilitate solution. In the case of reaction of terminal olefins, the solution is warmed to 0°C following the absorption of the first mole of oxygen to complete the reaction. With internal olefins, the oxidation is continued at −78°C until the absorption of oxygen stops. The hydroperoxides prepared by this method were 1-butyl, 1-octyl, 2-methyl-1-pentyl, 2-butyl, cyclopentyl, 2-methylcyclopentyl, cyclohexyl, and norbornyl hydroperoxides.

F. Reaction of Halides

Secondary and tertiary halides have been converted to hydroperoxides with concentrated hydrogen peroxide in the presence of a small amount of concentrated sulfuric acid [84,91]. Table IV shows the yields obtained from the reaction of a variety of alkyl halides. To be noted is the low yield associated with a primary halide such as benzyl chloride and the modest yields reported for the reaction with *sec*-butyl chloride and *tert*-butyl bromide. In the case of the reaction of *tert*-butyl chloride, a substantial fraction of di-*tert*-butyl peroxide is formed along with the expected *tert*-butyl hydroperoxide.

3-6. Preparation of Cumene Hydroperoxide [84]

$$C_6H_5C(CH_3)_2Cl + H_2O_2 \xrightarrow{H^+} C_6H_5C(CH_3)_2OOH + HCl \qquad (39)$$

With adequate precautions, 55 ml (2.4 moles) of 86% hydrogen peroxide is adjusted to pH 1 with a small quantity of concentrated sulfuric acid. While

TABLE IV
YIELDS ON TREATMENT OF ALKYL HALIDES WITH ACIDIFIED HYDROGEN PEROXIDE[a]

Alkyl halide	% Yield of hydroperoxide
Triphenylchloromethane	81
2-Chloro-2-phenylpropane	93
tert-Butyl chloride	65 (and 18% di-*tert*-butyl peroxide)
tert-Butyl bromide	48
Diphenylchloromethane	82
sec-Butyl chloride	37
Benzyl chloride	3.7

[a] From Ross and Hüttel [91].

stirring and cooling to between $-5°$ and 0°C, 77.3 gm (0.5 mole) of cumyl chloride is added over an 80 min period. At the same time, 65 gm of sodium bicarbonate is added portionwise. Stirring is continued for 2 hr, followed by cooling to -5°C for 3 hr. The two layers are separated. The organic phase is washed two times with 25 ml portions of water, dried with magnesium sulfate, filtered, and cautiously distilled through a packed column under reduced pressure. After a small fraction of α-methylstyrene is distilled, 70.6 gm (93%) of cumyl hydroperoxide is obtained, b.p. 49.5°–51°C (0.01 mm).

The use of sodium bicarbonate to absorb hydrogen chloride as it forms is not always necessary. If desired, the reaction may be carried out with benzene or ether as solvents for the alkyl halide [91].

G. Reaction of Alkyl Hydrogen Sulfates, Dialkyl Sulfates, Alkyl Methanesulfonates, and Related Compounds

Tertiary alkyl hydroperoxides have been prepared by the alkylation of 30% hydrogen peroxide with tertiary alkyl hydrogen sulfates [92–94]. If desired, the required monoester of sulfuric acid may be prepared *in situ*, just prior to the addition of hydrogen peroxide. In general, the method suffers from the fact that substantial amounts of dialkyl peroxide form along with the hydroperoxide. In fact, by raising the concentration of sulfuric acid in the system, the peroxide may form in yields up to 85%. The alkyl hydrogen sulfate may also react with an alkyl hydroperoxide to form dialkyl peroxides in yields up to 92%. Separation of the two products is accomplished either by distillation at reduced pressures or by extraction of the hydroperoxide with potassium hydroxide solutions.

3-7. *Preparation of tert-Butyl Hydroperoxide* [92]

$$CH_3-\overset{\overset{CH_3}{|}}{\underset{\underset{CH_3}{|}}{C}}-OH + H_2SO_4 \longrightarrow CH_3-\overset{\overset{CH_3}{|}}{\underset{\underset{CH_3}{|}}{C}}-O-SO_3H + H_2O \qquad (40)$$

$$CH_3-\overset{\overset{CH_3}{|}}{\underset{\underset{CH_3}{|}}{C}}-O-SO_3H + H_2O_2 \longrightarrow CH_3-\overset{\overset{CH_3}{|}}{\underset{\underset{CH_3}{|}}{C}}-O-OH + H_2SO_4 \qquad (41)$$

With all due precautions, at 5°C, to 140 gm (1 mole) of 70% aqueous sulfuric acid is added, with stirring, 74 gm (1 mole) of *tert*-butyl alcohol (m.p. 25.2°C).

In a second reactor, to 113 gm (1 mole) of 30% hydrogen peroxide maintained between 0° and 5°C is added with stirring the above reaction mixture which had been cooled to 0°C. The ice bath is removed and the reaction mixture is allowed to stand overnight at room temperature. The organic layer is separated, neutralized with a suspension of magnesium carbonate in 10 ml of water, washed with 20 ml of water, and dried over anhydrous magnesium sulfate. The crude yield is 69 gm containing 52 gm of *tert*-butyl hydroperoxide and 17 gm of di-*tert*-butyl peroxide.

The two products may be separated by fractional distillation through a 12 cm × 2.5 cm glass-helix-packed column. The peroxide passes over first at 12°–13°C (20 mm), followed by the hydroperoxide at 4.5–5°C (2 mm).

The *tert*-butyl hydroperoxide may also be separated by extracting it from the reaction mixture at 10°–15°C with 50% aqueous potassium hydroxide, followed by regeneration in the cold with dilute sulfuric acid.

The treatment of dialkyl sulfates with hydrogen peroxide to produce alkyl hydroperoxides, the Baeyer-Villiger reaction [95], has had only limited application [96,97]. The lower molecular weight starting materials are toxic and the hydroperoxides produced are unstable, and consequently low yields are obtained.

A more generally applicable method of synthesizing primary and secondary hydroperoxides involves the alkylation of hydrogen peroxide with alkyl methanesulfonates and alkyl tosylates [98–101].

While it is possible to produce dialkyl peroxides when two equivalents of an alkyl methanesulfonate are reacted with one equivalent of hydrogen peroxide [102], by use of an excess of hydrogen peroxide, hydroperoxides are formed. Yields are generally in the range of 38–45% for the preparation of primary hydroperoxides, and in the 10–25% range for the secondary compounds. Since the reaction is carried out in concentrated potassium hydroxide solution, the separation of hydroperoxide from peroxides is readily accomplished.

3-8. Preparation of n-Pentyl Hydroperoxide [98]

$$CH_3CH_2CH_2CH_2CH_2OSO_2CH_3 \xrightarrow[KOH]{H_2O_2} CH_3CH_2CH_2CH_2CH_2OOH + CH_3SO_3H \quad (42)$$

To a solution of 6.65 gm (0.040 mole) of *n*-pentyl methanesulfonate in 40 ml of methanol and 35 ml of water, cooled in an ice bath, is added in order: 20.0 gm (0.16 mole) of 27.7% hydrogen peroxide and 5 gm (0.045 mole) of 50% aqueous potassium hydroxide. The mixture is maintained in a constant water bath at 23°–25°C for 24 hr.

Then the mixture is placed in an ice bath and treated with 15.0 gm of 50% aqueous potassium hydroxide. The mixture is extracted with 25 ml of benzene. The benzene layer is discarded. The aqueous layer is cooled in ice and neutralized with concentrated hydrochloric acid. The product is extracted from the neutral mixture with four 15 ml portions of benzene. Since the benzene layer still contains some hydrogen peroxide, the product is taken up in 25% aqueous potassium hydroxide. The aqueous layer is neutralized again with concentrated hydrochloric acid and the product extracted therefrom with four 10 ml portions of benzene. The extract is dried with sodium sulfate and freed of benzene by distillation under reduced pressure. The residue is distilled at 41°–42°C (4 mm). Yield: 1.80 gm (43%).

In the case of the preparation of compounds such as *n*-propyl hydroperoxide, which have appreciable water solubility, the neutralized aqueous solution is extracted with peroxide-free ether, the ethereal extract dried with sodium sulfate and then directly distilled without further washing.

The higher *n*-alkyl hydroperoxides form sparingly soluble oily salts in the aqueous potassium hydroxide solution. The purification procedure needs to be modified to improve handling of these materials. After a 40 hr reaction time, the mixture is cooled to 0°C, 15 gm of 50% aqueous potassium hydroxide is added, and the basic solution is extracted three times with 15 ml portions of hexane. Then the aqueous layer is neutralized with hydrochloric acid. The product is extracted with three portions each of 10 ml of benzene. The extracts are combined, washed with water, dried over sodium sulfate, and distilled [98].

Table V gives the proportions of starting materials, solvents, reaction times, and yields for the preparation of alkyl hydroperoxides by this method.

The long-chain alkyl hydroperoxides were also prepared by a modification of this method which included the use of larger amounts of 30% hydrogen peroxide, powdered potassium hydroxide, and rather large volumes of

TABLE V

PREPARATION OF HYDROPEROXIDES FROM ALKYL METHANESULFONATES[a]

Methanesulfonate alkyl group	Weight charged (0.040 mole)	Methanol (ml)	Water (ml)	Time (hr)	Yield (%)
n-Propyl	5.52	20	0.0	20	—
n-Butyl	6.10	25	0.0	18	42
n-Pentyl	6.65	40	3.5	20	43
n-Hexyl	7.25	45	5.0	21	44
n-Heptyl	7.80	65	8.0	24	38
n-Octyl	8.35	90	7.0	40	38
n-Nonyl	8.90	100	8.0	40	39
n-Decyl	9.45	125	10.0	44	45
2-Butyl	0.08 mole[b]	50	0	20	19.5
2-Pentyl	0.08 mole[b]	80	7	20	20.8
3-Pentyl	0.08 mole[b]	80	7	20	17.0
3-Methylbutyl	0.08 mole[b]	80	7	20	52.0
Cyclopentyl	0.08 mole[b]	80	7	20	20.9
2-Hexyl	0.08 mole[b]	90	10	20	25.6
3-Hexyl	0.08 mole[b]	90	10	20	9.0
2-Heptyl	0.08 mole[b]	125	10	24	25.7
4-Heptyl	0.08 mole[b]	125	10	20	trace
2-Octyl	0.08 mole[b]	180	14	30	11.3

[a] From Williams and Mosher [98,99].

[b] In each case 0.08 mole of alkyl methanesulfonate was treated with 10 gm of a 50% aqueous potassium hydroxide solution and 40 gm of 30% hydrogen peroxide.

methanol as a reaction solvent. The reaction generally was carried out at higher temperatures. Yields were in the range of 70% [100].

4. HYDROLYSIS OF PERESTERS AND RELATED COMPOUNDS

The organic peresters, despite their formal resemblance to esters, cannot be considered reaction products of alcohols with peracids. Rather they are products of the reaction of hydroperoxides and carboxylic acids. Upon hydrolysis, peresters form hydroperoxides and acids rather than peracids and alcohols [103].

Consequently, from the preparative standpoint, the hydrolysis of peresters to afford hydroperoxides is of minor value, although it is of some interest for analytical purposes. *tert*-Butyl perbenzoate has been saponified in methanol–chloroform with sodium methylate to produce *tert*-butyl hydroperoxide and methyl benzoate [103].

TABLE VI

PROPERTIES OF HYDROPEROXIDES[a]

Hydroperoxide of (% purity)	B.p., °C (mm)	M.p. (°C)	n_D^t	d_4^t	Ref.
Cumene	53/0.1	—	1.5242^{20}	1.062^{20}	22
	49.5–51	—	—	—	84
Cyclohexane	—	−20	1.4638^{25}	1.018^{25}	18
	42–43 (0.1)	—	1.4645^{25}	1.019	68
	20 (0.01)	—	1.4624^{20}	1.0170^{20}	73
m-Diisopropylbenzene (dihydroperoxide)	—	64–65	—	—	27
p-Diisopropylbenzene (dihydroperoxide)	—	146.4–147.6	—	—	27
2-Pinane	100 (1)	—	1.4893^{20}	1.013^{20}	23
p-Menthane	—	—	1.4659^{20}	0.961^{20}	23
1-Tetralin	—	54–54.5	—	—	28
tert-Butyl (95%)	37 (18)	—	—	—	55
tert-Butyl (98.6%)	34–35 (20)	0.5–2.0	1.3980^{27}	0.897^{20}	68
	4.5–5 (2)	3.8–4.8	1.4013^{20}	0.896^{20}	92
Chloro-*tert*-butyl (91.5%)	51–61 (2)	—	1.4452	—	55
Cyclohexene 3-	50 (52.03)	—	—	—	56
Cyclopentene 3-	35 (0.01)	—	—	—	56
1-Methylcyclohexene	47–51 (0.01)	—	—	—	56
1,3-Dibromopropyl 2-	66 (0.1)	—	1.5493^{25}	—	58
1-Bromo-3-chloro-2-methyl-2-propyl	56–58 (0.25)	—	1.5199^{25}	—	58
1-Bromo-2-methyl-2-pentyl	51 (0.1)	—	—	—	58
1-Bromo-3-chloro-2-propyl	61 (0.1)	—	—	—	—
Δ^9-Octaline	—	59–60	—	—	59
Diphenylmethyl bis(*tert*-pentane) (83.2%)	—	94–96	—	—	67
	26 (3.5)		1.4161^{20}	0.903^{20}	93
	32–33 (4.3)		1.4132^{25}	0.902^{20}	68
	50–51 (20)		1.4154^{20}	—	78

2-Octane	58–59 (0.5)		1.4269^{25}	0.868^{20}	68
	58–59 (0.4)		1.4280^{20}	0.878^{22}	99
Benzyl (87.1%)	55–57 (0.1)	—	—	—	68
	54.5 (0.3)	—	1.5380^{20}	—	75
n-Heptane	51–53 (0.5–1)	—	1.4261^{20}	—	69,71
	46–47 (0.5)	—	1.4265^{20}	0.884^{20}	98
2-Heptane	35–41 (0.5–1)	—	1.4241^{20}	—	69,71
	32–33 (0.5)	—	1.4237^{20}	0.879^{22}	99
3-Heptane	38–42 (0.5–1)	—	1.4243^{20}	—	69,71
4-Heptane	36–40 (0.5–1)	—	1.4245^{20}	—	69,71
n-Hexane	42–45 (1–2)	—	1.4224^{20}	—	71
	42–43 (2)	—	1.4208^{20}	0.891	98
2-Hexane	34–36 (1–2)	—	1.4194^{20}	—	71
	29–30 (1)	—	1.4186^{20}	0.886^{22}	99
3-Hexane	30–35 (1–2)	—	1.4208^{20}	—	71
	48–50 (5)	—	1.4204^{20}	0.893^{22}	99
Indene 3-	—	47.5	1.5750^{55}	1.1555^{55}	73
Cyclododecane	—	27–28	—	—	75
n-Octane	60 (1.2)	—	1.4311^{20}	—	75
	54–55 (0.7)	—	1.4311^{20}	0.881	98
n-Decane	68 (0.5)	—	1.4378^{20}	—	75
n-Undecane	69 (5×10^{-3})	—	1.4415^{20}	—	75
	61–63 (0.3)	—	1.4378^{20}	0.871	98
2-Ethylhexane	60 (1.7)	—	1.4330^{20}	—	75
2-Phenylethane	55 (10^{-4})	—	1.5290^{20}	—	75
2-Phenylpropane	55 (10^{-3})	—	1.5241^{20}	—	75
3-Chloro-2,3-dimethyl-2-butane	72–73	—	—	—	81
3-Bromo-2,3-dimethyl-2-butane	93–94	—	—	—	81
1-Phenylethane	66 (0.5)	—	1.5251^{20}	—	83
1-Isopropyl-1-methylpropane	51–52 (2)	—	1.4350^{20}	0.895^{20}	78
Diphenylmethane	—	50–52, 46.5–47	—	—	78,88
		51–52	—	—	91
1-Ethyl-1-methylbutane	34 (0.1)	—	1.4300^{20}	—	79

(*cont.*)

TABLE VI (*cont.*)

Hydroperoxide of (% purity)	B.p., °C (mm)	M.p. (°C)	n_D^t	d_4^t	Ref.
4-Phenyldiphenylmethane	—	162.5–163	—	—	79
1-Methyl-1-phenylpropane	60 (0.01)	—	1.5230^{20}	—	79
2-Phenylpropane 2-	43–44 (0.005)	—	—	1.0545^{25}	84
1,1-Diphenylethane	—	83–84	—	—	85
1,1-Diphenyl-1-propane	—	79–79.5	—	—	85
1-Phenyl-2-methyl-2-propane	—	44.2–45.2	—	—	85
Chloro-*tert*-butane	48 (4.5)	—	—	—	86
Bromo-*tert*-butane	40 (10^{-3})	—	—	—	86
Triphenylmethane	—	86, 81–83	—	—	88,91
n-Butane	40–42 (8)	—	1.4057^{20}	0.907^{20}	98
n-Pentane	41–42 (4)	—	1.4146^{20}	0.897^{20}	98
n-Nonane	53–55 (0.3)	—	1.4330^{20}	0.878^{20}	98
2-Butane	41–42 (11)	—	1.4050^{20}	0.907^{22}	99
2-Pentane	46–47 (7)	—	1.4140^{20}	0.899^{22}	99
3-Pentane	51–52 (10)	—	1.4155^{20}	0.909^{22}	99
2-Methylbutane	47–48 (6)	—	1.4118^{20}	0.890^{22}	99
Cyclopentane	46–47 (4)	—	1.4533^{20}	1.026^{22}	99
Dodecane	—	33	—	—	100
Tetradecane	—	44–45	—	—	100
Hexadecane	—	54.5–56	—	—	100
Octadecane	—	61–62	—	—	100

[a] See Section 1, Introduction, for nomenclature used in this table.

An exceptional situation is the preparation of trifluoromethyl hydroperoxide (b.p. 11.3°C; m.p. −75.0°–74.0°C; d_{20} 1.460). By a complex reaction, trifluoromethyl fluoroformyl peroxide,

$$CF_3OOC(=O)F,$$

may be prepared from carbonyl fluoride and $CF_2(OF)_2$. Hydrolysis of the perester acyl fluoride afforded the hydroperoxide [104].

Table VI gives the physical properties of a selection of hydroperoxides.

5. MISCELLANEOUS METHODS

1. Autoxidation in the presence of potassium *tert*-butylene [105].
2. Synthesis of silyl hydroperoxides by oxidation of silylamines [106].
3. Photooxidation of azomethane and 2,2′-azoisobutane [107,108].
4. Oxidation of isobutylene chlorohydrine with acidified hydrogen peroxide [109].
5. Preparation of 2-cyanoisopropyl hydroperoxide by oxidation of 2,2′-azobis(isobutyronitrile) in the presence of hydroquinone [110].
6. Polyvinyltoluene hydroperoxide [111].
7. Liquid phase oxidation of alkylbenzenes in the presence of nitriles [112].

REFERENCES

1. *Chem. Abstr.* **56**, 61n (1962).
2. J. L. Bolland, *Quart. Rev., Chem. Soc.* **3**, 1 (1949).
3. C. E. Frank, *Chem. Rev.* **46**, 155 (1950).
4. R. Criegee, *in* "Methoden der organischen Chemie" (E. Mueller, ed.), 4th ed. Vol. 8, Part 3, p. 1. Thieme, Stuttgart, 1952.
5. A. G. Davies, "Organic Peroxides." Butterworth, London, 1961.
6. E. G. E. Hawkins, "Organic Peroxides." Van Nostrand-Reinhold, Princeton, New Jersey, 1961.
7. W. O. Lundberg, ed., "Autoxidation and Antioxidants," Vol. 1. Wiley (Interscience), New York, 1961.
8. G. Sosnovsky and J. H. Brown, *Chem. Rev.* **66**, 529 (1966).
9. V. J. Karnojitzki, *Chim. Ind., Genie Chim.* **100**, 24 (1968); *Chim. Ind. (Paris)* **93**, 56 (1965).
10. K. Maruyama, *Kagaku (Kyoto)* **21**, 1001 (1966); *Chem. Abstr.* **70**, 10702q (1969).
11. D. Swern, ed., "Organic Peroxides," Vol. I. Wiley (Interscience), New York, 1970.
12. R. Hiatt, *in* "Organic Peroxides" (D. Swern, ed.), Vol. II, p. 1. Wiley (Interscience), New York, 1971; also R. Hiatt, *in* "Organic Peroxides" (D. Swern ed.), Vol. III, p. 1. Wiley (Interscience), New York, 1972.

13. V. K. Krieble, U.S. Patent 2,895,950 (1959); *Chem. Abstr.* **53**, 18556 (1959); W. Karo and B. D. Halpern, U.S. Patent 3,125,480 (1964); W. Karo, U.S. Patents 3,180,777 (1965) and 3,239,477 (1966); J. Dickstein, E. Blommers, and W. Karo, *J. Appl. Polym. Sci.* **6**, 193 (1962); J. Dickstein, E. Blommers, and W. Karo, *in* "Symposium of Adhesives for Structural Applications" (M. J. Bodnar, ed.), p. 1. Wiley (Interscience), New York, 1967.
14. "Evaluation of Organic Peroxides from Half-Life Data," Bull. 30.30. Lucidol Division, Pennwalt Corporation.
15. S. Maira and A. A. Wahl, *Mod. Plast. Encycl.* p. 392 (1960).
16. R. Hiatt and W. M. J. Strachan, *J. Org. Chem.* **28**, 1893 (1963).
17. M. S. Kharasch, A. Fono, and W. Nudenberg, *J. Org. Chem.* **15**, 748 (1950).
18. A. Farkas and E. Panaglia, *J. Amer. Chem. Soc.* **72**, 3333 (1950).
19. N. Kornblum and H. E. DeLaMare, *J. Amer. Chem. Soc.* **74**, 3079 (1952).
20. W. Webster and B. H. M. Thompson, German Patent 1,004,611 (1957).
21. J. P. Fortuin and H. I. Waterman, German Patent 1,004,612 (1957).
22. G. P. Armstrong, R. H. Hall, and D. C. Quin, *J. Chem. Soc., London* p. 666 (1950); also R. H. Hall and D. C. Quin, British Patent 610,293 (1948).
23. G. S. Fisher and L. A. Goldblatt, U.S. Patent 2,735,870 (1956).
24. V. V. Fedorova and I. D. Sinovich, *Neftekhimiya* **4**, 772 (1964); *Chem. Abstr.* **62,** 2728f (1965).
25. Scholven-Chemie A.-G., Netherlands Patent 6,400,271 (1964); *Chem. Abstr.* **62,** 6434c (1965).
26. Scholven-Chemie A.-G., Netherlands Patent 6,400,048 (1964); *Chem. Abstr.* **62,** 7688b (1965).
27. Scientific Research Institute of Synthetic Alcohols and Organic Products, French Patent 1,415,211 (1955); *Chem. Abstr.* **64**, 9640c (1966).
28. H. B. Knight and D. Swern, *Org. Syn., Collect. Vol.* **4**, 895 (1963).
29. Scholven-Chemie A.-G., French Patent 1,374,846 (1964); *Chem. Abstr.* **62**, 7688f (1965).
30. J. E. van Lier and G. Kan, *J. Org. Chem.* **37**, 145 (1972).
31. F. F. Rust, *J. Amer. Chem. Soc.* **79**, 4000 (1957).
32. J. C. W. Chien, E. J. Vandenberg, and H. Jabloner, *J. Polym. Sci., Part A-1* **6**, 381 (1968).
33. N. A. Khan, *Oleagineux* **20**, 683 and 751 (1965); *Chem. Abstr.* **65**, 7482d (1966).
34. C. J. Norton, F. L. Dormish, N. F. Seppi, and P. M. Beazley, *Amer. Chem. Soc. Div. Petrol. Chem., Prepr.* **15**, E5–E18 (1970); *Chem. Abstr.* **76**, 4200g (1972).
35. G. Wagner, *J. Prakt. Chem.* [4] **27**, 297 (1965).
36. H. Boardman, *J. Amer. Chem. Soc.* **84**, 1376 (1962).
37. G. A. Russell, *J. Amer. Chem. Soc.* **79**, 3871 (1957).
38. H. S. Blanchard, *J. Amer. Chem. Soc.* **81**, 4548 (1959).
39. V. P. Shatalov, R. I. Zhilina, R. P. Furticheve, A. M. Antonova, E. N. Popova, and A. A. Semilutskaya, *Tr. Lab. Khim. Vysokomol. Soedin., Voronezh. Ges. Univ.* **2,** 50 (1963); *Chem. Abstr.* **62**, 7972c (1965).
40. Chemische Werke Huels A.-G., Netherlands Patent 6,504,559 (1965); *Chem. Abstr.* **64**, 9640d (1966).
41. K. A. Pecherskaya and N. T. Faldina, U.S.S.R. Patent 178,804 (1966).
42. F. List and L. Kuhnen, *Erdoel Kohle, Erdgas, Petrochem.* **20**, 192 (1967); *Chem. Abstr.* **67**, 4312w (1967).
43. Imperial Chemical Industries, Ltd., French Patent 1,530,986 (1968); *Chem. Abstr.* **71**, 49266w (1969).

44. W. R. Davie and J. C. E. Schult, U.S. Patent 3,160,668 (1964); *Chem. Abstr.* **62,** 7688d (1965).
45. N. P. Keier, I. L. Kotlyarevskii, J. M. Alikina, M. P. Terpugova, and Yu. M. Gridnev, U.S.S.R. Patent 192,758 (1967); *Chem. Abstr.* **69,** 27023w (1968).
46. Soc. Ital. Resine S.p.A., Italian Patent 663,421 (1964); *Chem. Abstr.* **63,** 11432c (1965).
47. D. Dimitrov and V. Vulkov, *God. Khim.-Tekhnol. Inst.* **11,** 53 (1964); *Chem. Abstr* **66,** 4641b (1967).
48. W. E. Weesner, U.S. Patent 2,792,424 (1957).
49. W. E. Weesner, U.S. Patent 2,792,425 (1957).
50. W. E. Weesner, U.S. Patent 2,792,426 (1957).
51. T. Bewley and W. Webster, French Patent 1,377,163 (1964); *Chem. Abstr.* **62,** 6434b (1965).
52. E. Jonkanski and M. Rolle, French Patent 1,377,237 (1964); *Chem. Abstr.* **62,** 6434d (1965).
53. B. Witkop and J. B. Patrick, *J. Amer. Chem. Soc.* **73,** 2188 (1951).
54. B. Witkop and J. B. Patrick, *J. Amer. Chem. Soc.* **73,** 2196 (1951).
55. E. R. Bell, F. H. Dickey, J. H. Raley, F. F. Rust, and W. E. Vaughan, *Ind. Eng. Chem.* **41,** 2597 (1949).
56. R. Criegee, H. Pilz, and H. Flygare, *Chem. Ber.* **72,** 1799 (1939).
57. W. J. Farrissey, *J. Org. Chem.* **29,** 391 (1964).
58. M. Schulz, A. Rieche, and K. Kirschke, East German Patent 55,328 (1966); *Chem. Abstr.* **68,** 21526v (1968).
59. G. O. Schenck and K.-H. Schulte-Elte, *Justus Liebigs Ann. Chem.* **618,** 185 (1958).
60. K. R. Kopecky and H. J. Reich, *Can. J. Chem.* **43,** 2265 (1965).
61. K. Gollnick and G. O. Schenk, *Pure Appl. Chem.* **9,** 507 (1964).
62. P. D. Bartlett and G. D. Mendenhall, *J. Amer. Chem. Soc.* **92,** 210 (1970).
63. H. H. Wasserman and J. R. Scheffer, *J. Amer. Chem. Soc.* **89,** 3073 (1967).
64. M. L. Kaplan and P. G. Kelleher, *Science* **169,** 1206 (1970).
65. B. N. Tyutyunnikov and A. S. Drozdov, *Maslob.-Zhir. Prom.* **35,** 21 (1969); *Chem. Abstr.* **70,** 69888h (1969).
66. W. P. Keaveney, M. G. Berger, and J. J. Pappas, *J. Org. Chem.* **32,** 1537 (1967).
67. J. C. Robertson and W. J. Verzino, Jr., *J. Org. Chem.* **35,** 545 (1970).
68. C. Walling and S. A. Buckler, *J. Amer. Chem. Soc.* **77,** 6032 (1955).
69. W. Pritzkow and K. A. Müller, *Justus Liebigs Ann. Chem.* **597,** 167 (1955).
70. C. Walling and S. A. Buckler, *J. Amer. Chem. Soc.* **77,** 6039 (1955).
71. W. Pritzkow and K. A. Müller, *Chem. Ber.* **89,** 2318 (1956).
72. H. Hock and F. Ernst, *Chem. Ber.* **92,** 2716 (1959).
73. H. Hock and F. Ernst, *Chem. Ber.* **92,** 2723 (1959).
74. R. C. Lamb, P. W. Ayers, M. K. Toney, and J. F. Garst, *J. Amer. Chem. Soc.* **88,** 4261 (1966).
75. G. Wilke and P. Heimbach, *Justus Liebigs Ann. Chem.* **652,** 7 (1962).
76. E. Hedaya and S. Winstein, *J. Amer. Chem. Soc.* **89,** 5314 (1967).
77. A. G. Davies and A. M. White, *Nature* (*London*) **170,** 668 (1952).
78. A. G. Davies, R. V. Foster, and A. M. White, *J. Chem. Soc., London* p. 1541 (1953).
79. A. G. Davies, R. V. Foster, and A. M. White, *J. Chem. Soc., London* p. 2200 (1954).
80. J. Hoffman, *Org. Syn.* **40,** 76 (1960).
81. K. R. Kopecky, J. H. van de Sande, and C. Mumford, *Can. J. Chem.* **46,** 25 (1968).
82. N. A. Milas, R. L. Peeler, Jr., and O. L. Mageli, *J. Amer. Chem. Soc.* **76,** 2322 (1954).

83. A. G. Davies and R. Feld, *J. Chem. Soc., London* p. 665 (1956).
84. H. Ross and R. Hüttel, *Chem. Ber.* **89**, 2641 (1956).
85. W. H. Richardson and V. F. Hodge, *J. Org. Chem.* **35**, 4012 (1970).
86. W. H. Richardson and V. F. Hodge, *J. Amer. Chem. Soc.* **93**, 3996 (1971).
87. T. I. Yurzhenko, M. R. Vilenskaya, E. I. Khutorskoi, and C. P. Mamchur, *Usp. Khim. Org. Perekisnykh Soedin. Antookisleniya, Dokl. Vses. Konf., 3rd, 1965* p. 59 (1965); *Chem. Abstr.* **72**, 21243h (1970).
88. A. G. Davies and R. Feld, *J. Chem. Soc., London* p. 4669 (1956).
89. A. G. Davies, D. G. Hare, and R. F. M. White, *J. Chem. Soc., London* p. 341 (1961); M. H. Abraham and A. G. Davies, *ibid.*, p. 429 (1959); A. G. Davies and B. P. Roberts, *J. Chem. Soc., B* p. 311 (1969).
90. H. C. Brown and M. M. Midland, *J. Amer. Chem. Soc.* **93**, 4078 (1971); H. C. Brown, M. M. Midland, and G. W. Kabalka, *ibid.*, p. 1024.
91. H. Ross and R. Hüttel, *Chem. Ber.* **89**, 2641 (1956).
92. N. A. Milas and D. M. Surgenor, *J. Amer. Chem. Soc.* **68**, 205 (1946).
93. N. A. Milas and D. M. Surgenor, *J. Amer. Chem. Soc.* **68**, 643 (1946).
94. A. E. Batog and M. K. Romantsevich, U.S.S.R. Patent 184,843 (1966); *Chem. Abstr.* **67**, 2762g (1967).
95. A. Baeyer and V. Villiger, *Ber. Deut. Chem. Ges.* **33**, 3387 (1900); **34**, 738 (1901).
96. A. D. Kirk, *Can. J. Chem.* **43**, 2236 (1965).
97. I. Vodnar and G. J. Kulcsar, *Rev. Roum. Chim.* **12**, 401 (1967); *Chem. Abstr.* **68**, 77640v (1968).
98. H. R. Williams and H. S. Mosher, *J. Amer. Chem. Soc.* **76**, 2984 (1954).
99. H. R. Williams and H. S. Mosher, *J. Amer. Chem. Soc.* **76**, 2987 (1954).
100. S. Wawzonek, P. D. Klimstra, and R. E. Kallia, *J. Org. Chem.* **25**, 621 (1960).
101. N. C. Deno, W. E. Billups, K. E. Kramer, and R. R. Lastomirsky, *J. Org. Chem.* **35**, 3080 (1970).
102. F. Welch, H. R. Williams, and H. S. Mosher, *J. Amer. Chem. Soc.* **77**, 551 (1955).
103. N. A. Milas and D. M. Surgenor, *J. Amer. Chem. Soc.* **68**, 642 (1946).
104. P. Bernstein, F. A. Hohorst, and D. D. DesMarteau, *Polym. Prepr., Amer. Chem. Soc., Div. Polym. Chem.* **12**, 378 (1971); *J. Amer. Chem. Soc.* **93**, 3882 (1971).
105. P. Musso and D. Maassen, *Justus Liebigs Ann. Chem.* **689**, 93 (1965); G. A. Russell and A. G. Bemis, *J. Amer. Chem. Soc.* **88**, 5491 (1966).
106. R. L. Dannley and G. Jalics, *J. Org. Chem.* **30**, 2417 (1965); A. K. Shubber and R. L. Dannley, *ibid.* **36**, 3784 (1971).
107. N. R. Subbaratnam and J. G. Calvert, *J. Amer. Chem. Soc.* **84**, 1113 (1962).
108. S. S. Thomas and J. G. Calvert, *J. Amer. Chem. Soc.* **84**, 4207 (1962).
109. W. H. Richardson, J. W. Peters, and W. P. Konopka, *Tetrahedron Lett.* No. 45, p. 5531 (1966).
110. L. Dulog and W. Vogt, *Tetrahedron Lett.* No. 42, p. 5169 (1966); No. 20, p. 1915 (1967).
111. S. R. Sandler, *Int. J. Appl. Radiat. Isotop.* **16**, 473 (1965).
112. K. Tanaka and T. Shimizu, *Japan Kokai* **73** (34), 84b (1973).

APPENDIX

1. NOMENCLATURE [1]

Polymers formed by the addition-condensation reaction are called poly, followed without space or hyphen by the starting monomer, ideally in parentheses. For example, poly(styrene) and poly(vinyl chloride). (In practice, parentheses are often omitted in polymer names where there is no possible ambiguity, e.g., polystyrene.) Condensation polymers are similarly named except the parentheses contain the connecting diradical(s) and the name of the parent compound, as for example poly(hexamethylene adipamide).

Polyamides, polyureas, and polyurethanes are named in a similar fashion. In some cases in the diacyl radical names the "o" is left out, as in adipyl rather than adipoyl, but this can be written both ways. The naming of some compounds is difficult. Copolymers are named by connecting the names of radicals with co. Reference [1] should be consulted for further information.

2. POLYMER ISOMERISM [2–4]

1. *Cis–trans isomerism* for olefinic-diene polymers means, for example,

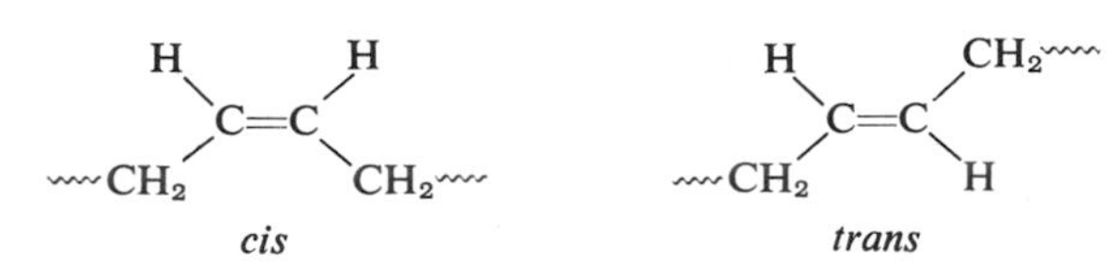

2. *Isotactic, syndiotactic, and atactic or heterotactic* refer to the relative chirility of successive monomer units:

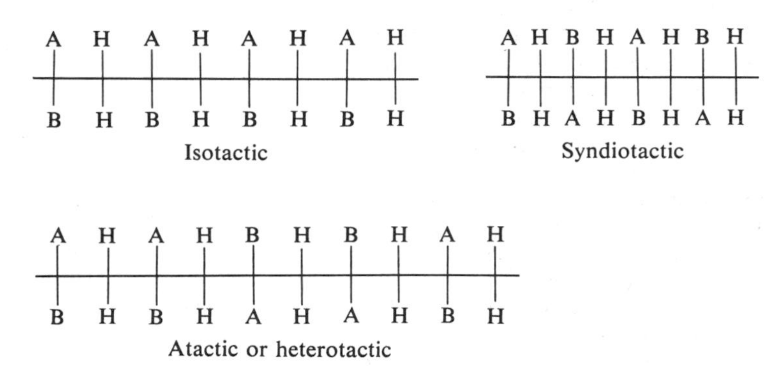

3. Monomer addition

A H
C=C
B H

may be *head-to-tail* or *head-to-head–tail-to-tail* as shown below:

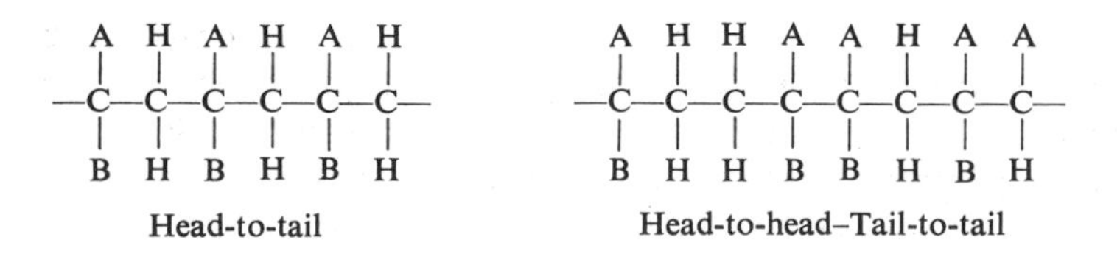

4. *Cross-linking or branch points:*

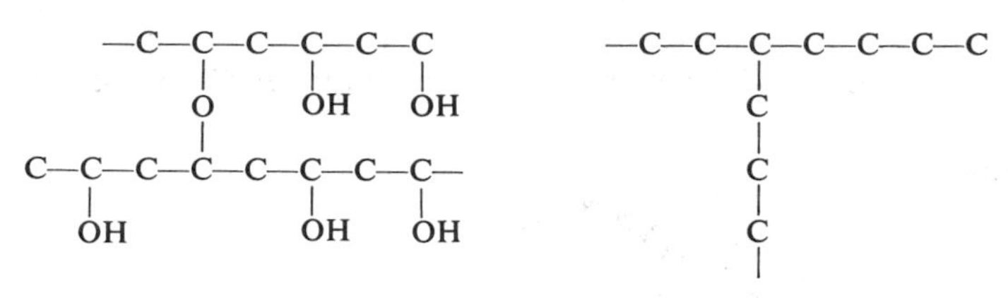

Cross-linked chains Branched chains

5. *Optical activity: Asymmetric* polymers have no symmetry elements in rotation or translation. *Dissymmetric* polymer structures cannot be superimposed on their mirror images.

Vinyl polymers may contain the asymmetric center in the main chain or the side group. Some monomers are asymmetric to start with, such as propylene oxide.

The presence of ortho-, meta-, or para-linked units in aromatic systems also leads to isomerism.

3. POLYMER CHARACTERIZATION [5,6]

As with monomeric organic compounds some of the routine structural tests are first used to establish the chemical structure, then others are used to determine the properties of the polymer.

A. Structure Proof

a. IR, UV, NMR, X-ray diffraction, etc. (4)
b. Elemental and functional group analyses
c. End-group analyses

B. Polymer Properties

a. Molecular weight determination
b. Melting (T_m), softening point, or glass-transition temperature (T_g)
c. Solubility characteristics
d. Melt viscosity
e. Solution viscosity

4. VISCOSITY

A. Viscosity Terms Commonly Used in Polymer Chemistry [7]

Symbols used:

η = viscosity (in centipoises)
η_s = viscosity of polymer solution
η_0 = viscosity of pure solvent
C = concentrations in gm/100 ml of solution (IUPAC recommendation is gm/ml)
t = flow time in viscometer corrected for kinetic energy factor inherent in instrument
t_0 = t of pure solvent
t_s = t of polymer solution

Name	Symbol	Definitions
1. Relative viscosity	η_{rel}	$\frac{\eta_s}{\eta_0} = \frac{t_s}{t_0}$
2. Specific viscosity	η_{sp}	$\frac{\eta_s - \eta_0}{\eta_0} = \frac{t_s - t_0}{t_0} = \eta_{rel} - 1$
3. Reduced viscosity (Viscosity number)	η_{red}	$\frac{\eta_{sp}}{C} = \frac{t_s - t_0}{Ct_0} = \frac{\eta_{rel} - 1}{C}$
4. Inherent viscosity	η_I (or η_{inh})	$\frac{2.303}{C} \log_{10} \eta_{rel} = \frac{\ln \eta_{rel}}{C}$
5a. Intrinsic viscosity[a]	$[\eta]$	$\lim_{C\to 0} \frac{\eta_{rel} - 1}{C} = \lim_{C\to 0} \eta_I$
5b. Limiting viscosity number (IUPAC recommended term)	LVN	$100\,[\eta]$
6. Fikentscher's K value	K	See below

[a] Experimentally, both the reduced viscosity and the inherent viscosity are plotted against C on the same graph paper and extrapolated to zero. If the intercepts coincide, this is taken as $[\eta]$; otherwise the two intercepts are averaged.

B. Fikentscher's *K* Value [8]

$$\frac{\log \eta_{rel}}{C} = \frac{75\,K^2}{1 + 1.5\,kC} + k,$$

where $1000\,k$ is *Fikentscher's K value* and η_{rel} is generally taken at 25.0° ± 0.1°C.

C = concentration in gm/100 ml up to 15 gm/100 ml, but usually below 5 gm/100 ml.

Table I will facilitate the conversion of K to $[\eta]$.

TABLE I

CONVERSION OF FIKENTSCHER'S K TO INTRINSIC VISCOSITY [4]

K Value	Intrinsic viscosity $[\eta]$
10	0.040
20	0.115
30	0.225
40	0.368
50	0.547
75	1.14
100	1.96
125	2.99
150	4.23
170	5.61
180	6.01
200	7.37

Fikentscher's K value may be used to compute the average K value for blends of polymers.

If P_1 is the percentage of a component with K value of K_1 and P_2 is the percentage of a component with K value of K_2 in a blend, the K value, K_B, of the blend may be computed by use of the equation

$$K_B = \frac{(P_1K_1 + P_2K_2)}{100}$$

where $P_1 + P_2 = 100$. Presumably, for multicomponent blends, analogous expressions may be developed.

5. MOLECULAR WEIGHT MEASUREMENTS [7]

A. Number Average Molecular Weight ($\overline{M}_n$)

$$\overline{M}_n = \frac{\sum_{i=1}^{\infty} n_i M_1}{\sum_{i=1}^{\infty} n_i} = \frac{n_1M_1}{\sum n_i} + \frac{n_2M_2}{\sum n_i} + \frac{n_3M_3}{\sum n_i} + \cdots$$

where n = number of molecules in sample

$n_1, n_2, n_3 \ldots$ = number of molecules of molecular weight M_1, M_2, M_3

$M_1, M_2, M_3 \ldots$ = molecular weights of individual molecules

The number average molecular weight is particularly sensitive to the number of small molecules present.

$\overline{M}_n$ may be determined by colligative methods such as osmometry or freezing point lowering.

B. Weight Average Molecular Weight ($\overline{M}_w$)

$$\overline{M}_w = \frac{\sum_{i=1}^{\infty} n_i M_i^2}{\sum_{i=1}^{\infty} n_i M_i} = \frac{n_1 M_1^2}{\sum n_i M_i} + \frac{n_2 M_2^2}{\sum n_i M_i} + \frac{n_3 M_3^2}{\sum n_i M_i} + \cdots$$

The weight average molecular weight is particularly sensitive to the number of large molecules.

$\overline{M}_w$ may be determined by light-scattering measurements.

[The viscosity molecular weight (see below) often is an approximation of the weight average molecular weight.]

C. Z-Average Molecular Weight ($\overline{M}_z$)

$$\overline{M}_z = \frac{\sum_{i=1}^{\infty} n_i M_i^3}{\sum_{i=1}^{\infty} n_i M_i^2} = \frac{n_1 M_1^3}{\sum n_i M_i^2} + \frac{n_2 M_2^3}{\sum n_i M_i^2} + \frac{n_3 M_3^3}{\sum n_i M_i^2} + \cdots$$

The Z average molecular weight is sensitive to the number of large molecules.

$\overline{M}_z$ may be determined by sedimentation equilibrium methods.

In general, the definitions of the average molecular weights take the form:

$$\overline{M} = \frac{\sum n_i M_i^j}{\sum n_i M_i^{j-1}}$$

where j is a positive integer greater than zero. Then, when:

$j = 1$, $\overline{M}$ is the number average mol. wt., $\overline{M}_n$
$j = 2$, $\overline{M}$ is the weight average mol. wt., $\overline{M}_w$
$j = 3$, $\overline{M}$ is the Z average mol. wt., $\overline{M}_z$
$j = 4$, $\overline{M}$ is the $(Z + 1)$ average mol. wt.

$\overline{m}_j$, where $j > 3$ seems to be unknown.

D. Viscosity Molecular Weight ($\overline{M}_v$)

The intrinsic viscosity is related to molecular weight by the equation

$$[\eta] = K'\overline{M}_v^a$$

where K' and a are parameters which have to be determined for each polymer–solvent system. The viscosity molecular weight frequently approximates the weight average molecular weight.

$\overline{M}_v$ has also been defined by the expression

$$\overline{M}_v = \left\{ \frac{\sum_{i=1}^{\infty} n_i M_i^{(1+a)}}{\sum_{i=1}^{\infty} n_i M_i} \right\}^{1/a}$$

where n_i is the number of molecules of molecular weight M_i and a is the same polymer–solvent interaction parameter as above. In good solvents a approaches 0.8; in poor solvents a approaches 0.5.

E. Degree of Polymerization ($\overline{DP}$)

$\overline{DP}$ is the "average" molecular weight (usually $\overline{M}_n$ or $\overline{M}_w$) divided by the molecular weight of the repeating structural unit.

6. CHAIN-TRANSFER EQUATION [7]

$$\frac{1}{\overline{DP}_n} = \left(\frac{1}{\overline{DP}_n}\right)_0 + C_t \frac{(S)}{(M)}$$

where $\overline{DP}_n$ = number average degree of polymerization
$(1/\overline{DP}_n)_0$ = reciprocal of $\overline{DP}_n$ without chain-transfer agent or solvent
C_t = chain-transfer constant (C_s for solvent, C_{RSH} for mercaptans)
(S) = concentration of chain-transfer agent
(M) = concentration of monomer.

Table II lists a number of typical chain-transfer agents. The activity of a chain-transfer reagent depends on the structure of the reagent, reaction temperature, concentration, monomer type, etc. Generally, the mercaptans have found wide acceptance. Halides must be used with great caution. For example, in some cases, violent reaction has been observed in polymerizations in the presence of traces of carbon tetrachloride.

TABLE II
TYPICAL CHAIN-TRANSFER AGENTS

sec-Butyl alcohol	Ethylene dichloride
tert-Butylbenzene	Ethyl thioglycolate
n-Butyl chloride	Methyl isobutyl ketone
n-Butyl iodide	Methylene chloride
Butyl mercaptan	Methyl ethyl ketone
Carbon tetrabromide	Pentaphenylethane
Carbon tetrachloride	Isopropylbenzene
Chlorobenzene	Tetrachloroethane
Chloroform	Thio-β-naphthol
Diethyl ketone	Thiophenol
Dodecyl mercaptan	Toluene
Ethylbenzene	Triphenylmethane
Ethylene dibromide	

7. COPOLYMERIZATION [7]

A. General

a. Definition of Reactivity Ratios

$$r_1 = \frac{k_{11}}{k_{12}}$$

$$r_2 = \frac{k_{22}}{k_{21}}$$

where r_1 and r_2 are the reactivity ratios of monomers M_1 and M_2 respectively. (These values are considered to be valid only for very low orders of monomer conversion. Note also that reactivity ratios deal with compositional factors not with rates of monomer conversion.) Also,

k_{11} = rate constant for the reaction of an M_1 free radical with monomer M_1
k_{12} = rate constant for the reaction of an M_1 free radical with monomer M_2
k_{22} = rate constant for the reaction of an M_2 free radical with monomer M_2
k_{21} = rate constant for the reaction of an M_2 free radical with monomer M_1

In general,

$$r_1 \times r_2 \leq 1$$

or

$$\frac{1}{r_1} \geq r_2$$

However, if

$$r_1 = r_2 = 1$$

the initial polymer has the same composition as that of the monomer mixture used, for all monomer ratios.

An *azeotropic monomer composition* is one in which the initially formed polymer has the same composition as the monomer solution. The necessary condition for azeotropic compositions to exist is

$$\frac{[M_1]}{[M_2]} = \frac{1 - r_2}{1 - r_1}$$

where $[M_1]$ and $[M_2]$ are the concentrations of the monomers used.

Alternating copolymers are said to form if r_1 and r_2 are low and r_1r_2 is "near zero."

Random copolymers are said to form if r_1 and r_2 are close to 1.

If r_1 is very large while r_2 is very small, copolymerization is strongly retarded as a rule when the concentration of M_1 is low.

b. Price-Alfrey Q-e Scheme

Price and Alfrey derived reactivity ratios from a monomer reactivity term Q and a polarity factor e. These are related to the reactivity ratios by

$$r_1 = \frac{(Q_1)}{(Q_2)} \exp[-e_1(e_1 - e_2)] \qquad r_2 = \frac{(Q_2)}{(Q_1)} \exp[-e_2(e_2 - e_1)]$$

$$r_1r_2 = \exp[-(e_1 - e_2)^2]$$

where Q_1 and e_1 are the monomer reactivity and polarity factor for M_1; Q_2 and e_2 are the monomer reactivity and polarity factor for M_2, respectively.

B. Compositional Requirements to Produce Copolymers of Predetermined Final Composition

The composition of the initially formed copolymer from a solution of two monomers is given by

$$M_0 = \frac{(P_0 - 1) + \sqrt{(1 - P_0)^2 + 4P_0r_1r_2}}{2r_1}$$

where M_0 is the mole ratio of Monomer 1 to Monomer 2 charged; P_0 is the mole ratio of Monomer 1 to Monomer 2 in *initially* formed copolymer; r_1r_2 is the product of the reactivity ratios of the two monomers and r_1 is the reactivity ratio of Monomer 1.

If the final, desired copolymer composition, P_t, is set equal to P_0, an M_0 can be computed.

If this composition M_0 is allowed to polymerize to low conversion, polymer of composition P_t is formed. If conditions are such that a fresh monomer solution of composition P_t can be added to the reaction vessel at the same rate that this polymer is formed, a monomer composition M_0 is maintained and only polymer of composition P_t is formed.

The indicated procedure therefore is to initiate a low level of monomers of composition M_0 followed by gradual addition of monomers of the desired final composition P_t. If the total charge of monomers of composition P_t greatly exceeds composition M_0, the final polymer composition will be essentially P_t, particularly if the polymerization is stopped before the last increment of added monomer is fully converted.

8. GLASS-TRANSITION TEMPERATURE OF COPOLYMERS [7]

For copolymers of known monomer composition:

$$\frac{1}{T_g} = \frac{F_1}{T_{g_1}} + \frac{F_2}{T_{g_2}} + \frac{F_3}{T_{g_3}} + \cdots + \frac{F_n}{F_{g_n}}$$

where T_g = glass-transition temperature of copolymer in degrees Kelvin

T_{g_1}, T_{g_2}, etc. = glass-transition temperature of constituent homopolymers in degrees Kelvin

F_1, F_2, etc. = weight fraction of polymers 1, 2, etc.

Table III lists some properties of polymers which are related to the glass-transition temperature.

TABLE III
POLYMER PROPERTIES DEPENDENT ON THE GLASS-TRANSITION TEMPERATURE

Physical state (transition from brittle glass to rubber)
Rate of thermal expansion
Thermal properties
Torsional modulus
Refractive index
Dissipation factor
Impact resistance—brittle point
Flow and heat distortion properties
Minimum film-forming temperature of polymer latex

REFERENCES

1. I.U.P.A.C. report in *J. Polym. Sci.* **8**, 257 (1952).
2. M. Goodman, *Stereochem.* **2**, 73 (1967).
3. A. D. Ketley, ed., "The Stereochemistry of Macromolecules," 3 vols. Dekker, New York, 1967.
4. F. A. Bovey, "High Resolution NMR of Macromolecules." Academic Press, New York, 1972.
5. P. W. Allen, ed., "Techniques of Polymer Characterization." Academic Press, New York, 1951.
6. R. G. Beaman and F. B. Cramer, *J. Polym. Sci.* **21**, 223 (1956).
7. W. Karo, Monomer–Polymer Catalog G-65-2 (1965). Reprinted in part by permission of the Monomer–Polymer Laboratories, a Division of Borden, Inc., Philadelphia, Pennsylvania.
8. H. Fikentscher, *Cellul.-Chem.* **13**, 58 (1932).

AUTHOR INDEX

Numbers in parentheses are reference numbers and indicate that an author's work is referred to although his name is not cited in the text. Numbers in italics show the page on which the complete reference is listed.

D

G

H

Q

R

S

T

Y

Z

SUBJECT INDEX

E

F

Q

R

A 4
B 5
C 6
D 7
E 8
F 9
G 0
H 1
I 2
J 3

ORGANIC CHEMISTRY

A SERIES OF MONOGRAPHS

1. Wolfgang Kirmse. CARBENE CHEMISTRY, 1964; 2nd Edition, 1971
2. Brandes H. Smith. BRIDGED AROMATIC COMPOUNDS, 1964
3. Michael Hanack. CONFORMATION THEORY, 1965
4. Donald J. Cram. FUNDAMENTALS OF CARBANION CHEMISTRY, 1965
5. Kenneth B. Wiberg (Editor). OXIDATION IN ORGANIC CHEMISTRY, PART A, 1965; Walter S. Trahanovsky (Editor). OXIDATION IN ORGANIC CHEMISTRY, PART B, 1973
6. R. F. Hudson. STRUCTURE AND MECHANISM IN ORGANO-PHOSPHORUS CHEMISTRY, 1965
7. A. William Johnson. YLID CHEMISTRY, 1966
8. Jan Hamer (Editor). 1,4-CYCLOADDITION REACTIONS, 1967
9. Henri Ulrich. CYCLOADDITION REACTIONS OF HETEROCUMULENES, 1967
10. M. P. Cava and M. J. Mitchell. CYCLOBUTADIENE AND RELATED COMPOUNDS, 1967
11. Reinhard W. Hoffman. DEHYDROBENZENE AND CYCLOALKYNES, 1967
12. Stanley R. Sandler and Wolf Karo. ORGANIC FUNCTIONAL GROUP PREPARATIONS, VOLUME I, 1968; VOLUME II, 1971; VOLUME III, 1972
13. Robert J. Cotter and Markus Matzner. RING-FORMING POLYMERIZATIONS, PART A, 1969; PART B, 1; B, 2, 1972
14. R. H. DeWolfe. CARBOXYLIC ORTHO ACID DERIVATIVES, 1970
15. R. Foster. ORGANIC CHARGE-TRANSFER COMPLEXES, 1969
16. James P. Snyder (Editor). NONBENZENOID AROMATICS, VOLUME I, 1969; VOLUME II, 1971

17. C. H. Rochester. ACIDITY FUNCTIONS, 1970

18. Richard J. Sundberg. THE CHEMISTRY OF INDOLES, 1970

19. A. R. Katritzky and J. M. Lagowski. CHEMISTRY OF THE HETEROCYCLIC *N*-OXIDES, 1970

20. Ivar Ugi (Editor). ISONITRILE CHEMISTRY, 1971

21. G. Chiurdoglu (Editor). CONFORMATIONAL ANALYSIS, 1971

22. Gottfried Schill. CATENANES, ROTAXANES, AND KNOTS, 1971

23. M. Liler. REACTION MECHANISMS IN SULPHURIC ACID AND OTHER STRONG ACID SOLUTIONS, 1971

24. J. B. Stothers. CARBON-13 NMR SPECTROSCOPY, 1972

25. Maurice Shamma. THE ISOQUINOLINE ALKALOIDS: CHEMISTRY AND PHARMACOLOGY, 1972

26. Samuel P. McManus (Editor). ORGANIC REACTIVE INTERMEDIATES, 1973

27. H.C. Van der Plas. RING TRANSFORMATIONS OF HETEROCYCLES, VOLUMES 1 AND 2, 1973

28. Paul N. Rylander. ORGANIC SYNTHESIS WITH NOBLE METAL CATALYSTS, 1973

29. Stanley R. Sandler and Wolf Karo. POLYMER SYNTHESES, VOLUME I, 1974

30. Robert T. Blickenstaff, Anil C. Ghosh, and Gordon C. Wolf. TOTAL SYNTHESIS OF STEROIDS, 1974

In preparation

Barry M. Trost and Lawrence S. Melvin, Jr. SULFUR YLIDES: EMERGING SYNTHETIC INTERMEDIATES